The Researcher's Guide to Heat Conduction

The Researcher's Guide to Heat Conduction

Edited by **Edgar Miller**

NY RESEARCH PRESS

New York

Published by NY Research Press,
23 West, 55th Street, Suite 816,
New York, NY 10019, USA
www.nyresearchpress.com

The Researcher's Guide to Heat Conduction
Edited by Edgar Miller

International Standard Book Number: 978-1-63238-441-6 (Hardback)

Printed in the United States of America.

Contents

Preface

This book was inspired by the evolution of our times; to answer the curiosity of inquisitive minds. Many developments have occurred across the globe in the recent past which has transformed the progress in the field.

This book consists of research-focused information regarding heat conduction and aims to serve as an efficient guide for comprehending the concepts of heat conduction. It presents various aspects of heat conduction theory like inverse heat conduction difficulties, coupled thermal and electromagnetic or mechanical effects and also, numerical methods for solving heat conduction equations. Other topics discussed in the book include inverse heat conduction difficulties, defining analytical explanations of nonlinear heat conduction difficulties, wavelike solutions, mixed effects of heat conduction in plasmas or in pyroelectric material elastic deformations. It also demonstrates numerical methods for solving heat conduction problems.

The book was developed from a mere concept to drafts to chapters and finally compiled together as a complete text to benefit the readers across all nations. To ensure the quality of the content we instilled two significant steps in our procedure. The first was to appoint an editorial team that would verify the data and statistics provided in the book and also select the most appropriate and valuable contributions from the plentiful contributions we received from authors worldwide. The next step was to appoint an expert of the topic as the Editor-in-Chief, who would head the project and finally make the necessary amendments and modifications to make the text reader-friendly. I was then commissioned to examine all the material to present the topics in the most comprehensible and productive format.

I would like to take this opportunity to thank all the contributing authors who were supportive enough to contribute their time and knowledge to this project. I also wish to convey my regards to my family who have been extremely supportive during the entire project.

Editor

Part 1

Inverse Heat Conduction Problems

1

Inverse Heat Conduction Problems

Krzysztof Grysa
Kielce University of Technology
Poland

1. Introduction

In the heat conduction problems if the heat flux and/or temperature histories at the surface of a solid body are known as functions of time, then the temperature distribution can be found. This is termed as a direct problem. However in many heat transfer situations, the surface heat flux and temperature histories must be determined from transient temperature measurements at one or more interior locations. This is an inverse problem. Briefly speaking one might say the inverse problems are concerned with determining causes for a desired or an observed effect.

The concept of an inverse problem have gained widespread acceptance in modern applied mathematics, although it is unlikely that any rigorous formal definition of this concept exists. Most commonly, by inverse problem is meant a problem of determining various quantitative characteristics of a medium such as density, thermal conductivity, surface loading, shape of a solid body etc. , by observation over physical fields in the medium or - in other words - a general framework that is used to convert observed measurements into information about a physical object or system that we are interested in. The fields may be of natural appearance or specially induced, stationary or depending on time, (Bakushinsky & Kokurin, 2004).

Within the class of inverse problems, it is the subclass of indirect measurement problems that characterize the nature of inverse problems that arise in applications. Usually measurements only record some indirect aspect of the phenomenon of interest. Even if the direct information is measured, it is measured as a correlation against a standard and this correlation can be quite indirect. The inverse problems are difficult because they ussually are extremely sensitive to measurement errors. The difficulties are particularly pronounced as one tries to obtain the maximum of information from the input data.

A formal mathematical model of an inverse problem can be derived with relative ease. However, the process of solving the inverse problem is extremely difficult and the so-called exact solution practically does not exist. Therefore, when solving an inverse problem the approximate methods like iterative procedures, regularization techniques, stochastic and system identification methods, methods based on searching an approximate solution in a subspace of the space of solutions (if the one is known), combined techniques or straight numerical methods are used.

2. Well-posed and ill-posed problems

The concept of well-posed or correctly posed problems was introduced in (Hadamard, 1923). Assume that a problem is defined as

$$\mathbf{A}u=g \tag{1}$$

where $u \in U$, $g \in G$, U and G are metric spaces and **A** is an operator so that $\mathbf{A}U \subset G$. In general u can be a vector that characterize a model of phenomenon and g can be the observed attribute of the phenomenon.

A well-posed problem must meet the following requirements:

- the solution of equation (1) must exist for any $\mathbf{g} \in \mathbf{G}$,
- the solution of equation (1) must be unique,
- the solution of equation (1) must be stable with respect to perturbation on the right-hand side, i.e. the operator $\mathbf{A}^{-1}$ must be defined throughout the space G and be continuous.

If one of the requirements is not fulfilled the problem is termed as an ill-posed. For ill-posed problems the inverse operator $\mathbf{A}^{-1}$ is not continuous in its domain $\mathbf{A}U \subset G$ which means that the solution of the equation (1) does not depend continuously on the input data $g \in G$, (Kurpisz & Nowak, 1995; Hohage, 2002; Grysa, 2010). In general we can say that the (usually approximate) solution of an ill-posed problem does not necessarily depend continuously on the measured data and the structure of the solution can have a tenuous link to the measured data. Moreover, small measurement errors can be the source for unacceptable perturbations in the solution. The best example of the last statement is numerical differentiation of a solution of an inverse problem with noisy input data. Some interesting remarks on the inverse and ill-posed problems can be found in (Anderssen, 2005).

Some typical inverse and ill-posed problems are mentioned in (Tan & Fox, 2009).

3. Classification of the inverse problems

Engineering field problems are defined by governing partial differential or integral equation(s), shape and size of the domain, boundary and initial conditions, material properties of the media contained in the field and by internal sources and external forces or inputs. As it has been mentioned above, if all of this information is known, the field problem is of a direct type and generally considered as well posed and solvable. In the case of heat conduction problems the governing equations and possible boundary and initial conditions have the following form:

$$\rho c \frac{\partial T}{\partial t} = \nabla \cdot (k \nabla T) + \dot{Q}_v, \quad (x,y,z) \in \Omega \subset R^3, \; t \in (0, t_f], \tag{2}$$

$$T(x,y,z,t) = T_b(x,y,z,t) \quad \text{for} \quad (x,y,z,t) \in S_D, \quad t \in (0, t_f], \tag{3}$$

$$-k \frac{\partial T(x,y,z,t)}{\partial n} = q_b(x,y,z,t) \quad \text{for} \quad (x,y,z,t) \in S_N, \quad t \in (0, t_f], \tag{4}$$

$$-k \frac{\partial T(x,y,z,t)}{\partial n} = h_c\left(T(x,y,z,t) - T_e(x,y,z,t)\right) \quad \text{for} \quad (x,y,z,t) \in S_R, \quad t \in (0, t_f], \tag{5}$$

$$T(x,y,z,0) = T_0(x,y,z) \quad \text{for} \quad (x,y,z) \in \Omega, \tag{6}$$

where $\nabla = (\partial / \partial x, \partial / \partial y, \partial / \partial z)$ stands for gradient differential operator in 3D; ρ denotes density of mass, [kg/m³]; c is the constant-volume specific heat, [J/kg K]; T is temperature, [K]; k denotes thermal conductivity, [W/m K]; $\dot{Q}_v$ is the rate of heat generation per unit volume, [W/m³], frequently termed as source function; $\partial / \partial n$ means differentiation along the outward normal; h_c denotes the heat transfer coefficient, [W/m² K]; T_b , q_b and T_0 are given functions and T_e stands for environmental temperature, t_f - final time. The boundary $\partial\Omega$ of the domain Ω is divided into three disjoint parts denoted with subscripts D for Dirichlet, N for Neumann and R for Robin boundary condition; $S_D \cup S_N \cup S_R = \partial\Omega$. Moreover, it is also possible to introduce the fourth-type or radiation boundary condition, but here this condition will not be dealt with.

The equation (2) with conditions (3) to (6) describes an initial-boundary value problem for transient heat conduction. In the case of stationary problem the equation (2) becomes a Poisson equation or – when the source function $\dot{Q}_v$ is equal to zero – a Laplace equation.

Broadly speaking, inverse problems may be subdivided into the following categories: inverse conduction, inverse convection, inverse radiation and inverse phase change (melting or solidification) problems as well as all combination of them (Özisik & Orlande, 2000). Here we have adopted classification based on the type of causal characteristics to be estimated:

1. Boundary value determination inverse problems,
2. Initial value determination inverse problems,
3. Material properties determination inverse problems,
4. Source determination inverse problems
5. Shape determination inverse problems.

3.1 Boundary value determination inverse problems

In this kind of inverse problem on a part of a boundary the condition is not known. Instead, in some internal points of the considered body some results of temperature measurements or anticipated values of temperature or heat flux are prescribed. The measured or anticipated values are called internal responses. They can be known on a line or surface inside the considered body or in a discrete set of points. If the internal responses are known as values of heat flux, on a part of the boundary a temperature has to be known, i.e. Dirichlet or Robin condition has to be prescribed. In the case of stationary problems an inverse problem for Laplace or Poisson equation has to be solved. If the temperature field depends on time, then the equation (2) becomes a starting point. The additional condition can be formulated as

$$T(x,y,z,t) = T_a(x,y,z,t) \quad \text{for} \quad (x,y,z) \in L \subset \Omega, \; t \in (0, t_f] \tag{7}$$

or

$$T(x_i,y_i,z_i,t_i) = T_{ik} \quad \text{for} \quad (x_i,y_i,z_i) \in \Omega, \; t_k \in (0, t_f], \; i=1,2,\ldots, I; \; k=1,2,..,K \tag{8}$$

with T_a being a given function and T_{ik} known from e.g. measurements. As examples of such problems can be presented papers (Reinhardt et al., 2007; Soti et al., 2007; Ciałkowski & Grysa, 2010) and many others.

3.2 Initial value determination inverse problems

In this case an initial condition is not known, i.e. in the condition (6) the function T_0 is not known. In order to find the initial temperature distribution a temperature field in the whole considered domain for fixed t>0 has to be known, i.e. instead of the condition (6) a condition like

$$T(x,y,z,t_{in}) = T_0(x,y,z) \quad \text{for} \quad (x,y,z) \in \Omega \text{ and } \ t_{in} \in (0,\, t_f] \tag{9}$$

has to be specified, compare (Yamamoto & Zou, 2001; Masood et al., 2002). In some papers instead of the condition (9) the temperature measurements on a part of the boundary are used, see e.g. (Pereverzyev et al., 2005).

3.3 Material properties determination inverse problems

Material properties determination makes a wide class of inverse heat conduction problems. The coefficients can depend on spatial coordinates or on temperature. Sometimes dependence on time is considered. In addition to the coefficients mentioned in part 3 also the thermal diffusivity, $a = k / \rho c$, [m/s²] is the one frequently being determined. In the case when thermal conductivity depends on temperature, Kirchhoff substitution is useful, (Ciałkowski & Grysa, 2010a). Also in the case of material properties determination some additional information concerning temperature and/or heat flux in the domain has to be known, usually the temperature measurements taken at the interior points, compare (Yang, 1998; Onyango et al., 2008; Hożejowski et al., 2009).

3.4 Source determination inverse problems

In the case of source determination, $\dot{Q}_v$, one can identify intensity of the source, its location or both. The problems are considered for steady state and for transient heat conduction. In many cases as an extra condition the temperature data are given at chosen points of the domain Ω , usually as results of measurements, see condition (8). As an additional condition can be also adopted measured or anticipated temperature and heat flux on a part of the boundary. A separate class of problems are those concerning moving sources, in particular those with unknown intensity. Some examples of such problems can be found in papers (Grysa & Maciejewska, 2005; Ikehata, 2007; Jin & Marin, 2007; Fan & Li, 2009).

3.5 Shape determination inverse problems

In such problems, in contrast to other types of inverse problems, the location and shape of the boundary of the domain of the problem under consideration is unknown. To compensate for this lack of information, more information is provided on the known part of the boundary. In particular, the boundary conditions are overspecified on the known part, and the unknown part of the boundary is determined by the imposition of a specific boundary condition(s) on it.

The shape determination inverse problems can be subivided into two class.

The first one can be considered as a design problem, e.g. to find such a shape of a part of the domain boundary, for which the temperature or heat flux achieves the intended values. The problems become then extremely difficult especially in the case when the boundary is multiply connected.

The second class is termed as Stefan problem. The Stefan problem consists of the determination of temperature distribution within a domain and the position of the moving interface between two phases of the body when the initial condition, boundary conditions and thermophysical properties of the body are known. The inverse Stefan problem consists of the determination of the initial condition, boundary conditions and thermophysical properties of the body. Lack of a portion of input data is compensated with certain additional information.

Among inverse problems, inverse geometric problems are the most difficult to solve numerically as their discretization leads to system of non-linear equations. Some examples of such problems are presented in (Cheng & Chang, 2003; Dennis et al., 2009; Ren, 2007).

4. Methods of solving the inverse heat conduction problems

Many analytical and semi-analytical approaches have been developed for solving heat conduction problems. Explicit analytical solutions are limited to simple geometries, but are very efficient computationally and are of fundamental importance for investigating basic properties of inverse heat conduction problems. Exact solutions of the inverse heat conduction problems are very important, because they provide closed form expressions for the heat flux in terms of temperature measurements, give considerable insight into the characteristics of inverse problems, and provide standards of comparison for approximate methods.

4.1 Analytical methods of solving the steady state inverse problems

In 1D steady state problems in a slab in which the temperature is known at two or more location, thermal conductivity is known and no heat source acts, a solution of the inverse problem can be easily obtained. For this situation the Fourier's law, being a differential equation to integrate directly, indicates that the temperature profile must be linear, i.e.

$$T(x) = ax + b = -qx / k + T_{con} , \tag{10}$$

with two unkowns, q (the steady-state heat flux) and T_{con} (a constant of integration). Suppose the temperature is measured at J locations, $\{x_1, x_2, ..., x_J\}$, below the upper surface (with x-axis directed from the surface downward) and the experimental temperature measurements are Y_j, $j = 1,2,...,J$. The steady-state heat flux and the integration constant can be calculated by minimizing the least square error between the computed and experimental temperatures. In order to generalize the analysis, assume that some of the sensors are more accurate than others, as indicated by the weighting factors, w_j, $j = 1,2,...,J$. A weighted least square criterion is defined as

$$I = \sum_{j=1}^{J} w_j^2 \left(Y_j - T(x_j)\right)^2 . \tag{11}$$

Differentiating equation (11) with respect to q and T_{con} gives

$$\sum_{j=1}^{J} w_j^2 \left(Y_j - T(x_j)\right) \frac{\partial T(x_j)}{\partial q} = 0 \text{ and } \sum_{j=1}^{J} w_j^2 \left(Y_j - T(x_j)\right) \frac{\partial T(x_j)}{\partial T_{con}} = 0 . \tag{12}$$

Equations (12) involve two sensitivity coefficients which can be evaluated from (10), $\partial T(x_j)/\partial q = -x_j/k$ and $\partial T(x_j)/\partial T_{con} = 1$, $j = 1,2,\ldots,J$, (Beck et al., 1985). Solving the system of equations (12) for the unknown heat flux gives

$$q = -k\frac{\left(\sum_{j=1}^{J} w_j^2\right)\left(\sum_{j=1}^{J} w_j^2 x_j Y_j\right) - \left(\sum_{j=1}^{J} w_j^2 x_j\right)\left(\sum_{j=1}^{J} w_j^2 Y_j\right)}{\left(\sum_{j=1}^{J} w_j^2\right)\left(\sum_{j=1}^{J} w_j^2 x_j^2\right) - \left(\sum_{j=1}^{J} w_j^2 x_j\right)^2}. \tag{13}$$

Note, that the unknown heat flux is linear in the temperature measurements.
Constants a and b in equation (10) could be developed by fitting a weighted least square curve to the experimental temperature data. Differentiating the curve according to the Fouriers'a law leads also to formula (13).
In the case of 2D and 3D steady state problems with constant thermophysical properties, the heat conduction equation becomes a Poisson equation. Any solution of the homogeneous (Laplace) equation can be expressed as a series of harmonic functions. An approximate solution, u, of an inverse problem can be then presented as a linear combination of a finite number of polynomials or harmonic functions plus a particular solution of the Poisson equation:

$$u = \sum_{k=1}^{K} \alpha_k H_k + T^{part} \tag{14}$$

where H_k's stand for harmonic functions, α_k denotes the k-th coefficient of the linear combination of the harmonic functions, $k = 1,2,\ldots,K$, and T^{part} stands for a particular solution of the Poisson equation. If the experimental temperature measurements Y_j, $j = 1,2,\ldots,J$, are known, coefficients of the combination, α_k, can be obtained by minimization an objective functional

$$I(u) = \iint_{\Omega}\left(\nabla^2 u - \dot{Q}_v\right)^2 d\Omega + w_1^2 \int_{S_D}\left(u - T_b\right)^2 dS + w_2^2 \int_{S_N}\left(k\frac{\partial u}{\partial n} + q_b\right)^2 dS$$
$$+ w_3^2 \int_{S_R}\left(k\frac{\partial v}{\partial n} + h_c v - h_c T_e\right)^2 dS + \sum_{j=1}^{J}\left(Y_j - u(\mathbf{x}_j)\right)^2 \tag{15}$$

where $\mathbf{x}_j \in \Omega$; w_1, w_2, w_3 – weights. Note that for harmonic functions the first integral vanishes.

4.2 Burggraf solution

Considering 1D transient boundary value inverse problem in a flat slab Burggraf obtained an exact solution in the case when the time-dependant temperature response was known at one internal point, (Burggraf, 1964). Assuming that $T(x^*,t) = T^*(t)$ and $q(x^*,t) = q^*(t)$ are known and are of class C^∞ in the considered domain, Burggraf found an exact solution to the inverse problem for a flat slab, a sphere and a circular cylinder in the following form:

$$T(x,t)=\sum_{n=0}^{\infty}\left[f_n(x)\frac{d^nT^*}{dt^n}-\frac{1}{a}g_n(x)\frac{d^nq^*}{dt^n}\right]. \tag{16}$$

with a standing for thermal diffusivity, $a=k/\rho c$, [m/s²]. The functions $f_n(x)$ and $g_n(x)$ have to fulfill the conditions

$$\frac{d^2f_0}{dx^2}=0,\ \ \frac{d^2f_n}{dx^2}=\frac{1}{a}f_{n-1},\ \ \frac{d^2g_0}{dx^2}=0,\ \ \frac{d^2g_n}{dx^2}=\frac{1}{a}g_{n-1},\ \ n=1,2,...$$

$$f_0(x^*)=1,\ \ f_n(x^*)=0,\ \ \left.\frac{df_n}{dx}\right|_{x=x^*}=0,\ \ n=0,1,...$$

$$g_0(x^*)=0,\ \ \left.\frac{dg_0}{dx}\right|_{x=x^*}=1\ \ g_n(x^*)=0,\ \ \left.\frac{dg_n}{dx}\right|_{x=x^*}=0,\ \ n=1,2,...$$

It is interesting that no initial condition is needed to determine the solution. This follows from the assumption that the functions $T^*(t)$ and $q^*(t)$ are defined for $t\in[0,\infty)$. The solutions of 1D inverse problems in the form of infinite series or polynomials was also proposed in (Kover'yanov, 1967) and in other papers.

4.3 Laplace transform approach

The Laplace transform approach is an integral technique that replaces time variable and the time derivative by a Laplace transform variable. This way in the case of 1D transient problems, the partial differential equation converts to the form of an ordinary differential equation. For the latter it is not difficult to find a solution in a closed form. However, in the case of inverse problems inverting of the obtained solutions to the time-space variables is practically impossible and usually one looks for approximate solutions, (Woo & Chow, 1981; Soti et al., 2007; Ciałkowski & Grysa, 2010). The Laplace transform is also useful when 2D inverse problems are considered (Monde et al., 2003)

The Laplace transform approach usually is applied for simple geometry (flat slab, halfspace, circular cylinder, a sphere, a rectangle and so on).

4.4 Trefftz method

The method known as "Trefftz method" was firstly presented in 1926, (Trefftz, 1926). In the case of any direct or inverse problem an approximate solution is assumed to have a form of a linear combination of functions that satisfy the governing partial linear differential equation (without sources). The functions are termed as Trefftz functions or T-functions. In the space of solutions of the considered equation they form a complete set of functions. The unknown coefficients of the linear combination are then determined basing on approximate fulfillment the boundary, initial and other conditions (for instance prescribed at chosen points inside the considered body), finally having a form of a system of algebraic equations (Ciałkowski & Grysa, 2010a).

T-functions usually are derived for differential equation in dimensionless form. The equation (2) with zero source term and constant material properties can be expressed in dimensionless form as follows:

$$\nabla^2 T(\xi,\tau) = \frac{\partial T(\xi,\tau)}{\partial \tau}, \quad (\xi,\tau) \in \Omega \times (0,\tau_f], \tag{17}$$

where ξ stands for dimensionless spatial location and $\tau = k/\rho c$ denotes dimensionless time (Fourier number). In further consideration we will use notation $\mathbf{x} = (x, y, z)$ and t for dimensionless coordinates.

For dimensionless heat conduction equation in 1D the set of T-functions read

$$v_n(x,t) = \sum_{k=0}^{[n/2]} \frac{x^{n-2k} t^k}{(n-2k)!k!} \cdot \quad n = 0,1,... \tag{18}$$

where $[n/2] = floor(n/2)$ stands for the greatest previous integer of $n/2$. T-functions in 2D are the products of proper T-functions for the 1D heat conduction equations:

$$V_m(x,y,t) = v_{n-k}(x,t) v_k(y,t), \quad n = 0,1,... \; ; \; k = 0,...,n; \; m = \frac{n(n+1)}{2} + k \tag{19}$$

The 3D T-functions are built in a similar way.

Consider an inverse problem formulated in dimensionless coordinates as follows:

$$\begin{aligned}
&\nabla^2 T = \partial T / \partial \tau && \text{in } \Omega \times (0,\tau_f], \\
&T = g_1 && \text{on } S_D \times (0,\tau_f], \\
&\partial T / \partial n = g_2 && \text{on } S_N \times (0,\tau_f], \\
&\partial T / \partial n + BiT = Big_3 && \text{on } S_R \times (0,\tau_f], \\
&T = g_4 && \text{on } S_{\text{int}} \times T_{\text{int}}, \\
&T = h && \text{on } \Omega \text{ for } t = 0,
\end{aligned} \tag{20}$$

where S_{int} stands for a set of points inside the considered region, $T_{\text{int}} \subset (0,\tau_f)$ is a set of moments of time, the functions g_i, i=1,2,3,4 and h are of proper class of differentiability in the domains in which they are determined and $S_D \cup S_N \cup S_R = \partial\Omega$. $Bi=h_c l/k$ denotes the Biot number (dimensionless heat transfer coefficient) and l stands for characteristic length. The sets S_{int} and T_{int} can be continuous (in the case of anticipated or smoothed or described by continuous functions input data) or discrete. Assume that g_1 in not known and g_4 describes results of measurements on $S_{\text{int}} \times T_{\text{int}}$. An approximate solution of the problem is expressed as a linear combination of the T-functions

$$T \approx u = \sum_{k=1}^{K} \alpha_k \theta_k \tag{21}$$

with θ_k standing for T-functions. The objective functional can be written down as

$$
\begin{aligned}
I(u) = & \int_{S_N \times (0,\tau_f)} \left(\partial u / \partial n - g_2\right)^2 dSdt \\
& + \int_{S_R x(0,\tau_f)} \left(\partial u / \partial n + Biu - Big_3\right)^2 dSdt \\
& + \int_{S_{\text{int}} \times T_{\text{int}}} \left(u - g_4\right)^2 dSdt + \int_{\Omega} \left(u - h\right)^2 d\Omega
\end{aligned}
\tag{22}
$$

In the contrary to the formula (15), the integral containing residuals of the governing equation fulfilling, $\iint_{\Omega\times(0,\tau_f)} \left(\left(\nabla^2 - \partial / \partial t\right)u\right)^2 d\Omega dt$, does not appear here because u, as a linear combination of T-functions, satisfies the equation $(20)_1$. Minimization of the functional $I(u)$ (being in fact a function of K unknown coefficients, $\alpha_1,...,\alpha_K$) leads to a system of K algebraic equations for the unknowns. The solution of this system leads to an approximate solution, (21), of the considered problem. Hence, for $(\mathbf{x},\tau) \in S_D \times (0,\tau_f)$ one obtains approximate form of the functions g_1.

It is worth to mention that approximate solution of the considered problem can also be obtained in the case when, for instance, the function h is unknown. In the formula (21) the last term is then omitted, but the minimization of the functional $I(u)$ can be done. The final result has physical meaning, because the approximate solution (21) consists of functions satisfying the governing partial differential equation.

The greater the number of T-functions in (21), the better the approximation of the solutions takes place. However, with increasing K, conditioning of the algebraic system of equation that results from minimization of I(u) can become worse. Therefore, the set S_{int} has to be chosen very carefully.

Since the system of algebraic equations for the whole domain may be ill-conditioned, a finite element method with the T-functions as base functions is often used to solve the problem.

4.5 Function specification method

The function specification method, originally proposed in (Beck, 1962), is particularly useful when the surface heat flux is to be determined from transient measurements at interior locations. In order to accomplish this, a functional form for the unknown heat flux is assumed. The functional form contains a number of unknown parameters that are estimated by employing the least square method. The function specification method can be also applied to other cases of inverse problems, but efficiency of the method for those cases is often not satisfactory.

As an illustration of the method, consider the 1D problem

$$
\begin{aligned}
& a\partial^2 T / \partial x^2 = \partial T / \partial t \quad \text{for } x \in (0,l) \text{ and } t \in (0, t_f], \\
& -k\partial T / \partial x = q(t) \quad \text{for } x = 0 \text{ and } t \in (0, t_f], \\
& k\partial T / \partial x = f(t) \quad \text{for } x = l \text{ and } t \in (0, t_f],
\end{aligned}
\tag{23}
$$

$$T = T_0(x) \qquad \text{for } x \in (0,l) \text{ and } t = 0\,.$$

For further analysis it is assumed that $q(t)$ is not known. Instead, some measured temperature histories are given at interior locations:

$$T(x_j, t_k) = U_{i,k}\,,\ \{x_j\}_{j=1,\dots,J} \subset (0,l)\,,\ \{t_k\}_{k=1,\dots,K} \subset (0,t_f)\,.$$

The heat flux is more difficult to calculate accurately than the surface temperature. When knowing the heat flux it is easy to determine temperature distribution. On the contrary, if the unknown boundary characteristics were assumed as temperature, calculating the heat flux would need numerical differentiating which may lead to very unstable results.

In order to solve the problem, it is assumed that the heat flux is also expressed in discrete form as a stepwise functions in the intervals (t_{k-1}, t_k) . It is assumed that the temperature distribution and the heat flux are known at times t_{k-1}, t_{k-2}, ... and it is desired to determine the heat flux q_k at time t_k. Therefore, the condition $(23)_2$ can be replaced by

$$q = -k\frac{\partial T}{\partial x} = \begin{cases} q_k = \text{const} & \text{for } t_{k-1} < t \le t_k \\ q(t) = \varphi(\text{t}) & \text{for } t > t_k \end{cases}$$

Now we assume that the unknown temperature field depends continuously on the unknown heat flux q. Let us denote $Z = \partial T / \partial q$ and differentiate the formulas (23) with respect to q. We arrive to a direct problem

$$\begin{aligned} &a\partial^2 Z / \partial x^2 = \partial Z / \partial t && \text{for } x \in (0,l) \text{ and } t \in (0, t_f], \\ &-k\partial Z / \partial x = 1 && \text{for } x = 0 \text{ and } t \in (0, t_f], \\ &k\partial Z / \partial x = 0 && \text{for } x = l \text{ and } t \in (0, t_f], \\ &Z = 0 && \text{for } x \in (0,l) \text{ and } t = 0\,. \end{aligned} \tag{24}$$

The direct problem (24) can be solved using different methods. Let us introduce now the sensitivity coefficients defined as

$$Z_{i,m}^k = \left.\frac{\partial T}{\partial q_k}\right|_{(x_i, t_m)} \equiv \frac{\partial T_{i,m}}{\partial q_k}\,. \tag{25}$$

The temperature $T_{i,k} = T(x_i, t_m)$ can be expanded in a Taylor series about arbitrary but known values of heat flux q_k^*. Neglecting the derivatives with order higher than one we obtain

$$T_{i,k} = T_{i,k}^* + \left.\frac{\partial T_{i,k}}{\partial q_k}\right|_{q_k = q_k^*} \left(q_k - q_k^*\right) = T_{i,k}^* + Z_{i,k}\left(q_k - q_k^*\right) \tag{26}$$

Making use of (24) and (25), solving (26) for heat flux component q_k and taking into consideration the temperature history only in one location, x_1, we arrive to the formula

$$q_k = q_k^* + \frac{U_{1,k} - T_{1,k}^*}{Z_{1,k}^k}, \quad k = 1,\ldots,K. \tag{27}$$

In the case when future temperature measurements are employed to calculate q_k, we use another formula (Beck et al, 1985, Kurpisz &Nowak, 1995), namely

$$q_k = q_k^* + \frac{\sum_{r=1}^{R}\left(U_{1,k+r-1} - T_{1,k+r-1}^*\right)Z_{1,k+r-1}^{k+r-1}}{\sum_{r=1}^{R}\left(Z_{1,k+r-1}^{k+r-1}\right)^2} \tag{28}$$

The case of many interior locations for temperature measurements is described e.g. in (Kurpisz &Nowak, 1995).

The detailed algorithm for 1D inverse problems with one interior point with measured temperature history is presented below:

1. Substitute $k=1$ and assume $q_k^* = 0$ over time interval $0 < t \le t_1$,
2. Calculate $T_{1,k+r-1}^*$ for $r = 1,2,\ldots,R$, $R \le K$, assuming $q_k = q_{k+1} = \ldots = q_{k+R-1}$; $T_{1,k+r-1}^*$ should be calculated, employing any numerical method to the following problem: differential equation $(23)_1$, boundary condition $(23)_2$ with q_k^* instead of $q(t)$, boundary condition $(23)_3$ and initial condition $T_{k-1}^* = T_{k-1}$, where T_{k-1} has been computed for the time interval $t_{k-2} < t \le t_{k-1}$ or is an initial condition $(23)_4$ when $k = 1$,
3. Calculate q_k from equation (27) or (28),
4. Determine the complete temperature distribution, using equation (26),
5. Substitute $k \to k+1$ and $q_k^* = q_{k-1}$ and repeat the calculations from step 2.

For nonlinear cases an iterative procedure should be involved for step 2 and 3.

4.6 Fundamental solution method

The fundamental solution method, like the Trefftz method, is useful to approximate the solution of multidimensional inverse problems under arbitrary geometry. The method uses the fundamental solution of the corresponding heat equation to generate a basis for approximating the solution of the problem.

Consider the problem described by equation $(20)_1$, Dirichlet and Neumann conditions $(20)_2$ and $(20)_3$ and initial condition $(20)_6$. The dimensionless time is here denoted as t. Let Ω be a simply connected domain in R^d, d = 2,3. Let $\{\mathbf{x}_i\}_{i=1}^{M} \subset \bar{\Omega}$ be a set of locations with noisy measured data $\tilde{Y}_i^{(k)}$ of exact temperature $T\left(\mathbf{x}_i, t_i^{(k)}\right) = Y_i^{(k)}$, $i = 1,2,\ldots,M$, $k = 1,2,\ldots,J_i$, where $t_i^{(k)} \in (0, t_f]$ are discrete times. The absolute error between the noisy measurement and exact data is assumed to be bounded for all measurement points at all measured times. The inverse problem is formulated as: reconstruct T and $\partial T / \partial n$ on $S_R \times (0, t_f)$ from $(20)_1$, $(20)_2$, $(20)_3$ and $(20)_6$ and the scattered noisy measurements $\tilde{Y}_i^{(k)}$, $i = 1,2,\ldots,M$, $k = 1,2,\ldots,J_i$. It is worth to mention that with reconstructed T and $\partial T / \partial n$ on $S_R \times (0, t_f)$ it is easy to identify heat transfer coefficient, h_c, on S_R.

The fundamental solution of $(20)_1$ in R^d is given by

$$F(\mathbf{x},t)=\frac{1}{(4\pi t)^{d/2}}\exp\left(-\frac{|\mathbf{x}^2|}{4t}\right)H(t) \tag{29}$$

where $H(t)$ is the Heaviside function. Assuming that $t^*>t_f$ is a constant, the function $\phi(\mathbf{x},t)=F(\mathbf{x},t+t^*)$ is a general solution of $(20)_1$ in the solution domain $\Omega\times(0,t_f)$.

We denote the measurement points to be $\left\{(\mathbf{x}_j,t_j)\right\}_{j=1}^{m}$, $m=\sum_{i=1}^{M}J_i$, so that a point at the same location but with different time is treated as two distinct points. In order to solve the problem one has to choose collocation points. They are chosen as

- $\left\{(\mathbf{x}_j,t_j)\right\}_{j=m+1}^{m+n}$ on the initial region $\bar{\Omega}\times\{0\}$,
- $\left\{(\mathbf{x}_j,t_j)\right\}_{j=m+n+1}^{m+n+p}$ on the surface $S_D\times(0,t_f]$, and
- $\left\{(\mathbf{x}_j,t_j)\right\}_{j=m+n+p+1}^{m+n+p+q}$ on the surface $S_N\times(0,t_f]$.

Here, n, p and q denote the total number of collocation points for initial condition $(20)_6$, Dirichlet boundary condition $(20)_2$ and Neumann boundary condition $(20)_3$, respectively. The only requirement on the collocation points are pairwisely distinct in the (d +1)-dimensional space $(\mathbf{x},t)$, (Hon & Wei, 2005, Chen et al., 2008).

To illustrate the procedure of choosing collocation points let us consider an inverse problem in a square (Hon & Wei, 2005): $\Omega=\left\{(x_1,x_2):0<x_1<1,\ 0<x_2<1\right\}$, $S_D=\left\{(x_1,x_2):x_1=1,\ 0<x_2<1\right\}$, $S_N=\left\{(x_1,x_2):0<x_1<1,\ x_2=1\right\}$, $S_R=\partial\Omega\setminus\left\{S_D\cup S_N\right\}$. Distribution of the measurement points and collocation points is shown in Figure 1.

An approximation $\tilde{T}$ to the solution of the inverse problem under the conditions $(20)_2$, $(20)_3$ and $(20)_6$ and the noisy measurements $\tilde{Y}_i^{(k)}$ can be expressed by the following linear combination:

$$\tilde{T}(\mathbf{x},t)=\sum_{j=1}^{n+m+p+q}\tilde{\lambda}_j\phi\left(\mathbf{x}-\mathbf{x}_j,t-t_j\right), \tag{30}$$

where $\phi(\mathbf{x},t)=F(\mathbf{x},t+t^*)$, F is given by (29) and $\tilde{\lambda}_j$ are unknown coefficients to be determined.

For this choice of basis functions ϕ, the approximated solution $\tilde{T}$ automatically satisfies the original heat equation $(20)_1$. Using the conditions $(20)_2$, $(20)_3$ and $(20)_6$, we then obtain the following system of linear equations for the unknown coefficients $\tilde{\lambda}_j$:

$$A\tilde{\lambda}=\tilde{b} \tag{31}$$

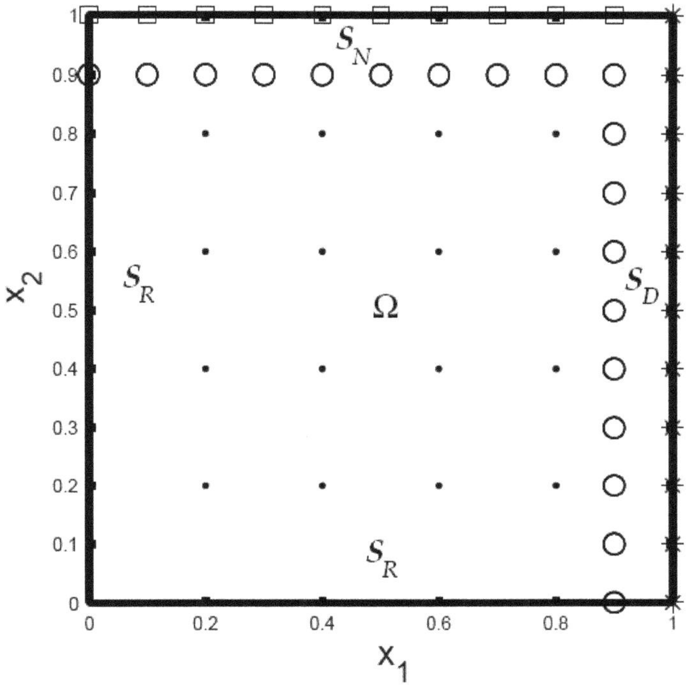

Fig. 1. Distribution of measurement points and collocation points. Stars represent collocation points matching Dirichlet data, squares represent collocation points matching Neumann data, dots represent collocation points matching initial data and circles denotes points with sensors for internal measurement.

where

$$A = \begin{pmatrix} \phi\left(\mathbf{x}_i - \mathbf{x}_j, t_i - t_j\right) \\ \dfrac{\partial \phi}{\partial n}\left(\mathbf{x}_k - \mathbf{x}_j, t_k - t_j\right) \end{pmatrix} \tag{32}$$

and

$$\tilde{b} = \begin{pmatrix} \tilde{Y}_i \\ h\left(\mathbf{x}_i, t_i\right) \\ g_1\left(\mathbf{x}_i, t_i\right) \\ g_2\left(\mathbf{x}_k, t_k\right) \end{pmatrix} \tag{33}$$

where $i = 1,2,\ldots,(n+m+p)$, $k = (n+m+p+1),\ldots,(m+n+p+q)$, $j = 1,2,\ldots,(n+m+p+q)$, respectively. The first m rows of the matrix A leads to values of measurements, the next n rows – to values of the right-hand side of the initial condition and, of course, time variable is then equal to zero, the next p rows leads to values of the right-hand side of the Dirichlet condition and the last q rows - to values of the right-hand side of Neumann condition.

The solvability of the system (31) depends on the non-singularity of the matrix A, which is still an open research problem.
Fundamental solution method belongs to the family of Trefftz method. Both methods, described in part 4.4 and 4.6, frequently lead to ill-conditioned system of algebraic equation. To solve the system of equations, different techniques are used. Two of them, namely single value decomposition and Tikhonov regularization technique, are briefly presented in the further parts of the chapter.

4.7 Singular value decomposition

The ill-conditioning of the coefficient matrix A (formula (32) in the previous part of the chapter) indicates that the numerical result is sensitive to the noise of the right hand side $\tilde{b}$ (formula (33)) and the number of collocation points. In fact, the condition number of the matrix A increases dramatically with respect to the total number of collocation points.
The singular value decomposition usually works well for the direct problems but usually fails to provide a stable and accurate solution to the system (31). However, a number of regularization methods have been developed for solving this kind of ill-conditioning problem, (Hansen, 1992; Hansen & O'Leary, 1993). Therefore, it seems useful to present the singular value decomposition method here.
Denote $N = n + m + p + q$. The singular value decomposition of the $N \times N$ matrix A is a decomposition of the form

$$A = W\Sigma V^T = \sum_{i=1}^{N} \mathbf{w}_i \sigma_i \mathbf{v}_i^T \tag{34}$$

with $W = (\mathbf{w}_1, \mathbf{w}_2, ..., \mathbf{w}_N)$ and $V = (\mathbf{v}_1, \mathbf{v}_2, ..., \mathbf{v}_N)$ satisfying $W^T W = V^T V = I_N$. Here, the superscript T denotes transposition of a matrix. It is known that $\Sigma = diag(\sigma_1, \sigma_2, ..., \sigma_N)$ has non-negative diagonal elements satisfying inequality

$$\sigma_1 \geq \sigma_2 \geq ... \geq \sigma_N \geq 0 \tag{35}$$

The values σ_i are called the singular values of A and the vectors $\mathbf{w}_i$ and $\mathbf{v}_i$ are called left and right singular vectors of A, respectively, (Golub & Van Loan, 1998). The more rapid is the decrease of singular values in (35), the less we can reconstruct reliably for a given noise level. Equivalently, in order to get good reconstruction when the singular values decrease rapidly, an extremely high signal-to-noise ratio in the data is required.
For the matrix A the singular values decay rapidly to zero and the ratio between the largest and the smallest nonzero singular values is often huge. Based on the singular value decomposition, it is easy to know that the solution for the system (31) is given by

$$\tilde{\lambda} = \sum_{i=1}^{N} \frac{\mathbf{w}_i^T \tilde{b}}{\sigma_i} \mathbf{v}_i \tag{36}$$

When there are small singular values, such approach leads to a very bad reconstruction of the vector $\tilde{\lambda}$. It is better to consider small singular values as being effectively zero, and to regard the components along such directions as being free parameters which are not determined by the data.

However, as it was stated above, the singular value decomposition usually fails for the inverse problems. Therefore it is better to use here Tikhonov regularization method.

4.8 Tikhonov regularization method

This is perhaps the most common and well known of regularization schemes, (Tikhonov & Arsenin, 1977). Instead of looking directly for a solution for an ill-posed problem (31) we consider a minimum of a functional

$$J\left[\tilde{\lambda}\right]=\left\|A\tilde{\lambda}-\tilde{b}\right\|^2+\alpha^2\left\|\tilde{\lambda}-\tilde{\lambda}_0\right\|^2 \tag{37}$$

with $\tilde{\lambda}_0$ being a known vector, $\|\,.\,\|$ denotes the Euclidean norm, and α^2 is called the regularization parameter. The necessary condition of minimum of the functional (37) leads to the following system of equation:

$$A^T\left(A\tilde{\lambda}-\tilde{b}\right)+\alpha^2\left(\tilde{\lambda}-\tilde{\lambda}_0\right)=0\,.$$

Hence

$$\tilde{\lambda}=\left(A^TA+\alpha^2I\right)^{-1}\left(A^T\tilde{b}+\alpha^2\tilde{\lambda}_0\right)$$

Taking into account (34) after transformation one obtains the following form of the functional J:

$$\begin{aligned}J\left[\tilde{\lambda}\right]&=\left\|W\Sigma V^T\tilde{\lambda}-WW^T\tilde{b}\right\|^2+\alpha^2\left\|VV^T\left(\tilde{\lambda}-\tilde{\lambda}_0\right)\right\|^2\\&=\left\|W(\Sigma\mathbf{y}-\mathbf{c})\right\|^2+\alpha^2\left\|V(\mathbf{y}-\mathbf{y}_0)\right\|^2=\left\|\Sigma\mathbf{y}-\mathbf{c}\right\|^2+\alpha^2\left\|\mathbf{y}-\mathbf{y}_0\right\|^2=J[\mathbf{y}]\end{aligned} \tag{38}$$

where $\mathbf{y}=V^T\tilde{\lambda}$, $\mathbf{y}_0=V^T\tilde{\lambda}$, $\mathbf{c}=W^T\tilde{b}$ and the use has been made from the properties $W^TW=V^TV=I_N$. Minimization of the functional $J[\mathbf{y}]$ leads to the following vector equation:

$$\Sigma^T(\Sigma\mathbf{y}-\mathbf{c})+\alpha^2(\mathbf{y}-\mathbf{y}_0)=0 \quad \text{or} \quad \left(\Sigma^T\Sigma\mathbf{y}+\alpha^2\mathbf{y}\right)=\Sigma^T\mathbf{c}+\alpha^2\mathbf{y}_0\,.$$

Hence

$$y_i=\frac{\sigma_i}{\sigma_i^2+\alpha^2}c_i+\frac{\alpha^2}{\sigma_i^2+\alpha^2}y_{0i}\;,\; i=1,\dots,N \quad \text{or} \quad \tilde{\lambda}=\sum_{i=1}^{N}\left(\frac{\sigma_i}{\sigma_i^2+\alpha^2}\mathbf{w}_i^T\tilde{b}\mathbf{v}_i+\frac{\alpha^2}{\sigma_i^2+\alpha^2}\tilde{\lambda}_0\right) \tag{39}$$

If $\tilde{\lambda}_0=\mathbf{0}$ the Tikhonov regularized solution for equation (31) based on singular value decomposition of the $N\times N$ matrix A can be expressed as

$$\tilde{\lambda}_\alpha=\sum_{i=1}^{N}\frac{\sigma_i}{\sigma_i^2+\alpha^2}\mathbf{w}_i^T\tilde{b}\mathbf{v}_i \tag{40}$$

The determination of a suitable value of the regularization parameter α^2 is crucial and is still under intensive research. Recently the L-curve criterion is frequently used to choose a good regularization parameter, (Hansen, 1992; Hansen & O'Leary, 1993). Define a curve L by

$$L = \left\{ \left(\log\left(\left\| \tilde{\lambda}_\alpha \right\|^2 \right), \log\left(\left\| A\tilde{\lambda}_\alpha - \tilde{b} \right\|^2 \right) \right) \right\} \quad (41)$$

A suitable regularization parameter α^2 is the one near the "corner" of the L-curve, (Hansen & O'Leary, 1993; Hansen, 2000).

4.9 The conjugate gradient method

The conjugate gradient method is a straightforward and powerful iterative technique for solving linear and nonlinear inverse problems of parameter estimation. In the iterative procedure, at each iteration a suitable step size is taken along a direction of descent in order to minimize the objective function. The direction of descent is obtained as a linear combination of the negative gradient direction at the current iteration with the direction of descent of the previous iteration. The linear combination is such that the resulting angle between the direction of descent and the negative gradient direction is less than 90º and the minimization of the objective function is assured, (Özisik & Orlande, 2000).
As an example consider the following problem in a flat slab with the unknown heat source $g_p(t)$ in the middle plane:

$$\partial^2 T / \partial x^2 + g_p(t)\delta(x-0.5) = \partial T / \partial t \quad \text{in } 0 < x < 1, \text{ for } t > 0$$

$$\partial T / \partial x = 0 \quad \text{at } x = 0 \text{ and at } x = 1, \text{ for } t > 0 \quad (42)$$

$$T(x,0) = 0 \quad \text{for } t = 0, \text{ in } 0 < x < 1$$

where $\delta(\cdot)$ is the Dirac delta function. Application of the conjugate gradient method can be organized in the following steps (Özisik & Orlande, 2000):

- The direct problem,
- The inverse problem,
- The iterative procedure,
- The stopping criterion,
- The computational algorithm.

The direct problem. In the direct problem associated with the problem (42) the source strength, $g_p(t)$, is known. Solving the direct problem one determines the transient temperature field $T(x,t)$ in the slab.

The inverse problem. For solution of the inverse problem we consider the unknown energy generation function $g_p(t)$ to be parameterized in the following form of linear combination of trial functions $C_j(t)$ (e.g. polynomials, B-splines, etc.):

$$g_p(t) = \sum_{j=1}^{N} P_j C_j(t) \tag{43}$$

P_j are unknown parameters, $j = 1, 2, ..., N$. The total number of parameters, *N*, is specified. The solution of the inverse problem is based on minimization of the ordinary least square norm, $S(\mathbf{P})$:

$$S(\mathbf{P}) = \sum_{i=1}^{I} [Y_i - T_i(\mathbf{P})]^2 = [\mathbf{Y} - \mathbf{T}(\mathbf{P})]^T [\mathbf{Y} - \mathbf{T}(\mathbf{P})] \tag{44}$$

where $\mathbf{P}^T = [P_1, P_2, ..., P_N]$, $T_i(\mathbf{P}) \equiv T(\mathbf{P}, t_i)$ states for estimated temperature at time t_i, $Y_i \equiv Y(t_i)$ denotes measured temperature at time t_i, *I* is a total number of measurements, $I \geq N$. The parameters estimation problem is solved by minimization of the norm (44).

The iterative procedure. The iterative procedure for the minimization of the norm ***S(P)*** is given by

$$\mathbf{P}^{k+1} = \mathbf{P}^k - \beta^k \mathbf{d}^k \tag{45}$$

where β^k is the search step size, $\mathbf{d}^k = \left[d_1^k, d_2^k, ..., d_N^k \right]$ is the direction of descent and *k* is the number of iteration. $\mathbf{d}^k$ is a conjugation of the gradient direction, $\nabla S\left(\mathbf{P}^k\right)$, and the direction of descent of the previous iteration, $\mathbf{d}^{k-1}$:

$$\mathbf{d}^k = \nabla S\left(\mathbf{P}^k\right) + \gamma^k \mathbf{d}^{k-1}. \tag{46}$$

Different expressions are available for the conjugation coefficient γ^k. For instance the Fletcher-Reeves expression is given as

$$\gamma^k = \frac{\sum_{j=1}^{N} \left[\nabla S\left(\mathbf{P}^k\right) \right]_j^2}{\sum_{j=1}^{N} \left[\nabla S\left(\mathbf{P}^{k-1}\right) \right]_j^2} \quad \text{for } k = 1, 2, ... \text{ with } \gamma^0 = 0. \tag{47}$$

Here

$$\left[\nabla S\left(\mathbf{P}^k\right) \right]_j = -2 \sum_{i=1}^{I} \frac{\partial T_i^k}{\partial P_j} \left[Y_i - T_i\left(\mathbf{P}^k\right) \right] \quad \text{for } j = 1, 2, ..., N. \tag{48}$$

Note that if $\gamma^k = 0$ for all iterations *k*, the direction of descent becomes the gradient direction in (46) and the *steepest-descent method* is obtained.

The search step β^k is obtained by minimizing the function $S\left(\mathbf{P}^{k+1}\right)$ with respect to β^k. It yields the following expression for β^k:

$$\beta^k = \frac{\sum_{i=1}^{I}\left[\left(\frac{\partial T_i}{\partial \mathbf{P}^k}\right)^T \mathbf{d}^k\right]\left[T_i\left(\mathbf{P}^k\right)-Y_i\right]}{\sum_{i=1}^{I}\left[\left(\frac{\partial T_i}{\partial \mathbf{P}^k}\right)^T \mathbf{d}^k\right]^2}, \text{ where } \left(\frac{\partial T_i}{\partial \mathbf{P}^k}\right)^T = \left[\frac{\partial T_i}{\partial P_1^k}, \frac{\partial T_i}{\partial P_2^k}, \ldots, \frac{\partial T_i}{\partial P_N^k}\right]. \tag{49}$$

The stopping criterion. The iterative procedure does not provide the conjugate gradient method with the stabilization necessary for the minimization of $S(\mathbf{P})$ to be classified as well-posed. Such is the case because of the random errors inherent to the measured temperatures. However, the method may become well-posed if the Discrepancy Principle is used to stop the iterative procedure, (Alifanov, 1994):

$$S\left(\mathbf{P}^{k+1}\right) < \varepsilon \tag{50}$$

where the value of the tolerance ε is chosen so that sufficiently stable solutions are obtained, i.e. when the residuals between measured and estimated temperatures are of the same order of magnitude of measurement errors, that is $\left|Y(t_i) - T(x_{meas}, t_i)\right| \approx \sigma_i$, where σ_i is the standard deviation of the measurement error at time t_i. For $\sigma_i = \sigma = const$ we obtain $\varepsilon = I\sigma$. Such a procedure gives the conjugate gradient method an *iterative regularization character*. If the measurements are regarded as errorless, the tolerance ε can be chosen as a sufficiently small number, since the expected minimum value for the $S(\mathbf{P})$ is zero.

The computation algorithm. Suppose that temperature measurements $\mathbf{Y} = [Y_1, Y_2, \ldots, Y_I]$ are given at times t_i, $i = 1, 2, \ldots, I$, and an initial guess $\mathbf{P}^0$ is available for the vector of unknown parameters **P**. Set $k = 0$ and then

Step 1. Solve the direct heat transfer problem (42) by using the available estimate $\mathbf{P}^k$ and obtain the vector of estimated temperatures $\mathbf{T}\left(\mathbf{P}^k\right) = [T_1, T_2, \ldots, T_I]$.

Step 2. Check the stopping criterion given by equation (50). Continue if not satisfied.

Step 3. Compute the gradient direction $\nabla S\left(\mathbf{P}^k\right)$ from equation (48) and then the conjugation coefficient γ^k from (47).

Step 4. Compute the direction of descent $\mathbf{d}^k$ by using equation (46).

Step 5. Compute the search step size β^k from formula (49).

Step 6. Compute the new estimate $\mathbf{P}^{k+1}$ using (45).

Step 7. Replace k by k+l and return to step 1.

4.10 The Levenberg-Marquardt method

The Levenberg-Marquardt method, originally devised for application to nonlinear parameter estimation problems, has also been successfully applied to the solution of linear ill-conditioned problems. Application of the method can be organized as for conjugate gradient. As an example we will again consider the problem (42).

The first two steps, **the direct problem** and **the inverse problem**, are the same as for the conjugate gradient method.

The iterative procedure. To minimize the least squares norm, (44), we need to equate to zero the derivatives of S(P) with respect to each of the unknown parameters $[P_1, P_2, ..., P_N]$, that is,

$$\frac{\partial S(\mathbf{P})}{\partial P_1} = \frac{\partial S(\mathbf{P})}{\partial P_2} = \ldots = \frac{\partial S(\mathbf{P})}{\partial P_N} = 0 \tag{51}$$

Let us introduce the *Sensitivity* or *Jacobian matrix*, as follows:

$$\mathbf{J}(\mathbf{P}) = \left[\frac{\partial \mathbf{T}^T(\mathbf{P})}{\partial \mathbf{P}}\right]^T = \begin{bmatrix} \frac{\partial T_1}{\partial P_1} & \frac{\partial T_1}{\partial P_2} & & \frac{\partial T_1}{\partial P_N} \\ \frac{\partial T_2}{\partial P_1} & \frac{\partial T_2}{\partial P_2} & & \frac{\partial T_2}{\partial P_N} \\ & & & \\ \frac{\partial T_I}{\partial P_1} & \frac{\partial T_I}{\partial P_2} & & \frac{\partial T_I}{\partial P_N} \end{bmatrix} \quad \text{or} \quad J_{ij} = \frac{\partial T_i}{\partial P_j} \tag{52}$$

where N = total number of unknown parameters, I= total number of measurements. The elements of the sensitivity matrix are called the *sensitivity coefficients*, (Özisik & Orlande, 2000). The results of differentiation (51) can be written down as follows:

$$-2\mathbf{J}^T(\mathbf{P})\left[\mathbf{Y} - \mathbf{T}(\mathbf{P})\right] = 0 \tag{53}$$

For linear inverse problem the sensitivity matrix is not a function of the unknown parameters. The equation (53) can be solved then in explicit form (Beck & Arnold, 1977):

$$\mathbf{P} = \left(\mathbf{J}^T\mathbf{J}\right)^{-1}\mathbf{J}^T\mathbf{Y} \tag{54}$$

In the case of a *nonlinear inverse problem*, the matrix **J** has some functional dependence on the vector **P**. The solution of equation (53) requires then an iterative procedure, which is obtained by linearizing the vector **T(P)** with a Taylor series expansion around the current solution at iteration *k*. Such a linearization is given by

$$\mathbf{T}(\mathbf{P}) = \mathbf{T}\left(\mathbf{P}^k\right) + \mathbf{J}^k\left(\mathbf{P} - \mathbf{P}^k\right) \tag{55}$$

where $\mathbf{T}\left(\mathbf{P}^k\right)$ and $\mathbf{J}^k$ are the estimated temperatures and the sensitivity matrix evaluated at iteration *k*, respectively. Equation (55) is substituted into (54) and the resulting expression is rearranged to yield the following iterative procedure to obtain the vector of unknown parameters **P** (Beck & Arnold, 1977):

$$\mathbf{P}^{k+1} = \mathbf{P}^k + [(\mathbf{J}^k)^T\mathbf{J}^k]^{-1}(\mathbf{J}^k)^T[\mathbf{Y} - \mathbf{T}(\mathbf{P}^k)] \tag{56}$$

The iterative procedure given by equation (56) is called the Gauss method. Such method is actually an approximation for the Newton (or Newton-Raphson) method. We note that equation (54), as well as the implementation of the iterative procedure given by equation (56), require the matrix $\mathbf{J}^T\mathbf{J}$ to be nonsingular, or

$$\left|\mathbf{J}^T\mathbf{J}\right| \neq 0 \tag{57}$$

where $|\,.\,|$ is the determinant.

Formula (57) gives the so called *Identifiability Condition,* that is, if the determinant of $\mathbf{J}^T\mathbf{J}$ is zero, or even very small, the parameters P_j , for $j = 1,2,...,N$, cannot be determined by using the iterative procedure of equation (56).

Problems satisfying $\left|\mathbf{J}^T\mathbf{J}\right| \approx 0$ are denoted *ill-conditioned.* Inverse heat transfer problems are generally very ill-conditioned, especially near the initial guess used for the unknown parameters, creating difficulties in the application of equations (54) or (56). The Levenberg-Marquardt method alleviates such difficulties by utilizing an iterative procedure in the form, (Özisik & Orlande, 2000):

$$\mathbf{P}^{k+1} = \mathbf{P}^k + [(\mathbf{J}^k)^T\mathbf{J}^k + \mu^k\Omega^k]^{-1}(\mathbf{J}^k)^T[\mathbf{Y} - \mathbf{T}(\mathbf{P}^k)] \tag{58}$$

where μ^k is a positive scalar named damping parameter and Ω^k is a diagonal matrix.

The purpose of the matrix term $\mu^k\Omega^k$ is to damp oscillations and instabilities due to the ill-conditioned character of the problem, by making its components large as compared to those of $\mathbf{J}^T\mathbf{J}$ if necessary. μ^k is made large in the beginning of the iterations, since the problem is generally ill-conditioned in the region around the initial guess used for iterative procedure, which can be quite far from the exact parameters. With such an approach, the matrix $\mathbf{J}^T\mathbf{J}$ is not required to be non-singular in the beginning of iterations and the Levenberg-Marquardt method tends to the *steepest descent method,* that is , a very small step is taken in the negative gradient direction. The parameter μ^k is then gradually reduced as the iteration procedure advances to the solution of the parameter estimation problem, and then the Levenberg-Marquardt method tends to the Gauss method given by (56).

The stopping criteria. The following criteria were suggested in (Dennis & Schnabel, 1983) to stop the iterative procedure of the Levenberg-Marquardt Method given by equation (58):

$$S\left(\mathbf{P}^{k+1}\right) < \varepsilon_1$$

$$\left\|\left(\mathbf{J}^k\right)[\mathbf{Y} - \mathbf{T}(\mathbf{P}^k)]\right\| < \varepsilon_2 \tag{59}$$

$$\left\|\mathbf{P}^{k+1} - \mathbf{P}^k\right\| < \varepsilon_3$$

where ε_1 , ε_2 and ε_3 are user prescribed tolerances and $\|\,.\,\|$ denotes the Euclidean norm.

The computational algorithm. Different versions of the Levenberg-Marquardt method can be found in the literature, depending on the choice of the diagonal matrix d and on the form chosen for the variation of the damping parameter μ^k (Özisik & Orlande, 2000). [l-91. Here

$$\Omega^k = diag[(\mathbf{J}^k)^T \mathbf{J}^k]. \tag{60}$$

Suppose that temperature measurements $\mathbf{Y} = [Y_1, Y_2, ..., Y_I]$ are given at times t_i, $i = 1,2,...,I$, and an initial guess $\mathbf{P}^0$ is available for the vector of unknown parameters **P**. Choose a value for μ^0, say, $\mu^0 = 0.001$ and set *k=0*. Then,

Step 1. Solve the direct heat transfer problem (42) with the available estimate $\mathbf{P}^k$ in order to obtain the vector $\mathbf{T}(\mathbf{P}^k) = [T_1, T_2, ..., T_I]$.

Step 2. Compute $S(\mathbf{P}^k)$ from the equation (44).

Step 3. Compute the sensitivity matrix $\mathbf{J}^k$ from (52) and then the matrix Ω^k from (60), by using the current value of $\mathbf{P}^k$.

Step 4. Solve the following linear system of algebraic equations, obtained from (58):

$$[(\mathbf{J}^k)^T \mathbf{J}^k + \mu^k \Omega^k] \Delta\mathbf{P}^k = (\mathbf{J}^k)^T [\mathbf{Y} - \mathbf{T}(\mathbf{P}^k)] \tag{61}$$

in order to compute $\Delta\mathbf{P}^k = \mathbf{P}^{k+1} - \mathbf{P}^k$.

Step 5. Compute the new estimate $\mathbf{P}^{k+1}$ as

$$\mathbf{P}^{k+1} = \mathbf{P}^k + \Delta\mathbf{P}^k \tag{62}$$

Step 6. Solve the exact problem (42) with the new estimate $\mathbf{P}^{k+1}$ in order to find $\mathbf{T}(\mathbf{P}^{k+1})$. Then compute $S(\mathbf{P}^{k+1})$.

Step 7. If $S(\mathbf{P}^{k+1}) \geq S(\mathbf{P}^k)$, replace μ^k by $10\mu^k$ and return to step 4.

Step 8. If $S(\mathbf{P}^{k+1}) \leq S(\mathbf{P}^k)$, accept the new estimate $\mathbf{P}^{k+1}$ and eplace μ^k by $0{,}1\mu^k$.

Step 9. Check the stopping criteria given by (59). Stop the iterative procedure if any of them is satisfied; otherwise, replace *k* by *k+1* and return to step 3.

4.11 Kalman filter method

Inverse problems can be regarded as a case of system identification problems. System identification has enjoyed outstanding attention as a research subject. Among a variety of methods successfully applied to them, the Kalman filter, (Kalman, 1960; Norton, 1986;Kurpisz. & Nowak, 1995), is particularly suitable for inverse problems.

The Kalman filter is a set of mathematical equations that provides an efficient computational (recursive) solution of the least-squares method. The Kalman filtering technique has been chosen extensively as a tool to solve the parameter estimation problem. The technique is simple and efficient, takes explicit measurement uncertainty incrementally (recursively), and can also take into account *a priori* information, if any.

The Kalman filter estimates a process by using a form of feedback control. To be precise, it estimates the process state at some time and then obtains feedback in the form of noisy measurements. As such, the equations for the Kalman filter fall into two categories: time update and measurement update equations. The time update equations project forward (in time) the current state and error covariance estimates to obtain the *a priori* estimates for the next time step. The measurement update equations are responsible for the feedback by

incorporating a new measurement into the *a priori* estimate to obtain an improved *a posteriori* estimate. The time update equations are thus predictor equations while the measurement update equations are corrector equations.
The standard Kalman filter addresses the general problem of trying to estimate $x \in \Re$ of a dynamic system governed by a linear stochastic difference equation, (Neaupane & Sugimoto, 2003)

4.12 Finite element method

The finite element method (FEM) or finite element analysis (FEA) is based on the idea of dividing the complicated object into small and manageable pieces. For example a two-dimensional domain can be divided and approximated by a set of triangles or rectangles (the elements or cells). On each element the function is approximated by a characteristic form.
The theory of FEM is well know and described in many monographs, e.g. (Zienkiewicz, 1977; Reddy & Gartling, 2001). The classic FEM ensures continuity of an approximate solution on the neighbouring elements. The solution in an element is built in the form of linear combination of shape function. The shape functions in general do not satisfy the differential equation which describes the considered problem. Therefore, when used to solve approximately an inverse heat transfer problem, usually leads to not satisfactory results.
The FEM leads to promising results when T-functions (see part 4.4) are used as shape functions. Application of the T-functions as base functions of FEM to solving the inverse heat conduction problem was reported in (Ciałkowski, 2001). A functional leading to the Finite Element Method with Trefftz functions may have other interpretation than usually accepted. Usually the functional describes mean-square fitting of the approximated temperature field to the initial and boundary conditions. For heat conduction equation the functional is interpreted as mean-square sum of defects in heat flux flowing from element to element, with condition of continuity of temperature in the common nodes of elements. Full continuity between elements is not ensured because of finite number of base functions in each element.
However, even the condition of temperature continuity in nodes may be weakened. Three different versions of the FEM with T-functions (FEMT) are considered in solving inverse heat conduction problems: (a) FEMT with the condition of continuity of temperature in the common nodes of elements, (b) no temperature continuity at any point between elements and (c) nodeless FEMT.
Let us discuss the three approaches on an example of a dimensionless 2D transient boundary inverse problem in a square $\Omega = \{(x,y): 0 < x < 1,\ 0 < y < 1\}$, for $t > 0$. Assume that for $y = 0$ the boundary condition is not known; instead measured values of temperature, Y_{ik}, are known at points $(1-\delta_b, y_i, t_k)$. Furthermore,

$$T(x,y,t)\big|_{t=0} = T_0(x,y),\ T(x,y,t)\big|_{x=0} = h_1(y,t),\ \frac{\partial T}{\partial y}(x,y,t)\bigg|_{y=1} = h_2(x,t),$$

$$\frac{\partial T}{\partial y}(x,y,t)\bigg|_{y=0} = h_3(x,t) \qquad (63)$$

(a) FEMT with the condition of continuity of temperature in the common nodes of elements (Figure 2). We consider time-space finite elements. The approximate temperature in a *j*-th element, $\tilde{T}^j(x,y,t)$, is a linear combination of the T-functions, $V_m(x,y,t)$:

$$T^j(x,y,t) \approx \tilde{T}^j(x,y,t) = \sum_{m=1}^{N} c_m^j V_m(x,y,t) = [C]^T [V(x,y,t)] \tag{64}$$

where N is the number of nodes in the *j-th* element and $[V(x, y, t)]$ is the column matrix consisting of the T-functions. The continuity of the solution in the nodes leads to the following matrix equation in the element:

$$[\bar{V}][C] = [T] \tag{65}$$

In (65) elements of matrix $[\bar{V}]$ stand for values of the T-functions, $V_m(x,y,t)$, in the nodal points, i.e. $\bar{V}_{rs} = V_s(x_r, y_r, t_r)$, $r,s = 1,2,\ldots,N$. The column matrix $[T] = [T^{1j}, T^{2j}, \ldots, T^{Nj}]^T$ consists of temperatures (mostly unknown) of the nodal points with T^{ij} standing for value of temperature in the *i*-th node, $i = 1,2,\ldots,N$. The unknown coefficients of the linear combination (63) are the elements of the column matrix [C]. Hence we obtain

$$[C] = [\bar{V}]^{-1}[T] \text{ and finally } \tilde{T}^j(x,y,t) = ([\bar{V}]^{-1}[T])^T [V(x,y,t)] \tag{66}$$

It is clear, that in each element the temperature $\tilde{T}^j(x,y,t)$ satisfies the heat conduction equation. The elements of matrix $([\bar{V}]^{-1}[T])^T$ can be calculated from minimization of the objective functional, describing the mean-square fitting of the approximated temperature field to the initial and boundary conditions.

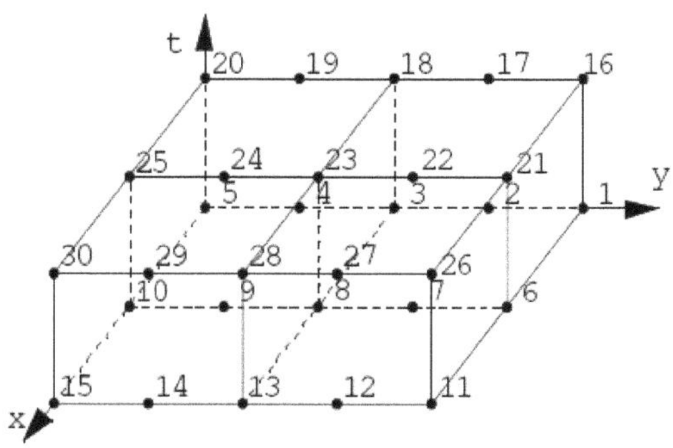

Fig. 2. Time-space elements in the case of temperature continuous in the nodes.

(b) No temperature continuity at any point between elements (Figure 3). The approximate temperature in a *j*-th element, $\tilde{T}^j(x,y,t)$, is a linear combination of the T-functions (63), too. In this case in order to ensure the physical sense of the solution we minimize inaccuracy of the temperature on the borders between elements. It means that the functional describing the mean-square fitting of the approximated temperature field to the initial and boundary conditions includes the temperature jump on the borders between elements. For the case

$$
\begin{aligned}
J = & \sum_i \int_{\Omega_i} \left(\tilde{T}_i(x,y,0) - T_0(x,y)\right)^2 d\Omega + \sum_i \int_0^{t_e} dt \int_{\Gamma_i} \left(\tilde{T}_i(0,y,t) - h_1(y,t)\right)^2 d\Gamma \\
& + \sum_i \int_0^{t_e} dt \int_{\Gamma_i} \left(\frac{\partial \tilde{T}_i}{\partial y}(x,1,t) - h_2(x,t)\right)^2 d\Gamma + \sum_i \int_0^{t_e} dt \int_{\Gamma_i} \left(\frac{\partial \tilde{T}_i}{\partial y}(x,0,t) - h_3(x,t)\right)^2 d\Gamma \\
& + \sum_{i,j} \int_0^{t_e} dt \int_{\Gamma_{i,j}} \left(\tilde{T}_i - \tilde{T}_j\right)^2 d\Gamma + \sum_i \sum_{k=1}^{I_{ITR}} \left(\tilde{T}_i(x_k, y_k, t_k) - Y_{ik}\right)^2 \Bigg|_{x=1-\delta_b}
\end{aligned}
\tag{67}
$$

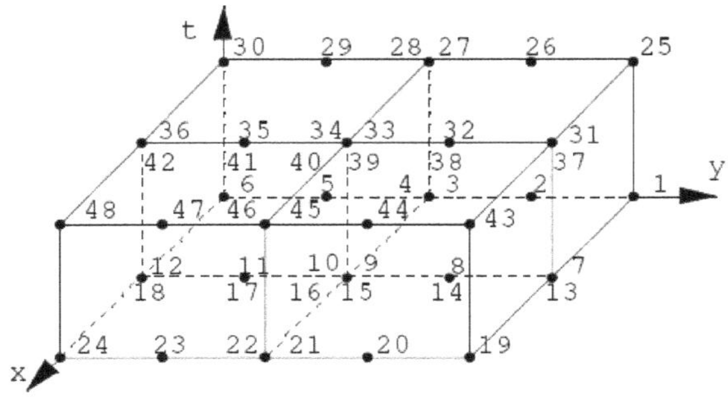

Fig. 3. Time-space elements in the case of temperature discontinuous in the nodes.

(c) Nodeless FEMT. Again, $\tilde{T}^j(x,y,t)$, is a linear combination of the T-functions. The time interval is divided into subintervals. In each subinterval the domain Ω is divided into J subdomains (finite elements) and in each subdomain Ω_j , $j=1, 2,..., J$ (with $\Gamma_i = \partial\Omega_i$) the temperature is approximated with the linear combination of the Trefftz functions according to the formula (64). The dimensionless time belongs to the considered subinterval. In the case of the first subinterval an initial condition is known. For the next subintervals initial condition is understood as the temperature distribution in the subdomain Ω_j at the final moment of time in the previous subinterval. The mean-square method is used to minimize the inaccuracy of the approximate solution on the boundary, at the initial moment of time and on the borders between elements. This way the unknown coefficients of the combination, c_m^j, can be calculated. Generally, the coefficients c_m^j depend on the time subinterval number, (Grysa & Lesniewska, 2009).

In (Ciałkowski et al., 2007) the FEM with Trefftz base functions (FEMT) has been compared with the classic FEM approach. The FEM solution of the inverse problem for the square considered was analysed. For the FEM the elements with four nodes and, consequently, the simplest set of base functions: (1, x, y, xy) have been applied.

Consider an inverse problem in a square (compare the paragraph before the equation (63)). Using FEM to solve the inverse problem gives acceptable solution only for the first row of elements. Even for exact values of the given temperature the results are encumbered with relatively high error. For the next row of the elements, the FEM solution is entirely not acceptable. When the distance δ_b greater than the size of the element, an instability of the numerical solution appears independently of the number of finite elements. Paradoxically, the greater number of elements, the sooner the instability appears even though the accuracy of solution in the first row of elements becomes better. The classic FEM leads to much worse results than the FEMT because the latter makes use of the Trefftz functions which satisfy the energy equation. This way the physical meaning of the results is ensured.

4.13 Energetic regularization in FEM

Three kinds of physical aspects of heat conduction can be applied to regularize an approximate solution obtained with the use of finite element method, (Ciałkowski et al., 2007). The first is minimization of heat flux jump between the elements, the second is minimization of the defect of energy dissipation on the border between elements and the third is the minimization of the intensity of entropy production between elements. Three kinds of regularizing terms for the objective functional are proposed:

- minimizing the heat flux inaccuracy between elements:

$$\sum_{i,j}\int_0^{t_e} dt \int_{\Gamma_{i,j}} \left(\frac{\partial \tilde{T}_i}{\partial n_i} - \frac{\partial \tilde{T}_j}{\partial n_j} \right)^2 d\Gamma \tag{68}$$

- minimizing numerical entropy production between elements:

$$\sum_{i,j}\int_0^{t_e} dt \int_{\Gamma_{i,j}} \left(\frac{1}{\tilde{T}_i}\frac{\partial \tilde{T}_i}{\partial n_i} - \frac{1}{\tilde{T}_j}\frac{\partial \tilde{T}_j}{\partial n_j} \right)^2 d\Gamma \text{ , and} \tag{69}$$

- minimizing the defect of energy of dissipation between elements:

$$\sum_{i,j}\int_0^{t_e} dt \int_{\Gamma_{i,j}} \left(\frac{\partial \tilde{T}_i}{\partial n_i}\ln \tilde{T}_i - \frac{\partial \tilde{T}_j}{\partial n_j}\ln \tilde{T}_j \right)^2 d\Gamma \tag{70}$$

with t_f being the final moment of the considered time interval, (Ciałkowski et al., 2007; Grysa & Leśniewska, 2009), and $\Gamma_{i,j}$ standing for the border between i-th and j-th element.

Notice that entropy production functional and energy dissipation functional are not quadratic functions of the coefficients of the base functions in elements. Hence, minimizing the objective functional leads to a non-linear system of algebraic equations. It seems to be the only disadvantage when compared with minimizing mean-square defects of heat flux (formula (68)); the latter leads to a system of linear equations.

4.14 Other methods

Many other methods are used to solve the inverse heat conduction problems. Many iterative methods for approximate solution of inverse problems are presented in monograph (Bakushinsky & Kokurin, 2004). Numerical methods for solving inverse problems of mathematical physics are presented in monograph (Samarski & Vabishchevich, 2007). Among other methods it is worth to mention boundary element method (Białecki et al., 2006; Onyango et al., 2008), the finite difference method (Luo & Shih, 2005; Soti et al., 2007), the theory of potentials method (Grysa, 1989), the radial basis functions method (Kołodziej et al., 2010), the artificial bee colony method (Hetmaniok et al., 2010), the Alifanov iterative regularization (Alifanov, 1994), the optimal dynamic filtration, (Guzik & Styrylska, 2002), the control volume approach (Taler & Zima, 1999), the meshless methods ((Sladek et al., 2006) and many other.

5. Examples of the inverse heat conduction problems

5.1 Inverse problems for the cooled gas turbine blade

Let us consider the following stationary problem concerning the gas turbine blade (Figure 4): find temperature distribution on the inner boundary Γ_i of the blade cross-section, $T|_{\Gamma_i}$, and heat transfer coefficient variation along Γ_i, with the condition

$$T_0 - \varepsilon_T \le T(s) \le T_0 + \varepsilon_T \tag{71}$$

where ε_T stands for temperature measurement tolerance and s is a normalized coordinate of a perimeter length (black dots in Figure 4 denote the beginning and the end of the inner and outer perimeter, coordinate is counted counterclockwise). Heat transfer coefficient distribution at the outer surface, $h_c|_{\Gamma_o}$, is known, T_{fo} = *1350* °*C*, T_{fi}=*780oC*, T_0 = *1100* °*C*, ε_T, standing for temperature measurement tolerance, does not exceed 1°C. Moreover, the inner and outer fluid temperature T_{fo} and T_{fi} are known, (Ciałkowski et al., 2007a). The unknowns: $T|_{\Gamma_i} = ?$, $h_c|_{\Gamma_i} = ?$ The solution has to be found in the class of functions fulfilling the energy equation

$$\nabla(k\nabla T) = 0 \tag{72}$$

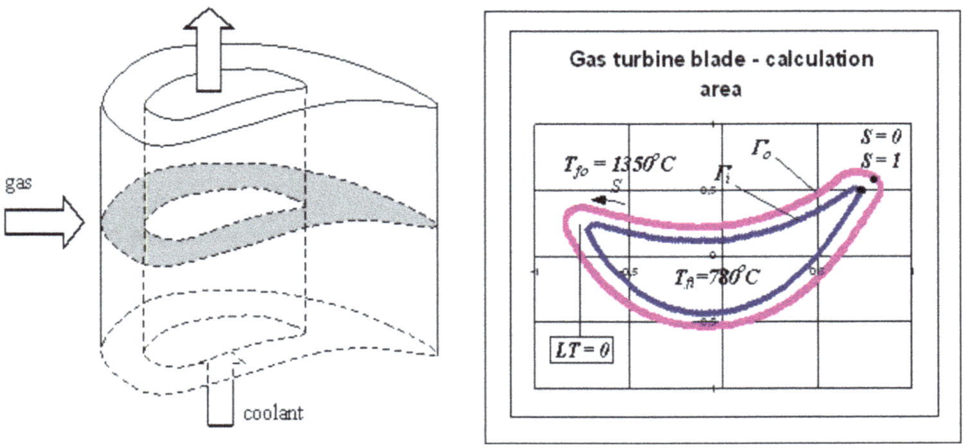

Fig. 4. An outline of a turbine blade.

with k assumed to be a constant. To solve the problem we use FEM with the shape functions belonging to the class of harmonic functions. It means that we can express an approximate solution of a stationary heat conduction problem in each element as a linear combination of the T-functions suitable for the equation (72). The functional with a term minimizing the heat flux inaccuracy between elements reads

$$I_{\delta}(T)=\sum_{ij}\left[\int_{\Gamma_{i,j}}(q_{+}-q_{-})^{2}\,d\Gamma+w\int_{\Gamma_{i,j}}(T_{+}-T_{-})^{2}\,d\Gamma\right] \text{ with } q=k\frac{\partial T}{\partial n} \tag{73}$$

In order to simplify the problem, temperature on the outer and inner surfaces was then approximated with 5 and 30 Bernstein polynomials, respectively, in order to simplify the problem. The area of the blade cross-section was divided into 99 rectangular finite elements with 16 nodes (12 on the boundary of each element and 4 inside). 16 harmonic (Trefftz) functions were used as base functions. All together 4x297 unknowns were introduced. Calculations were carried out with the use of PC with 1.6 GHz processor. Time of calculation was 1,5 hours using authors' own computer program in Fortran F90. The results are presented at Figures 5 and 6.

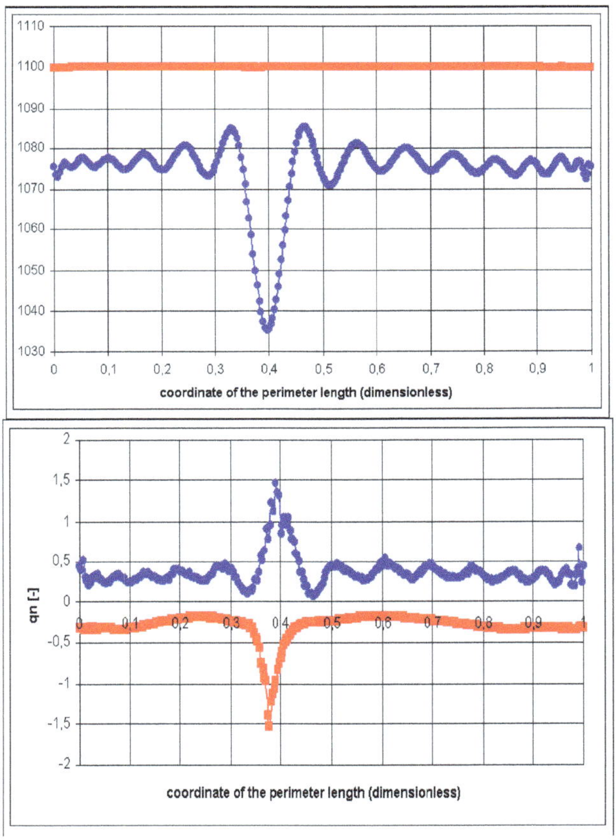

Fig. 5. Temperature [°C] (upper) and heat flux (lower) distribution on the outer (red squares) and inner (dark blue dots) surfaces of the blade.

Oscillations of temperature of the inner blade surface (Figure 5 left) is due to the number of Bernstein polynomials: it was too small. However, thanks to a small number of the polynomials a small number of unknown values of temperature could be taken for calculation. The same phenomenon appears in Figure 5 right for heat flux on the inner blade surface as well as in Figure 6 for the heat transfer coefficients values. The distance between peaks of the curves for the inner and outer surfaces in Figure 6 is a result of coordinate normalization of the inner and outer surfaces perimeter length. The normalization was done in such a way that only for $s = 0$ $(s = 1)$ points on both surfaces correspond to each other. The other points with the same value of the coordinate s for the outer and inner surface generally do not correspond to each other (in the case of peaks the difference is about 0,02).

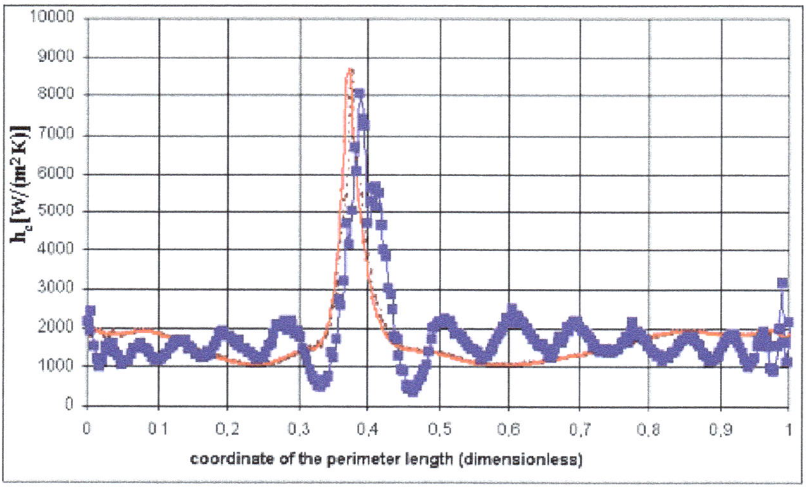

Fig. 6. Heat transfer coefficient over inner (dark blue squares) and outer (red dots – given; brown dots – calculated) surfaces of the blade.

5.2 Direct solution of a heat transfer coefficient identification problem

Consider a 1D dimensionless problem of heat conduction in a thermally isotropic flat slab (Grysa, 1982):

$$\partial^2 T / \partial x^2 = \partial T / \partial t \quad \text{for } x \in (0,1) \text{ and } t \in (0, t_f],$$

$$\partial T / \partial x = 0 \qquad \text{for } x = 0 \text{ and } t \in (0, t_f], \tag{74}$$

$$k\partial T / \partial x = -Bi\left[T(1,t) - T_f(t)\right] \quad \text{for } x = 1 \text{ and } t \in (0, t_f],$$

$$T = 0 \qquad \text{for} \quad x \in (0,1) \text{ and } t = 0 .$$

If the upper surface temperature (for x = 1) cannot be measured directly then in order to find the Biot number, temperature responses at some inner points of the slab or even temperature of the lower surface (x = 0) have to be known. Hence, the problem is ill-posed. Employing the Laplace transformation to the problem (74) we obtain

$$\bar{T}(x,s) = \bar{T}_f(s)\frac{Bi\cosh x\sqrt{s}}{\sqrt{s}\sinh\sqrt{s} + Bi\cosh\sqrt{s}} \quad \text{or}$$

$$\bar{T}_f(s)\frac{\cosh x\sqrt{s}}{s\cosh\sqrt{s}} = \frac{1}{s}\bar{T}(x,s) + \frac{1}{Bi}\bar{T}(x,s)\frac{\sinh\sqrt{s}}{\sqrt{s}\cosh\sqrt{s}} \tag{75}$$

The equation (75) is then used to find the formula describing the Biot number, Bi. Then, the inverse Laplace transformation yields:

$$Bi = \frac{2T(x,\tau) * \sum_{n=1}^{\infty}\exp\left(-\lambda_n^2 t\right)}{T_f(t) * \left[1 + 2\sum_{n=1}^{\infty}\frac{(-1)^n}{\lambda_n}\cos(x\lambda_n)\exp\left(-\lambda_n^2 t\right)\right] - H(t) * \theta(x,t)} \tag{76}$$

Here asterisk denotes convolution, $H(\tau)$ is the Heaviside function and $\lambda_n = \pi(2n-1)/2$, $n = 1,2,\ldots$.

If the temperature is known on the boundary $x = 0$ (e.g. from measurements), values of Bi (because of noisy input data having form of a function of time) can be calculated from formula (76). Of course, formula (76) is obtained with the assumption that Bi = const. Therefore, the results have to be averaged in the considered time interval.

6. Final remarks

It is not possible to present such a broad topic like inverse heat conduction problems in one short chapter. Many interesting achievements were discussed very briefly, some were omitted. Little attention was paid to stochastic methods. Also, the non-linear issues were only mentioned when discussing some methods of solving inverse problems. For lack of space only few examples could be presented.

The inverse heat conduction problems have been presented in many monographs and tutorials. Some of them are mentioned in references, e.g. (Alifanov, 1994; Bakushinsky & Kokurin, 2004; Beck & Arnold, 1977; Grysa, 2010; Kurpisz & Nowak, 1995; Özisik & Orlande, 2000; Samarski & Vabishchevich, 2007; Duda & Taler, 2006; Hohage, 2002; Bal, 2004; Tan & Fox, 2009).

7. References

Alifanov, O. M. (1994), *Inverse heat transfer problems*, Springer-Verlag, ISBN 0-387-53679-5, New York

Anderssen, R. S. (2005), Inverse problems: A pragmatist's approach to the recovery of information from indirect measurements, *Australian and New Zealand Industrial and Applied Mathematics Journal* Vol.46, pp. C588--C622, ISSN 1445-8735

Bakushinsky, A. B. & Kokurin M. Yu. (2004), *Iterative Methods for Approximate Solution of Inverse Problems*, Springer, ISBN 1-4020-3121-1, Dordrecht, The Netherlands

Bal. G, (2004), *Lecture Notes, Introduction to Inverse Problems*, Columbia University, New York, Date of acces: June 30, 2011, Available from: http://www.columbia.edu/~gb2030/COURSES/E6901/LectureNotesIP.pdf

Beck, J. V. & Arnold, K. J. (1977) *Parameter Estimation in Engineering and Science*, Wiley, ISBN 0471061182, New York

Beck, J. V. (1962), Calculation of surface heat flux from an internal temperature history, *ASME Paper* 62-HT-46

Beck, J. V., Blackwell B. & St. Clair, Jr, R. St. (1985), *Inverse heat conduction*, A Wiley-Interscience Publication, ISBN 0-471-08319-4, New York - Chichester - Brisbane - Toronto - Singapure

Bialecki, R., Divo E. & Kassab, A. (2006), Reconstruction of time-dependent boundary heat flux by a BEM-based inverse algorithm, *Engineering Analysis with Boundary Elements*, Vol.30, No.9, September 2006, pp. 767-773, ISSN 0955-7997

Burggraf, O. R. (1964), An Exact Solution of the Inverse Problem in Heat Conduction Theory and Application, *Journal of Heat Transfer*, Vol.86, August 1964, pp.373-382, ISSN 0022-1481

Chen, C. S., Karageorghis, A. & Smyrlis Y.S. (2008), *The Method of Fundamental Solutions - A Meshless Method*, Dynamic Publishers, Inc., ISBN 1890888-04-4, Atlanta, USA

Cheng, C.H. & Chang, M.H. (2003), Shape design for a cylinder with uniform temperature distribution on the outer surface by inverse heat transfer method, *International Journal of Heat and Mass Transfer*, Vol.46, No.1, (January 2003), pp. 101-111, ISSN 0017-9310

Ciałkowski, M. J. (2001), New type of basic functions of FEM in application to solution of inverse heat conduction problem, *Journal of Thermal Science*, Vol.11, No.2, pp. 163-171, ISSN 1003-2169

Ciałkowski, M. J., Frąckowiak, A. & Grysa, K. (2007), Solution of a stationary inverse heat conduction problem by means of Trefftz non-continuous method, *International Journal of Heat and Mass Transfer* Vol.50, No.11-12, pp.2170–2181, ISSN 0017-9310

Cialkowski, M. J., Frąckowiak, A. & Grysa, K. (2007a), Physical regularization for inverse problems of stationary heat conduction, *Journal of Inverse and Ill-Posed Problems*, Vol.15, No.4, pp. 347–364. ISSN 0928-0219

Ciałkowski, M. J. & Grysa, K. (2010), A sequential and global method of solving an inverse problem of heat conduction equation, *Journal of Theoretical and Applied Mechanics*, Vol.48, No.1, pp. 111-134, ISSN 1429-2955

Ciałkowski, M. J. & Grysa, K. (2010a), Trefftz method in solving the inverse problems, *Journal of Inverse and Ill-posed Problems,* Vol.18, No.6, pp. 595 - 616, ISSN 0928-0219

Dennis, B. H., Dulikravich, G. S. Egorov, I. N., Yoshimura, S. & Herceg, D. (2009), Three-Dimensional Parametric Shape Optimization Using Parallel Computers, *Computational Fluid Dynamics Journal,* Vol.17, No.4, pp.256–266, ISSN 0918-6654

Dennis, J. & Schnabel, R. (1983), *Numerical Methods for Unconstrained Optimization and Nonlinear Equations,* Prentice Hall, ISBN 0-89871-364-1

Duda, P. & Taler, J. (2006), *Solving Direct and Inverse Heat Conduction Problems,* Springer, ISBN 354033470X

Fan, Y. & Li, D.-G. (2009), Identifying the Heat Source for the Heat Equation with Convection Term, *International Journal of Mathematical Analysis,* Vol.3, No.27, pp. 1317–1323, ISSN 1312-8876

Golub, G. & Van Loan, C.(1998), *Matrix Computations.*: The Johns Hopkins University Press, ISBN 0-8018-5413-X, Baltimore, USA

Guzik, A. & Styrylska, T (2002), An application of the generalized optimal dynamic filtration method for solving inverse heat transfer problems, *Numerical Heat Transfer,* Vol.42, No.5, October 2002, pp.531-548, ISSN 1040-7782

Grysa, K. (1982), Methods of determination of the Biot number and the heat transfer coefficient, *Journal of Theoretical and Applied Mechanics,* 20, 1/2, 71-86, ISSN 1429-2955

Grysa, K. (1989), On *the exact and approximate methods of solving inverse problems of temperature fields,* Rozprawy 204, Politechnika Poznańska, ISBN 0551-6528, Poznań, Poland

Grysa, K. & Lesniewska, R. (2009), Different Finite Element Approaches For The Inverse Heat Conduction Problems, *Inverse Problems in Science and Engineering,* Vol.18, No.1 pp. 3-17, ISSN 1741-5977

Grysa, K. & Maciejewska, B. (2005), Application of the modified finite elements method to identify a moving heat source, In: *Numerical Heat Transfer 2005,* Vol.2, pp. 493-502, ISBN 83-922381-2-5, EUTOTERM 82, Gliwice-Cracow, Poland, September 13-16, 2005

Grysa, K. (2010), *Trefftz functions and their Applications in Solving Inverse Problems,* Politechnika Świętokrzyska, PL ISSN 1897-2691, (in Polish)

Hadamard, J. (1923), *Lectures on the Cauchy's Problem in Linear Partial Differential Equations,* Yale University Press, New Haven, recent edition: Nabu Press, 2010, ISBN 9781177646918

Hansen, P. C. (1992), Analysis of discrete ill-posed problems by means of the *L*-curve, *SIAM Review,* Vol.34, No.4, pp. 561–580, ISSN 0036-1445

Hansen, P. C. (2000), The L-curve and its use in the numerical treatment of inverse problems, In: *Computational Inverse Problems in Electrocardiology,* P. Johnston (Ed.), 119-142, Advances in Computational Bioengineering. Available, WIT Press, from http://www.sintef.no/project/eVITAmeeting/2005/Lcurve.pdf

Hansen, P.C. & O'Leary, D.P. (1993), The use of the *L*-curve in the regularization of discrete ill-posed problems, *SIAM Journal of Scientific Computing,* Vol.14, No.6, pp. 1487–1503, ISSN 1064-8275

Hetmaniok, E., Słota, D. & Zielonka A. (2010), Solution of the inverse heat conduction problem by using the ABC algorithm, *Proceedings of the 7th international conference on Rough sets and current trends in computing RSCTC'10,* pp.659-668, ISBN:3-642-13528-5, Springer-Verlag, Berlin, Heidelberg

Hohage, T. (2002), *Lecture Notes on Inverse Problems*, University of Goettingen. Date of acces : June 30, 2011, Available from http://www.mathematik.uni-stuttgart.de/studium/infomat/Inverse-Probleme-Kaltenbacher-WS0607/ip.pdf

Hon, Y.C. & Wei, T. (2005), The method of fundamental solutions for solving multidimensional inverse heat conduction problems, *Computer Modeling in Engineering & Sciences,* Vol.7, No.2, pp. 119-132, ISSN 1526-1492

Hożejowski, L., Grysa, K., Marczewski, W. & Sendek-Matysiak, E. (2009), Thermal diffusivity estimation from temperature measurements with a use of a thermal probe, *Proceedings of the International Conference Experimental Fluid Mechanics 2009,* pp. 63-72, ISBN 978-80-7372-538-9, Liberec, Czech Republic, November 25.-27, 2009

Ikehata, M. (2007), An inverse source problem for the heat equation and the enclosure method, *Inverse Problems,* Vol. 23, No 1, pp. 183–202, ISSN 0266-5611

Jin, B. & Marin, L. (2007), The method of fundamental solutions for inverse source problems associated with the steady-state heat conduction, *International Journal for Numerical Methods in Engineering,* Vol.69, No.8, pp. 1570–1589, ISSN 0029-5981

Kalman, R. E. (1960), A New Approach to Linear Filtering and Prediction Problems, *Transactions of the ASME - Journal of Basic Engineering,* Vol.82, pp. 35-45, ISSN 0021-9223

Kołodziej, J. A., Mierzwiczak, M. & Ciałkowski M. J. (2010), Application of the method of fundamental solutions and radial basis functions for inverse heat source problem in case of steady-state, *International Communications in Heat and Mass Transfer,* Vol.37, No.2, February 2010, pp.121-124, ISSN 0735-1933

Kover'yanov, A. V. (1967), Inverse problem of nonsteady state thermal conductivity, *Teplofizika vysokikh temperatur,* Vol.5 No.1, pp.141-148, ISSN 0040-3644

Kurpisz, K. & Nowak, A. J. (1995), *Inverse Thermal Problems,* Computational Mechanics Publications, ISBN 1 85312 276 9, Southampton, UK

Lorentz, G. G. (1953), *Bernstein Polynomials.* University of Toronto Press, ISBN 0-8284-0323-6, Toronto,

Luo, J. & Shih, A. J. (2005), Inverse Heat Transfer Solution of the Heat Flux Due to Induction Heating, *Journal of Manufacturing Science and Engineering,* Vol.127, No.3, pp.555-563, ISSN 1087-1357

Masood, K., Messaoudi, S. & Zaman, F.D. (2002), Initial inverse problem in heat equation with Bessel operator, *International Journal of Heat and Mass Transfer,* Vol.45, No.14, pp. 2959–2965, ISSN 0017-9310

Monde, M., Arima, H., Liu, W., Mitutake, Y. & Hammad, J.A. (2003), An analytical solution for two-dimensional inverse heat conduction problems using Laplace transform, *International Journal of Heat and Mass Transfer,* Vol.46, No.12, pp. 2135–2148, ISSN 0017-9310

Neaupane, K. M. & Sugimoto, M. (2003), An inverse Boundary Value problem using the Extended Kalman Filter, *ScienceAsia*, Vol.29, pp.121-126, ISSN 1513-1874

Norton, J. P. (1986), *An Introduction to identification*, Academic Press, ISBN 0125217307, London

Onyango, T.T.M., Ingham, D.B. &. Lesnic, D. (2008), Reconstruction of heat transfer coefficients using the boundary element metod. *Computers and Mathematics with Applications*, Vol.56 No.1, pp. 114–126, ISSN: 0898-1221

Özisik, M. N. & Orlande, H. R. B. (2000), *Inverse Heat Transfer: Fundamentals and Applications*, Taylor $ Francis, ISBN 1-56032-838-X, New York, USA

Pereverzyev, S.S., Pinnau R. & Siedow N., (2005), Initial temperature reconstruction for nonlinear heat equation: application to a coupled radiative-conductive heat transfer problem, *Inverse Problems in Science and Engineering*, Vol.16, No.1, pp. 55-67, ISSN 1741-5977

Reddy, J. N. & Gartling, D. K. (2001) *The finite element method in heat transfer and fluid dynamics*, CRC Press, ISBN 084932355X, London, UK

Reinhardt, H.-J., Hao, D. N., Frohne, J. & Suttmeier, F.-T. (2007), Numerical solution of inverse heat conduction problems in two spatial dimensions, *Journal of Inverse and Ill-posed Problems*, Vol.15, No. 5, pp. 181-198, ISSN: 0928-0219

Ren, H.-S. (2007), Application of the heat-balance integral to an inverse Stefan problem, *International Journal of Thermal Sciences*, Vol.46, No.2, (February 2007), pp. 118–127, ISSN 1290-0729

Samarski, A. A. & Vabishchevich, P. N. (2007), *Numerical methods for solving inverse problems of mathematical physics*, de Gruyter, ISBN 978-3-11-019666-5, Berlin, Germany

Sladek, J., Sladek, V. & Hon, Y. C. (2006) Inverse heat conduction problems by meshless local Petrov–Galerkin method, *Engineering Analysis with Boundary Elements*, Vol.30, No.8, August 2006, pp. 650–661, ISSN 0955-7997

Soti, V., Ahmadizadeh, Y., Pourgholi, Y. R. & Ebrahimi M. (2007), Estimation of heat flux in one-dimensional inverse heat conduction problem, *International Mathematical Forum*, Vol.2, No. 10, pp. 455 - 464, ISSN 1312-7594

Taler, J. & Zima, W. (1999), Solution of inverse heat conduction problems using control volume approach, *International Journal Of Heat and Mass Transfer*, Vol.42, No. 6, pp.1123-1140, ISSN 0017-9310

Tan, S. M. & Fox, C. (2009), *Physics 707 Inverse Problems*. The University of Auckland. Date of acces : June 30, 2011, Available from http://home.comcast.net/~SzeMengTan/InverseProblems/chap1.pdf

Tikhonov, A. N. & Arsenin, V. Y. (1977), *On the solution of ill-posed problems*, John Wiley and Sons, ISBN 0-470-99124-0, New York, USA

Trefftz, E. (1926), Ein Gegenstuek zum Ritz'schen Verfahren. *Proceedings of the 2nd International Congress of Applied Mechanics*, pp.131–137, Orell Fussli Verlag, Zurich,

Woo, K. C. & Chow, L. C. (1981), Inverse Heat Conduction by Direct Inverse Laplace Transform, *Numerical Heat Transfer*, Vol.4, pp.499-504, ISSN 1040-7782

Yamamoto, M. & Zou, J. (2001), Simultaneous reconstruction of the initial temperature and heat radiative coefficient, *Inverse Problems* Vol.17, No.4, pp. 1181–1202, ISSN 0266-5611

Yang C. 1998, A linear inverse model for the temperature-dependent thermal conductivity determination in one-dimensional problems, *Applied Mathematical Modelling*, Vol.22 No.1-2, pp. 1-9, ISSN: 0307-904X

Zienkiewicz, O. (1977), *The Finite Element Method,* McGraw-Hill, ISBN 0-07-084072-5, London

2

Assessment of Various Methods in Solving Inverse Heat Conduction Problems

M. S. Gadala and S. Vakili
Department of Mechanical Engineering, The University of British Columbia, Vancouver, BC, Canada

1. Introduction

In an inverse heat conduction problem (IHCP), the boundary conditions, initial conditions, or thermo-physical properties of material are not fully specified, and they are determined from measured internal temperature profiles. The challenge is that the effect of changes in boundary conditions are normally damped or lagged, i.e. the varying magnitude of the interior temperature profile lags behind the changes in boundary conditions and is generally of lesser magnitude. Therefore, such a problem would be a typically ill-posed and would normally be sensitive to the measurement errors. Also, in the uniqueness and stability of the solution are not generally guaranteed (Beck et al., 1985; Alifanov, 1995; Ozisik, 2000).
Inverse heat conduction problems, like most of the inverse problems encountered in science and engineering may be reformulated as an optimization problem. Therefore, many available techniques of solving the optimization problems are available as methods of solving the IHCPs. However, the corresponding objective function of the inverse problems can be highly nonlinear or non-monotonic, may have a very complex form, or in many practical applications, its analytical expression may be unknown. The objective function usually involves the squared difference between measured and estimated unknown variables. If **Y** and **T** are the vectors of the measured and estimated temperatures, then the objective function will be in the form of

$$U = [\mathbf{Y} - \mathbf{T}]^T [\mathbf{Y} - \mathbf{T}] \quad (1)$$

However, normally there is need for another term, called "regularization" in order to eliminate the oscillations in the results and make the solution more stable. The effect of this term and the strategy of choosing it will be discussed in details in the subsequent chapters.
The above equation is only valid, if the measured temperatures and the associated errors have the following statistical characteristics (Beck & Arnold, 1977):

- The errors are additive, i.e.

$$Y_i = T_i + \varepsilon_i \quad (2)$$

where ε_i is the random error associated with the ith measurement.

- The temperature errors have zero mean.
- The errors have constant variance.

- The errors associated with different measurements are uncorrelated.
- The measurement errors have a normal (Gaussian) distribution.
- The statistical parameters describing the errors, such as their variance, are known.
- Measured temperatures are the only variables that contain measurement errors. Measured time, positions, dimensions, and all other quantities are all accurately known.
- There is no more prior information regarding the quantities to be estimated. If such information is available, it should be utilized to improve the estimates.

While classical methods, such as the least square regularization method (Beck et al., 1985; Beck et al., 1996), the sequential function specification method (Alifanov, 1995; Beck et al., 1996; Blanc et al., 1998), the space marching method (Al-Khalidy, 1998), conjugate gradient method (Abou khachfe & Jarny, 2001; Huang & Wang, 1999), steepest descent method (Huang et al., 2003), and the model reduction algorithm (Battaglia, 2002; Girault et al., 2003) are vastly studied in the literature, and applied to the problems in thermal engineering (Bass, 1980; Osman, 1190; Kumagai et al., 1995; Louahia-Gualous et al., 2003; Kim & Oh, 2001; Pietrzyk & Lenard, 1990; Alifanov et al., 2004; Gadala & Xu, 2006), there are still some unsolved problems:

- The solution often shows some kinds of overshoot and undershoot, which may result in non-physical answers.
- Very high heat flux peak values such as those experienced in jet impingement cooling are normally damped and considerably underestimated.
- Results are very sensitive to the quality of input. Measurement errors are intrinsic in laboratory experiments, so we need a more robust approach in solving the inverse problem.
- The time step size that can be used with these methods is bounded from below, and cannot be less than a specific limit (Beck et al., 1985). This causes temporal resolutions that are not sufficient for some real world applications, where changes happen very fast.

More recent optimization techniques may be used in the solution of the IHCPs to aid in stability, solution time, and to help in achieving global minimum solutions. Some of these techniques are briefly reviewed in the following section:

Genetic algorithm

This technique has been widely adopted to solve inverse problems (Raudensky et al., 1995; Silieti et al., 2005; Karr et al., 2000). Genetic algorithms (GAs) belong to the family of computational techniques originally inspired by the living nature. They perform random search optimization algorithms to find the global optimum to a given problem. The main advantage of GAs may not necessarily be their computational efficiency, but their robustness, i.e. the search process may take much longer than the conventional gradient-based algorithms, but the resulting solution is usually the global optimum. Also, they can converge to the solution when other classical methods become unstable or diverge. However, this process can be time consuming since it needs to search through a large tree of possible solutions. Luckily, they are inherently parallel algorithms, and can be easily implemented on parallel structures.

Neural networks

Artificial neural networks can be successfully applied in the solution of inverse heat conduction problems (Krejsa et al., 1999; Shiguemori et al., 2004; Lecoeuche et al., 2006).

They are capable of dealing with significant non-linearities and are known to be effective in damping the measurement errors.

Self-learning finite elements

This methodology combines neural network with a nonlinear finite element program in an algorithm which uses very basic conductivity measurements to produce a constitutive model of the material under study. Through manipulating a series of neural network embedded finite element analyses, an accurate constitutive model for a highly nonlinear material can be evolved (Aquino & Brigham, 2006; Roudbari, 2006). It is also shown to exhibit a great stability when dealing with noisy data.

Maximum entropy method

This method seeks the solution that maximizes the entropy functional under given temperature measurements. It converts the inverse problem to a non-linear constrained optimization problem. The constraint is the statistical consistency between the measured and estimated temperatures. It can guarantee the uniqueness of the solution. When there is no error in the measurements, maximum entropy method can find a solution with no deterministic error (Kim & Lee, 2002).

Proper orthogonal decomposition

Here, the idea is to expand the direct problem solution into a sequence of orthonormal basis vectors, describing the most essential features of spatial and temporal variation of the temperature field. This can result in the filtration of the noise in the field under study (Ostrowski et al., 2007).

Particle Swarm Optimization (PSO)

This is a population based stochastic optimization technique, inspired by social behavior of bird flocking or fish schooling. Like GA, the system is initialized with a population of random solutions and searches for optima by updating generations. However, unlike GA, PSO has no evolution operators such as crossover and mutation. In PSO, the potential solutions, called particles, fly through the problem space by following the current optimum particles. Compared to GA, the advantages of PSO are the ease of implementation and that there are few parameters to adjust. Some researchers showed that it requires less computational expense when compared to GA for the same level of accuracy in finding the global minimum (Hassan et al., 2005).

In this chapter, in addition to the classical function specification method, we will study the genetic algorithm, neural network, and particle swarm optimization techniques in more details. We will investigate their strengths and weaknesses, and try to modify them in order to increase their efficiency and effectiveness in solving inverse heat conduction problems.

2. Function specification methods

As mentioned above, in order to stabilize the solution to the ill-posed IHCP, it is very common to include more variables in the objective function. A common choice in inverse heat transfer problems is to use a scalar quantity based on the boundary heat fluxes, with a weighting parameter α, which is normally called the regularization parameter. The regularization term can be linked to the values of heat flux, or their first or second derivatives, with respect to time or space. Previous research (Gadala & Xu, 2006) has shown that using the heat flux values (zeroth-order regularization) is the most suitable choice. The objective function then will be

$$F(\mathbf{q}) = \sum_{i=1}^{N} (\mathbf{T}_m^i - \mathbf{T}_c^i)^T (\mathbf{T}_m^i - \mathbf{T}_c^i) + \alpha \sum_{i=1}^{N} \mathbf{q}^{iT} \mathbf{q}^i \tag{3}$$

where $\mathbf{T_m^i}$ and $\mathbf{T}_c^\mathbf{i}$ are the vectors of expected (measured) and calculated temperatures at the ith time step, respectively, each having J spatial components; a is the regularization coefficient; and q^i is the boundary heat flux. It is important to notice that in the inverse analysis, the number of spatial components is equal in the measured and calculated temperature vectors; i.e. the spatial resolution of the recovered boundary heat flux vector is determined by the number of embedded thermocouples.

Due to the fact that inverse problems are generally ill-posed, the solution may not be unique and would be in general sensitive to measurement errors. To decrease such sensitivity and improve the simulation, a number of future time steps (n_{FTS}) are utilized in the analysis of each time step. This means that in addition to the measured temperature at the present time step T^i, the measured temperatures at future time steps, $T^{i+1}, T^{i+2}, \ldots., T^{i+n_{FTS}}$, are also used to approximate the heat flux q^i. In this process, a temporary assumption would be usually considered for the values of $q^{i+1}, q^{i+2}, \ldots., q^{i+n_{FTS}}$. The simplest and the most widely used one is to assume $q^{i+k} = q^i$ for $1 \le k \le n_{FTS}$, which is also used in our work. In this chapter, a combined function specification-regularization method is used, which utilizes both concepts of regularization, and future time steps (Beck & Murio, 1986).

Mathematically we may express T_c^k, the temperature at the kth time step and at location c as an implicit function of the heat flux history and initial temperature:

$$\mathrm{T}_c^k = f(\mathrm{q}^1, \mathrm{q}^2, \ldots, \mathrm{q}^k, T_c^0) \tag{4}$$

and the following equation is valid

$$\mathrm{T}_c^k = \mathrm{T}_c^{k*} + \frac{\partial \mathrm{T}_c^1}{\partial \mathrm{q}^1}(\mathrm{q}^1 - \mathrm{q}^{1^*}) + \frac{\partial \mathrm{T}_c^2}{\partial \mathrm{q}^2}(\mathrm{q}^2 - \mathrm{q}^{2^*}) + \ldots + \frac{\partial \mathrm{T}_c^k}{\partial \mathrm{q}^k}(\mathrm{q}^k - \mathrm{q}^{k^*}) \tag{5}$$

The values with a '*' superscript in the above may be considered as initial guess values.

The first derivative of temperature $\mathbf{T}_c^i$ with respect to heat flux q^i is called the sensitivity matrix:

$$\mathbf{X}^i = \frac{\partial \mathbf{T}_c^i}{\partial \mathbf{q}^i} = \begin{bmatrix} a_{11}(i) & a_{12}(i) & \ldots & a_{1J}(i) \\ a_{21}(i) & a_{22}(i) & \ldots & a_{2J}(i) \\ \vdots & \vdots & \ddots & \vdots \\ a_{L1}(i) & a_{L2}(i) & \ldots & a_{LJ}(i) \end{bmatrix} \tag{6}$$

$$a_{rs}(i) = \frac{\partial T_{cr}^i}{\partial q_s^i} \tag{7}$$

The optimal solution for Eq. (3) may be obtained by setting $\partial F / \partial \mathbf{q} = 0$, which results in the following set of equations (note that $\partial F / \partial \mathbf{q}$ should be calculated with respect to each component q^i, with i=1, 2...N):

$$\left\{ \sum_{i=1}^{N} \left(\frac{\partial \mathbf{T}_c^i}{\partial \mathbf{q}^j} \right)^T_{\mathbf{q}^j=\mathbf{q}^{j^*}} \left(\frac{\partial \mathbf{T}_c^i}{\partial \mathbf{q}^j} \right)_{\mathbf{q}^j=\mathbf{q}^{j^*}} + \alpha \mathbf{I} \right\} (\mathbf{q}^j - \mathbf{q}^{j^*}) = \sum_{i=1}^{N} \left(\frac{\partial T_c^i}{\partial \mathbf{q}^j} \right)^T_{\mathbf{q}^j=\mathbf{q}^{j^*}} \left(\mathbf{T}_m^i - \mathbf{T}_c^{i*} \right) - \alpha\, \mathbf{q}^{j^*} \qquad j = 1, 2, \ldots, N \tag{8}$$

where $\mathbf{q}^{j^*}$ is the initial guess of heat fluxes, $\mathbf{T}_c^{i*}$ is the calculated temperature vector with the initial guess values.
Recalling equations (6) and (7), equation (8) may be rearranged and written in the following form:

$$(\mathbf{X}^T_{\mathbf{q}=\mathbf{q}^*} \mathbf{X}_{\mathbf{q}=\mathbf{q}^*} + \alpha\, \mathbf{I})(\mathbf{q} - \mathbf{q}^*) = \mathbf{X}^T \Delta \mathbf{T} - \alpha\, \mathbf{q}^* \tag{9}$$

where $\mathbf{X}$ is the total sensitivity matrix for multi-dimensional problem and has the following form:

$$\mathbf{X} = \begin{bmatrix} \mathbf{X}^1 & 0 & 0 & 0 \\ \mathbf{X}^2 & \mathbf{X}^1 & 0 & 0 \\ \vdots & \vdots & \ddots & 0 \\ \mathbf{X}^N & \ldots & \mathbf{X}^2 & \mathbf{X}^1 \end{bmatrix} \tag{10}$$

and

$$\Delta \mathbf{T} = \left(\mathbf{T}_m^1 - \mathbf{T}_c^{1*} \quad \mathbf{T}_m^2 - \mathbf{T}_c^{2*} \quad \ldots \quad \mathbf{T}_m^N - \mathbf{T}_c^{N*} \right)^T \tag{11}$$

By solving Eq. (9), the heat flux update will be calculated and added to the initial guess. In this chapter, a fully sequential approach with function specification is used. First, the newly calculated $\mathbf{q}^{1''}$ is used for all time steps in the computation window after the first iteration, i.e., constant function specification is used for this computation window. Then, the computation window moves one time step at the next sequence after obtaining a convergent solution in the current sequence.
One important consideration in calculating the sensitivity values is the nonlinearity. The whole sensitivity matrix is independent of the heat flux only if the thermal properties of the material are not changing with temperature. For most materials, the thermophysical properties are temperature dependent. In such case, all properties should be updated at the beginning of each time step, which is time consuming especially for large size models. Moreover, such changes in properties would not be very large and would not significantly change the magnitude of the sensitivity coefficients. Also, updating the material properties at the beginning of each time step would be based on the temperatures $\mathbf{T}^{k*}$ obtained from the initially given values of heat flux $\mathbf{q}^*$, which is essentially an approximation. So, we may

update the sensitivity matrix every M steps (in our numerical experiments, M=10). The results obtained under this assumption were very close to those obtained by updating the values at each step, so the assumption is justified.
To obtain an appropriate level of regularization, the number of future time steps (or more accurately, the size of look-ahead time window, i.e. the product of the number of future time steps and time step size) and the value of the regularization parameter must be chosen with respect to the errors involved in the temperature readings. The residual principle (Alifanov, 1995; Woodbury & Thakur, 1996) has been used to determine these parameters based on the accuracy of thermocouples in the relative temperature range.

3. Genetic algorithm

Genetic algorithm is probably the most popular stochastic optimization method. It is also widely used in many heat transfer applications, including inverse heat transfer analysis (Gosselin et al., 2009). Figure 1 shows a flowchart of the basic GA. GA starts its search from a randomly generated population. This population evolves over successive generations (iterations) by applying three major operations. The first operation is "Selection", which mimics the principle of "Survival of the Fittest" in nature. It finds the members of the population with the best performance, and assigns them to generate the new members for future generations. This is basically a sort procedure based on the obtained values of the objective function. The number of elite members that are chosen to be the parents of the next generation is also an important parameter. Usually, a small fraction of the less fit solutions are also included in the selection, to increase the global capability of the search, and prevent a premature convergence. The second operator is called "Reproduction" or "Crossover", which imitates mating and reproduction in biological populations. It propagates the good features of the parent generation into the offspring population. In numerical applications, this can be done in several ways. One way is to have each part of the array come from one parent. This is normally used in binary encoded algorithms. Another method that is more popular in real encoded algorithms is to use a weighted average of the parents to produce the children. The latter approach is used in this chapter. The last operator is "Mutation", which allows for global search of the best features, by applying random changes in random members of the generation. This operation is crucial in avoiding the local minima traps. More details about the genetic algorithm may be found in (Davis, 1991; Goldberg, 1989).
Among the many variations of GAs, in this study, we use a real encoded GA with roulette selection, intermediate crossover, and uniform high-rate mutation (Davis, 1991). The crossover probability is 0.2, and the probability of adjustment mutation is 0.9. These settings were found to be the most effective based on our experience with this problem. A mutation rate of 0.9 may seem higher than normal. This is because we start the process with a random initial guess, which needs a higher global search capability. However, if smarter initial guesses are utilized, a lower rate of mutation may be more effective. Genes in the present application of GA consist of arrays of real numbers, with each number representing the value of the heat flux at a certain time step, or a spatial location.

4. Particle Swarm Optimization

We start by giving a description of the basic concepts of the algorithm. Then a brief description of the three variations of the PSO algorithm that are used in this study is given.

Finally we investigate some modifications in PSO algorithm to make it a more robust and efficient solver for the inverse heat conduction problem.

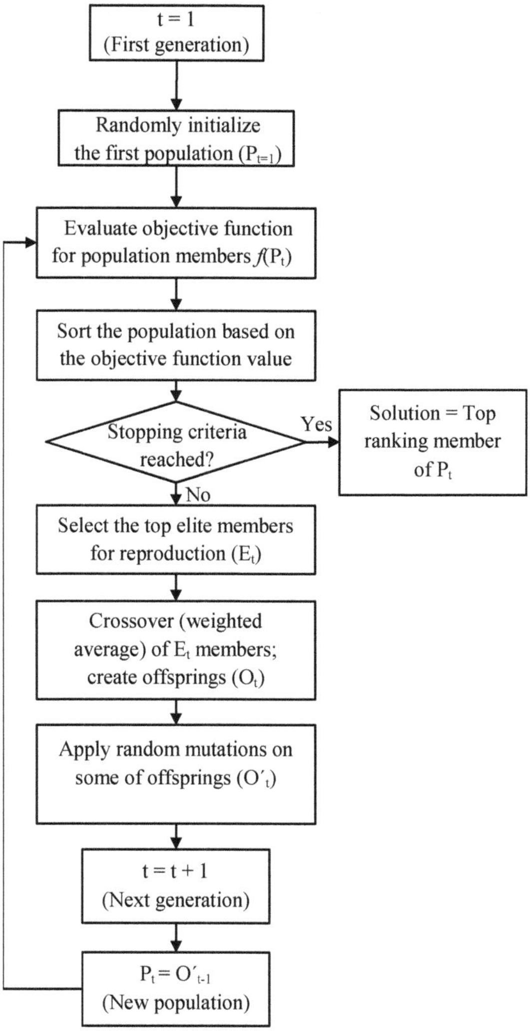

Fig. 1. Flowchart of a General Implementation of Genetic Algorithm (GA).

4.1 Basic concepts

Particle swarm optimization (PSO) is a high-performance stochastical search algorithm that can also be used to solve inverse problems. The method is based on the social behavior of species in nature, e.g., a swarm of birds or a school of fish (Eberhart & Kennedy, 1995).

In the basic PSO algorithm, if a member of the swarm finds a desirable position, it will influence the traveling path of the rest of the swarm members. Every member searches in its vicinity, and not only learns from its own experience (obtained in the previous iterations), but also benefits from the experiences of the other members of the swarm, especially from the experience of the best performer. The original PSO algorithm includes the following components (Clerc, 2006):

- *Particle Position Vector* $\boldsymbol{x}$: For each particle, this vector stores its current location in the search domain. These are the values for which the value of the objective function is calculated, and the optimization problem is solved.
- *Particle Velocity Vector* $\boldsymbol{v}$: For every particle, this vector determines the magnitude and direction of change in the position of that particle in the next iteration. This is the factor that causes the particles to move around the search space.
- *Best Solution of a Particle* $\boldsymbol{p}$: For each particle, this is the position that has produced the lowest value of the objective function (the best solution with the lowest error in our case). So if f is the objective function that is supposed to be minimized; i is the index for each particle, and m is the iteration counter, then:

$$p_i^m = \underset{0 \le s \le m}{\arg\min} \left(f\left(x_i^s\right) \right) \tag{12}$$

- *Best Global Solution* $\boldsymbol{g}$: This is the best single position found by all particles of the swarm, i.e., the single p point that produces the lowest value for the objective function, among all the swarm members. In other words, if n is the swarm size, then:

$$g^m = \underset{0 \le s \le m, 1 \le k \le n}{\arg\min} \left(f\left(x_k^s\right) \right) \tag{13}$$

The number of particles in the swarm (n) needs to be specified at the beginning. Fewer particles in the swarm results in lower computational effort in each iteration, but possibly higher number of iterations is required to find the global optimum. On the other hand, a larger population will have a higher computational expense in each iteration, but is expected to require less iterations to reach the global optimum point. Earlier studies have shown that a smaller population is normally preferred (Alrasheed et al., 2008; Karray & de Silva, 2004). This was also observed in our study; however, its effect seems to be insignificant.

The steps involved in the basic PSO algorithm are detailed below (Clerc, 2006):

1. Randomly initialize the positions and velocities for all of the particles in the swarm.
2. Evaluate the fitness of each swarm member (objective function value at each position point).
3. At iteration m, the velocity of the particle i, is updated as:

$$v_i^{m+1} = c_0 v_i^m + c_1 r_1 \left(p_i^m - x_i^m \right) + c_2 r_2 \left(g^m - x_i^m \right) \tag{14}$$

where x_i^m and v_i^m are the position and velocity of particle i at the m-th iteration, respectively; p_i^m and g^m are the best positions found up to now by this particle (local memory) and by the whole swarm (global memory) so far in the iterations, respectively; c_0 is called the inertia coefficient or the self-confidence parameter and is usually between zero and one; c_1 and c_2 are the acceleration coefficients that pull the particles toward the local and global best positions; and r_1 and r_2 are random vectors in the range of (0,1). The ratio between these three parameters controls the effect of the previous velocities and the trade-off between the global and local exploration capabilities.

1. Update the position of each particle using the updated velocity and assuming unit time:

$$x_i^{m+1} = x_i^m + v_i^{m+1} \tag{15}$$

2. Repeat (2) - (4) until a convergence criterion (an acceptable fitness value or a certain maximum number of iterations) is satisfied.

There are some considerations that must be taken into account when updating velocity of particles (step 3 of the above algorithm). First, we need a value for the maximum velocity. A rule of thumb requires that, for a given dimension, the maximum velocity, $v_{i,\max}$, should be equal to one-half the range of possible values for the search space. For example, if the search space for a specific dimension is the interval [0, 100], we will take a maximum velocity of 50 for this dimension. If the velocity obtained from Equation (14) is higher than $v_{i,\max}$, then we will substitute the maximum velocity instead of v_i^{m+1}. The reason for having this maximum allowable velocity is to prevent the swarm from "*explosion*" (divergence). Another popular way of preventing divergence is a technique called "*constriction*", which dynamically scales the velocity update (Clerc, 2006). The first method was used in a previous research by the authors (Vakili & Gadala, 2009). However, further investigation showed that a better performance is obtained when combining the constriction technique with limiting the maximum velocity. In this chapter, the velocity updates are done using constriction and can be written as:

$$v_i^{m+1} = K\left(v_i^m + c_1 r_1\left(p_i^m - x_i^m\right) + c_2 r_2\left(g^m - x_i^m\right)\right) \tag{16}$$

where K is the constriction factor, and is calculated as (Clerc, 2006):

$$K = \frac{2}{\left|2 - \varphi - \sqrt{\varphi^2 - 4\varphi}\right|} \tag{17}$$

where $\varphi = c_1 + c_2$. Here, following the recommendations in (Clerc, 2006), the initial values for c_1 and c_2 are set to 2.8 and 1.3, respectively. These values will be modified in subsequent iterations, as discussed below.

As mentioned above, the relation between the self-confidence parameter, c_0, and the acceleration coefficients, c_1 and c_2, determines the trade-off between the local and global search capabilities. When using the constriction concept, the constriction factor is responsible for this balance. As we progress in time through iterations, we get closer to the best value. Thus, a reduction in the value of the self-confidence parameter will limit the global exploration, and a more localized search will be performed. In this study, if the value of the best objective function is not changed in a certain number of iterations (10 iterations in our case), the value of K is multiplied by a number less than one (0.95 for our problems) to reduce it (i.e. $K^{new} = 0.95K^{old}$). These numbers are mainly based on the authors' experience, and the performance is not very sensitive to their exact values. Some other researchers have used a linearly decreasing function to make the search more localized after the few initial

iterations (Alrasheed et al., 2008). These techniques are called *"dynamic adaptation"*, and are very popular in the recent implementations of PSO (Fan & Chang, 2007).
Also, in updating the positions, one can impose a lower and an upper limit for the values, usually based on the physics of the problem. If the position values fall outside this range, several treatments are possible. In this study, we set the value to the limit that has been passed by the particle. Other ideas include substituting that particle with a randomly chosen other particle in the swarm, or penalizing this solution by increasing the value of the objective function.
Figure 2 shows a flowchart of the whole process. Figure 3 gives a visual representation of the basic velocity and position update equations.

4.2 Variations

Unfortunately, the basic PSO algorithm may get trapped in a local minimum, which can result in a slow convergence rate, or even premature convergence, especially for complex problems with many local optima. Therefore, several variants of PSO have been developed to improve the performance of the basic algorithm (Kennedy et al., 2001). Some variants try to add a chaotic acceleration factor to the position update equation, in order to prevent the algorithm from being trapped in local minima (Alrasheed et al., 2007). Others try to modify the velocity update equation to achieve this goal. One of these variants is called the Repulsive Particle Swarm Optimization (RPSO), and is based on the idea that repulsion between the particles can be effective in improving the global search capabilities and finding the global minimum (Urfalioglu, 2004; Lee et al., 2008). The velocity update equation for RPSO is

$$v_i^{m+1} = c_0 v_i^m + c_1 r_1 \left(p_i^m - x_i^m\right) + c_2 r_2 \left(p_j^m - x_i^m\right) + c_3 r_3 v_r \tag{18}$$

where p_j^m is the best position of a randomly chosen other particle among the swarm, c_3 is an acceleration coefficient, r_3 is a random vector in the range of (0,1), and v_r is a random velocity component. Here c_2 is -1.43, and c_3 is 0.5. These values are based on recommendations in (Clerc, 2006). The newly introduced third term on the right-hand side of Eq. 18., with always a negative coefficient (c_2), causes a repulsion between the particle and the best position of a randomly chosen other particle. Its role is to prevent the population from being trapped in a local minimum. The fourth term generates noise in the particle's velocity in order to take the exploration to new areas in the search space.
Once again, we are gradually decreasing the weight of the self-confidence parameter. Note that the third term on the right-hand side of Eq. (1), i.e., the tendency toward the global best position, is not included in a repulsive particle swarm algorithm in most of the literature.
The repulsive particle swarm optimization technique does not benefit from the global best position found. A modification to RPSO that also uses the tendency towards the best global point is called the "Complete Repulsive Particle Swarm Optimization" or CRPSO (Vakili & Gadala, 2009). The velocity update equation for CPRSO will be:

$$v_i^{m+1} = c_0 v_i^m + c_1 r_1 \left(p_i^m - x_i^m\right) + c_2 r_2 \left(g^m - x_i^m\right) + c_3 r_3 \left(p_j^m - x_i^m\right) + c_4 r_4 v_r \tag{19}$$

In CRPSO, by having both an attraction toward the particle's best performance, and a repulsion from the best performance of a random particle, we are trying to create a balance between the local and global search operations.

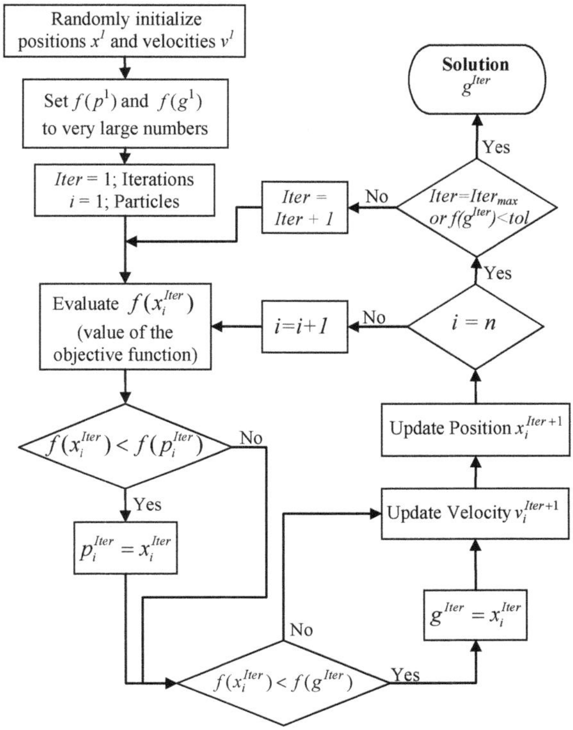

Fig. 2. Flowchart of the basic particle swarm optimization procedure.

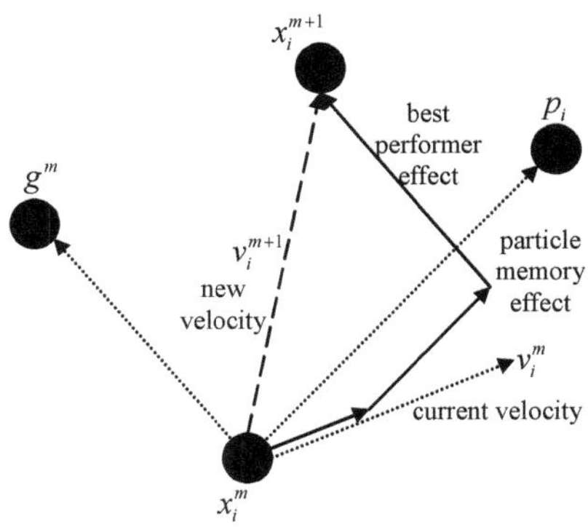

Fig. 3. Velocity and position updates in the basic particle swarm optimization algorithm.

Both RPSO and CRPSO were previously tested in solving inverse heat conduction problems by the authors (Vakili & Gadala, 2009). It was found that CRPSO shows better performance than the basic and repulsive PSO algorithms. In handling the noisy data, however, RPSO was the most efficient variation, followed closely by CRPSO. It was concluded then that the CRPSO variation is the suitable choice for IHCPs. Also, in (Vakili & Gadala, 2011), several modifications were done on the formulation of inverse heat conduction problem as an optimization problem, as well as on the implementation of the PSO algorithm.

5. Artificial Neural Networks

Artificial Neural Networks (ANN) are motivated by the efficiency of brain in performing computations. These networks are made of a large number of processing units (neurons) that are interconnected through weighted connections, similar to synapses in brain. In order for the network to perform the expected tasks, it should first go through a "learning" process. There are two main categories of learning: supervised, or unsupervised. In supervised learning, the network learning is achieved by practicing on pre-designed training sets, while in unsupervised learning, the network is presented with a set of patterns, and learns to group these patterns into certain categories. The supervised learning is useful in function fitting and prediction, while unsupervised learning is more applicable to pattern recognition and data clustering. Since the learning process in our application is a supervised one, we focus on this type of learning process.

While there are several major classes of neural networks, in this chapter, we have studied only two of them, which are introduced in this section.

5.1 Feedforward Multilayer Perceptrons (FMLP)

In a feedforward network, the nodes are arranged in layers, starting from the input layer, and ending with the output layer. In between these two layers, a set of layers called hidden layers, are present, with the nodes in each layer connected to the ones in the next layer through some unidirectional paths. See Fig. 4 for a presentation of the topology. It is common to have different number of elements in the input and output vectors. These vectors can occur either concurrently (order is not important), or sequentially (order is important). In inverse heat conduction applications, normally the order of elements is important, so sequential vectors are used.

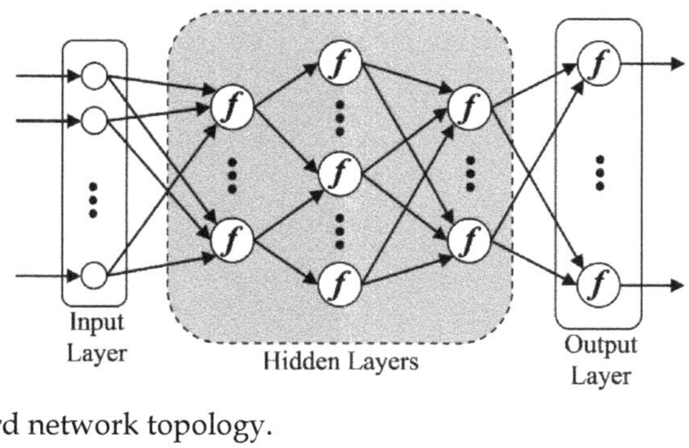

Fig. 4. A feedforward network topology.

5.2 Radial Basis Function Networks (RBFN)

The basic RBFN includes only an input layer, a single hidden layer, and an output layer. See Fig. 5 for a visual representation. The form of the radial basis function can be generally given by

$$f_i(\mathbf{x}) = r_i\left(\frac{\|\mathbf{x} - \mathbf{v}_i\|}{\sigma_i}\right) \tag{20}$$

in which $\mathbf{x}$ is the input vector, and $\mathbf{v}_i$ is the vector denoting the center of the receptive field unit f_i with σ_i as its unit width parameter. The most popular form of this function is the Gaussian kernel function, given as

$$f_i(\mathbf{x}) = \exp\left(\frac{-\|\mathbf{x}-\mathbf{v}_i\|^2}{2\sigma_i^2}\right) \tag{21}$$

These networks normally require more neurons than the feedforward networks, but they can be designed and trained much faster. However, in order to have a good performance, the training set should be available in the beginning of the process.

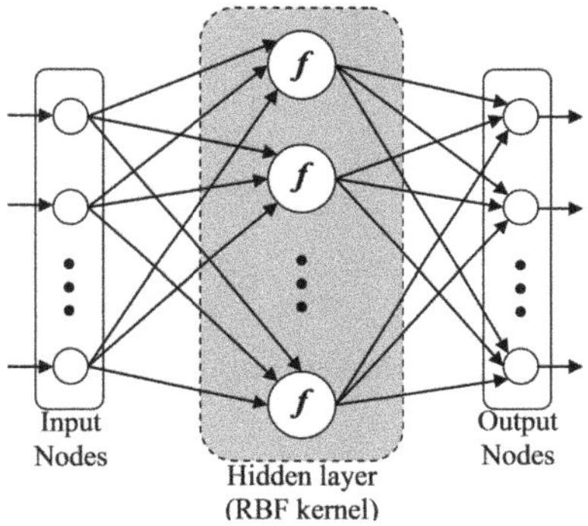

Fig. 5. An RBF network topology.

5.3 Implementation in inverse heat conduction problem

In order to use the artificial neural networks in the inverse heat conduction problem, we first started with a direct heat conduction finite element code, and applied several sets of heat fluxes in the boundary. The resulting temperatures in locations inside the domain, which correspond to the thermocouple locations in the experiments, were obtained. The neural network was then trained using the internal temperature history as an input, and the corresponding applied heat flux as the target. The assumption was that this way, the neural network should be able to act as an inverse analysis tool, and given a set of measured thermocouple readings, be able to reproduce the heat fluxes.

The obtained results, however, were far from satisfactory. It seemed that the relationship between the actual values of temperatures and heat fluxes is a complicated one, which is very hard for the neural networks to understand and simulate, at least when using a reasonably small number of layers. Thus, we decided to reformulate the problem, and use the change in the temperature in each time step as the input. In this formulation, neural networks performed much better, and a good quality was achieved in the solution in a reasonable amount of time.

Further investigations showed that if the time step size is varying, we can use a derivative of temperature with respect to the heat flux as the input, i.e. divide the temperature change by the time step size. The results were again satisfactory, however, more bookkeeping is needed, which complicates the implementation and makes the algorithm more prone to coding errors. This practice is not normally recommended, unless it can result in a considerable reduction in the solution time.

6. Test cases

A block containing nine thermocouples is modeled for each pass of water jet cooling of a steel strip. The length of the block is 114.3 mm (9 sections of each 12.7 mm). The width and thickness are 12.7 mm and 6.65 mm, respectively. To model the thermocouple hole, a cylinder of radius 0.5 mm and height of 5.65 mm is taken out of the block. Isoparametric eight-node brick elements are used to discretize the domain. Fig. 6(a) shows the whole domain, and Fig. 6(b) is a close-up view of one of the TC holes.

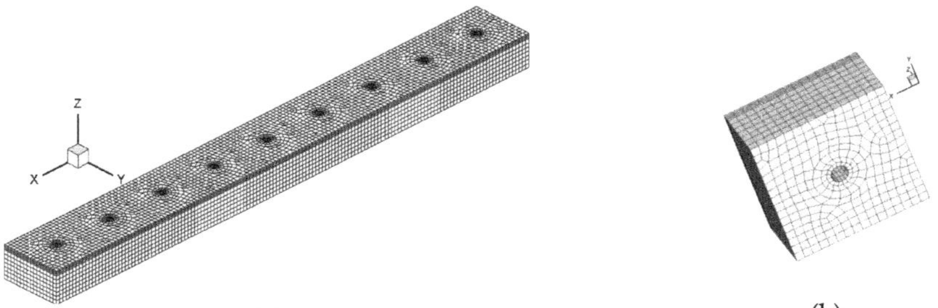

(a) (b)

Fig. 6. (a) The whole block consisting of nine thermocouple zones; (b) A close-up view of the TC hole from bottom.

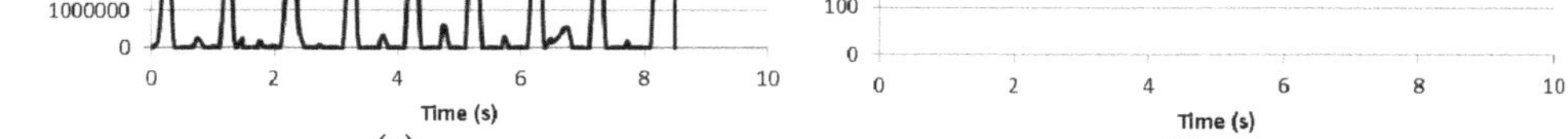

(a) (b)

Fig. 7. Time history of cooling on a run-out table; (a) Surface heat fluxes; (b) Internal temperatures.

The boundary condition on the top surface is prescribed heat flux which is chosen to resemble the one in water cooling of steel strips. Figure 7(a) shows the applied heat fluxes on top of one of the thermocouple locations for the whole cooling process, while Figure 7(b) shows the history of the temperature drop at the corresponding thermocouple location. Figure 8(a) shows a close-up of the applied heat flux at five of the nine thermocouple locations. It is very similar to the actual heat flux values on a run-out table with two rows of staggered circular jets, impinging on the third and seventh locations (Vakili & Gadala, 2010). Figure 8(b) is a close-up view of the temperature history at five of the nine thermocouple locations inside the plate, obtained from direct finite element simulation. The other boundaries are assumed to be adiabatic. The density, ρ, is 7850 kg/m^3, C_p is 475 J/kgK, and the thermal conductivity, k, is first assumed to be constant and equal to 40 $W/m.°C$ and later changed to be depending on temperature, as will be discussed in section 7.4. These are the physical properties of the steel strips that are used in our controlled cooling experiment. Results are obtained at the top of the cylindrical hole, which is the assumed position of a thermocouple. Inverse analysis is conducted to obtain the transient heat flux profile at the top surface of the plate.

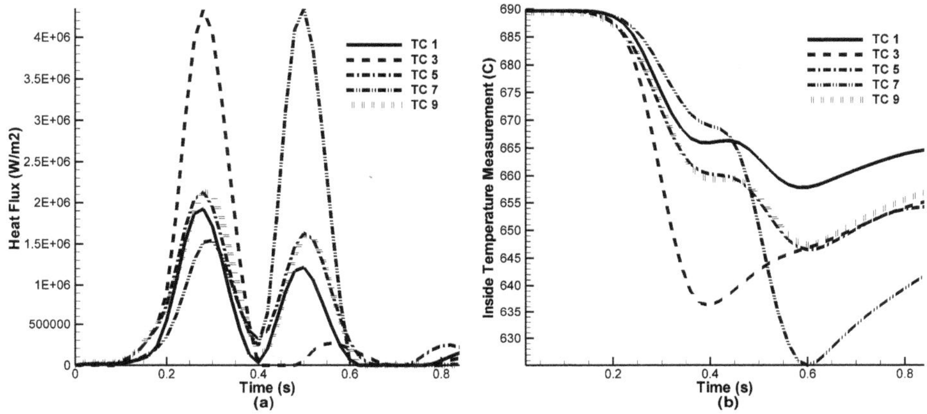

Fig. 8. (a) The applied heat flux on the top surface; (b) The thermocouple readings used for inverse analysis.

7. Results and discussion

As the first step in our study of these techniques, we investigate their capability in solving a general inverse heat conduction problem. We start by applying the artificial neural networks to the inverse heat conduction problem. This is different from GA and PSO, since those methods perform a stochastical search and are similar in many aspects, while the artificial neural networks are more like a correlation between the inputs and outputs. Fig. 9 shows the result of the application of the radial basis function neural networks for the whole history of the heat fluxes on the runout table. Temperatures start at 700 °C and go down to 176 °C. The heat flux vs. time profile is plotted in Fig. 9. As can be seen from this figure, neural networks are generally capable of dealing with the whole range of the cooling history.

However, this method has limitations, as observed in Fig. 9, and in more details in Fig. 10. The latter figure shows a close-up view of each of the 17 peaks of heat flux that happen during the cooling process on the run-out table, i.e. the peaks in Fig. 9. The circles are the expected heat flux, and the plusses are the result of NNs. The top left sub-figure is the first peak heat flux in time, and then it moves to the right, and then to the next row. Note that each even sub-figure (2nd, 4th, and so on) is a very smaller peak which is associated with the second row of jets. These peaks are not very obvious in Fig. 9, due to the scaling. Going through the subfigures of heat fluxes, it is apparent that the success or failure of NNs is not that much related to the temperature range, or the magnitude of heat fluxes, but on the actual shape of the heat flux profile. If the heat flux has a clear thin peak and two tails before and after the peak, the NN is doing a good job. However, the existence of other details in the heat flux profile reduces the quality of the NN predictions. Also, considering the ill-posed nature of the problem, and all the complications that are involved, we can generally say that in most cases (about 75% of the cases) it does a decent job. Of course, there is the possibility of slightly improving the results by trying to modify the performance parameters of the NN, but overall we can say that NNs are more useful in getting a general picture of the solution, rather than producing a very accurate and detailed answer to the IHCP.

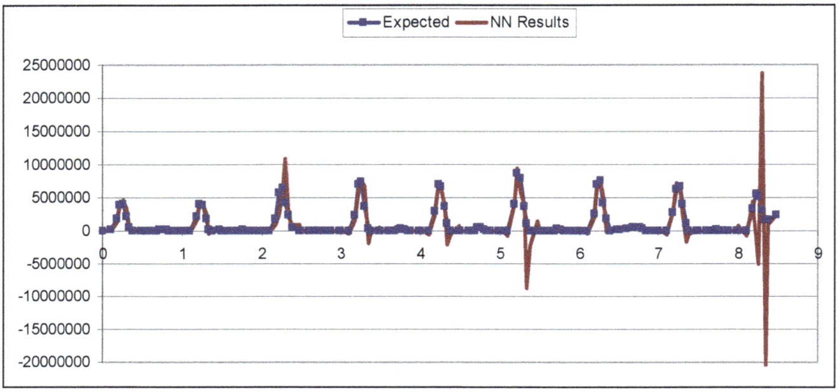

Fig. 9. Time History of Heat Fluxes in a Typical Run-Out Table Application; Expected Results (Squares) vs. the RBF Network Results (Line).

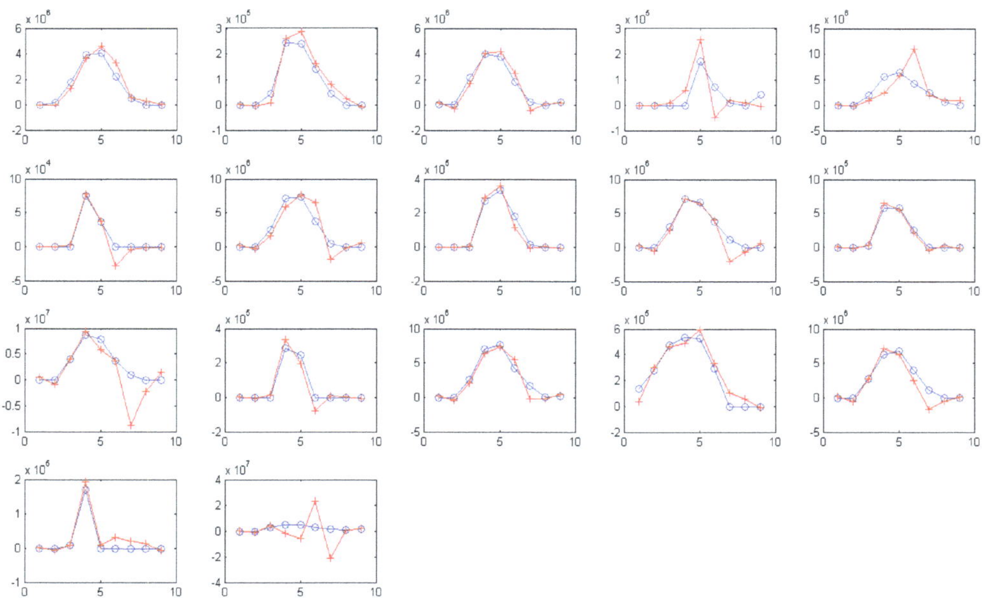

Fig. 10. Individual Heat Flux Peaks vs. Time from a Typical Run-Out Table Application; Expected Results (Circles) vs. the RBF Network Results (Pluses)

On the other hand, GA and PSO algorithms show reasonably good predictions of the details of the missing boundary conditions. Notice that we still need to have some form of regularization for these methods to work properly. The figures for the results of GA and PSO are not presented here for the sake of brevity, but can be found in (Vakili & Gadala, 2009). They will be used, however, for comparisons in the next sections.

7.1 Time step size

One of the main problems of the classical approaches, such as the sequential function specification method studied in this chapter, is their instability when small time steps are used. Unlike direct problems where the stability requirement gives the upper limit of the time step size, in inverse problems the time step is bounded from below. Fig. 11(a)(Vakili & Gadala, 2009) shows the oscillation in the results obtained by the function specification method and a time step size of 0.01 (s), which corresponds to the onset of instability. For time steps smaller than this, the whole process diverges. PSO, GA, and NNs successfully produce, however, the results for the same time step size as presented in Fig. 11(b) for PSO. Note that the oscillations here are not due to the instability caused by the time step size, and can be improved by performing more iterations. It is, however, important to mention that the time requirements for these techniques are much higher than those of the classical function specification approaches.

7.2 Efficiency

In this section, we compare the solution time required for GA, the three variations of PSO, and feed forward and radial basis function neural networks. We assume that there is no noise in the solution, and we compare the time that is required to get to certain accuracy in the heat flux predictions. Table 1 compares the solution time for different inverse analysis algorithms. The fastest solution technique is the gradient-based function specification method. The stochastical methods such as GA and PSO variants suffer a high computational cost. RBF neural networks perform much faster than GA and PSO, but they are still slower than the gradient-based methods, such as function specification.

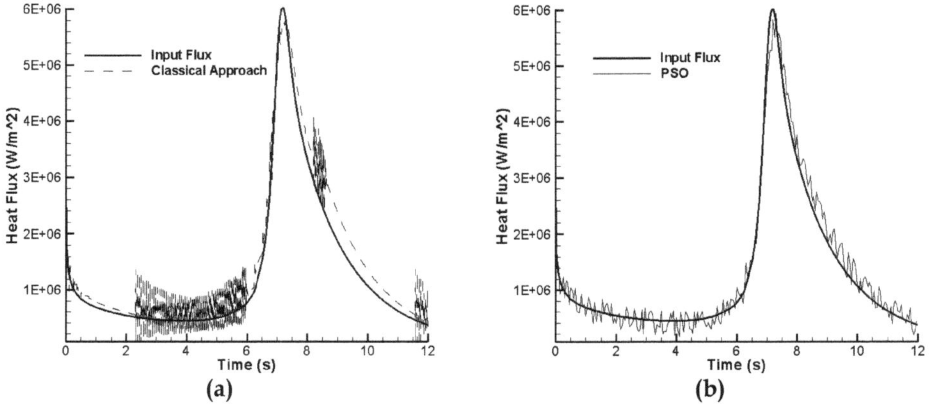

Fig. 11. Heat flux vs. time: (a) classical approach, (b) PSO (Vakili & Gadala, 2009).

	Function Specification Method	GA	PSO	RPSO	CRPSO	FMLP	RBFN
Solution Time (s)	1406	8430	6189	5907	6136	7321	2316

Table 1. Comparison of the solution time for different inverse analysis algorithms.

A more detailed comparison between the efficiency of GA and PSO variations can be found in (Vakili & Gadala, 2009).

7.3 Noisy domain solution

To investigate the behavior of different inverse algorithm variations in dealing with noise in the data, a known boundary condition is first applied to the direct problem. The temperature at some internal point(s) will be calculated and stored. Then random errors are imposed onto the calculated exact internal temperatures with the following equation:

$$T_m = T_{exact} + \sigma \cdot r \tag{22}$$

where T_m is the virtual internal temperature that is used in the inverse calculations instead of the exact temperature, T_{exact}; r is a normally distributed random variable with zero mean and unit standard deviation; and σ is the standard deviation. Virtual errors of 0.1% and 1% of the temperature magnitude are investigated here.

We start by studying the effectiveness of the neural networks in handling noisy domains. Generally, the stability of the neural networks is on the same order as other inverse methods. It may be possible to tune the parameters to make it a little bit more stable, but generally it does not look promising in terms of noise resistance, since such modifications exist for almost all other methods. Fig. 12 - Fig. 13 show the results of the RBF network

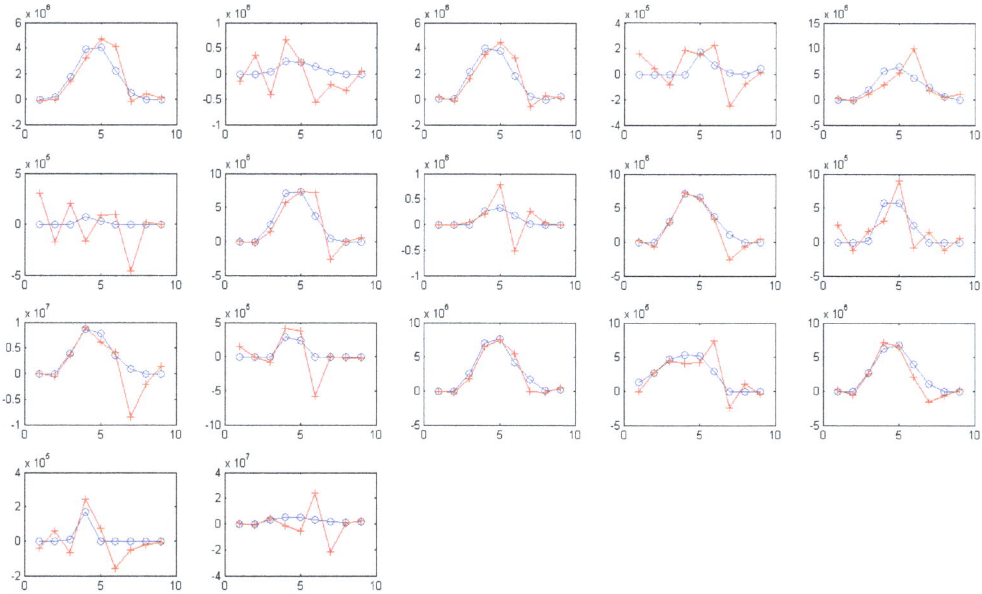

Fig. 12. Individual heat flux peaks vs. time from a typical run-out tale application; Expected results (blue circles) vs. the RBF network results (red pluses); Artificial noise added: c = ±0.1%.

(red pluses) versus the expected results (blue circles) for individual heat flux peaks during the cooling history of the plate. The amount of added noise in these figures is ±0.1%and ±1%, respectively.

There are several ways to make an inverse algorithm more stable when dealing with noisy data. For example, (Gadala & Xu, 2006) have shown that increasing the number of "*future time steps*" in their sequential function specification algorithm resulted in greater stability. They have also demonstrated that increasing the regularization parameter, a, improves the ability of the algorithm to handle noisy data. However, the latter approach was shown to greatly increase the required number of iterations, and in many cases the solution may diverge. In this work, we first examine the effect of the regularization parameter, and then investigate an approach unique to the PSO method, to improve the effectiveness of the inverse algorithm in dealing with noise.

Fig. 14 shows the effect of varying the regularization parameter value on the reconstructed heat flux, using the basic particle swarm optimization technique. Stable and accurate results are obtained for a range of values of $a = 10^{-12}$ to 10^{-10}. These results are very close to those reported in (Gadala & Xu, 2006), i.e., the proper values of a are very similar for the sequential specification approach and PSO.

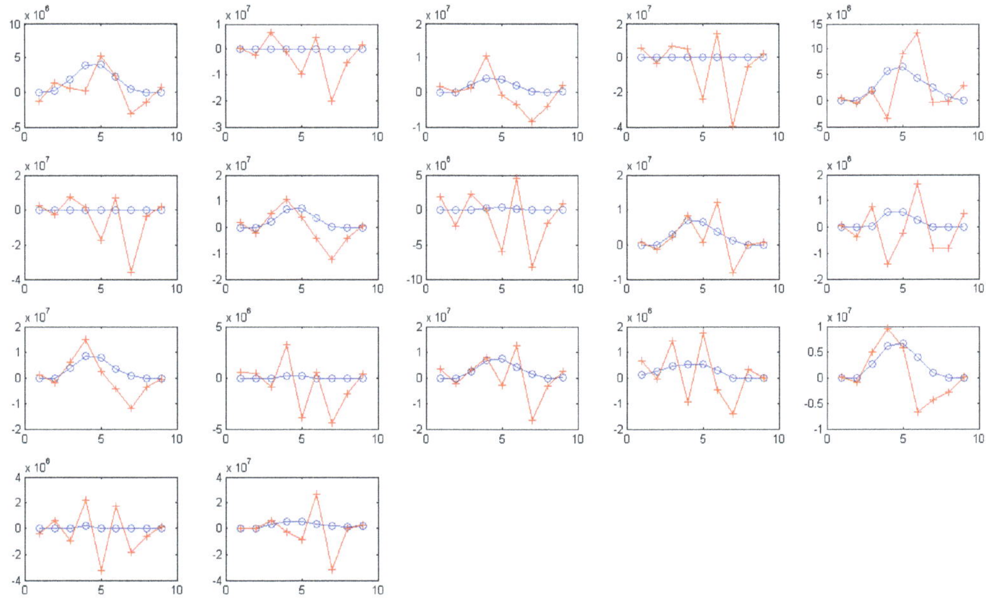

Fig. 13. Individual heat flux peaks vs. time from a typical run-out tale application; Expected results (blue circles) vs. the RBF network results (red pluses); Artificial noise added: c = ±1%.

Another factor that can affect the performance of a PSO inverse approach in dealing with noisy data is the value of the self-confidence parameter, c_0, or the ratio between this parameter and the acceleration coefficients. The acceleration coefficients are set to the default value of 1.42. The initial value of the self-confidence parameter, c_0, is changed from the default value of 0.7. The results are shown below.

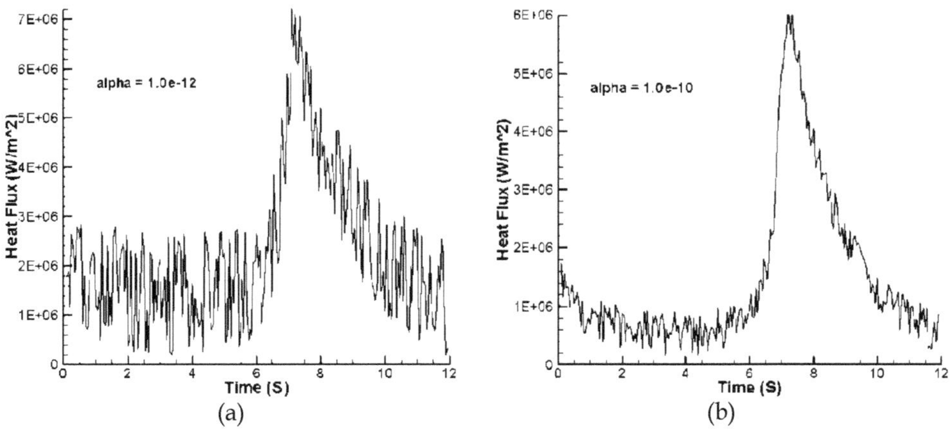

Fig. 14. Effect of Regularization Parameter; a: $\alpha = 10^{-12}$; b: $\alpha = 10^{-10}$

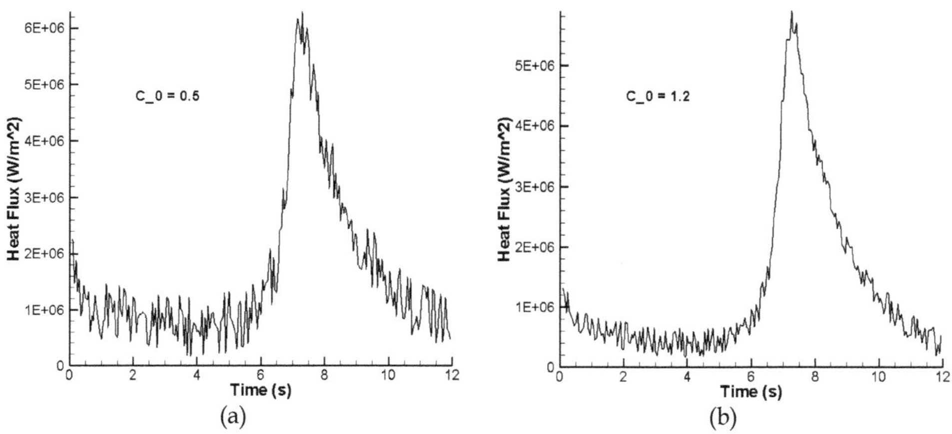

Fig. 15. Effect of Self-Confidence Parameter; (a) c0=0.5; (b) c0=1.2

As can be seen in Fig. 15 (for $a = 10^{-10}$), increasing the value of the self-confidence parameter results in better handling of the noisy data. This trend was observed for values up to approximately 1.3, after which the results become worse, and diverge. One possible explanation is that increasing the ratio of the self-confidence parameter with respect to the acceleration coefficients results in a more global search in the domain, and therefore increases the capability of the method to escape from the local minima caused by the noise, and find values closer to the global minimum. This effect was observed to be weaker in highly noisy domains. However, in the presence of a moderate amount of noise, increasing the self-confidence ratio results in more effectiveness. As can be seen in Table 2, the best effectiveness is normally obtained by RPSO, closely followed by CRPSO. Considering the higher efficiency of CRPSO, it is still recommended for the inverse heat conduction analysis.

C_0	0.7	0.8	0.95	1.1	1.2
PSO	8.105e+4	7.532e+4	7.079e+4	6.823e+4	6.257e+4
RPSO	7.577e+4	7.064e+4	6.685e+4	6.346e+4	5.816e+4
CRPSO	7.611e+4	6.739e+4	6.117e+4	5.999e+4	5.822e+4

Table 2. Effect of the Self-Confidence Parameter on the L2 Norm of Error in the Solution

Table 3 shows the value of L_2 norm of error in the solution, for ±1% added noise, and for different algorithms. It can be seen that the RBF neural networks perform better than the function specification method, and somewhere between the genetic algorithm and PSO variants. The most noise resistant algorithms are PSO variants, and the least stable algorithm is the gradient-based function specification method.

	Function Specification Method	GA	PSO	RPSO	CRPSO	FMLP	RBFN
L_2 Norm of Error	9.14e4	6.61e4	5.24e4	4.82e4	5.02e4	8.91e4	5.91e4

Table 3. The L2 Norm of Error in the Solution in a Noisy Domain for Different Algorithms

7.4 Effect of non-linearity

In many applications of inverse heat conduction, the thermophysical properties change with temperature. This results in nonlinearity of the problem. In other words, a same drop in the temperature values can be caused by different values of heat flux. So, a neural network that is trained with the relationship between the temperature change values and heat flux magnitudes may not be correctly capable of recognizing this nonlinear pattern, and as a result the performance will suffer. To investigate this effect, two kinds of expressions are used for thermal conductivity in this study. In one, we assume a constant thermal conductivity of *W/m.°C*, while in the other a temperature-dependent expression is used:

$$k = 60.571 - 0.03849 \times T \quad W/m.^\circ C \tag{23}$$

As expected, the nonlinearity will weaken the performance of both feedforward and radial basis function neural networks. The effect is seen as the training of the network stalls after a number of epochs. In order to deal with this, increasing the number of hidden layers, increasing the number of neurons in each layer, and choosing different types of transfer function were investigated. However, none of these methods showed a significant improvement in the behavior of the network. The other methods of solving the inverse problem are much less sensitive to the effect of nonlinearity. Table 4 compares the error in the solution for both the linear and nonlinear cases, if the same numbers of iterations, generations, and epochs are used for different methods of solving the inverse heat conduction. As it can be seen, the neural networks methods perform very poorly in the nonlinear cases, while the other methods, either gradient based or stochastical, are immune to the problems caused by nonlinearity. Basically, neural networks, at least in the form that is used in this chapter, see nonlinearity as a kind of noise. It should be noted that neural networks can be useful in making rough estimates of the answer, or combined with some other techniques employed as an inverse solver for nonlinear cases (Aquino & Brigham, 2006), but on their own, are not a suitable choice for an accurate prediction of the boundary conditions in a nonlinear inverse heat conduction problem.

	Function Specification Method	GA	PSO	RPSO	CRPSO	FMLP	RBFN
Linear	1.81e2	7.62e2	3.85e2	3.42e2	3.17e2	9.90e2	5.35e2
Nonlinear	2.14e2	7.71e2	4.46e2	5.12e2	4.26e2	3.57e4	2.76e4

Table 4. The L_2 norm of error in the solution in an exact domain for different algorithms.

8. Conclusion

In this chapter, we introduced a gradient-based inverse solver to obtain the missing boundary conditions based on the readings of internal thermocouples. The results show that the method is very sensitive to measurement errors, and becomes unstable when small time steps are used. Then, we tried to find algorithms that are capable of solving the inverse heat conduction problem without the shortcomings of the gradient-based methods.

The artificial neural networks are capable of capturing the whole thermal history on the run-out table, but are not very effective in restoring the detailed behavior of the boundary conditions. Also, they behave poorly in nonlinear cases and where the boundary condition profile is different.

GA and PSO are more effective in finding a detailed representation of the time-varying boundary conditions, as well as in nonlinear cases. However, their convergence takes longer. A variation of the basic PSO, called CRPSO, showed the best performance among the three versions. The effectiveness of PSO was also studied in the presence of noise. PSO proved to be effective in handling noisy data, especially when its performance parameters were tuned. The proper choice of the regularization parameter helped PSO deal with noisy data, similar to the way it helps the classical function specification approaches. An increase in the self-confidence parameter was also found to be effective, as it increased the global search capabilities of the algorithm. RPSO was the most effective variation in dealing with noise, closely followed by CRPSO. The latter variation is recommended for inverse heat conduction problems, as it combines the efficiency and effectiveness required by these problems.

9. References

Abou khachfe, R., and Jarny, Y. (2001). Determination of Heat Sources and Heat Transfer Coefficient for Two-Dimensional Heat flow–numerical and Experimental Study. *International Journal of Heat and Mass Transfer*, Vol. 44, No. 7 , pp.1309-1322.

Alifanov, O. M., Nenarokomov, A. V., Budnik, S. A., Michailov, V. V., and Ydin, V. M. (2004). Identification of Thermal Properties of Materials with Applications for Spacecraft Structures. *Inverse Problems in Science and Engineering*, Vol. 12, No. 5 , pp.579-594.

Alifanov, O.M., (1995). *Inverse Heat Transfer Problems*, Berlin ; New York: Springer-Verlag.

Al-Khalidy, N., (1998). A General Space Marching Algorithm for the Solution of Two-Dimensional Boundary Inverse Heat Conduction Problems. *Numerical Heat Transfer, Part B*, Vol. 34, , pp.339-360.

Alrasheed, M.R., de Silva, C.W., and Gadala, M.S. (2008).Evolutionary Optimization in the Design of a Heat Sink, in: *Mechatronic Systems: Devices, Design, Control, Operation and Monitoring*, C. W. de Silva, Taylor & Francis.

Alrasheed, M.R., de Silva, C.W., and Gadala, M.S. (2007).A Modified Particle Swarm Optimization Scheme and its Application in Electronic Heat Sink Design, ASME/Pacific Rim Technical Conference and Exhibition on Packaging and Integration of Electronic and Photonic Systems, MEMS, and NEMS.

Aquino, W., and Brigham, J. C. (2006). Self-Learning Finite Elements for Inverse Estimation of Thermal Constitutive Models. *International Journal of Heat and Mass Transfer*, Vol. 49, No. 15-16 , pp.2466-2478.

Bass, B. R., (1980). Application of the Finite Element Method to the Nonlinear Inverse Heat Conduction Problem using Beck's Second Method. *Journal of Engineering and Industry*, Vol. 102, , pp.168-176.

Battaglia, J. L., (2002). A Modal Approach to Solve Inverse Heat Conduction Problems. *Inverse Problems in Engineering*, Vol. 10, No. 1 , pp.41-63.

Beck, J. V., Blackwell, B., and Haji-Sheikh, A. (1996). Comparison of some Inverse Heat Conduction Methods using Experimental Data. *International Journal of Heat and Mass Transfer*, Vol. 39, , pp.3649-3657.

Beck, J. V., and Murio, D. A. (1986). Combined Function Specification-Regularization Procedure for Solution of Inverse Heat Conduction Problem. *AIAA Journal*, Vol. 24, No. 1 , pp.180-185.

Beck, J.V., Blackwell, B., and Clair Jr, C.R.S. (1985). *Inverse Heat Conduction: Ill-Posed Problem*, New York: Wiley-Interscience Publication.

Beck, J.V., and Arnold, K.J. (1977). *Parameter Estimation in Engineering and Science*, New York: Wiley.

Blanc, G., Raynaud, M., and Chau, T. H. (1998). A Guide for the use of the Function Specification Method for 2D Inverse Heat Conduction Problems. *Revue Generale De Thermique*, Vol. 37, No. 1 , pp.17-30.

Clerc, M., (2006). *Particle Swarm Optimization*, ISTE.

Davis, L., (1991). *Handbook of Genetic Algorithms*, Thomson Publishing Group.

Eberhart, R., and Kennedy, J. (1995).A New Optimizer using Particle Swarm Theory, Proceedings of the Sixth International Symposium on Micro Machine and Human Science pp. 39-43.

Fan, S.K.S., and Chang, J.M. (2007).A Modified Particle Swarm Optimizer using an Adaptive Dynamic Weight Scheme, in: *Digital Human Modeling,* V. G. Duffy, Springer Berlin / Heidelberg. pp. 56-65, .

Gadala, M. S., and Xu, F. (2006). An FE-Based Sequential Inverse Algorithm for Heat Flux Calculation during Impingement Water Cooling. *International Journal of Numerical Methods for Heat and Fluid Flow,* Vol. 16, No. 3 , pp.356-385.

Girault, M., Petit, D., and Videcoq, E. (2003). The use of Model Reduction and Function Decomposition for Identifying Boundary Conditions of A Linear Thermal System. *Inverse Problems in Science and Engineering,* Vol. 11, No. 5 , pp.425-455.

Goldberg, D.E., (1989). *Genetic Algorithms in Search, Optimization and Machine Learning,* Addison-Wesley Longman Publishing Co., Inc. Boston, MA, USA.

Gosselin, L., Tye-Gingras, M., and Mathieu-Potvin, F. (2009). Review of Utilization of Genetic Algorithms in Heat Transfer Problems. *International Journal of Heat and Mass Transfer,* Vol. 52, No. 9-10 , pp.2169-2188.

Hassan, R., Cohanim, B., de Weck, O., and Venter, G. (2005).A Comparison of Particle Swarm Optimization and the Genetic Algorithm, Proceedings of the 46th AIAA/ASME/ASCE/AHS/ASC Structures, Structural Dynamics and Materials Conference.

Huang, C. H., Yuan, I. C., and Herchang, A. (2003). A Three-Dimensional Inverse Problem in Imaging the Local Heat Transfer Coefficients for Plate Finned-Tube Heat Exchangers. *International Journal of Heat and Mass Transfer,* Vol. 46, No. 19 , pp.3629-3638.

Huang, C. H., and Wang, S. P. (1999). A Three-Dimensional Inverse Heat Conduction Problem in Estimating Surface Heat Flux by Conjugate Gradient Method. *International Journal of Heat and Mass Transfer,* Vol. 42, No. 18 , pp.3387-3403.

Karr, C. L., Yakushin, I., and Nicolosi, K. (2000). Solving Inverse Initial-Value, Boundary-Value Problems Via Genetic Algorithm. *Engineering Applications of Artificial Intelligence,* Vol. 13, No. 6 , pp.625-633.

Karray, F.O., and de Silva, C.W. (2004). *Soft Computing and Intelligent System Design - Theory, Tools, and Applications,* New York: Addison Wesley.

Kennedy, J., Eberhart, R.C., and Shi, Y. (2001). *Swarm Intelligence,* Morgan Kaufmann.

Kim, H. K., and Oh, S. I. (2001). Evaluation of Heat Transfer Coefficient during Heat Treatment by Inverse Analysis. *Journal of Material Processing Technology,* Vol. 112, , pp.157-165.

Kim, S. K., and Lee, W. I. (2002). Solution of Inverse Heat Conduction Problems using Maximum Entropy Method. *International Journal of Heat and Mass Transfer,* Vol. 45, No. 2 , pp.381-391.

Krejsa, J., Woodbury, K. A., Ratliff, J. D., and Raudensky, M. (1999). Assessment of Strategies and Potential for Neural Networks in the IHCP. *Inverse Problems in Engineering,* Vol. 7, No. 3 , pp.197-213.

Kumagai, S., Suzuki, S., Sano, Y.R., and Kawazoe, M. (1995).Transient Cooling of Hot Slab by an Impinging Jet with Boiling Heat Transfer, ASME/JSME Thermal Engineering Conference pp. 347-352.

Lecoeuche, S., Mercere, G., and Lalot, S. (2006). Evaluating Time-Dependent Heat Fluxes using Artificial Neural Networks. *Inverse Problems in Science and Engineering*, Vol. 14, No. 2 , pp.97-109.

Lee, K. H., Baek, S. W., and Kim, K. W. (2008). Inverse Radiation Analysis using Repulsive Particle Swarm Optimization Algorithm. *International Journal of Heat and Mass Transfer*, Vol. 51, , pp.2772-2783.

Louahia-Gualous, H., Panday, P. K., and Artioukhine, E. A. (2003). Inverse Determination of the Local Heat Transfer Coefficients for Nucleate Boiling on a Horizontal Pipe Cylinder. *Journal of Heat Transfer*, Vol. 125, , pp.1087-1095.

Osman, A. S., (1190). Investigation of Transient Heat Transfer Coefficients in Quenching Experiments. *Journal of Heat Transfer*, Vol. 112, , pp.843-848.

Ostrowski, Z., R. A. Bialstrokecki, and A. J. Kassab. "Solving Inverse Heat Conduction Problems using Trained POD-RBF Network Inverse Method." *Inverse Problems in Science and Engineering* (2007).

Ozisik, M.N., (2000). *Inverse Heat Transfer: Fundamentals and Applications*, New York: Taylor & Francis.

Pietrzyk, M., and Lenard, J. G. (1990). A Study of Heat Transfer during Flat Rolling. *International Journal of Numerical Methods in Engineering*, Vol. 30, , pp.1459-1469.

Raudensky, M., Woodbury, K. A., Kral, J., and Brezina, T. (1995). Genetic Algorithm in Solution of Inverse Heat Conduction Problems. *Numerical Heat Transfer, Part B: Fundamentals*, Vol. 28, No. 3 , pp.293-306.

Roudbari, S., (2006). Self Adaptive Finite Element Analysis, Cornell University, .

Shiguemori, E. H., Da Silva, J. D. S., and De Campos Velho, H. F. (2004). Estmiation of Initial Condition in Heat Conduction by Neural Networks. *Inverse Problems in Science and Engineering*, Vol. 12, No. 3 , pp.317-328.

Silieti, M., Divo, E., and Kassab, A. J. (2005). An Inverse Boundary Element method/genetic Algorithm Based Approach for Retrieval of Multi-Dimensional Heat Transfer Coefficients within Film Cooling holes/slots. *Inverse Problems in Science and Engineering*, Vol. 13, No. 1 , pp.79-98.

Urfalioglu, O., (2004).Robust Estimation of Camera Rotation, Translation and Focal Length at High Outlier Rates, The First Canadian Conference on Computer and Robot Vision pp. 464-471.

Vakili, S., and Gadala, M. S. (2011). A Modified Sequential Particle Swarm Optimization Algorithm with Future Time Data for Solving Transient Inverse Heat Conduction Problems. *Numerical Heat Transfer, Part A: Applications*, Vol. 59, No. 12 , pp.911-933.

Vakili, S., and Gadala, M. S. (2010). Boiling Heat Transfer of Multiple Impinging Round Jets on a Hot Moving Plate. *Submitted to Heat Transfer Engineering* .

Vakili, S., and Gadala, M. S. (2009). Effectiveness and Efficiency of Particle Swarm Optimization Technique in Inverse Heat Conduction Analysis. *Numerical Heat Transfer, Part B: Fundamentals*, Vol. 56, No. 2 , pp.119-141.

Woodbury, K. A., and Thakur, S. K. (1996). Redundant Data and Future Times in the Inverse Heat Conduction Problem. *Inverse Problems in Science and Engineering*, Vol. 2, No. 4 , pp.319-333.

3

Experimental and Numerical Studies of Evaporation Local Heat Transfer in Free Jet

Hasna Louahlia Gualous
Caen Basse Normandie University/LUSAC
France

1. Introduction

Jet impingement heat transfer has been used extensively in many industrial applications for cooling because it provides high local heat transfer coefficients at low flow rates. Several experimental and theoretical studies on liquid jet impingement heat transfer have been reported in the literature (Louahlia & Baonga, 2008, Chen et al., 2002, Lin & Ponnappan, 2004, Liu & Zhu, 2004, Pan & Webb, 1995). Numerous studies are conducted in average heat transfer, but local heat transfer analysis for steady and unsteady states has not been much attention. Jet impingement heat transfer is influenced by different physical parameters such as: (i) the velocity turbulent fluctuations (Oliphant et al. 1998, Stevens & Webb, 1989), (ii) the difference between the temperatures of inlet jet and heat exchange surface (Siba et al. 2003, MA et al. 1997), (iii) the surface geometry and the jet orientation (MA et al. 1997b, Elison & Webb, 1994), (iv) the liquid flow rate and Prandtl number (Elison & Webb, 1994, Fabbri et al. 2003, Stevens & Webb, 1993), and (v) the nozzle diameter (Stevens & Webb, 1993, 1992).

2. Hydrodynamic characteristics of the jet impinging on a horizontal surface

When a liquid jet impinges on a horizontal surface, three distinct regions can be identified as shown in Figure 1. The first zone is the free jet region where the flow is accelerated because of the gravitational force. The second zone is the impingement region where the interaction between the jet and the heat exchange surface produces a strong deceleration of the flow. After this zone, the liquid wets the surface and flows in a parallel direction to the heat exchange surface. Heat transfer efficiency in each zone is related to the flow velocity and its structure. In the impingement zone, jet diameter could be measured using fluresecence induced laser (Baonga et al. 2006) combined to the images processing. In this method, liquid impinging the heat exchange surface is illuminated by a laser sheet in the axial direction as shown by Figure 2. Rhodamine B with low concentration must be used as the fluerescent substance added to the liquid jet. In this case, fluorescent substance becomes visible when liquid jet is illuminated with light. A CCD camera can be used to record the flow video images. Video images are treated in order to extract the profiles of the jet as shown by Figure 1.

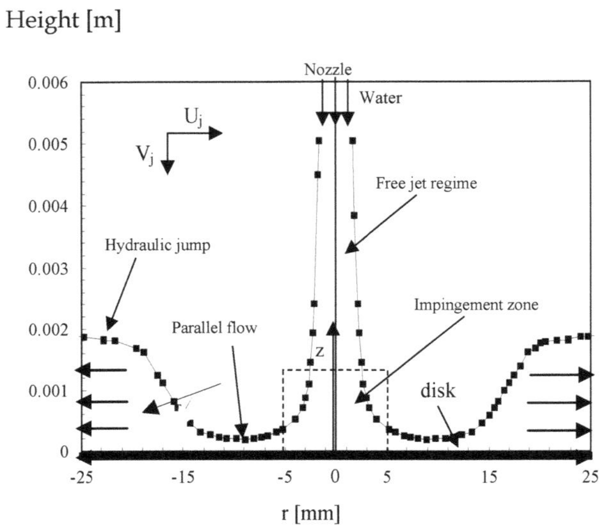

Fig. 1. Schematic of flow developing from nozzle to heated disk.

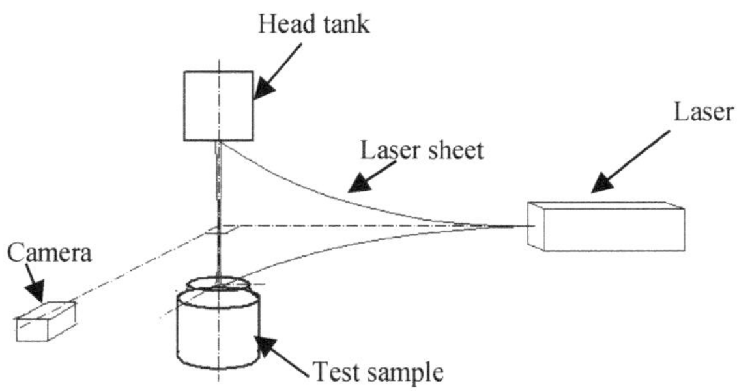

Fig. 2. Flow visualization system.

2.1 Axial flow structure

For inlet Reynolds number ranging from 1520 to 5900 (the corresponding values of the inlet mean velocity are in the range of 3.24 to 12.5 m/s), Figure 3 shows effect of the jet flow rate on the distribution of the jet diameter along the axial direction. The nozzle diameter is of 4 mm. The nozzle-heat exchange surface spacing is of 13 mm. Reynolds number is calculated as follow :

$$\mathrm{Re} = \frac{4\dot{m}}{\pi d_i \mu_L} \tag{1}$$

where: d_i is the inner diameter of the nozzle, μ_L is the dynamic viscosity, $\dot{m}$ is the total mass flow rate of the jet. Physical properties are used at the inlet jet temperature measured at the nozzle exit.

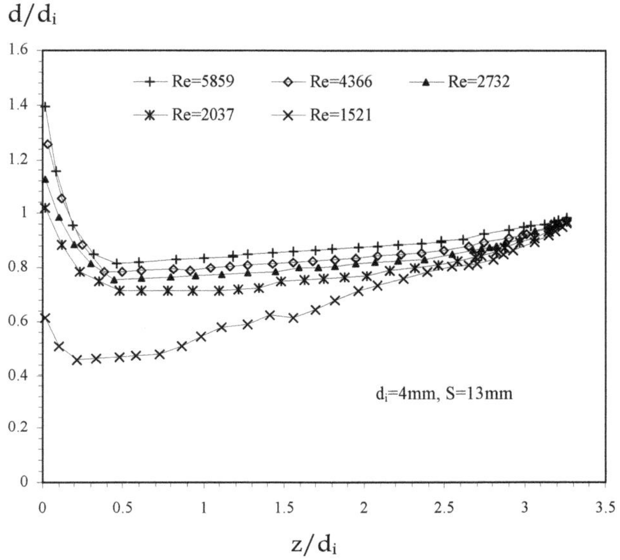

Fig. 3. Evolution of the jet diameter along the z direction.

It can be seen from figure 3 that for the same axial position (z), the jet diameter increases with inlet Reynolds number because gravitational force increases with flow velocity and becomes higher than surface tension force at the jet free surface. For lower Reynolds number (Re=1521), it shows that instability starts and waves appears on the jet free surface because capillarity force increases and becomes non-negligible compared to gravitational force.

Along the falling jet, no evaporation has been produced and the mass flow rate is conserved. In this case, axial distribution of the flow velocity can be deduced from the following equation:

$$\dot{m} = \rho_L V_j(z)\,\pi \frac{d^2(z)}{4} \tag{2}$$

At each axial position (z), $V_j(z)$ is the average velocity of the jet, $d(z)$ is the jet diameter, ρ_L is the jet density. Figure 4 shows evolution of $V_j(z)/V_{j,inlet}$ from the injection zone to the heat exchange surface for various inlets Reynolds numbers. $V_{j,inlet}$ refers liquid velocity of the jet at the nozzle exit. For each Reynolds number, velocity is high near the impingement zone where the jet diameter is low. The free jet is accelerated after the nozzle exit because the gravity force effect is very pronounced. After this zone, the jet velocity decelerates quickly because liquid flow is retained on the heat exchange surface under the effect of the capillarity force and the wall friction.

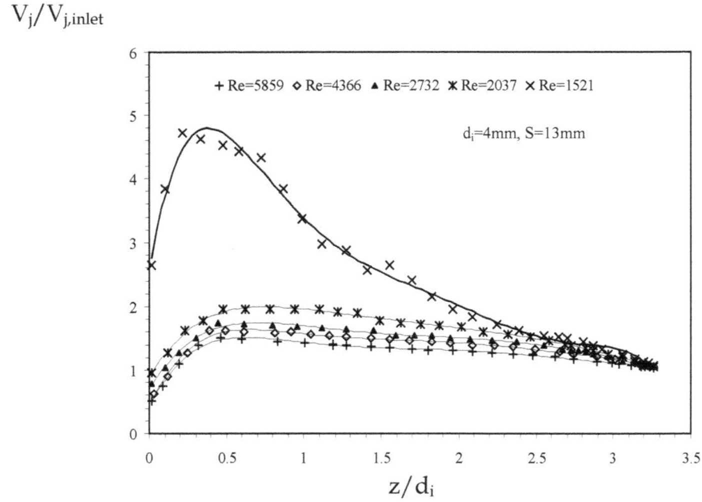

Fig. 4. Dimensionless axial velocity of the jet.

2.2 Wall parallel flow structure

Turning now to the characterisation of the local liquid layer depth near the heat exchange surface and the velocity profile along the radial direction where the heat transfer occurs.

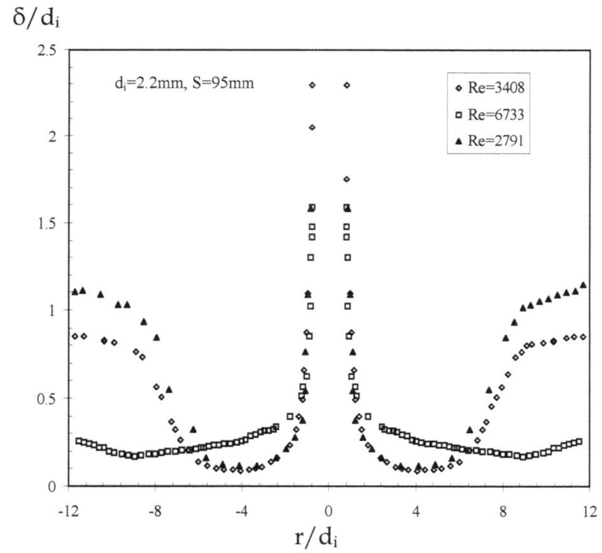

Fig. 5. Local evolution of the dimensionless liquid layer depth.

Figure 5 shows an example of the local liquid layer depth ($\delta(r)$) measured for three values of the inlet Reynolds number (Re=6733, Re=3408, and Re=2791). The nozzle diameter is of 2.2 mm for theses experiments. The jet inlet temperature is of 32°C and the nozzle-heat exchange surface spacing is of 95 mm. Figure 5 shows three distinct zones: the impingement zone, the zone where the liquid layer depth is approximately uniform, and the final zone where a hydraulic jump is formed. The radius, at which the liquid layer depth increases, is termed as the hydraulic jump radius. For higher Reynolds number, hydraulic jump is not appeared on the heat exchange surface because it is certainly higher than the radius of the heat exchange surface. Location of hydraulic jump on the surface is an interest physical phenomenon. In the previous work, some authors (Stevens & Webb, 1992, 1993, Liu et al. 1991, 1989, Watson, 1964) show the influence of the jet mass flow rate on the hydraulic jump radius that is defined at the radius location where the liquid layer depth attains a highest value in the parallel flow (Figure 6a).

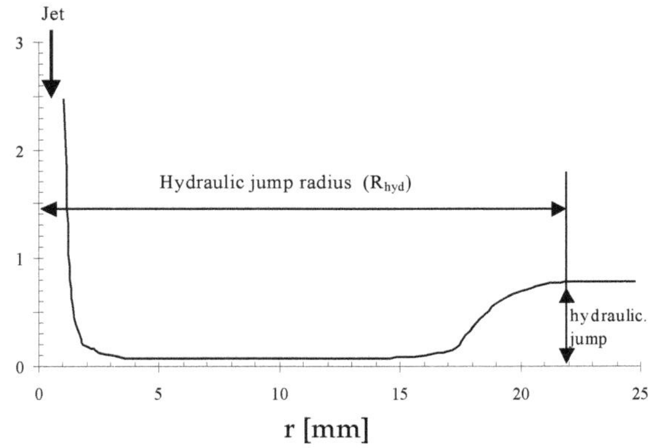

(a)

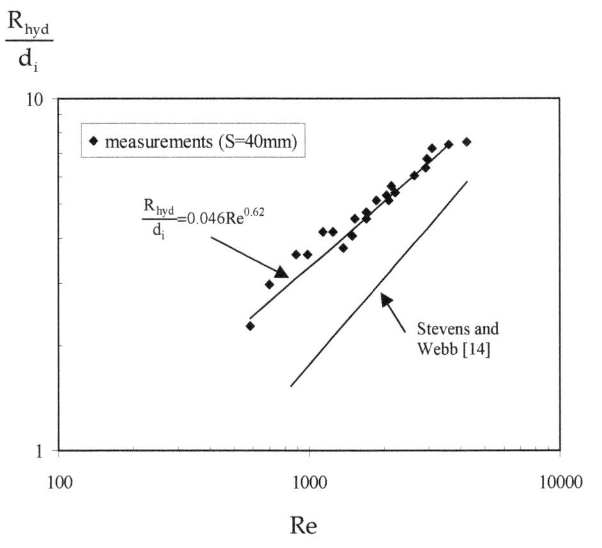

(b)

Fig. 6. a- Schematic of the hydraulic jump radius, b- Dimensionless hydraulic jump radius.

For Reynolds number ranging from 700 to 5000, Figure 6b shows dimensionless hydraulic jump radius as a function of Reynolds number. It shows that the hydraulic jump radius increases with the Reynolds number because flow is accelerated in the radial direction and the hydraulic jump is moved far from the stagnation zone. The difference between the present results and the experimental data of Stevens and Webb can be due to the uncertainty in the data of Stevens and Webb estimated of ±0.5 cm. The present results are defined with a maximum uncertainty of 2% and revealed an approximation dependence of the hydraulic jump radius on the Reynolds number as $Re^{0.62}$:

$$\frac{R_{hyd}}{d_i} = 0.046\,Re^{0.62} \tag{3}$$

Equation (3) estimates hydraulic jump radius with a maximum uncertainty of ±7%.
Distribution of the liquid velocity along the radial direction is determined by assuming conservation of the mass flow rate of liquid jet. For parallel flow:

$$\dot{m} = \rho_L U_j(r)\, 2\,\pi\, r\, \delta(r) \tag{4}$$

Where ρ_L is the jet density, $U_j(r)$ is the jet average velocity in the radial direction, r is the radial coordinate, $\delta(r)$ is the liquid layer depth on the surface.
Figure 7 shows profiles of dimensionless velocity and shows for each inlet Reynolds number, radial velocity profiles reaches a maximum value which is very pronounced for higher Reynolds number.

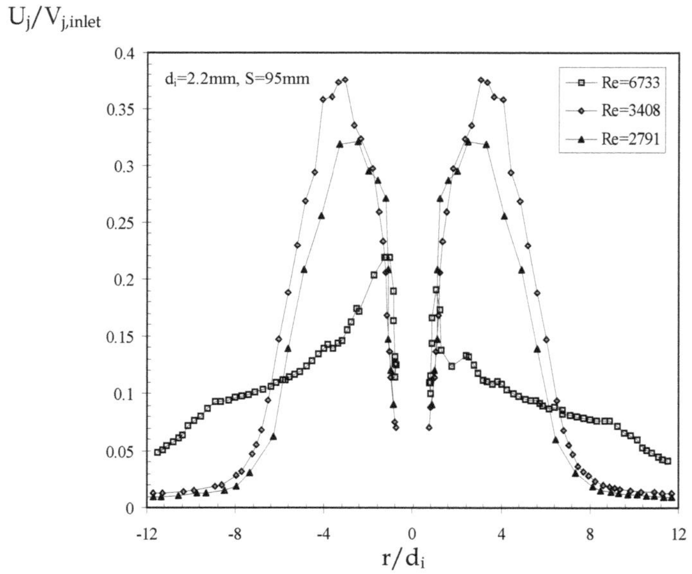

Fig. 7. Local evolution of the dimensionless radial velocity.

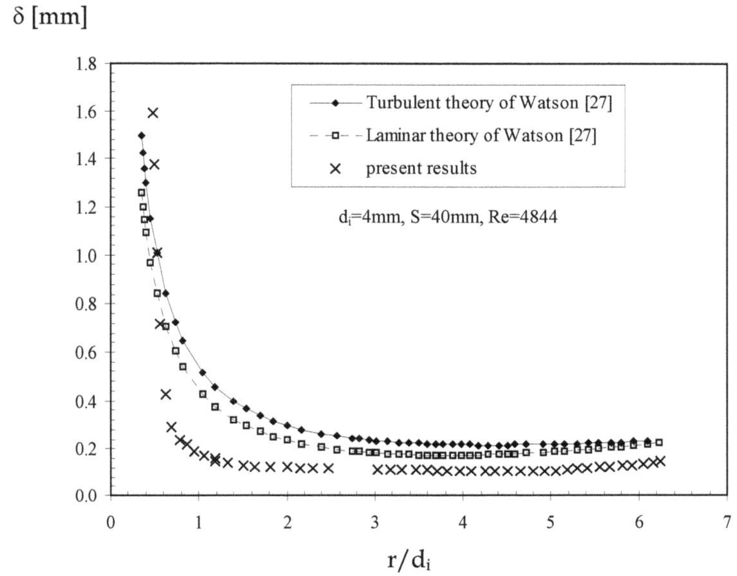

(a)

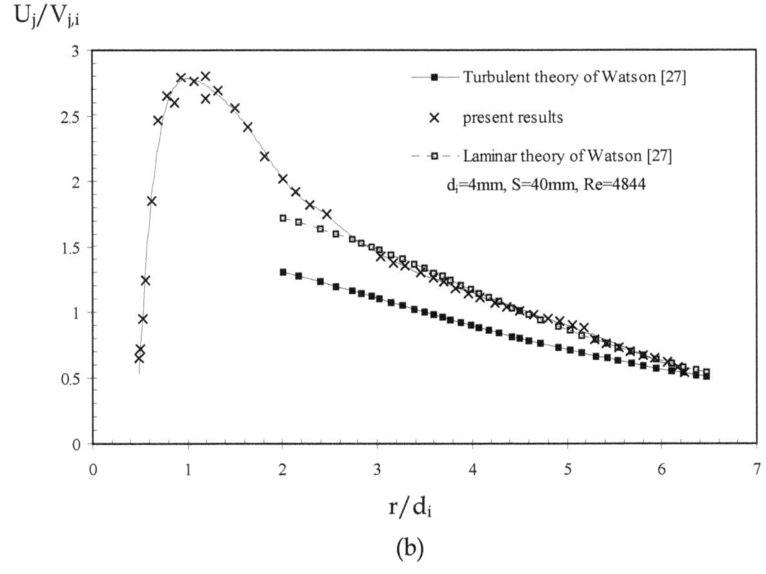

(b)

Fig. 8. Comparison of the experimental results with Watson's theory: (a) liquid layer depth (b) dimensionless radial mean velocity.

For the same radial position, Figure 7 shows effect of the hydraulic jump on the flow velocity. It shows that in the zone of the hydraulic jump, radial velocity is the lowest and approximately uniform for Re=3408 and Re=2791. For all data, the maximum dimensionless velocity is obtained for radius ranging from 2 to 4 times nozzle diameter. In the previous

work, Stevens and Webb (1989) found this maximum at r/d_i of 2.5 for the horizontal impinging jet on the vertical surface. Figure 7 also indicates that in the parallel flow, radial velocity is not uniform and it is lower than inlet jet velocity at the nozzle exit. The present results contradicts the assumption of some authors (Liu et al. 1989, Liu et al. 1991) assuming that the flow is fully developed before the hydraulic jump, and the free surface velocity is equal to the exit average jet velocity.
Experimental results are compared with the laminar and the turbulent theories predictions defined by Watson (1964) in figures 8a and b. It shows that laminar theory provides the best agreement with experimental data but sub-estimates the liquid layer depth. However, the turbulent theory underestimates liquid velocity along the radial direction and sub-estimates the liquid layer depth.
For all experiences showed in this section, it can be seen that when a circular liquid free jet strikes a flat plate, it spreads radially in very thin film along the heated surface, and the hydraulic jump that is associated with a Rayleigh-Taylor instability, can be appeared. Three distinct regions are identified and flow velocity is varied along the jet. Therefore, local distribution of heat flux and heat transfer coefficient is variable following the liquid layer depth and flow velocity.
There has been little information available in the published literature on local heat transfer for cooling using evaporation of impinging free liquid jet. The reason is that the liquid film spreads radially on the heated surface in very thin film, and determination of local heat flux on the wetted surface requires measurement of the temperature profiles along the axial and radial directions without perturbing the flow. Therefore, inverse heat conduction problem (IHCP) has been solved in order to determine locally distribution of thermal boundary conditions at the wetted surface using only temperatures measured inside the wall.

3. Determination of the thermal boundary conditions

In the previous work (Chen et al., 2001, Martin & Dulkravich, 1998, Louahlia-Gualous et al., 2003, Louahlia & El Omari, 2006), IHCP is used to estimate the thermal boundary conditions in various applications of science and engineering when direct measurements are difficult. IHCP could determine the precise results with numerical computations and simple instrumentation inside the wall.
In this study, experiments were investigates using a disk heated at its lower surface. The disk is 50 mm in diameter and 8 mm thick (Figure 9). It is thermally insulated with Teflon on all faces except the cooling face in order to prevent the heat loss. Liquid jet impactes perpendicularly in the center of the heat exchange surface (top surface of the disk). Temperatures inside the experimental disk are measured using 7 Chromel-Alumel thermocouples of 200 μm diameter (uncertainty of ±0.2°C). As shown in Figure 9, thermocouples are placed at 0.6 mm below the wetted surface at radial intervals of 3.5 mm.
The experimental disk is heated continually and the wall temperatures are monitored. When thermal steady state is reached, the heat exchange surface is quickly cooled with the liquid jet. Time-dependent local wall temperatures are recorded, until the experimental disk reaches a new steady state. The local surface temperature and heat flux are determined by solving IHCP using these measurements.

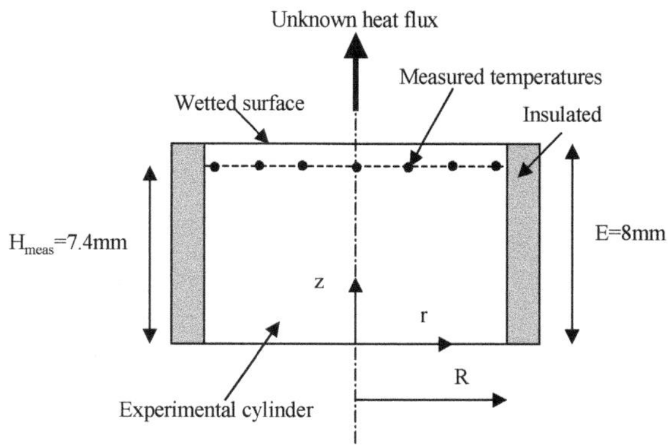

Fig. 9. Physical model.

Physical model of a unsteady heat conduction process is given by the following system of equations:

$$\frac{\rho C_p}{\lambda}\frac{\partial T(r,z,t)}{\partial t} = \frac{\partial^2 T(r,z,t)}{\partial r^2} + \frac{1}{r}\frac{\partial T(r,z,t)}{\partial r} + \frac{\partial^2 T(r,z,t)}{\partial z^2}, \tag{4}$$

where $0 < r < R$, $0 < z < E$

$$\frac{\partial T}{\partial r}(0,z,t) = 0 \text{, where } 0 < t \le t_f,\ 0 \le z \le E \tag{5}$$

$$\frac{\partial T}{\partial r}(R,z,t) = 0 \text{, where } 0 < t \le t_f,\ 0 \le z \le E \tag{6}$$

$$T(r,z,0) = T_0 \text{, where: } 0 \le r \le R,\ 0 \le z \le E \tag{7}$$

$$\lambda \frac{\partial T}{\partial z}(r,E,t) = Q_w(r,E,t) \text{, where: } 0 < t \le t_f,\ 0 \le r \le R \tag{8}$$

$$T(r,0,t) = f(r,t) \text{, where: } 0 < t \le t_f,\ 0 \le r \le R \tag{9}$$

Distribution of local heat flux $Q_w(r,E,t)$ at the heat exchange surface (z=E) is unknown. It is estimated by solving the IHCP using temperatures $T_{meas}(r_n, z_n, t)$ measured at nodes (r_n, z_n) inside the disk (Figure 9). Solution of the inverse problem is based on the minimization of the residual functional defined as:

$$J(C(T), \lambda(T)) = \sum_{n=1}^{N} \int_{t_0}^{t_f} \left[T(X_n, t; C(T), \lambda(T)) - f_n(t)\right]^2 dt \to \min \tag{10}$$

where $T(r_n, z_n; Q_w)$ are temperatures at the sensor locations computed from the direct problem (4-9). Minimization is carried out by using conjugate gradient algorithm (Alifanov et al., 1995). Heat flux $Q_w(r,E,t)$ is approximated in the form of a cubic B-spline and the IHCP is reduced to the estimation of a vector of B-Spline parameters. Conjugate gradient procedure is iterative. For each iteration, successive improvements of desired parameters are built. Descent parameter is computed using a linear approximation as follows:

$$\gamma^{it} = -\frac{\int_0^{t_f} \sum_{n=1}^{N_{meas}} \left[T^{it}(r_n,z_n,t;Q_w) - T_{meas}(r_n,z_n,t)\right] \theta^{it}(r_n,z_n,t;\delta Q_w)\,dt}{\int_0^{t_f} \sum_{n=1}^{N_{meas}} \theta^{it}(r_n,z_n,t;\delta Q_w)^2 dt} \tag{11}$$

Variation of temperature at the sensor locations $\theta^{it}(r_n,z_n,t;\delta Q_w)$ resulting from the variation of heat flux $\delta Q_w(r,E,t)$ is determined by solving variational problem. Variation of functional $J(Q_w)$ resulting from temperature variation is given by:

$$\delta J(Q_w,\delta Q_w) = \sum_{n=1}^{N_{meas}} \int_{t_0}^{t_f} \left[T(r_n,z_n,t,q_w) - T_{meas}(r_n,z_n,t)\right]\theta(r_n,z_n,t)dt \tag{12}$$

where $\theta^{it}(r_n,z_n,t;\delta Q_w)$ is determined at the sensor locations (r_n,z_n) by solving variational problem that defined by the following equations:

$$\frac{\rho C_p}{\lambda}\frac{\partial\theta(r,z,t)}{\partial t} = \frac{\partial^2\theta(r,z,t)}{\partial r^2} + \frac{1}{r}\frac{\partial\theta(r,z,t)}{\partial r} + \frac{\partial^2\theta(r,z,t)}{\partial z^2} \tag{13}$$

where $0 \le r \le R$, $0 \le z \le E$, $0 < t \le t_f$

$$\frac{\partial\theta}{\partial r}(0,z,t) = 0, \text{ where } 0 < t \le t_f,\ 0 \le z \le E \tag{14}$$

$$\frac{\partial\theta}{\partial r}(R,z,t) = 0, \text{ where } 0 < t \le t_f,\ 0 \le z \le E \tag{15}$$

$$\theta(r,z,0) = 0, \text{ where: } 0 \le r \le R,\ 0 \le z \le E \tag{16}$$

$$\lambda\frac{\partial\theta}{\partial z}(r,E,t) = 0, \text{ where: } 0 < t \le t_f,\ 0 \le r \le R \tag{17}$$

$$\theta(r,0,t) = 0, \text{ where: } 0 < t \le t_f,\ 0 \le r \le R \tag{18}$$

3.1 Lagrangian functional and adjoint problem

Using Lagrange multiplier method, Lagrangian functional is defined as:

$$J(C(T),\lambda(T)) = \sum_{n=1}^{N} \int_{t_0}^{t_f} \left[T(X_n,t;C(T),\lambda(T)) - f_n(t)\right]^2 dt \to \min$$

$$+\int_0^{t_f}\int_0^R \psi(r,z,t)\left[\frac{\lambda}{r}\frac{\partial}{\partial r}\left(r\frac{\partial T}{\partial r}\right)+\lambda\frac{\partial^2 T}{\partial z^2}-\rho C_p\frac{\partial T}{\partial t}\right]dr\,dz\,dt$$

$$+\int_0^{t_f}\int_0^R \xi(r,t)\left[T(r,0,t)-f(r,t)\right]dr\,dt+\int_0^{t_f}\int_0^E \omega(z,t)\,\lambda\frac{\partial T}{\partial \phi}(R,z,t)\,dz\,dt$$

$$+\int_0^{t_f}\int_0^R \gamma(r,t)\left[\lambda\frac{\partial T}{\partial z}(r,E,t)+Q_w(r,E,t)\right]dr\,dt+\int_0^{t_f}\int_0^E \mu(z,t)\left[\lambda\frac{\partial T}{\partial r}(0,z,t)\right]dz\,dt$$

$$+\int_0^R\int_0^E \eta(r,z)\left[T(r,z,0)-T_0\right]dr\,dz \tag{19}$$

Let $\psi(r,z,t)$, $\xi(r,t)$, $\omega(z,t)$, $\mu(z,t)$, $\eta(r,z)$ and $\gamma(r,t)$ be the Lagrange multipliers. The necessary condition of the optimization problem is obtained from the following equation:

$$\delta L(Q_w,\delta Q_w)=0 \tag{20}$$

where $\delta L(Q_w,\delta Q_w)$ is the variation of Lagrangian functional. Equation (19) requires that all coefficients of the temperature variation $\theta(r,z,t)$ be equal to 0. To satisfy this condition the necessary conditions of optimization are defined in the form of adjoint problem.

$$-\frac{\rho C_p}{\lambda}\frac{\partial\psi(r,z,t)}{\partial t}=\frac{\partial^2\psi}{\partial r^2}-\frac{1}{r}\frac{\partial\psi}{\partial r}+\frac{1}{r^2}\psi+\frac{\partial^2\psi}{\partial z^2}+S(r,z,t) \tag{21}$$

where: $$S(r,z,t)=-\sum_{n=1}^{N_{meas}}\left\{\delta(r,r_n;z,z_n)\times\left[T(r_n,z_n,t;Q_w)-T_{meas}(r_n,z_n,t)\right]\right\},$$

$$0\le r\le R,\ 0\le z\le E,\ 0<t\le t_f$$

$$\frac{\partial\psi}{\partial r}(0,z,t)=\frac{\psi}{r}(0,z,t)\text{, where } 0<t\le t_f,\ 0\le z\le E \tag{22}$$

$$\frac{\partial\psi}{\partial r}(R,z,t)=\frac{\psi}{r}(R,z,t)\text{, where } 0<t\le t_f,\ 0\le z\le E \tag{23}$$

$$\psi(r,z,t_f)=0\text{, where: } 0\le r\le R,\ 0\le z\le E \tag{24}$$

$$\lambda\frac{\partial\psi}{\partial z}(r,E,t)=0\text{, where: } 0<t\le t_f,\ 0\le r\le R \tag{25}$$

$$\psi(r,0,t)=0\text{, where: } 0<t\le t_f,\ 0\le r\le R \tag{26}$$

where $\psi(r,z,t)$ is the Lagrange multiplier,

$$J(C(T),\lambda(T))=\sum_{n=1}^{N}\int_{t_0}^{t_f}\left[T(X_n,t;C(T),\lambda(T))-f_n(t)\right]^2 dt\rightarrow \min$$

is the Dirac Function, $S(r,z,t)$ is the deviation between temperature measurements and computed temperatures. $S(r,z,t)$ is equal to 0 everywhere in the physical domain except at sensor locations (r_n,z_n).
The Dirac function is defined by

$$J(C(T),\lambda(T))=\sum_{n=1}^{N}\int_{t_0}^{t_f}\left[T(X_n,t;C(T),\lambda(T))-f_n(t)\right]^2 dt\rightarrow \min \tag{27}$$

where $\delta(0)=1$, $\delta(r)=0$ for $r\neq 0$ and $\delta(z)=0$ for $z\neq 0$
If the direct problem and the adjoint problem are verified, variation of the Lagrangian functional becomes:

$$\delta\mathbf{L}(Q_w,\delta Q_w)=-\int_0^{t_f}\int_0^{R}\psi(r,E,t)\,\delta Q_w(r,E,t)\,dr\,dt \tag{28}$$

Vector gradient can be verified by the following equation:

$$J'_{Q_w}(r,E,t)\ \ =-\psi(r,E,t) \tag{29}$$

3.2 Gradient vector computation

Variation of functional $J(Q_w)$ can be approximated in the form:

$$\delta J(Q_w,\delta Q_w)=-\int_0^{t_f}\int_0^{E}\int_0^{R}\left[\frac{\rho C_p}{\lambda}\frac{\partial\psi(r,z,t)}{\partial t}+\frac{\partial^2\psi}{\partial r^2}-\frac{1}{r}\frac{\partial\psi}{\partial r}+\frac{1}{r^2}\psi+\frac{\partial^2\psi}{\partial z^2}\right]\theta(r,z,t)\,dr\,dz\,dt \tag{30}$$

Integration by parts gives, the variation of functional becomes using Eqs (21-26):

$$\delta J(Q_w,\delta Q_w)=\int_0^{t_f}\int_0^{R_2}\left[-\lambda\frac{\partial\psi}{\partial z}\theta(r,t)+\lambda\psi(r,t)\frac{\partial\theta(r,t)}{\partial z}\right]drdt \tag{31}$$

Substituting Eqs. (25) and (17) into Eq. (31), $\delta J(q_w,\delta q_w)$ becomes:

$$\delta J(Q_w,\delta Q_w)=-\int_0^{t_f}\int_0^{R}\psi(r,E,t)\delta Q_w(r,E,t)\,dr\,dt$$

$$=\delta\mathbf{L}(Q_w,\delta Q_w) \tag{32}$$

Variation of functional is defined as:

$$\delta J(Q_w, \delta Q_w) = \int_0^{t_f} \int_0^{R_2} J'_{Q_w}(r,E,t)\delta Q_w(r,E,t)\,dr\,dt \tag{33}$$

Equations (32) and (33) imply that:

$$J'_{Q_w}(r,E,t) = -\psi(r,E,t) \tag{34}$$

Vector gradient can verified the following equation:

$$J'_{Q_w}(r,E,t) = -\psi(r,E,t) \tag{35}$$

3.3 Algorithm

The following iterative procedure is adopted to solve the inverse heat conduction problem:

i. solution of the direct problem,
ii. calculation of the residual functional,
iii. solution of the adjoint problem,
iv. calculation of the components of the functional gradient,
v. calculation of the parameter in descent direction,
vi. calculation of the component of descent direction,
vii. solution of the variational problem to determine the descent parameter,
viii. the new value of the heat flux density is corrected.

If the convergence criteria is not satisfied the iterative procedure is repeated until the functional is minimized. The minimal value of the functional depends on the temperature measurement errors.

The direct problem, adjoint problem, and variational problem are solved using the control volume method (Patankar, 1980) and the implicit fractional-step time scheme proposed by (Brian, 1961).

3.4 Regularization

The inverse problem is ill-posed and numerical solution depends on the fluctuation occurring in the measurements. The iterations are stopped at the optimal value of the residual functional which satisfies the criteria:

$$J(Q_w) = \frac{1}{2}\int_0^{t_f} \sum_{n=1}^{N_{meas}} \sigma^2(r_n, z_n, t)\,dt \tag{36}$$

Here, $\sigma^2(r_n, z_n, t)$ is the standard deviation of measurement errors for the temperatures measured at locations (r_n, z_n).

4. Inverse estimation of the boundary conditions

4.1 Numerical verification of the solution procedure

The numerical procedure is verified by using a known heat flux varying with time and the radius of the disk. Heat flux is imposed at the top surface of the disk (z = E) as shown in Figure 10 by the continuous curve. The bottom surface (z=0) is assumed to be at the constant temperature of T(r,0,t)= 40°C. For each numerical application, time step size is chosen with respect to delta Fourier number condition defined by the following equation:

$$\Delta Fo = \frac{\lambda}{\rho C_p} \frac{\Delta t}{(E - H_{meas})^2} \geq 0.001 \quad (37)$$

The delta Fourier number is based on the sensor depth, thermal characteristics of the solid, and time step (Williams & Beck, 1995, Beck & Brown, 1996).

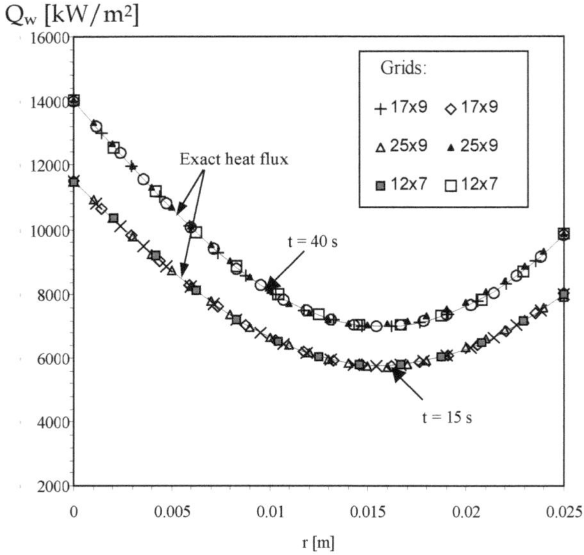

Fig. 10. Heat flux variation with radius on the top surface. Verification of the IHCP: solid line ("measurements"), symbols ("estimations using inverse method").

In order to validate inverse estimation procedure, it is assumed that temperatures calculated from the direct problem at the measurement points are used as the measured temperatures ($T_{meas}(r_n, z_n, t) = f_n(r_n, z_n, t)$) for solving ICHP. Figure 10 shows that the estimated heat flux is closed with the exact heat flux for different times. This validation is carried out for the number of approximation parameters equal to 9x9. The maximum deviation between the computed temperatures and the simulated measured temperatures is of ±0.03°C. The evolution of the residual functional $J(Q_w)$ is a function of the number of iterations that are continued till the convergence criteria is satisfied.

4.2 Inverse estimation of evaporation local heat transfer for jet impingement

4.2.1 Evaporation local heat transfer for unsteady state

For inlet Reynolds number of 7600, Figure 11 shows an example of temporal temperatures measured for different radial locations at 0.6 mm below the heat exchange surface. During experiments, heat flux imposed inside the experimental disk is 45 W, the nozzle-heat exchange surface spacing is 30 mm, and the liquid inlet temperature is 42°C. At the steady state, wall temperatures are 78°C. When the heat exchange surface is wetted, the wall temperatures decrease continually and reach a stable value during a short period. Temperature at the stagnation zone is lower than the temperature measured far from the impingement zone. IHCP is solved using temperatures measured at H_{meas} = 7.4mm (Figure 11) in order to estimate the local surface temperature and heat flux. These local thermal characteristics are estimated using the temperatures measured at the bottom surface (z=0) as the boundary condition to solve the direct problem.

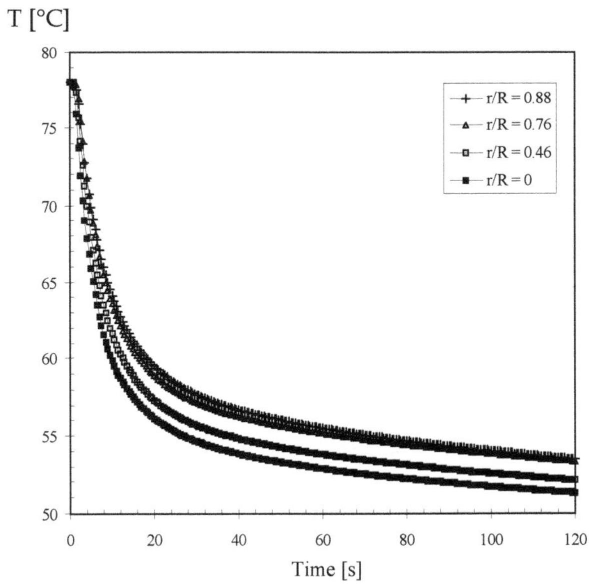

Fig. 11. Temperatures measured inside the solid at z = H_{meas}.

Figures 12 and 13 show, respectively, the unsteady evolution of the predicted surface heat flux and temperature at different radial locations on the cooling surface (z = E =8 mm). Surface temperature is low in the stagnation and in impingement zone where heat flux is high. The difference between the wall and liquid temperatures is high at the moment when the liquid jet impinges the heat exchange surface. After this, heat flux decreases with time and follows the same trend for each radial location. Heat flux decreases after the impingement zone because liquid spreads along the radial direction as a very thin film. The experimental data for each radial location and inlet Reynolds number, follows the same trend. For brevity, theses curves are not shown in this figure.

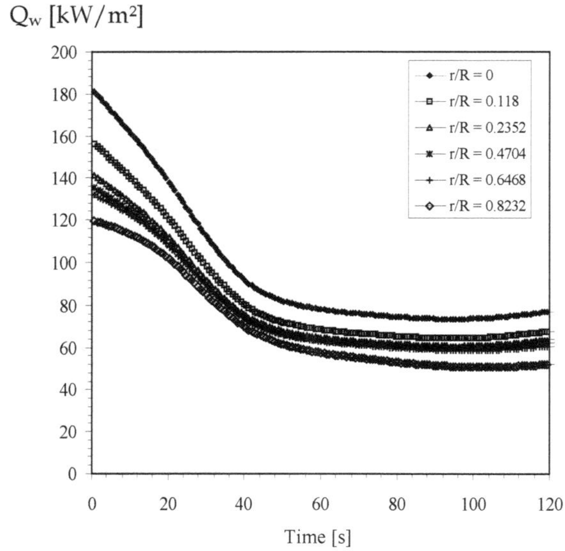

Fig. 12. Heat flux inversely predicted at the top surface.

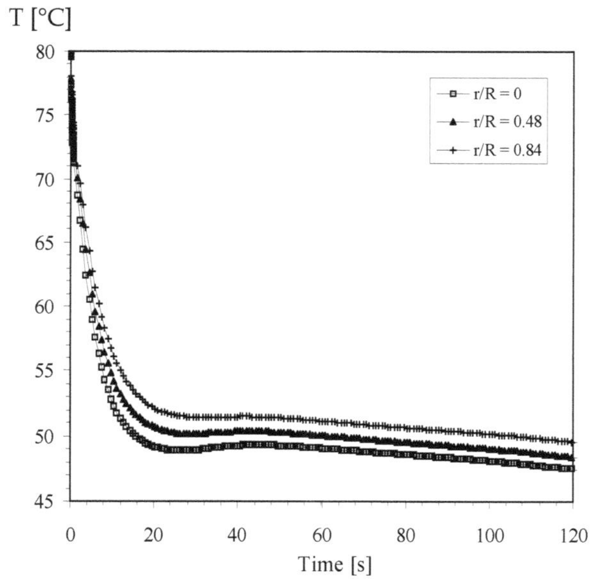

Fig. 13. Temperatures inversely predicted at the top surface.

For both sides of the disk, radial distributions of the surface heat flux and heat transfer coefficients are presented in Figures 14a and 14b for different times. Local heat flux and heat transfer coefficients are not uniform along the radial direction, and they are high in the impingement zone.

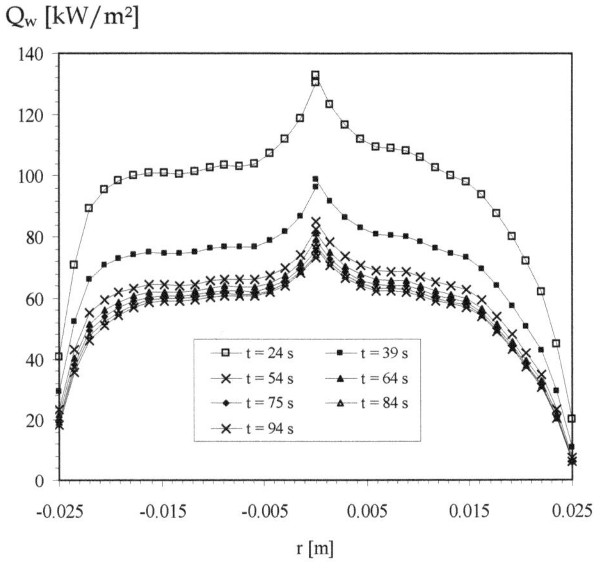

(a)

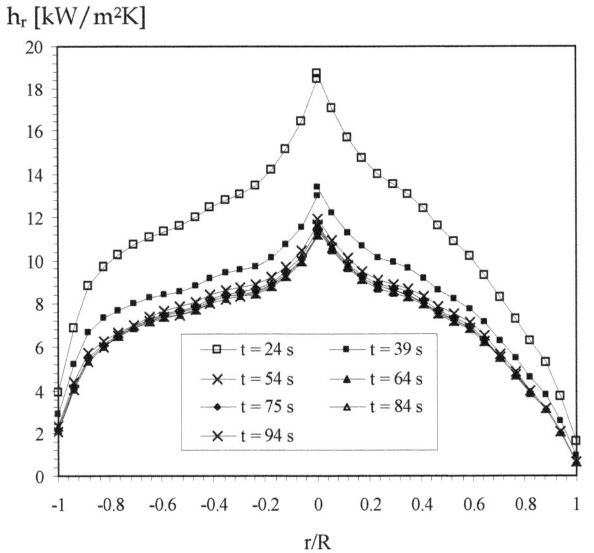

(b)

Fig. 14. Radial distribution inversely predicted at the top surface (z = E) : (a) heat flux and (b) heat transfer coefficient.

After the impingement zone, heat transfer decreases because the liquid jet covers the entire heat exchange surface. Therefore, local liquid flow rate decreases in spite of the decrease of the film thickness. When the radius r becomes higher than approximately 0.018 mm, heat transfer is reduced because of the hydraulic jump formation where the velocity of the flow becomes relatively negligible. At each time, the local heat flux and heat transfer coefficient follow the same trend. Beyond 64s, the curves of the heat flux and those of the heat transfer coefficient are independent on the time because of the steady state.

4.2.2 Evaporation local heat transfer for steady state

For steady state, Figure 15 shows the local distributions of the surface temperature and heat transfer coefficient. For each radial location, the local heat transfer coefficient is determined from the surface heat flux and temperature as follows:

$$h_r = \frac{Q_{w,r}}{T_{s,r} - T_e} \tag{38}$$

where h_r is the local heat transfer coefficient, $Q_{w,r}$ is the local heat flux, $T_{s,r}$ is the local surface temperature, and T_e is the liquid temperature at the nozzle exit.

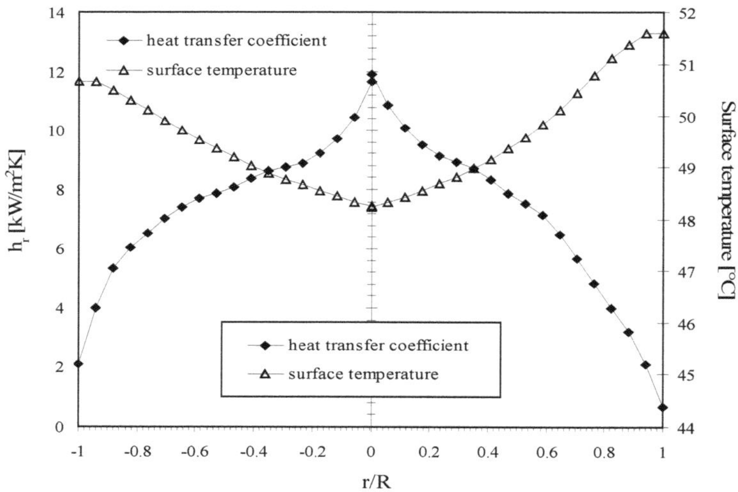

Fig. 15. Local thermal characteristics for steady state.

The surface temperature is low in the stagnation zone compared to all the zones of the heat exchange surface. The maximum heat transfer coefficient is occurred in the stagnation point. For different flow rates, Figure 16 illustrates the unsteady evolution of the surface temperatures for two radial locations. The first one is at the stagnation point where the surface temperature is low. The second is far from the impingement zone (at r=0.82R), where the heat transfer coefficient is deteriorated because of the hydraulic jump. The surface temperature in this zone is higher than in the stagnation point. It is shown that the surface temperature is less influenced by the flow rate at the stagnation zone than for r=0.82R where the film thickness is small. The normalized heat transfer coefficient is determined as the fraction of the local heat transfer coefficient and h_0 that is defined at the stagnation zone (Figure 17). For each tested flow rate, the heat transfer coefficient decreases from h_0 to 50% of h_0 at radial location approximately equal to 0.6R.

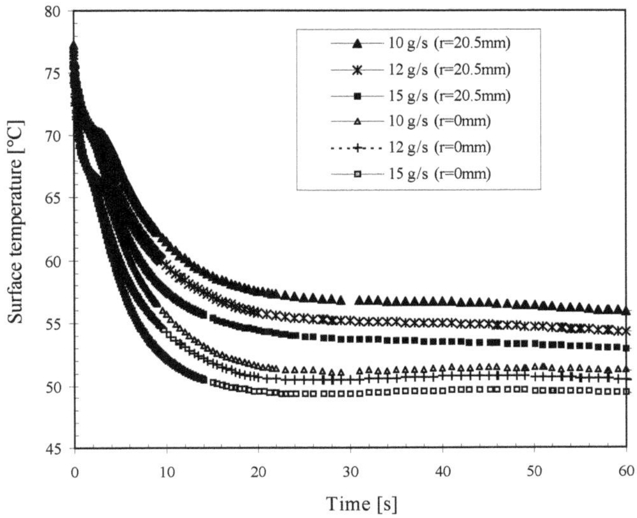

Fig. 16. Local surface temperatures inversely predicted at the top surface.

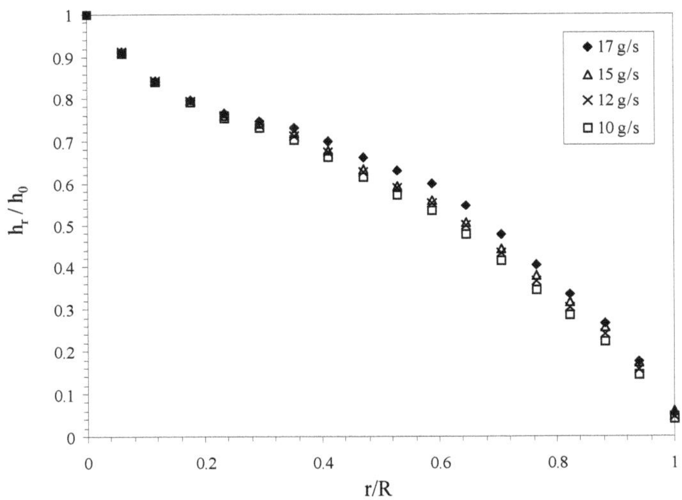

Fig. 17. The normalized heat transfer coefficient distribution as a function of water jet flow rate.

5. Conclusion

Various theoretical and experimental investigations on convective local heat transfer have been published in the literature where local heat transfer coefficient is determined from total heat flux or using direct estimation (Fourier's law). In this case, heat flux is assumed to be dissipated only in the axial direction and constant along the heat exchange surface.

In this work, local heat transfer is analyzed by solving inverse heat conduction problem and using only sensors responses placed inside the experimental disk. Iterative regularization method is used to solve the inverse problem under analysis. Solution procedure is based on the conjugate gradient method used to minimize the residual functional and the residual discrepancy principal as the regularizing stopping criterion.
For each radial location, local heat transfer coefficient is determined using local heat flux and surface temperature. The heat flux and heat transfer coefficient are high in the impingement zone and decrease after this zone because liquid flow spreads along the radial direction as a very thin film. At each time, surface temperature is low in the stagnation zone and the highest heat transfer coefficient occurs in the stagnation zone and falls off with the radial location because local flow rate decreases. For different tested flow rates, the heat transfer coefficient decreases from h_0 to 50% of h_0 at the radial location approximately equal to 0.6R.

6. References

Alifanov, O.M., Artyukhin, E.A. and Rumyantsev, S.V., (1995). Extreme Methods for Solving Ill-Posed Problems with Applications to Inverse Heat Transfer Problems, Begell House, New York.

Baonga, J.B., Louahlia-Gualous, H. & Imbert, M. (2006). Experimental study of the hydrodynamic and heat transfer of free liquid jet impinging a flat circular heated disk, *Applied Thermal Engineering*, Vol. 26, pp. 1125-1138.

Brian, P.L.T., (1961). A finite-difference method of higher order accuracy for solution of three dimensional transient heat conduction, *A.I.Ch.E. Journal*, vol. 7, pp. 367-370.

Chen, R.H., Chow, L.C. & Navedo, J.E. (2002). Effects of spray characteristics on critical heat flux in subcooled water spray cooling, *Int. J. Heat and Mass Transfer*, Vol. 45, pp. 4033-4043.

Chen, H.T., Lin, S.Y. and Fang, L.Ch., (2001). Estimation of surface temperature in two dimensional inverse heat conduction problems, *Int. J. of Heat and Mass Transfer*, vol. 44, pp. 1455-1463.

Elison, B. & Webb, B.W. (2003). Local heat transfer to impinging liquid jets in the initially laminar, transitional, and turbulent regimes, *Int. J. Heat Mass Transfer*, Vol. 37, pp. 1207-1217.

Fabbri, M., Jiang, Sh. & Dhir, V.K. (2003). Experimental investigation of single-phase micro jets impingement cooling for electronic applications, *Proc. Of Heat Transfer Conference ASME*, pp. 1-10.

Liu, X., Lienhard J.H. & Lombara, J.S. (1991). Convective heat transfer by impingement of circular liquid jets, *J. of Heat Transfer, Transaction of the ASME*, Vol. 113, pp. 571-582.

Liu, X. & Lienhard J.H. (1989). Liquid jet impingement heat transfer on a uniform flux surface, *National Heat Transfer Conference*, Vol. 106, pp. 523-530.

Lin, L. & Ponnappan, R. (2004). Critical heat flux of Multi-nozzle spray cooling, *J. of Heat Transfer, Transaction of the ASME*, Vol. 126, pp. 482-485.

Liu, Z.H. & Zhu, Q.Z. (2004). Prediction of critical heat flux for convective boiling of saturated water jet impinging on the stagnation zone, *J. of Heat Transfer, Transaction of the ASME*, Vol. 124, pp. 1125-1130

Louahlia-Gualous, H., Panday, P.K. and Artioukhine, E., (2003). Inverse determination of the local heat transfer coefficients of nucleate boiling on a horizontal cylinder, *Trans. ASME, J. Heat Transfer*, vol. 125, pp. 1087-1095.

Louahlia-Gualous, H. & Baonga, J.B. (2008). Experimental study of unsteady local heat transfer for impinging miniature jet, *Heat Transfer Engineering*, Vol. 29, N°. 2, pp.782-792.

Louahlia-Gualous H. & El Omari, L. (2006). Local heat transfer for the evaporation of a laminar falling liquid film on a cylinder - Experimental, numerical and inverse heat conduction analysis, *Numerical Heat Transfer, Part A: Applications*, Vol. 50 N° 7, p. 667-688.

MA, C.F., Zheng, Q., Lee, S.C. & Gomi, T. (1997). Impingement heat transfer and recovery effect with submerged jets of large Prandtl number liquid-I. Unconfined circular jets, *Int. J. Heat Mass Transfer*, Vol. 40, No. 6 pp. 1481-1490.

MA, C.F., Zheng, Q., Sun, H., Wu, K., Gomi, T. & Webb, B.W. (1997). Local characteristics of impingement heat transfer with oblique round free-surface jets of large Prandtl number liquid, *Int. J. Heat Mass Transfer*, Vol. 40, No. 10, pp. 2249-2259

Martin, T.J. & Dulkravich, G.S., (1998) Inverse determination of steady heat convection coefficient distributions, *J. of Heat Transfer, Transaction of the ASME*, Vol. 120, pp. 328-334.

Oliphant, K., Webb, B.W., McQuay, M.Q., (1998). An experimental comparison of liquid jet array and spray impingement cooling in the non-boiling regime, *Exp. Thermal and Fluid Science*, Vol. 18, pp. 1-10.

Pan, Y. & Webb, B.W., (1995). Heat transfer characteristics of array of free-surface liquid jets, *Transaction of the ASME*, Vol. 177, pp. 878-884

Patankar, S.V., (1980). Numerical Heat Transfer and Fluid Flow, Mc Graw Hill, New York.

Siba, E.A., Ganesa-Pillai, M., Harris, K.T. & Haji-Sheikh, A. (2003) Heat transfer in a high turbulence air jet impinging over a flat circular disk, *J. of Heat Transfer, Transaction of the ASME*, Vol. 125, pp. 257-265.

Stevens, J. & Webb, B.W. (1989). Local heat transfer coefficients under an axisymmetric, single-phase liquid jet, *National Heat Transfer Conference*, Vol. 11, pp. 113-119.

Stevens, J., & Webb, B.W. (1993). Measurements of flow structure in the radial layer of impinging free surface liquid jets, *Int. J. Heat Mass Transfer*, Vol. 36, N°.15, pp. 3751-3758.

Stevens, J., & Webb, B.W. (1992). Measurements of the free surface flow structure under an impinging, free liquid jet, *Journal of Heat Transfer, Transaction of ASME*, Vol. 114, pp. 79-84.

Stevens,J. & Webb, B.W. (1989). Local heat transfer coefficients under an axisymmetric, single-phase liquid jet, *American society Mechanical Engineers. Heat Transfer Division*, Vol. 111 pp. 113-119.

Stevens, J. & Webb, B.W., (1991). Local heat transfer coefficients under an axisymmetric, single-phase liquid jet, *Journal of Heat Transfer*, Vol. 113, pp. 71-78.

Watson, E.J. (1964). The radial spread of a liquid over horizontal plane, *J. Fluid Mech*. Vol. 20, pp. 481-500.

4

Identifiability of Piecewise Constant Conductivity

Semion Gutman[1] and Junhong Ha[2]
[1]*University of Oklahoma*
[2]*Korea University of Technology and Education*
[1]*USA*
[2]*South Korea*

1. Introduction

Consider the heat conduction in a nonhomogeneous insulated rod of a unit length, with the ends kept at zero temperature at all times. Our main interest is in the identification and identifiability of the discontinuous conductivity (thermal diffusivity) coefficient $a(x),\ 0 \le x \le 1$. The **identification** problem consists of finding a conductivity $a(x)$ in an admissible set K for which the temperature $u(x,t)$ fits given observations in a prescribed sense.
Under a wide range of conditions one can establish the continuity of the objective function $J(a)$ representing the best fit to the observations. Then the existence of the best fit to data conductivity follows if the admissible set K is compact in the appropriate topology. However, such an approach usually does not guarantee the uniqueness of the found conductivity $a(x)$. Establishing such a uniqueness is referred to as the **identifiability** problem. For an extensive survey of heat conduction, including inverse heat conduction problems see (Beck et al., 1985; Cannon, 1984; Ramm, 2005)
From physical considerations the conductivity coefficients $a(x)$ are assumed to be in

$$A_{\text{ad}} = \{a \in L^\infty(0,1) : 0 < \nu \le a(x) \le \mu\}. \tag{1}$$

The temperature $u(a) = u(x,t;a)$ inside the rod satisfies

$$\begin{array}{ll} u_t - (a(x)u_x)_x = f(x,t), & Q = (0,1) \times (0,T), \\ u(0,t) = q_1(t),\ u(1,t) = q_2(t), & t \in (0,T), \\ u(x,0) = g(x), & x \in (0,1), \end{array} \tag{2}$$

where $g \in H = L^2(0,1),\ q_1, q_2 \in C^1[0,\infty)$. Suppose that one is given an observation $z(t) = u(p,t;a)$ of the heat conduction process (2) for $t_1 < t < t_2$ at some observation point $0 < p < 1$. From the series solution for (2) and the uniqueness of the Dirichlet series expansion (see Section 5), one can, in principle, recover all the eigenvalues of the associated Sturm-Lioville problem. If one also knows the eigenvalues for the heat conduction process with the same coefficient a and different boundary conditions, then classical results of Gelfand and Levitan (Gelfand & Levitan, 1955) show that the conductivity $a(x)$ can be uniquely identified from the knowledge of the two spectral sequences.
Alternatively, the conductivity is identifiable if the entire spectral function is known (i.e. the eigenvalues and the values of the derivatives of the normalized eigenfunctions at $x = 0$). However, such results have little practical value, since the observation data $z(t)$ always

contain some noise, and therefore one cannot hope to adequately identify more than just a few first eigenvalues of the problem.

A different approach is taken in (Duchateau, 1995; Kitamura & Nakagiri, 1977; Nakagiri, 1993; Orlov & Bentsman, 2000; Pierce, 1979). These works show that one can identify a constant conductivity a in (2) from the measurement $z(t)$ taken at one point $p \in (0,1)$. These works also discuss problems more general than (2), including problems with a broad range of boundary conditions, non-zero forcing functions, as well as elliptic and hyperbolic problems. In (Elayyan & Isakov, 1997; Kohn & Vogelius, 1985) and references therein identifiability results are obtained for elliptic and parabolic equations with discontinuous parameters in a multidimensional setting. A typical assumption there is that one knows the normal derivative of the solution at the boundary of the region for every Dirichlet boundary input. For more recent work see (Benabdallah et al., 2007; Demir & Hasanov, 2008; Isakov, 2006).

In our work we examine piecewise constant conductivities $a(x)$, $x \in [0,1]$. Suppose that the conductivity a is known to have sufficiently separated points of discontinuity. More precisely, let $a \in PC(\sigma)$ defined in Section 2. Let $u(x,t;a)$ be the solution of (2). The eigenfunctions and the eigenvalues for (2) are defined from the associated Sturm-Liouville problem (5).

In our approach the identifiability is achieved in two steps:

First, given finitely many equidistant observation points $\{p_m\}_{m=1}^{M-1}$ on interval $(0,1)$ (as specified in Theorem 5.5), we extract the first eigenvalue $\lambda_1(a)$ and a constant nonzero multiple of the first eigenfunction $G_m(a) = C(a)\psi_1(p_m;a)$ from the observations $z_m(t;a) = u(p_m,t;a)$. This defines the M-tuple

$$\mathcal{G}(a) = (\lambda_1(a), G_1(a), \cdots, G_{M-1}(a)) \in \mathbb{R}^M. \tag{3}$$

Second, the Marching Algorithm (see Theorem 5.5) identifies the conductivity a from $\mathcal{G}(a)$.

We start by recalling some basic properties of the eigenvalues and the eigenfunctions for (2) in Section 2. Our main identifiability result is Theorem 5.5. It is discussed in Section 5. The continuity properties of the solution map $a \to \mathcal{G}(a)$ are established in Section 4, and the continuity of the identification map $\mathcal{G}^{-1}(a)$ is proved in Section 8. Computational algorithms for the identification of $a(x)$ from noisy data are presented in Section 10.

This exposition outlines main results obtained in (Gutman & Ha, 2007; 2009). In (Gutman & Ha, 2007) the case of distributed measurements is considered as well.

2. Properties of the eigenvalues and the eigenfunctions

The admissible set A_{ad} is too wide to obtain the desired identifiability results, so we restrict it as follows.

Definition 2.1. (i) $a \in \mathcal{PS}_N$ if function a is piecewise smooth, that is there exists a finite sequence of points $0 = x_0 < x_1 < \cdots < x_{N-1} < x_N = 1$ such that both $a(x)$ and $a'(x)$ are continuous on every open subinterval (x_{i-1}, x_i), $i = 1, \cdots, N$ and both can be continuously extended to the closed intervals $[x_{i-1}, x_i]$, $i = 1, \cdots, N$. For definiteness, we assume that a and a' are continuous from the right, i.e. $a(x) = a(x+)$ and $a'(x) = a'(x+)$ for all $x \in [0,1)$. Also let $a(1) = a(1-)$.

(ii) Define $\mathcal{PS} = \cup_{N=1}^{\infty} \mathcal{PS}_N$.

(iii) Define $\mathcal{PC} \subset \mathcal{PS}$ as the class of piecewise constant conductivities, and $\mathcal{PC}_N = \mathcal{PC} \cap \mathcal{PS}_N$. Any $a \in \mathcal{PC}_N$ has the form $a(x) = a_i$ for $x \in [x_{i-1}, x_i)$, $i = 1, 2, \cdots, N$.

(iv) Let $\sigma > 0$. Define

$$\mathcal{PC}(\sigma) = \{a \in \mathcal{PC} \ : \ x_i - x_{i-1} \geq \sigma, \quad i = 1, 2, \cdots, N\},$$

where $x_1, x_2, \cdots, x_{N-1}$ are the discontinuity points of a, and $x_0 = 0,\ x_N = 1$.

Note that $a \in \mathcal{PC}(\sigma)$ attains at most $N = [[1/\sigma]]$ distinct values $a_i,\ 0 < \nu \le a_i \le \mu$. For $a \in \mathcal{PS}_N$ the governing system (2) is given by

$$\begin{cases} u_t - (a(x)u_x)_x = f(x,t), & x \neq x_i, \quad t \in (0,T), \\ u(0,t) = q_1(t),\ u(1,t) = q_2(t), & t \in (0,T), \\ u(x_i+,t) = u(x_i-,t), & t \in (0,T), \\ a(x_i+)u_x(x_i+,t) = a(x_i-)u_x(x_i-,t), & t \in (0,T), \\ u(x,0) = g(x), & x \in (0,1). \end{cases} \tag{4}$$

The associated Sturm-Liouville problem for (4) is

$$\begin{cases} (a(x)\psi(x)')' = -\lambda\psi(x), \ x \neq x_i, \\ \psi(0) = \psi(1) = 0, \\ \psi(x_i+) = \psi(x_i-), \\ a(x_i+)\psi_x(x_i+) = a(x_i-)\psi_x(x_i-). \end{cases} \tag{5}$$

For convenience we collect basic properties of the eigenvalues and the eigenfunctions of (5). Additional details can be found in (Birkhoff & Rota, 1978; Evans, 2010; Gutman & Ha, 2007).

Theorem 2.2. *Let $a \in \mathcal{PS}$. Then*

(i) The associated Sturm-Liouville problem (5) has infinitely many eigenvalues

$$0 < \lambda_1 < \lambda_2 < \cdots \to \infty.$$

The eigenvalues $\{\lambda_k\}_{k=1}^{\infty}$ and the corresponding orthonormal set of eigenfunctions $\{\psi_k\}_{k=1}^{\infty}$ satisfy

$$\lambda_k = \int_0^1 a(x)[\psi_k'(x)]^2 dx, \tag{6}$$

$$\lambda_k = \inf\left\{ \frac{\int_0^1 a(x)[\psi'(x)]^2 dx}{\int_0^1 [\psi(x)]^2 dx} \ :\ \psi \perp span\{\psi_1, \ldots, \psi_{k-1}\} \subset H_0^1(0,1) \right\}. \tag{7}$$

The normalized eigenfunctions $\{\psi_k\}_{k=1}^{\infty}$ form a basis in $L^2(0,1)$. Eigenfunctions $\{\psi_k/\sqrt{\lambda_k}\}_{k=1}^{\infty}$ form an orthonormal basis in

$$V_a = \{\psi \in H_0^1(0,1)\ :\ \int_0^1 a(x)[\psi'(x)]^2 dx < \infty\}.$$

(ii) Each eigenvalue is simple. For each eigenvalue λ_k there exists a unique continuous, piecewise smooth normalized eigenfunction $\psi_k(x)$ such that $\psi_k'(0+) > 0$, and the function $a(x)\psi_k'(x)$ is continuous on $[0,1]$.

(iii) Eigenvalues $\{\lambda_k\}_{k=1}^{\infty}$ satisfy Courant min-max principle

$$\lambda_k = \min_{V_k} \max\left\{ \frac{\int_0^1 a(x)[\psi'(x)]^2 dx}{\int_0^1 [\psi(x)]^2 dx} \ :\ \psi \in V_k \right\},$$

where V_k varies over all subspaces of $H_0^1(0,1)$ of finite dimension k.

(iv) Eigenvalues $\{\lambda_k\}_{k=1}^{\infty}$ *satisfy the inequality*

$$\nu\pi^2k^2 \leq \lambda_k \leq \mu\pi^2k^2.$$

(v) First eigenfunction ψ_1 *satisfies* $\psi_1(x) > 0$ *for any* $x \in (0,1)$.

(vi) First eigenfunction ψ_1 *has a unique point of maximum* $q \in (0,1)$: $\psi_1(x) < \psi_1(q)$ *for any* $x \neq q$.

Proof. (i) See (Evans, 2010).

(ii) On any subinterval (x_i, x_{i+1}) the coefficient $a(x)$ has a bounded continuous derivative. Therefore, on any such interval the initial value problem $(a(x)v'(x))' + \lambda v = 0$, $v(x_i) = A$, $v'(x_i) = B$ has a unique solution. Suppose that two eigenfunctions $w_1(x)$ and $w_2(x)$ correspond to the same eigenvalue λ_k. Then they both satisfy the condition $w_1(0) = w_2(0) = 0$. Therefore their Wronskian is equal to zero at $x = 0$. Consequently, the Wronskian is zero throughout the interval (x_0, x_1), and the solutions are linearly dependent there. Thus $w_2(x) = Cw_1(x)$ on (x_0, x_1), $w_2(x_1-) = Cw_1(x_1-)$ and $w_2'(x_1-) = Cw_1'(x_1-)$. The linear matching conditions imply that $w_2(x_1+) = Cw_1(x_1+)$ and $w_2'(x_1+) = Cw_1'(x_1+)$. The uniqueness of solutions implies that $w_2(x) = Cw_1(x)$ on (x_1, x_2), etc. Thus $w_2(x) = Cw_1(x)$ on $(0,1)$ and each eigenvalue λ_k is simple. In particular λ_1 is a simple eigenvalue. The uniqueness and the matching conditions also imply that any solution of $(a(x)v'(x))' + \lambda v = 0$, $v(0) = 0$, $v'(0) = 0$ must be identically equal to zero on the entire interval $(0,1)$. Thus no eigenfunction $\psi_k(x)$ satisfies $\psi_k'(0) = 0$. Assuming that the eigenfunction ψ_k is normalized in $L^2(0,1)$ it leaves us with the choice of its sign for $\psi_k'(0)$. Letting $\psi_k'(0) > 0$ makes the eigenfunction unique.

(iii) See (Evans, 2010).

(iv) Suppose $a(x) \leq b(x)$ for $x \in [0,1]$. The min-max principle implies $\lambda_k(a) \leq \lambda_k(b)$. Since the eigenvalues of (7) with $a(x) = 1$ are π^2k^2 the required inequality follows.

(v) Recall that $\psi_1(x)$ is a continuous function on $[0,1]$. Suppose that there exists $p \in (0,1)$ such that $\psi_1(p) = 0$. Let $w_l(x) = \psi_1(x)$ for $0 \leq x < p$, and $w_l(x) = 0$ for $p \leq x \leq 1$. Let $w_r(x) = \psi_1(x) - w_l(x)$, $x \in [0,1]$. Then w_l, w_r are continuous, and, moreover, $w_l, w_r \in H_0^1(0,1)$. Also

$$\int_0^1 w_l(x)w_r(x)dx = 0, \quad \text{and} \quad \int_0^1 a(x)w_l'(x)w_r'(x)dx = 0.$$

Suppose that w_l is not an eigenfunction for λ_1. Then

$$\int_0^1 a(x)[w_l'(x)]^2dx > \lambda_1 \int_0^1 [w_l(x)]^2dx.$$

Since

$$\int_0^1 a(x)[w_r'(x)]^2dx \geq \lambda_1 \int_0^1 [w_r(x)]^2dx$$

we have

$$\lambda_1 = \frac{\int_0^1 a(x)[\psi_1'(x)]^2dx}{\int_0^1 [\psi_1(x)]^2dx} = \frac{\int_0^1 a(x)([w_l'(x)]^2 + [w_r'(x)]^2)dx}{\int_0^1 ([w_l(x)]^2 + [w_r(x)]^2)dx} >$$

$$\frac{\int_0^1 (\lambda_1 [w_l(x)]^2 + \lambda_1 [w_r(x)]^2) dx}{\int_0^1 ([w_l(x)]^2 + [w_r(x)]^2) dx} = \lambda_1.$$

This contradiction implies that w_l (and w_r) must be an eigenfunction for λ_1. However, $w_l(x) = 0$ for $p \le x \le 1$, and as in (ii) it implies that $w_l(x) = 0$ for all $x \in [0,1]$ which is impossible. Since $\psi_1'(0) > 0$ the conclusion is that $\psi_1(x) > 0$ for $x \in (0,1)$.

(vi) From part (ii), any eigenfunction ψ_k is continuous and satisfies

$$(a(x)\psi_k'(x))' = -\lambda_k \psi_k(x)$$

for $x \ne x_i$. Also function $a(x)\psi_k'(x)$ is continuous on $[0,1]$ because of the matching conditions at the points of discontinuity x_i, $i = 1, 2, \cdots, N-1$ of a. The integration gives

$$a(x)\psi_k'(x) = a(p)\psi_k'(p) - \lambda_k \int_p^x \psi_k(s) ds,$$

for any $x, p \in (0,1)$.

Let $p \in (0,1)$ be a point of maximum of ψ_k. If $p \ne x_i$ then $\psi_k'(p) = 0$. If $p = x_i$, then $\psi_k'(x_i-) \ge 0$ and $\psi_k'(x_i+) \le 0$. Therefore $\lim_{x\to p} a(x)\psi_k'(x) = 0$, and $\psi_k'(p+) = \psi_k'(p-) = 0$ since $a(x) \ge \nu > 0$. In any case for such point p we have

$$a(x)\psi_k'(x) = -\lambda_k \int_p^x \psi_k(s) ds, \quad x \in (0,1). \tag{8}$$

Since $\psi_1(x) > 0$, $a(x) > 0$ on $(0,1)$ equation (8) implies that $\psi_1'(x) > 0$ for any $0 \le x < p$ and $\psi_1'(x) < 0$ for any $p < x \le 1$. Since the derivative of ψ_1 is zero at any point of maximum, we have to conclude that such a maximum p is unique.

□

3. Representation of solutions

First, we derive the solution of (4) with $f = q_1 = q_2 = 0$. Then we consider the general case.

Theorem 3.1. *(i) Let $g \in H = L^2(0,1)$. For any fixed $t > 0$ the solution $u(x,t)$ of*

$$\begin{array}{ll} u_t - (a(x)u_x)_x = 0, & Q = (0,1) \times (0,T), \\ u(0,t) = 0,\ u(1,t) = 0, & t \in (0,T), \\ u(x,0) = g(x), & x \in (0,1) \end{array} \tag{9}$$

is given by

$$u(x,t;a) = \sum_{k=1}^{\infty} \langle g, \psi_k \rangle e^{-\lambda_k t} \psi_k(x),$$

and the series converges uniformly and absolutely on $[0,1]$.

(ii) For any $p \in (0,1)$ function

$$z(t) = u(p,t;a), \quad t > 0$$

is real analytic on $(0,\infty)$.

Proof. (i) Note that the eigenvalues and the eigenfunctions satisfy

$$\nu \|\psi_k'\|^2 \le \int_0^1 a(x)[\psi_k'(x)]^2 dx = \lambda_k \|\psi_k\|^2 = \lambda_k.$$

Thus

$$\|\psi_k'\| \le \frac{\sqrt{\lambda_k}}{\sqrt{\nu}},$$

and

$$|\psi_k(x)| \le \int_0^x |\psi_k'(s)| ds \le \|\psi_k'\| \le \frac{\sqrt{\lambda_k}}{\sqrt{\nu}}.$$

Bessel's inequality implies that the sequence of Fourier coefficients $\langle g, \psi_k \rangle$ is bounded. Therefore, denoting by C various constants and using the fact that the function $s \to \sqrt{s}e^{-\sigma s}$ is bounded on $[0, \infty)$ for any $\sigma > 0$ one gets

$$|\langle g, \psi_k \rangle e^{-\lambda_k t} \psi_k(x)| \le C \frac{\sqrt{\lambda_k}}{\sqrt{\nu}} e^{-\frac{\lambda_k t}{2}} e^{-\frac{\lambda_k t}{2}} \le C e^{-\frac{\lambda_k t}{2}}.$$

From (iv) of Theorem 2.2 $\lambda_k \ge \nu \pi^2 k^2$. Thus

$$\sum_{k=1}^{\infty} |\langle g, \psi_k \rangle e^{-\lambda_k t} \psi_k(x)| \le C \sum_{k=1}^{\infty} e^{-\frac{\nu \pi^2 k^2 t}{2}} \le C \sum_{k=1}^{\infty} \left(e^{-\frac{\nu \pi^2 t}{2}} \right)^k < \infty.$$

By Weierstrass M-test the series converges absolutely and uniformly on $[0, 1]$.

(ii) Let $t_0 > 0$ and $p \in (0, 1)$. From (i), the series $\sum_{k=1}^{\infty} \langle g, \psi_k \rangle e^{-\lambda_k t_0} \psi_k(p)$ converges absolutely. Therefore $\sum_{k=1}^{\infty} \langle g, \psi_k \rangle e^{-\lambda_k s} \psi_k(p)$ is analytic in the part of the complex plane $\{s \in \mathbb{C} \; : \; Re\, s > t_0\}$, and the result follows.

□

Next we establish a representation formula for the solutions $u(x, t; a)$ of (4) under more general conditions. Suppose that $u(x, t; a)$ is a strong solution of (4), i.e. the equation and the initial condition in (4) are satisfied in $H = L^2(0, 1)$. Let

$$\Phi(x, t; a) = \frac{q_2(t) - q_1(t)}{\int_0^1 \frac{1}{a(s)} ds} \int_0^x \frac{1}{a(s)} ds + q_1(t). \tag{10}$$

Then $v(x, t; a) = u(x, t; a) - \Phi(x, t; a)$ is a strong solution of

$$\begin{cases} v_t - (a v_x)_x = -\Phi_t + f, \; 0 < x < 1, \; 0 < t < T, \\ v(0, t) = 0, \; 0 < t < T, \\ v(1, t) = 0, \; 0 < t < T, \\ v(x, 0) = g(x) - \Phi(x, 0), \; 0 < x < 1. \end{cases} \tag{11}$$

Accordingly, the weak solution u of (4) is defined by $u(x, t; a) = v(x, t; a) + \Phi(x, t; a)$ where v is the weak solution of (11). For the existence and the uniqueness of the weak solutions for such evolution equations see (Evans, 2010; Lions, 1971).

Let $V = H_0^1(0, 1)$ and $X = C[0, 1]$.

Theorem 3.2. *Suppose that* $T > 0$, $a \in \mathcal{PS}$, $g \in H$, $q_1, q_2 \in C^1[0, T]$ *and* $f(x, t) = h(x) r(t)$ *where* $h \in H$ *and* $r \in C[0, T]$. *Then*

(i) There exists a unique weak solution $u \in C((0, T]; X)$ *of (4).*

(ii) *Let* $\{\lambda_k, \psi_k\}_{k=1}^{\infty}$ *be the eigenvalues and the eigenfunctions of (5). Let* $g_k = \langle g, \psi_k \rangle$, $\phi_k(t) = \langle \Phi(\cdot, t), \psi_k \rangle$ *and* $f_k(t) = \langle f(\cdot, t), \psi_k \rangle$ *for* $k = 1, 2, \cdots$. *Then the solution* $u(x, t; a)$, $t > 0$ *of (4) is given by*

$$u(x,t;a) = \Phi(x,t;a) + \sum_{k=1}^{\infty} B_k(t;a)\psi_k(x), \tag{12}$$

where

$$B_k(t;a) = e^{-\lambda_k t}(g_k - \phi_k(0;a)) + \int_0^t e^{-\lambda_k(t-\tau)}(f_k(\tau) - \phi_k'(\tau;a))d\tau \tag{13}$$

for $k = 1, 2, \cdots$.

(iii) *For each* $t > 0$ *and* $a \in \mathcal{PS}$ *the series in (12) converges in X. Moreover, this convergence is uniform with respect to t in* $0 < t_0 \le t \le T$ *and* $a \in \mathcal{PS}$.

Proof. Under the conditions specified in the Theorem the existence and the uniqueness of the weak solution $v \in C([0,T]; H) \cap L^2([0,T]; V)$ of (11) is established in (Evans, 2010; Lions, 1971). By the definition $u = v + \Phi$. Thus the existence and the uniqueness of the weak solution u of (4) is established as well.
Let $\{\psi_k\}_{k=1}^{\infty}$ be the orthonormal basis of eigenfunctions in H corresponding to the conductivity $a \in \mathcal{PS}$. Let $B_k(t) = \langle v(\cdot, t), \psi_k \rangle$. To simplify the notation the dependency of B_k on a is suppressed. Then $v = \sum_{k=1}^{\infty} B_k(t)\psi_k$ in H for any $t \ge 0$, and

$$B_k'(t) + \lambda_k B_k(t) = -\phi_k'(t) + f_k(t), \quad B_k(0) = g_k - \phi_k(0).$$

Therefore $B_k(t)$ has the representation stated in (13).
Let $0 < t_0 < T$. Our goal is to show that v defined by $v = \sum_{k=1}^{\infty} B_k(t)\psi_k$ is in $C([t_0, T]; X)$. For this purpose we establish that this series converges in $X = C[0,1]$ uniformly with respect to $t \in [t_0, T]$ and $a \in A_{\text{ad}}$.
Note that V is continuously embedded in X. Furthermore, since $0 < \nu \le a(x) \le \mu$ the original norm in V is equivalent to the norm $\|\cdot\|_{V_a}$ defined by $\|w\|_{V_a}^2 = \int_0^1 a|w'|^2 dx$. Thus it is enough to prove the uniform convergence of the series for v in V_a. The uniformity follows from the fact that the convergence estimates below do not depend on a particular $t \in [t_0, T]$ or $a \in A_{\text{ad}}$.
By the definition of the eigenfunctions ψ_k one has $\langle a\psi_k', \psi_j' \rangle = \lambda_k \langle \psi_k, \psi_j \rangle$ for all k and j. Thus the eigenfunctions are orthogonal in V_a. In fact, $\{\psi_k/\sqrt{\lambda_k}\}_{k=1}^{\infty}$ is an orthonormal basis in V_a, see (Evans, 2010). Therefore the series $\sum_{k=1}^{\infty} B_k(t)\psi_k$ converges in V_a if and only if $\sum_{k=1}^{\infty} \lambda_k |B_k(t)|^2 = \|v(\cdot, t; a)\|_{V_a}^2 < \infty$ for any $t > 0$. This convergence follows from the fact that the function $s \to \sqrt{s}e^{-\sigma s}$ is bounded on $[0, \infty)$ for any $\sigma > 0$, see (Gutman & Ha, 2009). □

4. Continuity of the solution map

In this section we establish the continuous dependence of the eigenvalues λ_k, eigenfunctions ψ_k and the solution u of (4) on the conductivities $a \in \mathcal{PS} \subset A_{\text{ad}}$, when A_{ad} is equipped with the $L^1(0,1)$ topology. For smooth a see (Courant & Hilbert, 1989).

Theorem 4.1. *Let* $a \in \mathcal{PS}$, $\mathcal{PS} \subset A_{\text{ad}}$ *be equipped with the* $L^1(0,1)$ *topology, and* $\{\lambda_k(a)\}_{k=1}^{\infty}$ *be the eigenvalues of the associated Sturm-Liouville system (5). Then the mapping* $a \to \lambda_k(a)$ *is continuous for every* $k = 1, 2, \cdots$.

Proof. Let $a, \hat{a} \in \mathcal{PS}$, $\{\lambda_k, \psi_k\}_{k=1}^{\infty}$ be the eigenvalues and the eigenfunctions corresponding to a, and $\{\hat{\lambda}_k, \hat{\psi}_k\}_{k=1}^{\infty}$ be the eigenvalues and the eigenfunctions corresponding to $\hat{a}$. According

to Theorem 2.2 the eigenfunctions form a complete orthonormal set in H. Since $\int_0^1 a\psi_j'\psi' dx = \lambda_j \int_0^1 \psi_j \psi dx$ for any $\psi \in H_0^1(0,1)$ we have $\int_0^1 a\psi_i'\psi_j' dx = 0$ for $i \neq j$.
Let $W_k = \text{span}\{\psi_j\}_{j=1}^k$. Then W_k is a k-dimensional subspace of $H_0^1(0,1)$, and any $\psi \in W_k$ has the form $\psi(x) = \sum_{j=1}^k \alpha_j \psi_j(x)$, $\alpha_j \in \mathbb{R}$. From the min-max principle (Theorem 2.2(iii))

$$\hat{\lambda}_k \leq \max_{\psi \in W_k} \frac{\int_0^1 \hat{a}(x)[\psi'(x)]^2 dx}{\int_0^1 [\psi(x)]^2 dx}.$$

Note that

$$\max_{\psi \in W_k} \frac{\int_0^1 a(x)[\psi'(x)]^2 dx}{\int_0^1 [\psi(x)]^2 dx} = \max \left\{ \frac{\sum_{j=1}^k \alpha_j^2 \lambda_j}{\sum_{j=1}^k \alpha_j^2} : \alpha_j \in \mathbb{R},\ j = 1, 2, \cdots, k \right\} = \lambda_k.$$

Therefore

$$\begin{aligned} \hat{\lambda}_k &\leq \max_{\psi \in W_k} \frac{\int_0^1 a(x)[\psi'(x)]^2 dx}{\int_0^1 [\psi(x)]^2 dx} + \max_{\psi \in W_k} \frac{\int_0^1 (\hat{a}(x) - a(x))[\psi'(x)]^2 dx}{\int_0^1 [\psi(x)]^2 dx} \\ &\leq \lambda_k + \|a - \hat{a}\|_{L^1} \max_{\alpha_j} \frac{\| \sum_{j=1}^k \alpha_j \psi_j' \|_\infty^2}{\sum_{j=1}^k \alpha_j^2}, \end{aligned}$$

where $\| \cdot \|_\infty$ is the norm in $L^\infty(0,1)$. Estimates from Theorem 3.1 and the Cauchy-Schwarz inequality give

$$\frac{| \sum_{j=1}^k \alpha_j \psi_j'(x)|^2}{\sum_{j=1}^k \alpha_j^2} \leq \frac{\sum_{j=1}^k \alpha_j^2 \sum_{j=1}^k |\psi_j'(x)|^2}{\sum_{j=1}^k \alpha_j^2} \leq \frac{\lambda_k^2 k}{\nu^2} \leq \frac{(\mu \pi^2 k^2)^2 k}{\nu^2} = C(k).$$

Therefore

$$|\lambda_k - \hat{\lambda}_k| \leq C(k) \|a - \hat{a}\|_{L^1}$$

and the desired continuity is established. □

The following theorem is established in (Gutman & Ha, 2007).

Theorem 4.2. *Let $a \in \mathcal{PS}$, $\mathcal{PS} \subset A_{\text{ad}}$ be equipped with the $L^1(0,1)$ topology, and $\{\psi_k(x;a)\}_{k=1}^\infty$ be the unique normalized eigenfunctions of the associated Sturm-Liouville system (5) satisfying the condition $\psi_k'(0+;a) > 0$. Then the mapping $a \to \psi_k(a)$ from $\mathcal{PS}$ into $X = C[0,1]$ is continuous for every $k = 1, 2, \cdots$.*

Theorem 4.3. *Let $a \in \mathcal{PS} \subset A_{\text{ad}}$ equipped with the $L^1(0,1)$ topology, and $u(a)$ be the solution of the heat conduction process (4), under the conditions of Theorem 3.2. Then the mapping $a \to u(a)$ from $\mathcal{PS}$ into $C([0,T];X)$ is continuous.*

Proof. According to Theorem 3.2 the solution $u(x,t;a)$ is given by $u(x,t;a) = v(x,t;a) + \Phi(x,t;a)$, where $v(x,t;a) = \sum_{k=1}^\infty B_k(t;a)\psi_k(x)$ with the coefficients $B_k(t;a)$ given by (13). Let

$$v^N(x,t;a) = \sum_{k=1}^N B_k(t;a)\psi_k(x).$$

By Theorems 4.1 and 4.2 the eigenvalues and the eigenfunctions are continuously dependent on the conductivity a. Therefore, according to (13), the coefficients $B_k(t,a)$ are continuous as functions of a from $\mathcal{PS}$ into $C([0,T];X)$. This implies that $a \to v^N(a)$ is continuous. By Theorem 3.2 the convergence $v^N \to v$ is uniform on A_{ad} as $N \to \infty$ and the result follows. □

5. Identifiability of piecewise constant conductivities from finitely many observations

Series of the form $\sum_{k=1}^{\infty} C_k e^{-\lambda_k t}$ are known as Dirichlet series. The following lemma shows that a Dirichlet series representation of a function is unique. Additional results on Dirichlet series can be found in Chapter 9 of (Saks & Zygmund, 1965).

Lemma 5.1. *Let* $\mu_k > 0,\ k = 1,2,\ldots$ *be a strictly increasing sequence, and* $0 \le T_1 < T_2 \le \infty$. *Suppose that either*

(i) $\sum_{k=1}^{\infty} |C_k| < \infty$,

or

(ii) $\gamma > 0,\ \mu_k \ge \gamma k^2,\ k = 1,2,\ldots,$ *and* $\sup_k |C_k| < \infty$.

Then

$$\sum_{k=1}^{\infty} C_k e^{-\mu_k t} = 0 \quad \textit{for all} \quad t \in (T_1, T_2)$$

implies $C_k = 0$ *for* $k = 1,2,\ldots$.

Proof. In both cases the series $\sum_{k=1}^{\infty} C_k e^{-\mu_k z}$ converges uniformly in $Re\ z > 0$ region of the complex plane, implying that it is an analytic function there. Thus

$$\sum_{k=1}^{\infty} C_k e^{-\mu_k t} = 0 \quad \text{for all} \quad t > 0.$$

Suppose that some coefficients C_k are nonzero. Without loss of generality we can assume $C_1 \ne 0$. Then

$$0 = e^{\mu_1 t} \sum_{k=1}^{\infty} C_k e^{-\mu_k t} = C_1 + \sum_{k=2}^{\infty} C_k e^{(\mu_1 - \mu_k)t} \to C_1, \quad t \to \infty,$$

which is a contradiction. □

Remark. According to Theorem 3.1 for each fixed $p \in (0,1)$ the solution $z(t) = u(p,t;a)$ of (4) is given by a Dirichlet series. The series coefficients $C_k = \langle g, v_k \rangle v_k(p)$ are square summable, therefore they form a bounded sequence. The growth condition for the eigenvalues stated in (iv) of Theorem 2.2 shows that Lemma 5.1(ii) is applicable to the solution $z(t)$.
Functions $a \in \mathcal{PC}_N$ have the form $a(x) = a_i$ for $x \in [x_{i-1}, x_i),\ i = 1,2,\cdots,N$. Assuming $f = q_1 = q_2 = 0$, in this case the governing system (4) is

$$\begin{array}{lll} u_t - a_i u_{xx} = 0, & x \in (x_{i-1}, x_i), & t \in (0,T), \\ u(0,t) = u(1,t) = 0, & t \in (0,T), & \\ u(x_i+,t) = u(x_i-,t), & t \in (0,T), & \\ a_{i+1} u_x(x_i+,t) = a_i u_x(x_i-,t), & t \in (0,T), & \\ u(x,0) = g(x), & x \in (0,1), & \end{array} \tag{14}$$

where $g \in L^2(0,1)$ and $i = 1, 2, \cdots, N-1$. The associated Sturm-Liouville problem is

$$\begin{aligned} &a_i \psi''(x) = -\lambda \psi(x), \quad x \in (x_{i-1}, x_i), \\ &\psi(0) = \psi(1) = 0, \\ &\psi(x_i+) = \psi(x_i-), \\ &a_{i+1}\psi'(x_i+) = a_i \psi'(x_i-) \end{aligned} \tag{15}$$

for $i = 1, 2, \cdots, N-1$.

The central part of the identification method is the Marching Algorithm contained in Theorem 5.5. Recall that it uses only the M-tuple $\mathcal{G}(a)$, see (3). That is we need only the first eigenvalue λ_1 and a nonzero multiple of the first eigenfunction ψ_1 of (15) for the identification of the conductivity $a(x)$.

Suppose that $p^* \in (x_{i-1}, x_i)$. Then ψ_1 can be expressed on (x_{i-1}, x_i) as

$$\psi_1(x) = A\cos\left(\sqrt{\frac{\lambda_1}{a_i}}(x - p^*) + \gamma\right), \quad -\frac{\pi}{2} < \gamma < \frac{\pi}{2}$$

with $A > 0$. The range for γ in the above representation follows from the fact that $\psi_1(p^*) = A\cos\gamma > 0$ by Theorem 2.2(5).

The identifiability of piecewise constant conductivities is based on the following three Lemmas, see (Gutman & Ha, 2007).

Lemma 5.2. *Suppose that $\delta > 0$. Assume $Q_1, Q_3 \geq 0$, $Q_2 > 0$ and $0 < Q_1 + Q_3 < 2Q_2$. Let*

$$\Gamma = \left\{(A, \omega, \gamma) \;:\; A > 0,\ 0 < \omega < \frac{\pi}{2\delta},\ -\frac{\pi}{2} < \gamma < \frac{\pi}{2}\right\}.$$

Then the system of equations

$$A\cos(\omega\delta - \gamma) = Q_1, \quad A\cos\gamma = Q_2, \quad A\cos(\omega\delta + \gamma) = Q_3$$

has a unique solution $(A, \omega, \gamma) \in \Gamma$ given by

$$\omega = \frac{1}{\delta}\arccos\frac{Q_1 + Q_3}{2Q_2}, \quad \gamma = \arctan\left(\frac{Q_1 - Q_3}{2Q_2\sin\omega\delta}\right),$$

$$A = \frac{Q_2}{\cos\gamma}.$$

Lemma 5.3. *Suppose that $\delta > 0$, $0 < p \leq x_1 < p + \delta < 1$, $0 < \omega_1, \omega_2 < \pi/2\delta$. Let $w(x)$, $v(x)$, $x \in [p, p+\delta]$ be such that*

$$\begin{aligned} w(x) &= A_1\cos\omega_1 x + B_1\sin\omega_1 x, \\ v(x) &= A_2\cos\omega_2 x + B_2\sin\omega_2 x. \end{aligned}$$

Suppose that

$$\begin{aligned} &v(x_1) = w(x_1), \quad \omega_1^2 v'(x_1) = \omega_2^2 w'(x_1), \\ &v'(x_1) > 0, \quad v(x_1) > 0. \end{aligned}$$

Then

(i) Conditions $v(p+\delta) = w(p+\delta)$, $v'(p+\delta) \geq 0$ and $\omega_1 \leq \omega_2$ imply $\omega_1 = \omega_2$.

(ii) Conditions $v(p+\delta) = w(p+\delta)$, $w'(p+\delta) \geq 0$ *and* $\omega_1 \geq \omega_2$ *imply* $\omega_1 = \omega_2$.

Lemma 5.4. *Let* $\delta > 0$, $0 < \eta \leq 2\delta$, $\omega_1 \neq \omega_2$ *with* $0 < \omega_1\delta, \omega_2\delta < \pi/2$. *Also let* $A, B > 0$, $0 \leq p < p + \eta \leq 1$ *and*

$$w(x) = A\cos[\omega_1(x-p) + \gamma_1],$$
$$v(x) = B\cos[\omega_2(x-p-\eta) + \gamma_2]$$

with $|\gamma_1|, |\gamma_2| < \pi/2$. *Then system*

$$w(q) = v(q), \tag{16}$$
$$\omega_2^2 w'(q) = \omega_1^2 v'(q), \tag{17}$$
$$w(q) > 0, \quad v(q) > 0 \tag{18}$$

admits at most one solution q *on* $[p, p+\eta]$. *This unique solution* q *can be computed as follows:*
If $\gamma_1 \geq 0$ *then*

$$q = p + \frac{1}{\omega 1}\left[\arctan\left(\omega_1\sqrt{\left|\frac{B^2 - A^2}{A^2\omega_2^2 - B^2\omega_1^2}\right|}\right) - \gamma_1\right]. \tag{19}$$

If $\gamma_2 \leq 0$ *then*

$$q = p + \eta + \frac{1}{\omega 2}\left[-\arctan\left(\omega_2\sqrt{\left|\frac{B^2 - A^2}{A^2\omega_2^2 - B^2\omega_1^2}\right|}\right) - \gamma_2\right]. \tag{20}$$

Otherwise compute q_1 *and* q_2 *according to formulas (19) and (20) and discard the one that does not satisfy the conditions of the Lemma.*

By the definition of $a \in \mathcal{PC}$ there exist $N \in \mathbb{N}$ and a finite sequence $0 = x_0 < x_1 < \cdots < x_{N-1} < x_N = 1$ such that a is a constant on each subinterval (x_{n-1}, x_n), $n = 1, \cdots, N$. Let $\sigma > 0$. The following Theorem is our main result.

Theorem 5.5. *Given* $\sigma > 0$ *let an integer* M *be such that*

$$M \geq \frac{3}{\sigma} \quad \text{and} \quad M > 2\sqrt{\frac{\mu}{\nu}}.$$

Suppose that the initial data $g(x) > 0$, $0 < x < 1$ *and the observations* $z_m(t) = u(p_m, t; a), p_m = m/M$ *for* $m = 1, 2, \cdots, M-1$ *and* $0 \leq T_1 < t < T_2$ *of the heat conduction process (14) are given. Then the conductivity* $a \in A_{\text{ad}}$ *is identifiable in the class of piecewise constant functions* $\mathcal{PC}(\sigma)$.

Proof. The identification proceeds in two steps. In step I the M-tuple $\mathcal{G}(a)$ is extracted from the observations $z_m(t)$. In step II the Marching Algorithm identifies $a(x)$.
Step I. *Data extraction.*
By Theorem 3.1 we get

$$z_m(t) = \sum_{k=1}^{\infty} g_k e^{-\lambda_k t}\psi_k(p_m), \quad m = 1, 2, \cdots, M-1, \tag{21}$$

where $g_k = \langle g, \psi_k \rangle$ for $k = 1, 2, \cdots$. By Theorem 2.2(5) $\psi_1(x) > 0$ on interval $(0, 1)$. Since g is positive on $(0, 1)$ we conclude that $g_1\psi_1(p_m) > 0$. Since $z_m(t)$ is represented by a Dirichlet

series, Lemma 5.1 assures that all nonzero coefficients (and the first term, in particular) are defined uniquely.
An algorithm for determining the first eigenvalue λ_1, and the coefficient $g_1\psi_1(p_m)$ from (21) is given in Section 10. Repeating this process for every m one gets the values of

$$G_m = g_1\psi_1(p_m) > 0, \quad p_m = m/M \tag{22}$$

for $m = 1, 2, \cdots, M-1$. This determines the M-tuple $\mathcal{G}(a)$, see (3). Because of the zero boundary conditions we let $G_0 = G_M = 0$.
Step II. *Marching Algorithm.*
The algorithm marches from the left end $x = 0$ to a certain observation point $p_{l-1} \in (0,1)$ and identifies the values a_n and the discontinuity points x_n of the conductivity a on $[0, p_{l-1}]$. Then the algorithm marches from the right end point $x = 1$ to the left until it reaches the observation point $p_{l+1} \in (0,1)$ identifying the values and the discontinuity points of a on $[p_{l+1}, 1]$. Finally, the values of a and its discontinuity are identified on the interval $[p_{l-1}, p_{l+1}]$.
The overall goal of the algorithm is to determine the number $N-1$ of the discontinuities of a on $[0,1]$, the discontinuity points x_n, $n = 1, 2, \cdots, N-1$ and the values a_n of a on $[x_{n-1}, x_n]$, $n = 1, 2, \cdots, N$ ($x_0 = 0$, $x_N = 1$). As a part of the process the algorithm determines certain functions $H_n(x)$ defined on intervals $[x_{n-1}, x_n]$, $n = 1, 2, \cdots N$. The resulting function $H(x)$ defined on $[0,1]$ is a multiple of the first eigenfunction v_1 over the entire interval $[0,1]$. An illustration of the Marching Algorithm is given in Figure 1.

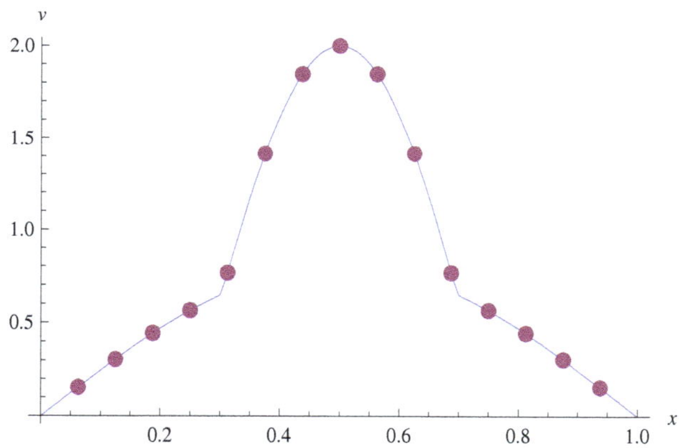

Fig. 1. Conductivity identification by the Marching Algorithm. The dots are a multiple of the first eigenfunction at the observation points p_m. The algorithm identifies the values of the conductivity a and its discontinuity points

(i) Find l, $0 < l < M$ such that $G_l = \max\{G_m : m = 1, 2, \cdots, M-1\}$ and $G_m < G_l$ for any $0 \le m < l$.

(ii) Let $i = 1$, $m = 0$.

(iii) Use Lemma 5.2 to find A_i, ω_i and γ_i from the system

$$\begin{cases} A_i \cos(\omega_i\delta - \gamma_i) = G_m, \\ A_i \cos\gamma_i = G_{m+1}, \\ A_i \cos(\omega_i\delta + \gamma_i) = G_{m+2}. \end{cases} \tag{23}$$

Let

$$H_i(x) = A_i \cos(\omega_i(x - p_{m+1}) + \gamma_i).$$

(iv) If $m + 3 \geq l$ then go to step (vii). If $H_i(p_{m+3}) \neq G_{m+3}$, or $H_i(p_{m+3}) = G_{m+3}$ and $H_i'(p_{m+3}) \leq 0$ then a has a discontinuity x_i on interval $[p_{m+2}, p_{m+3})$. Proceed to the next step (v).

If $H_i(p_{m+3}) = G_{m+3}$ and $H_i'(p_{m+3}) > 0$ then let $m := m + 1$ and repeat this step (iv).

(v) Use Lemma 5.2 to find A_{i+1}, ω_{i+1} and γ_{i+1} from the system

$$\begin{cases} A_{i+1} \cos(\omega_{i+1}\delta - \gamma_{i+1}) = G_{m+3}, \\ A_{i+1} \cos \gamma_{i+1} = G_{m+4}, \\ A_{i+1} \cos(\omega_{i+1}\delta + \gamma_{i+1}) = G_{m+5}. \end{cases} \tag{24}$$

Let

$$H_{i+1}(x) = A_{i+1} \cos(\omega_{i+1}(x - p_{m+4}) + \gamma_{i+1}).$$

(vi) Use formulas in Lemma 5.4 to find the unique discontinuity point $x_i \in [p_{m+2}, p_{m+3})$. The parameters and functions used in Lemma 5.4 are defined as follows. Let $p = p_{m+2}$, $\eta = \delta$. To avoid a confusion we are going to use the notation Ω_1, Ω_2, Γ_1, Γ_2 for the corresponding parameters ω_1, ω_2, γ_1, γ_2 required in Lemma 5.4. Let $\Omega_1 = \omega_i$, $\Omega_2 = \omega_{i+1}$. For $w(x)$ use function $H_i(x)$ recentered at $p = p_{m+2}$, i.e. rewrite $H_i(x)$ in the form

$$w(x) = H_i(x) = A \cos(\Omega_1(x - p_{m+2}) + \Gamma_1), \quad |\Gamma_1| < \pi/2.$$

For $v(x)$ use function H_{i+1} recentered at $p + \eta = p_{m+3}$, i.e.

$$v(x) = H_{i+1}(x) = B \cos(\Omega_2(x - p_{m+3}) + \Gamma_2), \quad |\Gamma_2| < \pi/2.$$

Let $i := i + 1$, $m := m + 3$. If $m < l$ then return to step (iv). If $m \geq l$ then go to the next step (vii).

(vii) Do steps (ii)-(vi) in the reverse direction of x, advancing from $x = 1$ to $x = p_{l+1}$. Identify the values and the discontinuity points of a on $[p_{l+1}, 1]$, as well as determine the corresponding functions $H_i(x)$.

(viii) Using the notation introduced in (vi) let $H_j(x)$ be the previously determined function H on interval $[p_{l-2}, p_{l-1}]$. Recenter it at $p = p_{l-1}$, i.e. $w(x) = H_j(x) = A\cos(\Omega_1(x - p_{l-1}) + \Gamma_1)$. Let $H_{j+1}(x)$ be the previously determined function H on interval $[p_{l+1}, p_{l+2}]$. Recenter it at p_{l+1}: $v(x) = H_{j+1}(x) = B\cos(\Omega_2(x - p_{l+1}) + \Gamma_2)$. If $\Omega_1 = \Omega_2$ then stop, otherwise use Lemma 5.4 with $\eta = 2\delta$, and the above parameters to find the discontinuity $x_j \in [p_{l-1}, p_{l+1}]$. Stop.

The justification of the Marching Algorithm is given in (Gutman & Ha, 2007). □

6. Identifiability of piecewise constant conductivity with one discontinuity

The Marching Algorithm of Theorem 5.5 requires measurements of the system at possibly large number of observation points. Our next Theorem shows that if a piecewise constant conductivity a is known to have just one point of discontinuity x_1, and its values a_1 and a_2 are known beforehand, then the discontinuity point x_1 can be determined from just one measurement of the heat conduction process.

Theorem 6.1. *Let $p \in (0,1)$ be an observation point, $g(x) > 0$ on $(0,1)$, and the observation $z_p(t) = u(x_p, t; a)$, $t \in (T_1, T_2)$ of the heat conduction process (14) be given. Suppose that the conductivity $a \in A_{\text{ad}}$ is piecewise constant and has only one (unknown) point of discontinuity $x_1 \in (0,1)$. Given positive values $a_1 \neq a_2$ such that $a(x) = a_1$ for $0 \leq x < x_1$ and $a(x) = a_2$ for $x_1 \leq x < 1$ the point of discontinuity x_1 is constructively identifiable.*

Proof. Arguing as in the previous Theorem

$$z_p(t) = \sum_{k=1}^{\infty} g_k e^{-\lambda_k t} \psi_k(p), \quad 0 \leq T_1 < t < T_2,$$

where $g_k = \langle g, \psi_k \rangle$ for $k = 1, 2, \cdots$. Since $g_1\psi_1(p) > 0$ the uniqueness of the Dirichlet series representation implies that one can uniquely determine the first eigenvalue λ_1 and the value of $G_p = g_1\psi_1(p)$.

Without loss of generality one can assume that $a_1 > a_2$. In this case we show that the first eigenvalue λ_1 is strictly increasing as a function of the discontinuity point $x_1 \in [0,1]$. Indeed, suppose that

$$0 \leq x_1^a < x_1^b \leq 1,$$

that is

$$a(x) = \begin{cases} a_1, & 0 < x < x_1^a \\ a_2, & x_1^a < x < 1 \end{cases} \quad \text{and} \quad b(x) = \begin{cases} a_1, & 0 < x < x_1^b \\ a_2, & x_1^b < x < 1 \end{cases}.$$

By Theorem 2.2(i)

$$\lambda_1^b = \frac{\int_0^1 b(x)[\psi'_{1,b}(x)]^2 dx}{\int_0^1 [\psi_{1,b}(x)]^2 dx} > \frac{\int_0^1 a(x)[\psi'_{1,b}(x)]^2 dx}{\int_0^1 [\psi_{1,b}(x)]^2 dx} \geq \inf_{\psi \in H_0^1(0,1)} \frac{\int_0^1 a(x)[\psi'(x)]^2 dx}{\int_0^1 [\psi(x)]^2 dx} = \lambda_1^a$$

provided that the derivative $\psi'_{1,b}(x)$ of the first eigenfunction $\psi_{1,b}(x)$ is not identically zero on (x_1^a, x_1^b). But, from $(b(x)\psi'_{1,b}(x))' = -\lambda_1^b \psi_{1,b}(x)$, the assumption $\psi'_{1,b}(x) = 0$ on (x_1^a, x_1^b) implies $\psi_{1,b}(x) = 0$ on (x_1^a, x_1^b). However, this is impossible, since $\psi_{1,b}(x) > 0$ on $(0,1)$. Thus there exists a unique conductivity of the type sought in the Theorem for which its first eigenvalue is equal to λ_1, i.e. a is identifiable.

Now the unique discontinuity point x_1 of a can be determined as follows. Let

$$\omega_1 = \sqrt{\frac{\lambda_1}{a_1}}, \quad \omega_2 = \sqrt{\frac{\lambda_1}{a_2}}.$$

Then the first eigenfunction ψ_1 is given by

$$\psi_1(x) = \begin{cases} A \sin \omega_1 x, & 0 < x < x_1, \\ B \sin \omega_2 (1 - x), & x_1 < x < 1 \end{cases} \tag{25}$$

for some $A, B > 0$. The matching conditions at x_1 give

$$A \sin \omega_1 x_1 = B \sin \omega_2 (1 - x_1) \quad \text{and} \quad \frac{A}{\omega_1} \cos \omega_1 x_1 = \frac{B}{\omega_2} \cos \omega_2 (1 - x_1).$$

Since $\psi_1(x_1) > 0$ we have $0 < \omega_1 x_1 < \pi$ and $0 < \omega_2(1 - x_1) < \pi$. Therefore x_1 satisfies

$$\frac{1}{\omega_1} \cot \omega_1 x = \frac{1}{\omega_2} \cot \omega_2 (1 - x).$$

The existence and the uniqueness of the solution x_1 of the above nonlinear equation follows from the monotonicity and the continuity of the cotangent functions. Practically, the value of x_1 can be found by a numerical method. □

7. Identifiability with non-zero boundary conditions

Let $a \in \mathcal{PS}$, and $u(x, t; a)$ be the unique solution of the heat conduction process (4). Next Theorem describes some conditions under which the identifiability for (4) is possible.

Theorem 7.1. *Given $\sigma > 0$ let an integer M be such that*

$$M \geq \frac{3}{\sigma} \quad \text{and} \quad M > 2\sqrt{\frac{\mu}{\nu}}.$$

Suppose that the observations $z_m(t; a) = u(p_m, t; a)$ for $p_m = m/M$, $m = 1, 2, \cdots, M-1$ and $t > 0$ of the heat conduction process (4) are given. Then the conductivity $a \in A_{\text{ad}}$ is identifiable in the class of piecewise constant functions $\mathcal{PC}(\sigma)$ in each one of the following four cases.

(i) $f = 0,\ q_1 = 0,\ q_2 = 0,\ g > 0,\ g \in L^2(0,1)$.

(ii) $g = 0,\ q_1 = 0,\ q_2 = 0,\ f(x,t) = h(x)r(t) \neq 0,\ h > 0,\ h \in L^2(0,1),\ r \in C[0,\infty)$.

(iii) $g = 0,\ f = 0,\ q_2 = 0,\ q_1 \neq 0,\ q_1(0) = 0,\ q_1 \in C^1[0,\infty)$.

(iv) $g = 0,\ f = 0,\ q_1 = 0,\ q_2 \neq 0,\ q_2(0) = 0,\ q_2 \in C^1[0,\infty)$.

Proof. Case (i) is considered in Theorem 5.5. In case (ii) of the Theorem let

$$y_m(t) = \sum_{k=1}^{\infty} \langle h, \psi_k \rangle \psi_k(p_m) e^{-\lambda_k t}. \tag{26}$$

Then $y_m(t)$ is the solution of (4) with $g = h$, $f = 0$ and zero boundary conditions, observed at $p_m \in (0,1)$. It is shown in Theorem 3.2 that such a solution is a continuous function for $t > 0$. Furthermore, using the estimate $|\psi_k(x)| \leq \sqrt{\lambda_k}/\sqrt{\nu}$ established in Theorem 3.1, and the Cauchy-Schwarz inequality we get

$$\int_0^\infty |y_m(t)| dt \leq \sum_{k=1}^{\infty} \frac{1}{\lambda_k} |h_k| |\psi_k(p_m)| \leq \frac{1}{\sqrt{\nu}} \sum_{k=1}^{\infty} \frac{|h_k|}{\sqrt{\lambda_k}} \leq C\|h\| < \infty. \tag{27}$$

Therefore $y_m(t) \in L^1[0,\infty)$.
Returning to the observation $z_m(t)$, Theorem 3.2 shows that it is given by

$$z_m(t) = u(p_m, t) = \int_0^t \left[\sum_{k=1}^{\infty} \langle h, \psi_k \rangle \psi_k(p_m) e^{-\lambda_k (t-\tau)} \right] r(\tau)\, d\tau.$$

That is

$$z_m(t) = \int_0^t y_m(t - \tau) r(\tau)\, d\tau.$$

Since $y_m(t) \in L^1[0,\infty)$ and $r(t)$ is continuous and bounded on $[0,\infty)$, Titchmarsh Theorem (Titchmarsh, 1962), Theorem 152, Chap. XI, p. 325, implies that this Volterra integral equation is uniquely solvable for $y_m(t)$.
Since $h > 0$ is assumed to be in $L^2(0,1)$, one has $C(a) = \langle h, \psi_1(a)\rangle \neq 0$. The uniqueness of the Dirichlet series representation (26) and rest of the argument is the same as in the proof of case (i).
In case (iii) of the Theorem function $\Phi(x,t;a)$ has the form $\Phi(x,t;a) = q_1(t)\xi(x;a)$, where

$$\xi(x;a) = 1 - \frac{1}{\int_0^1 \frac{1}{a(s)}\,ds} \int_0^x \frac{1}{a(s)}\,ds.$$

Note that $\xi(x;a)$ is bounded, continuous and strictly positive on $(0,1)$. Thus $\xi \in L^2(0,1)$. Let $\xi_k = \langle \xi(x;a), \psi_k(x;a)\rangle$ for $k = 1,2,....$ Then $\phi_k(t;a) = q_1(t)\xi_k$, $\phi_k(0;a) = 0$ and $\phi_k'(t;a) = q_1'(t)\xi_k$.
Let

$$y_m(t) = -\sum_{k=1}^{\infty} \xi_k \psi_k(p_m) e^{-\lambda_k t}. \tag{28}$$

Arguing as in case (ii), we conclude that $y_m(t)$ is continuous on $[0,\infty)$ and $y_m(t) \in L^1[0,\infty)$. Also, by Theorem 3.2

$$z_m(t) = u(p_m,t) = -\int_0^t \left[\sum_{k=1}^{\infty} \xi_k \psi_k(p_m) e^{-\lambda_k(t-\tau)}\right] q_1'(\tau)\,d\tau.$$

That is

$$z_m(t) = \int_0^t y_m(t-\tau) q_1'(\tau)\,d\tau.$$

Since $y_m(t) \in L^1[0,\infty)$ and $q_1'(t)$ is continuous and bounded on $[0,\infty)$, Titchmarsh Theorem (Titchmarsh, 1962), Theorem 152, Chap. XI, p. 325, implies that this Volterra integral equation is uniquely solvable for $y_m(t)$.
Since $\xi_1 > 0$ and $\psi_1(p_m) > 0$, the uniqueness of the Dirichlet series representation (28) implies that the M-tuple $\mathcal{G}(a)$ is recoverable from the observations $z_m(t)$. In this case $C(a) = \langle \xi(x;a), \psi_1(x;a)\rangle$. Finally, the Marching Algorithm identifies the unknown conductivity a.
Case (iv) of the Theorem is treated in the same way as case (iii). □

8. Continuity of the identification map

The Marching Algorithm establishes the identifiability of the conductivity $a \in \mathcal{PC}(\sigma)$ from the data $\mathcal{G}(a)$. In other words, the inverse mapping $\mathcal{G}^{-1}$ is well defined on $\mathcal{G}(\mathcal{PC}(\sigma))$. To prove our main result that the identifiability map $\mathcal{G}^{-1}$ is continuous, first we show that the set $\mathcal{PC}(\sigma) \subset A_{ad}$ is compact in $L^1(0,1)$. A proof of this result can be found in (Gutman & Ha, 2009).

Theorem 8.1. *Let A_{ad} be equipped with the $L^1(0,1)$ topology. Let $N \in \mathbb{N}$ and $\sigma > 0$. Then*

(i) Set $\mathcal{PC}_N \subset A_{ad}$ is compact.

(ii) Set $\mathcal{PC}(\sigma) \subset A_{ad}$ is compact.

Theorem 8.2. *Let A_{ad} be equipped with the $L^1(0,1)$ topology, and the data map $\mathcal{G} : \mathcal{PC}(\sigma) \to \mathbb{R}^M$ be defined as in (3). Then the identifiability map $\mathcal{G}^{-1} : \mathcal{G}(\mathcal{PC}(\sigma)) \to \mathcal{PC}(\sigma)$ is continuous.*

Proof. Theorem 7.1 shows that in every case specified there the data map $a \to \mathcal{G}(a)$ is defined everywhere on $\mathcal{PC}(\sigma)$ and that the conductivity a is identifiable from $\mathcal{G}(a)$, i.e. $\mathcal{G}$ is invertible on $\mathcal{G}(\mathcal{PC}(\sigma))$. By Theorem 8.1 the set $\mathcal{PC}(\sigma)$ is compact in $L^1(0,1)$. Thus the Theorem would be established if the injective map $a \to \mathcal{G}(a)$ were shown to be continuous.
Recall that $\mathcal{G}(a) = (\lambda_1(a), G_1(a), \cdots, G_{M-1}(a)) \in \mathbb{R}^M$. The continuity of $a \to \lambda_1(a)$ was established in Theorem 4.1. In every case of Theorem 7.1 the data G_m has the form $G_m(a) = C(a)\psi_1(p_m; a)$, where p_m are the observation points. By Theorem 4.2 the mapping $a \to \psi_1(\cdot; a)$ is continuous from $\mathcal{PC}(\sigma) \subset L^1(0,1)$ into $C[0,1]$. Thus the evaluation maps $a \to \psi_1(p_m; a) \in \mathbb{R}$ are continuous for every $p_m \in [0,1]$.
To see that $a \to C(a)$ is continuous we have to examine it separately for each case of Theorem 7.1. In case (i) $C(a) = \langle g, \psi_1(a)\rangle$, where $g \in L^2(0,1)$ is a fixed initial condition. The continuity of the inner product and of $a \to \psi_1(\cdot; a)$ imply the continuity of $C(a)$. In case (ii) $C(a) = \langle h, \psi_1(a)\rangle$ for an $h \in L^2(0,1)$ and the continuity of $C(a)$ follows. In cases (iii) and (iv) the continuity of $C(a)$ is established similarly. □

9. Identifiability with a known heat flux

Let Π be the set of piecewise constant functions on $[0,1]$ with finitely many discontinuity points,

$$\Pi = \{a(x) : 0 < \nu \le a(x) \le \mu,\ a(x) = a_j,\ x \in [x_{j-1}, x_j),\ j = 1, 2, ..., n\} \tag{29}$$

with $x_0 = 0$ and $x_n = 1$.
Consider the following heat conduction problem in an inhomogeneous bar of the unit length with a conductivity $a \in \Pi$:

$$\begin{cases} u_t = (a(x)u_x)_x, & (x,t) \in Q = (0,1) \times (0,\infty), \\ u(0,t) = g(t),\ u(1,t) = 0, & t \in (0,\infty), \\ u(x,0) = 0, & x \in (0,1). \end{cases} \tag{30}$$

Suppose that the extra data $f(t) = a(0)u_x(0,t) \not\equiv 0$, i.e., the heat flux through the left end of the bar, is known.
The inverse problem (IP) for (29)-(30) is:
IP: *Given $f(t)$ and $g(t)$ for all $t > 0$, find $a(x)$.*
In this Section we establish the identifiability for the IP. Additional details including a fast computational algorithm can be found in (Gutman & Ramm, 2010) and (Hoang & Ramm, 2009).
The main idea of the proof is to apply a "layer peeling" argument. Suppose that two conductivities $a, b \in \Pi$ satisfy (30) with the same data $f(t)$ and $g(t)$ for $t > 0$. Let both a and b have no discontinuities on an interval $[0, y]$, $0 < y \le 1$. Then we can show that $a(x) = b(x)$ for $x \in [0, y]$. A repeated application of this argument shows that $a = b$ on the entire interval $[0,1]$. See (Hoang & Ramm, 2009) for further refinements of this result, in particular for the data f, g available only on a finite interval $(0, T)$.
The main tool for the uniqueness proof is Property C (completeness of the products of solutions for (30)). We will use the following Property C result established in (Hoang & Ramm, 2009).

Theorem 9.1. *Let $PC[0,1]$ be the set of piecewise-constant functions on $[0,1]$. Let $q_1, q_2 \in PC[0,1]$ be two positive functions. Suppose that $\psi_1(x,k)$ and $\psi_2(x,k)$ satisfy*

$$-\psi_j''(x,k) + k^2 q_j^2(x)\psi_j(x,k) = 0, \quad \psi_j(1,k) = 1, \quad \psi_j'(1,k) = 0, \quad j = 1, 2. \tag{31}$$

Then the set of products $\{\psi_1(x,k)\psi_2(x,k)\}_{k>0}$ *is dense in* $PC[0,1]$. *That is, if* $h \in PC[0,1]$ *and*

$$\int_0^1 h(x)\psi_1(x,k)\psi_2(x,k)dx = 0 \tag{32}$$

for any $k > 0$, *then* $h = 0$.

Theorem 9.2. *Problem IP has at most one solution* $a \in \Pi$.

Proof. Following Hoang & Ramm (2009), problem (30) is restated in terms of the Laplace transform

$$v(x,s;a) = (\mathcal{L}u)(x,s;a) = \int_0^\infty u(x,t;a)e^{-st}dt, \quad s > 0.$$

Let $G(s) = \mathcal{L}(g(t))$ and $F(s) = \mathcal{L}(f(t))$. Thus (30) with the extra condition $a(0)u_x(0,t) = f(t)$ becomes

$$\begin{aligned} &(a(x)v')' - sv = 0, \quad 0 < x < 1, \\ &v(0,s;a) = G(s), \quad a(0)v'(0,s;a) = F(s), \\ &v(1,s;a) = 0. \end{aligned} \tag{33}$$

Let

$$k = \sqrt{s}, \quad \psi(x,k) = a(x)v'(x,s;a), \quad \text{and} \quad q(x) = \sqrt{\frac{1}{a(x)}}.$$

Then, using $k^2 v(x,s;a) = \psi'(x,k)$, system (33) is rewritten as

$$\begin{aligned} &-\psi''(x,k) + k^2 q^2(x)\psi(x,k) = 0, \quad 0 < x < 1, \\ &\psi(0,k) = F(k^2), \quad \psi'(0,k) = k^2 G(k^2), \quad \psi'(1,k) = 0. \end{aligned} \tag{34}$$

Let $\psi_1(x,k)$ and $\psi_2(x,k)$ be the solutions of (34) for two positive piecewise-constant functions $q_1(x)$ and $q_2(x)$ correspondingly. That is,

$$\begin{aligned} &-\psi_1''(x,k) + k^2 q_1^2(x)\psi_1(x,k) = 0, \quad 0 < x < 1, \\ &\psi_1(0,k) = F(k^2), \quad \psi_1'(0,k) = k^2 G(k^2), \quad \psi_1'(1,k) = 0, \end{aligned} \tag{35}$$

and

$$\begin{aligned} &-\psi_2''(x,k) + k^2 q_2^2(x)\psi_2(x,k) = 0, \quad 0 < x < 1, \\ &\psi_2(0,k) = F(k^2), \quad \psi_2'(0,k) = k^2 G(k^2), \quad \psi_2'(1,k) = 0. \end{aligned} \tag{36}$$

Multiply equation (35) by $\psi_2(x,k)$ and integrate it over $[0,1]$. Then use an integration by parts and the boundary conditions in (35) and (36) to obtain

$$k^2 \int_0^1 q_1^2 \psi_1 \psi_2 dx = \psi_1'\psi_2|_0^1 - \int_0^1 \psi_1'\psi_2' dx = -k^2 G(k^2)F(k^2) - \int_0^1 \psi_1'\psi_2' dx. \tag{37}$$

Similarly,

$$k^2 \int_0^1 q_2^2 \psi_1 \psi_2 dx = -k^2 G(k^2)F(k^2) - \int_0^1 \psi_1'\psi_2' dx. \tag{38}$$

Subtracting (38) from (37) gives

$$\int_0^1 (q_1^2 - q_2^2)\psi_1\psi_2 dx = 0$$

for any $k > 0$.
Given nonzero F and G, consider (35) as an initial value problem for ψ_1 at $x = 0$. Its solution $\psi_1(x,k)$ must satisfy $\psi_1(1,k) \neq 0$, because of the condition $\psi_1'(1,k) = 0$. The same goes for $\psi_2(x,k)$. Now we can conclude that the set of products $\{\psi_1(x,k)\psi_2(x,k)\}_{k>0}$ is dense in $PC[0,1]$ by Theorem 9.1. Therefore $q_1 = q_2$. Thus (34) has a unique solution $q \in PC[0,1]$. Consequently (33) has a unique solution $a \in \Pi$, and the Theorem is proved. □

10. Computational algorithms

The main objective of this research is the development of a theoretical framework for the parameter identifiability described in previous sections. Nevertheless, from a practical perspective it is desirable to develop an algorithm for such an identifiability incorporating the new insights gained in the theoretical part. The main new element of it is the separation of the identification process into the following two parts. First, the observation data is used to recover the M-tuple $\mathcal{G}(a)$, i.e. the first eigenvalue of (5), and a multiple of the first eigenfunction at the observation points p_m, see (3). In the second step this input is used to recover the conductivity distribution. We emphasize that only one (first) eigenvalue and the eigenfunction are needed for the identification. For other methods for inverse heat conduction problems see (Beck et al., 1985) and the references therein.
Before considering noise contaminated observation data $z_m(t)$, let us assume that $z_m(t)$ are known precisely on an interval $I = (t_0, T)$, $t_0 \geq 0$. In case (i) of Theorem 7.1 the observations are given by the Dirichlet series

$$z_m(t) = \sum_{k=1}^{\infty} \langle g, \psi_k \rangle e^{-\lambda_k t} \psi_k(p_m). \tag{39}$$

We have not implemented yet other cases of Theorem 7.1.
In principle, functions $z_m(t)$ are analytic for $t > 0$. Therefore they can be uniquely extended to $(0, \infty)$ from I. Then the first eigenvalue λ_1 and the data sequence $\{G_m =< g, \psi_1 > \psi_1(p_m)\}_{m=1}^{M-1}$ can be recovered from the Dirichlet series (39) representing $z_m(t)$ by

$$\lambda_1 = -\frac{1}{h} \lim_{t\to\infty} \ln \frac{z_m(t+h)}{z_m(t)}, \quad G_m = \lim_{t\to\infty} e^{\lambda_1 t} z_m(t), \tag{40}$$

where $h > 0$.
The second step of the algorithm, i.e. the identification of the conductivity a is accomplished by the Marching Algorithm. Numerical experiments show that it provides the perfect identification only if $\mathcal{G}(a)$ is known precisely. However, even for noiseless data $z_m(t)$, the numerical identification of $\mathcal{G}(a)$ from the Dirichlet series (39) representing $z_m(t)$ can only be accomplished with a significant error. This numerical evidence is presented in (Gutman & Ha, 2009).
Hence a different algorithm is needed for the practically important case of noise contaminated data. It should also take into account the severe ill-posedness of the identification of data from Dirichlet series, see (Acton, 1990). Our numerical experiments confirm that even the second eigenvalue of the associated Sturm-Liouville problem cannot be reliably identified even for

noiseless data. It is the distinct advantage of the proposed algorithm that it uses only the first eigenvalue λ_1 for the conductivity identification. In what follows LMA refers to the Levenberg-Marquardt algorithm for the nonlinear least squares minimization, and BA to the Brent algorithm for a single variable nonlinear minimization, see (Press et al., 1992) for details. First, consider a simple regression type algorithm for the identification of the M-tuple $\mathcal{G}(a)$. In step 1, for each observation data $z_m(t)$ we find λ and c to best fit $z_m(t)$ in the objective function $\Psi(\lambda, c; m)$ defined by (41). In step 2 the obtained eigenvalues $\lambda^{(m)}$ are averaged over the middle third of the observation points, since such data would presumably be less affected by noise. The result of the averaging is the sought eigenvalue λ_1. In step 3, the averaged eigenvalue λ_1 is kept fixed, and the functions $\Psi(\lambda_1, c; m)$ are minimized in variable c only. The resulting values G_m form the M-tuple $\mathcal{G}(a)$.

Regression Algorithm for λ_1 identification.
Let the data consist of the observations $z_m(t_j),\ j = 1, 2, \dots J,\ m = 1, 2, \dots, M-1$.

(i) Let $\lambda,\ c \in \mathbb{R}$ and

$$\Psi(\lambda, c; m) = \sum_{j=1}^{J} (ce^{-\lambda t_j} - z_m(t_j))^2. \tag{41}$$

Let

$$\Psi(\lambda, c_m(\lambda); m) = \min_{c \in \mathbb{R}} \Psi(\lambda, c; m).$$

Note that such a minimizer $c_m(\lambda)$ can be found directly by

$$c_m(\lambda) = \frac{\sum_{j=1}^{J} z_m(j) e^{-\lambda t_j}}{\sum_{j=1}^{J} e^{-2\lambda t_j}}.$$

For each $m = 1, \dots, M-1$ apply BA to find a $\lambda^{(m)}$ such that

$$\Psi(\lambda^{(m)}, c_m(\lambda^{(m)}); m) = \min_{\lambda \in \mathbb{R}} \Psi(\lambda, c_m(\lambda); m).$$

(ii) Let $k = card\{[[M/3]], \dots, [[2M/3]]\}$ and

$$\lambda_1 = \frac{1}{k} \sum_{m=[[M/3]]}^{[[2M/3]]} \lambda_1^{(m)}.$$

(iii) Keep λ_1 fixed. For each $m = 1, \dots, M-1$ find $G_m = c_m(\lambda_1)$ such that

$$\Psi(\lambda_1, G_m; m) = \min_{c \in \mathbb{R}} \Psi(\lambda_1, c; m).$$

(iv) Let $\mathcal{G}(a) = \{\lambda_1, G_1, \dots, G_{M-1}\}$.

One may assume that fitting the data $z_m(t)$ using two exponents as in (43) could result in a better estimate for the eigenvalue λ_1. To examine this assumption let us consider a more complicated algorithm which we call the *LMA Algorithm for λ_1 identification*. This algorithm proceeds as follows (see details below).

(i). This step is the same as step (i) in the regression algorithm above, i.e. we minimize the functions $\Psi(\lambda, c; m)$ in both λ and c for $m = 1, \ldots, M-1$. Call the minimizers by $\mu^{(m)}$ and $c_m(\mu^{(m)})$ respectively.
(ii). Apply the *LMA* to minimize $\Phi(\mu, \nu, c, b; m)$ defined in (43). Use the initial guess $\mu^{(m)}$, $4\mu^{(m)}$, $c_m(\lambda)$, 0 for the variables μ, ν, c, b correspondingly. Call the results of these minimizations for the variable μ by $\lambda_1^{(m)}$. The initial value $4\mu^{(m)}$ for the second eigenvalue is used because of Theorem 2.2(iii). A direct application of the *LMA* without the initial values obtained in Step (i) did not produce consistent results. Now the data $z_m(t)$ is approximated by the first two terms of the Dirichlet series (39). Thus, for each m there is an estimate $\lambda_1^{(m)}$ for the first eigenvalue λ_1.
(iii). Let λ_1 be an average of the computed values $\lambda_1^{(m)}$. We used the middle third of the indices m since the maximum of our initial data $g(x)$ was attained in the middle of the interval $[0, 1]$. Hence these observations were relatively less affected by the noise.
(iv-v). Repeat the minimizations of Steps (i) and (ii), but keep λ_1 frozen. Let G_m be the values of the coefficients c that minimize $\Phi(\lambda_1, \nu, c, b; m)$. This is the best fit to the data $z_m(t)$ by the first two terms of the Dirichlet series (39) with the fixed first eigenvalue λ_1. By now the first part of the identification algorithm is completed, since we have recovered the first eigenvalue λ_1 and a multiple G_m of the first eigenfunction $\psi_1(p_m)$, $m = 1, 2, \ldots, M-1$.

LMA Algorithm for λ_1 identification.
Let the data consist of the observations $z_m(t_j)$, $j = 1, 2, \ldots J$, $m = 1, 2, \ldots, M-1$.

(i) Let λ, $c \in \mathbb{R}$ and

$$\Psi(\lambda, c; m) = \sum_{j=1}^{J} (ce^{-\lambda t_j} - z_m(t_j))^2. \tag{42}$$

Let

$$\Psi(\lambda, c_m(\lambda); m) = \min_{c \in \mathbb{R}} \Psi(\lambda, c; m).$$

Note that such a minimizer $c_m(\lambda)$ can be found directly by

$$c_m(\lambda) = \frac{\sum_{j=1}^{J} z_m(j) e^{-\lambda t_j}}{\sum_{j=1}^{J} e^{-2\lambda t_j}}.$$

For each $m = 1, \ldots, M-1$ apply *BA* to find a $\mu^{(m)}$ such that

$$\Psi(\mu^{(m)}, c_m(\mu^{(m)}); m) = \min_{\lambda \in \mathbb{R}} \Psi(\lambda, c_m(\lambda); m).$$

(ii) Let

$$\Phi(\mu, \nu, c, b; m) = \sum_{j=1}^{J} (ce^{-\mu t_j} + be^{-\nu t_j} - z_m(t_j))^2. \tag{43}$$

Apply the *LMA* to minimize $\Phi(\mu, \nu, c, b; m)$ using the initial guess $\mu^{(m)}, 4\mu^{(m)}, c_m(\mu^{(m)}), 0$ for the variables μ, ν, c, b correspondingly. Let

$$\Phi(\lambda_1^{(m)}, \nu_m, c_m, b_m; m) = \min_{\mu, \nu, c, b} \Phi(\mu, \nu, c, b; m).$$

(iii) Let $k = card\{[[M/3]], \ldots, [[2M/3]]\}$ and

$$\lambda_1 = \frac{1}{k} \sum_{m=[[M/3]]}^{[[2M/3]]} \lambda_1^{(m)}.$$

(iv) Find $c_m(\lambda_1),\ m = 1, 2, \ldots, M$ (as in Step 1) such that

$$\Psi(\lambda_1, c_m(\lambda_1); m) = \min_{c \in \mathbb{R}} \Psi(\lambda_1, c; m).$$

(v) Apply the LMA to minimize $\Phi(\lambda_1, \nu, c, b; m)$ in variables ν, c, b using the initial guess $4\lambda_1, c_m(\lambda_1), 0$ for the variables ν, c, b correspondingly. Let

$$\Phi(\lambda_1, \nu_m, G_m, b_m; m) = \min_{\nu, c, b} \Phi(\lambda_1, \nu, c, b; m).$$

(vi) Let $\mathcal{G}(a) = \{\lambda_1, G_1, \ldots, G_{M-1}\}$.

The second part of the algorithm identifies the conductivity $\bar{a}$ from the M-tuple $\mathcal{G}(a)$. As we have already mentioned the Marching Algorithm provides a perfect identification for noiseless data, otherwise one has to find $\bar{a}$ by a nonlinear minimization.

Identification of piecewise constant conductivity.
The data is the M-tuple $\mathcal{G}(a) = \{\lambda_1, G_1, \ldots, G_{M-1}\}$.

(i) Fix $N > 0$. Form the objective function $\Pi(a)$ by

$$\Pi(a) = \min_{c \in \mathbb{R}} \sum_{m=1}^{M} (cG_m - \psi_1(p_m; a))^2, \tag{44}$$

for the conductivities $a \in A_N \subset A_{\text{ad}}$ having at most $N - 1$ discontinuity points on the interval $[0, 1]$.

(ii) Use Powell's minimization method in $K = 2N - 1$ variables ($N - 1$ discontinuity points and N conductivity values) to find

$$\Pi(\bar{a}) = \min_{a \in A_N} \Pi(a).$$

The minimizer $\bar{a}$ is the sought conductivity.

The function $\psi_1(p_m; a)$ in step (i) of the above algorithm is the first normalized eigenfunction of the Sturm-Liouville problem (5) corresponding to the conductivity $a \in A_N$. Powell's minimization method, a shooting method for the computation of the eigenvalues and the eigenfunctions, and numerical experiments are presented in (Gutman & Ha, 2009).

11. Conclusions

While in most parameter estimation problems one can hope only to achieve the best fit to data solution, sometimes it can be shown that such an identification is unique. In such case it is said that the sought parameter is identifiable within a certain class. In our recent work (Gutman & Ha, 2007; 2009) we have shown that piecewise constant conductivities $a \in \mathcal{PC}(\sigma)$ are identifiable from observations $z_m(t; a)$ of the heat conduction process (2) taken at finitely many points p_m.

Let $\mathcal{G}(a) = \{\lambda_1(a), G_1(a), \cdots, G_{M-1}(a)\}$, where he values $G_m(a)$ are a constant nonzero multiple of the first eigenfunction $\psi_1(a)$. In principle, if $\mathcal{G}(a)$ is known, then the identification of the conductivity a can be accomplished by the Marching Algorithm. Theorem 7.1 shows under what conditions the M-tuple $\mathcal{G}(a)$ can be extracted from the observations $z_m(t)$, thus assuring the identifiability of a.

It is shown in Theorem 8.2 that the Marching Algorithm not only provides the unique identification of the conductivity a, but that the identification is also continuous (stable). This result is based on the continuity of eigenvalues, eigenfunctions, and the solutions with respect to the $L^1(0,1)$ topology in the set of admissible parameters A_{ad}, see Section 4.

Numerical experiments show that, because of the ill-posedness of the identification of eigenvalues from a Dirichlet series representation, one can only identify $\mathcal{G}(a)$ with some error. Thus the Marching Algorithm would not be practically useful. In Section 10 we presented algorithms for the identification of conductivities from noise contaminated data. Its main novel point is, in agreement with the theoretical developments, the separation of the identification process into two separate parts. In part one the first eigenvalue and a multiple of the first eigenfunction are extracted from the observations. In the second part a general minimization method is used to find a conductivity which corresponds to the recovered eigenfunction.

The first eigenvalue and the eigenfunction in part one of the algorithm are found from the Dirichlet series representation of the solution of the heat conduction process. The numerical experiments in (Gutman & Ha, 2009) confirm that even for noiseless data the second eigenvalue cannot be reliably found. These experiments showed that in our tests a simple regression type algorithm identified λ_1 better than a more complex Levenberg-Marquardt algorithm. The last part of the algorithm employs Powell's nonlinear minimization method because it does not require numerical computation of the gradient of the objective function. The numerical experiments show that the conductivity identification was achieved with a 15-18% relative error for various noise levels in the observations.

12. References

Acton, F. S. (1990). *Numerical methods that work. Rev. and updated ed.*, Washington: The Mathematical Association of America.

Beck, J. V., Blackwell, B. & Clair, C. R. (1985). *Inverse heat conduction. Ill-posed problems.*, A Wiley-Interscience Publication. New York etc.: John Wileyand Sons, Inc. XVII.

Benabdallah, A., Dermenjian, Y. & Rousseau, J. L. (2007). Carleman estimates for the one-dimensional heat equation with a discontinuous coefficient and applications to controllability and an inverse problem, *Journal of Mathematical Analysis and Applications* 336(2): 865 – 887.

Birkhoff, G. & Rota, G.-C. (1978). *Ordinary Differential Equations*, 3rd edn, Wiley, New York.

Cannon, J. R. (1984). *The one-dimensional heat equation. Foreword by Felix E. Browder.*, Encyclopedia of Mathematics and Its Applications, Vol. 23. Menlo Park, California etc.: Addison-Wesley Publishing Company; Cambridge etc.: Cambridge University Press.

Courant, R. & Hilbert, D. (1989). *Methods of mathematical physics. Volume I. Transl. and rev. from the German Original. Reprint of the 1st Engl. ed. 1953.*, Wiley Classics Edition; A Wiley-Interscience Publication. New York etc.: John Wiley & Sons.

Demir, A. & Hasanov, A. (2008). Identification of the unknown diffusion coefficient in a linear parabolic equation by the semigroup approach, *Journal of Mathematical Analysis and Applications* 340(1): 5 – 15.

Duchateau, P. (1995). Monotonicity and invertibility of coefficient-to-data mappings for parabolic inverse problems, *SIAM Journal on Mathematical Analysis* 26(6): 1473–1487. URL: *http://link.aip.org/link/?SJM/26/1473/1*

Elayyan, A. & Isakov, V. (1997). On uniqueness of recovery of the discontinuous conductivity coefficient of a parabolic equation, *SIAM Journal on Mathematical Analysis* 28(1): 49–59. URL: *http://link.aip.org/link/?SJM/28/49/1*

Evans, L. C. (2010). *Partial differential equations. 2nd ed.*, Graduate Studies in Mathematics 19. Providence, RI: American Mathematical Society.

Gelfand, I. M. & Levitan, B. M. (1955). On the determination of a differential equation from its spectral function, *Amer. math. Soc. Transl. Ser.* 2(2): 253–304.

Gutman, S. & Ha, J. (2007). Identifiability of piecewise constant conductivity in a heat conduction process, *SIAM Journal on Control and Optimization* 46(2): 694–713. URL: *http://link.aip.org/link/?SJC/46/694/1*

Gutman, S. & Ha, J. (2009). Parameter identifiability for heat conduction with a boundary input., *Math. Comput. Simul.* 79(7): 2192–2210.

Gutman, S. & Ramm, A. G. (2010). Inverse problem for a heat equation with piecewise-constant conductivity, *J. Appl. Math. and Informatics* 28(3–4): 651–661.

Hoang, N. S. & Ramm, A. G. (2009). An inverse problem for a heat equation with piecewise-constant thermal conductivity, *Journal of Mathematical Physics* 50(6): 063512. URL: *http://link.aip.org/link/?JMP/50/063512/1*

Isakov, V. (2006). *Inverse problems for partial differential equations. 2nd ed.*, Applied Mathematical Sciences 127. New York, NY: Springer.

Kitamura, S. & Nakagiri, S. (1977). Identifiability of spatially-varying and constant parameters in distributed systems of parabolic type., *SIAM J. Control Optimization* 15: 785–802.

Kohn, R. V. & Vogelius, M. (1985). Determining conductivity by boundary measurements ii. interior results, *Communications on Pure and Applied Mathematics* 38(5): 643–667. URL: *http://dx.doi.org/10.1002/cpa.3160380513*

Lions, J. (1971). *Optimal control of systems governed by partial differential equations.*, Berlin-Heidelberg-New York: Springer-Verlag.

Nakagiri, S.-i. (1993). Review of Japanese work of the last ten years on identifiability in distributed parameter systems., *Inverse Probl.* 9(2): 143–191.

Orlov, Y. & Bentsman, J. (2000). Adaptive distributed parameter systems identification with enforceable identifiability conditions and reduced-order spatial differentiation., *IEEE Trans. Autom. Control* 45(2): 203–216.

Pierce, A. (1979). Unique identification of eigenvalues and coefficients in a parabolic problem., *SIAM J. Control Optimization* 17: 494–499.

Press, W. H., Teukolsky, S. A., Vetterling, W. T. & Flannery, B. P. (1992). *Numerical recipes in FORTRAN. The art of scientific computing. 2nd ed.*, Cambridge: Cambridge University Press.

Ramm, A. G. (2005). *Inverse Problems. Mathematical and analytical techniques with applications to engineering.*, New York, NY: Springer.

Saks, S. & Zygmund, A. (1965). *Analytic functions*, Warsaw: PWN - Polish Scientific Publishers. IX.

Titchmarsh, E. (1962). *Introduction to the theory of Fourier integrals*, Oxford University Press.

Part 2

Non-Fourier and Nonlinear Heat Conduction, Time Varying Heat Sources

5

Time Varying Heat Conduction in Solids

Ernesto Marín Moares
Centro de Investigación en Ciencia Aplicada y Tecnología Avanzada (CICATA)
Unidad Legaria, Instituto Politécnico Nacional (IPN)
México

1. Introduction

People experiences heat propagation since ancient times. The mathematical foundations of this phenomenon were established nearly two centuries ago with the early works of Fourier [Fourier, 1952]. During this time the equations describing the conduction of heat in solids have proved to be powerful tools for analyzing not only the transfer of heat, but also an enormous array of diffusion-like problems appearing in physical, chemical, biological, earth and even economic and social sciences [Ahmed & Hassan, 2000]. This is because the conceptual mathematical structure of the non-stationary heat conduction equation, also known as the heat diffusion equation, has inspired the mathematical formulation of several other physical processes in terms of diffusion, such as electricity flow, mass diffusion, fluid flow, photons diffusion, etc [Mandelis, 2000; Marín, 2009a]. A review on the history of the Fourier´s heat conduction equations and how Fourier´s work influenced and inspired others can be found elsewhere [Narasimhan, 1999].
But although Fourier´s heat conduction equations have served people well over the last two centuries there are still several phenomena appearing often in daily life and scientific research that require special attention and carefully interpretation. For example, when very fast phenomena and small structure dimensions are involved, the classical law of Fourier becomes inaccurate and more sophisticated models are then needed to describe the thermal conduction mechanism in a physically acceptable way [Joseph & Preziosi, 1989, 1990]. Moreover, the temperature, the basic parameter of Thermodynamics, may not be defined at very short length scales but only over a length larger than the phonons mean free paths, since its concept is related to the average energy of a system of particles [Cahill, et al., 2003; Wautelet & Duvivier, 2007]. Thus, as the mean free path is in the nanometer range for many materials at room temperature, systems with characteristic dimensions below about 10 nm are in a nonthermodynamical regime, although the concepts of thermodynamics are often used for the description of heat transport in them. To the author´s knowledge there is no yet a comprehensible and well established way to solve this very important problem about the definition of temperature in such systems and the measurement of their thermal properties remains a challenging task. On the other hand there are some aspects of the heat conduction through solids heated by time varying sources that contradict common intuition of many people, being the subject of possible misinterpretations. The same occurs with the understanding of the role of thermal parameters governing these phenomena.

Thus, this chapter will be devoted to discuss some questions related to the above mentioned problems starting with the presentation of the equations governing heat transfer for different cases of interest and discussing their solutions, with emphasis in the role of the thermal parameters involved and in applications in the field of materials thermal characterization.

The chapter will be distributed as follows. In the next section a brief discussion of the principal mechanisms of heat transfer will be given, namely those of convection, radiation and conduction. Emphasis will be made in the definition of the heat transfer coefficients for each mechanism and in the concept of the overall heat transfer coefficient that will be used in later sections. Section 2 will be devoted to present the general equation governing non-stationary heat propagation, namely the well known (parabolic) Fourier's heat diffusion equation, in which further discussions will be mainly based. The conditions will be discussed under which this equation can be applied. The modified Fourier's law, also known as Cattaneo's Equation [Cattaneo, 1948], will be presented as a useful alternative when the experimental conditions are such that it becomes necessary to consider a relaxation time or build-up time for the onset of the thermal flux after a temperature gradient is suddenly imposed on the sample. Cattaneo's equation leads them to the hyperbolic heat diffusion equation. Due to its intrinsic importance it will be discussed with some detail. In Section 3 three important situations involving time varying heat sources will be analyzed, namely: (i) a sample periodically and uniformly heated at one of its surfaces, (ii) a finite sample exposed to a finite duration heat pulse, and (iii) a finite slab with superficial continuous uniform thermal excitation. In each case characteristic time and length scales will be defined and discussed. Some apparently paradoxical behaviors of the thermal signals and the role playing by the characteristic thermal properties will be explained and physical implications in practical fields of applications will be presented too. In Section 4 our conclusions will be drawn.

2. Heat transfer mechanisms

Any temperature difference within a physical system causes a transfer of heat from the region of higher temperature to the one of lower. This transport process takes place until the system has become uniform temperature throughout. Thus, the flux of heat, Φ (units of W), should be some function of the temperatures, T_1 and T_2, of both the regions involved (we will suppose that $T_2 > T_1$). The mathematical form of the heat flux depends on the nature of the transport mechanism, which can be convection, conduction or radiation, or a coupling of them. The dependence of the heat flux on the temperature is in general non linear, a fact that makes some calculations quite difficult. But when small temperature variations are involved, things become much simpler. Fortunately, this is the case in several practical situations, for example when the sun rays heat our bodies, in optical experiments with low intensity laser beams and in the experiments that we will describe here later.

Heat convection takes place by means of macroscopic fluid motion. It can be caused by an external source (forced convection) or by temperature dependent density variations in the fluid (free or natural convection). In general, the mathematical analysis of convective heat transfer can be difficult so that often the problems can be solved only numerically or graphically [Marín, et al., 2009]. But convective heat flow, in its most simple form, i.e. heat transfer from surface of wetted area A and temperature T_2, to a fluid with a temperature T_1 , for small temperature differences, $\Delta T=T_2-T_1$, is given by the (linear with temperature) Newton's law of cooling,

$$\Phi_{conv}=h_{conv}\Delta T \quad (1)$$

The convective heat transfer coefficient, h_{conv}, is a variable function of several parameters of different kinds but independent on ΔT.
On the other hand heat radiation is the continuous energy interchange by means of electromagnetic waves. For this mechanism the net rate of heat flow, Φ_{rad}, radiated by a body surrounded by a medium at a temperature T_1, is given by the Stefan-Boltzmann Law.

$$\Phi_{rad}=\sigma\ e\ A(T_2^4-T_1^4) \quad (2)$$

where σ is the Stefan-Boltzmann constant, A is the surface area of the radiating object and e is the total emissivity of its surface having absolute temperature T_2.
A glance at Eq. (2) shows that if the temperature difference is small, then one should expand it as Taylor series around T_1 obtaining a linear relationship:

$$\Phi_{rad}=4\sigma e\ A\ T_1^3\ (T_2-T_1)=h_{rad}\Delta T \quad (3)$$

If we compare this equation with Eq. (1) we can conclude that in this case $h_{rad}=4\sigma e\ A\ T_1^3$ can be defined as a radiation heat transfer coefficient.
On the other hand, heat can be transmitted through solids mainly by electrical carriers (electrons and holes) and elementary excitations such as spin waves and phonons (lattice waves). The stationary heat conduction through the opposite surfaces of a sample is governed by Fourier's Law

$$\Phi_{cond}=-kA\nabla T \quad (4)$$

The thermal conductivity, k (W/mK), is expressed as the quantity of heat transmitted per unit time, t, per unit area, A, and per unit temperature gradient. For one-dimensional steady state conduction in extended samples of homogeneous and isotropic materials and for small temperature gradients, Fourier's law can be integrated in each direction to its potential form. In rectangular coordinates it reads:

$$\Phi_{cond}=kA\frac{T_2-T_1}{x_2-x_1}=\frac{K A\Delta T}{L}=\frac{\Delta T}{R_{cond}}=h_{cond}\Delta T \quad (5)$$

Here T_1 and T_2 represent two planar isotherms at positions x_1 and x_2, respectively, $L=x_2-x_1$, and

$$R_{cond}=\frac{L}{Ak}=\frac{1}{h_{cond}} \quad (6)$$

is the thermal resistance against heat conduction (thermal resistance for short) of the sample. The Eq. (5) is often denoted as Ohm's law for thermal conduction following analogies existing between thermal and electrical phenomena. Comparing with Eq. (1) we see that the parameter h_{cond} has been incorporated in Eq. (6) as the conduction heat transfer coefficient.
Using

$$H=h_{conv}+h_{rad}=1/R \quad (7)$$

heat transfer scientists define the dimensionless Biot number as:

$$B_i = \frac{H}{h_{cond}} = \frac{R_{cond}}{R} \tag{8}$$

as the fraction of material thermal resistance that opposes to convection and radiation heat looses.

3. The heat diffusion equation

Eq. 4 represents a very simple empirical law that has been widely used to explain heat transport phenomena appearing often in daily life, engineering applications and scientific research. In terms of the heat flux density ($q=\Phi/A$) it lauds:

$$\vec{q} = -k\vec{\nabla}T \tag{9}$$

When combined with the law of energy conservation for the heat flux

$$\frac{\partial E}{\partial t} = -\mathrm{div}(\vec{q}) + Q \tag{10}$$

where Q represents the internal heat source and

$$\partial E/\partial t = \rho c \partial T/\partial t \tag{11}$$

is the temporal change in internal energy, E, for a material with density ρ and specific heat c, and assuming constant thermal conductivity, Fourier's law leads to another important relationship, namely the non-stationary heat diffusion equation also called second Fourier's law of conduction. It can be written as:

$$\nabla^2 T - \frac{1}{\alpha}\frac{\partial T}{\partial t} = -\frac{Q}{k} \tag{12}$$

with

$$a = k/\rho c \tag{13}$$

as the thermal diffusivity.
Fourier's law of heat conduction predicts an infinite speed of propagation for thermal signals, i.e. a behavior that contradicts the main results of Einstein´s theory of relativity, namely that the greatest known speed is that of the electromagnetic waves propagation in vacuum. Consider for example a flat slab and apply at a given instant a supply of heat to one of its faces. Then according to Eq. (9) there is an instantaneous effect at the rear face. Loosely speaking, according to Eq. (9), and also due to the intrinsic parabolic nature of the partial differential Eq. (12), the diffusion of heat gives rise to infinite speeds of heat propagation. This conclusion, named by some authors the paradox of instantaneous heat propagation, is not physically reasonable.
This contradiction can be overcome using several models, the most of them inspired in the so-called CV model.
This model takes its name from the authors of two pioneering works on this subject, namely that due to Cattaneo [Cattaneo, 1948] and that developed later and (apparently) independently by Vernotte [Vernotte, 1958]. The CV model introduces the concept of the

relaxation time, τ, as the build-up time for the onset of the thermal flux after a temperature gradient is suddenly imposed on the sample.
Suppose that as a consequence of the temperature existing at each time instant, t, the heat flux appears only in a posterior instant, $t + \tau$. Under these conditions Fourier's Law adopts the form:

$$\vec{q}(r, t+\tau) = -k\vec{\nabla}T(r,t) \tag{14}$$

For small τ (as it should be, because otherwise the first Fourier´s law would fail when explaining every day phenomena) one can expand the heat flux in a Taylor Serie around $\tau = 0$ obtaining, after neglecting higher order terms:

$$\vec{q}(r, t+\tau) = \vec{q}(r,t) + \tau\frac{\partial \vec{q}(r,t)}{\partial t} \tag{15}$$

Substituting Eq. (15) in Eq. (14) leads to the modified Fourier´s law of heat conduction or CV equation that states:

$$\tau\frac{\partial \vec{q}}{\partial t} + \vec{q} = -k\vec{\nabla}\mathrm{T}. \tag{16}$$

Here the time derivative term makes the heat propagation speed finite. Eq. (16) tells us that the heat flux does not appear instantaneously but it grows gradually with a build-up time, τ. For macroscopic solids at ambient temperature this time is of the order of 10^{-11}s so that for practical purposes the use of Eq. (1) is adequate, as daily experience shows.
Substituting Eq. (17) into the energy conservation law (Eq. (10)) one obtains:

$$\nabla^2 T - \frac{1}{\alpha}\frac{\partial T}{\partial t} - \frac{1}{u^2}\frac{\partial^2 T}{\partial t^2} = -\frac{1}{k}\left(Q + \tau\frac{\partial Q}{\partial t}\right). \tag{17}$$

Here $u = (\alpha/\tau)^{1/2}$ represents a (finite) speed of propagation of the thermal signal, which diverges only for the unphysical assumption of $\tau = 0$.
Eq. (16) is a hyperbolic instead of a parabolic (diffusion) equation (Eq. (12)) so that the wave nature of heat propagation is implied and new (non-diffusive) phenomena can be advised. Some of them will be discussed in section (3.1).
The CV model, although necessary, has some disadvantages, among them: i). The hyperbolic differential Eq. (7) is not easy to be solved from the mathematical point of view and in the majority of the physical situations has non analytical solutions. ii) The relaxation time of a given system is in general an unknown variable. Therefore care must be taken in the interpretation of its results. Nevertheless, several examples can be found in the literature.
As described with more detail elsewhere [Joseph & Preziosi, 1989, 1990] other authors [Band & Meyer, 1948], proposed exactly the same Eq. (7) to account for dissipative effects in liquid He II, where temperature waves propagating at velocity u were predicted [Tisza, 1938; Landau, 1941; Peshkov, 1944)] and verified. Due to these early works the speed u is often called the second sound velocity. More recently Tzou reported on phenomena such as thermal wave resonance [Tzou, 1991] and thermal shock waves generated by a moving heat source [Tzou, 1989]. Very rapid heating processes must be explained using the CV model too, such as those taking place, for example, during the absorption of energy coming from ultra short laser pulses [Marín, et al., 2005] and during the gravitational collapse of some stars [Govender, et al., 2001]. In the field of nanoscience and nanotechnology thermal time

constants, τ_c, characterizing heat transfer rates depend strongly on particle size and on its thermal diffusivity. One can assume that for spherical particles of radius R, these times scale proportional to R^2/α [Greffet, 2004; Marín, 2010; Wolf, 2004]. As for condensed matter the order of magnitude of a is 10^{-6}m^2/s, for spherical particles having nanometric diameters, for example between 100 and 1 nm, we obtain for these times values ranging from about 10 ns to 1ps, which are very close to the above mentioned relaxation times. At these short time scales Fourier's laws do not work in their initial forms.
In the next sections some interesting problems involving time varying heat sources will be discussed assuming that the conditions for the parabolic approach are well fulfilled, and, when required, these conditions will be deduced.

4. Some non-stationary problems on heat conduction

While the parabolic Fourier´s law of heat conduction (4) describes stationary problems, with the thermal conductivity as the relevant thermophysical parameter, time varying heat conduction phenomena, which appear often in praxis, are described by the heat diffusion equation (12), being the thermal diffusivity the important parameter in such cases. Thermal conductivity can be measured using stationary methods based in Eq. (4), whose principal limitation is that precise knowledge of the amount of heat flowing through the sample and of the temperature gradient in the normal direction to this flow is necessary, a difficult task when small specimens are investigated. Therefore the use of non-stationary or dynamic methods becomes many times advantageous that allow, in general, determination of the thermal diffusivity. Thus knowledge of the specific heat capacity (per unit volume) is necessary if the thermal conductivity is to be obtained as well, as predicted by Eq. (13). Although this can be a disadvantage, often available specific heat data are used, so that it is not always necessary to determine experimentally it in order to account for the thermal conductivity. This is because specific heat capacity is less sensitive to impurities and structure of materials and comparatively independent of temperature above the Debye temperature than thermal conductivity and diffusivity. More precise, C is nearly a constant parameter for solids. In a plot of thermal conductivity versus thermal diffusivity we can see that solid materials typically fall along the line $C \approx 3\times10^6$ J/m^{3}K at room temperature. This experimentally proved fact is a consequence of the well known Dulongs and Petit's classical law for the molar specific heat of solids and of the consideration that the volume occupied per atom is about 1.4×10^{-29} m^3 for almost all solids. In other words, the almost constant value of C can be explained by taking into account its definition as the product of the density (ρ) and the specific heat (c). The specific heat is defined as the change in the internal energy per unit of temperature change; thus, if the density of a solid increases (or decreases) the solid can store less (or more) energy. Therefore, as the density increases, the specific heat must decrease and then the product $C=\rho c$ stays constant and, according to Eq. 13, the behavior of the thermal conductivity is similar to that of the diffusivity. In accurate work, however, particularly on porous materials and composites, it is highly recommendable to measure also C. This is because some materials have lower-than-average volumetric specific heat capacity. Sometimes this happens because the Debye temperature lies well above room temperature and heat absorption is not classical. Deviations are observed in porous materials too, whose conductivity is limited partially by the gas entrapped in the porosity, in low density solids, which contain fewer atoms per unit volume so that ρC becomes low, and in composites due to heat fluxes through series and parallel arrangements of layers and through embedded regions from different materials that strongly modified their effective thermal properties values, as has been described elsewhere [Salazar, 2003]. Although there are several methods for measurement of C their applications are often limited because they involve temperature variations that can affect thermal properties during measurement, in particular in the vicinity of phase transitions and structural changes. Fortunately there is another parameter involved in non-stationary problems and that can be also measured using dynamic techniques. While thermal diffusivity is defined as the ratio between the thermal conductivity and the specific heat capacity, this new parameter, named as thermal effusivity, e, but also called thermal contact parameter by some authors [Boeker & van Grondelle, 1999], is related to their product, as follows:

$$\varepsilon = (k\,C)^{1/2} \tag{18}$$

It is worth to be noticed that while the two expressions contain the same parameters, they are quite different. Diffusivity is related to the speed at which thermal equilibrium can be reached, while effusivity is related to the heated body surface temperature and it is the property that determines the contact temperature between two bodies in touch to one another, as will be seen below. Measuring both quantities provides the thermal conductivity without the need to know the specific heat capacity (note that Eqs. (13) and (18) lead to $k=\varepsilon a^{1/2}$). Dynamic techniques for thermal properties measurement can be divided in three classes, namely those involving pulsed, periodical and transient heat sources. There are also phenomena encounter in daily life that also involve these kinds of heating sources. This section will be devoted to analyze and discuss the solution of the heat diffusion equation in the presence of these sources. In each case characteristic time and length scales will be presented, the role playing by the characteristic thermal properties will be discussed as well as physical implications in practical fields of applications.

4.1 A sample periodically and uniformly heated at one of its surfaces

Consider an isotropic homogeneous semi-infinite solid, whose surface is heated uniformly (in such a way that the one-dimensional approach used in what follows is valid) by a source of periodically modulated intensity $I_0(1+\cos(\omega t))/2$, where I_0 is the intensity, $\omega=2\pi f$ is the angular modulation frequency, and t is the time (this form of heating can be achieved in praxis using a modulated light beam whose energy is partially absorbed by the sample and converted to heat [Almond & Patel, 1996] but other methods can be used as well, e.g. using joule´s heating [Ivanov et al., 2010]). The temperature distribution $T(x,t)$ within the solid can be obtained by solving the homogeneous (parabolic) heat diffusion equation, which can be written in one dimension as

$$\frac{\partial^2 T(x,t)}{\partial x^2} - \frac{1}{\alpha}\frac{\partial T(x,t)}{\partial t} = 0 \tag{19}$$

The solution of physical interest for most applications (for example in photothermal (PT) techniques [Almond & Patel, 1996]) is the one related to the time dependent component. If we separate this component from the spatial distribution, the temperature can be expressed as:

$$T(x,t) = Re[\theta(x)exp(i\omega t)] \tag{20}$$

Substituting Eq. (20) into Eq. (19) leads to

$$\frac{\partial^2 \theta(x,t)}{\partial x^2} - q^2 \theta(x) = 0 \tag{21}$$

where

$$q = \sqrt{\frac{i\omega}{\alpha}} = \frac{(1+i)}{\mu} \tag{22}$$

is the thermal wave number and μ represents the thermal diffusion length defined as

$$\mu = \sqrt{\frac{2\alpha}{\omega}} \tag{23}$$

Using the boundary condition

$$k \frac{\partial \theta(x,t)}{\partial x}\bigg|_{x=0} = Re\left[\frac{I_0}{2} exp(i\omega t)\right], \tag{24}$$

the Eq. (21) can be solved and Eq. (19) leads to

$$T(x,t) = \frac{I_0}{2\varepsilon\sqrt{\omega}} exp\left(-\frac{x}{\mu}\right) Re\left\{exp\left[-i\left(\frac{x}{\mu} + \omega t + \frac{\pi}{4}\right)\right]\right\} \tag{25}$$

This solution represents a mode of heat propagation through which the heat generated in the sample is transferred to the surrounding media by diffusion at a rate determined by the thermal diffusivity. Because this solution has a form similar to that of a plane attenuated wave it is called a thermal wave. Although it is not a real wave because it does not transport energy as normally waves do [Salazar, 2006], the thermal wave approach has demonstrated to be useful for the description of several experimental situations, as will be seen later.

Suppose that we have an alternating heat flux, related to a periodic oscillating temperature field. The analogy between thermal and electrical phenomena described in Section 1 when dealing with Fourier´s law can be followed to define the thermal impedance Z_t as the temperature difference between two faces of a thermal conductor divided by the heat flux crossing the conductor. Then the thermal impedance becomes the ratio between the change in thermal wave amplitude and the thermal wave flux. At the surface of the semi-infinite medium treated with above one gets,

$$Z_t = \frac{T(x=0,t) - T_{amb}}{-k\frac{dT(x,t)}{dx}\big|_{x=0}} \tag{26}$$

where T_{amb} is the ambient (constant) temperature (it was settled equal to zero for simplicity). Substituting Eq. (25) in Eq. (26) one obtains after a straightforward calculation [Marín, 2009b]:

$$Z_t = \frac{1-i}{\varepsilon\sqrt{\omega}} = \frac{1}{\varepsilon\sqrt{\omega}} exp\left(-i\frac{\pi}{4}\right) \tag{27}$$

Note that, contrary to thermal resistance (see Eq. (6)), which depends on thermal conductivity, in the thermal impedance definition the thermal effusivity becomes the relevant parameter.

Using Eq. (27) the Eq. (25) can be rewritten as:

$$T(x,t) = \frac{I_0}{2} Z_t exp\left(-\frac{x}{\mu}\right) cos\left(\frac{x}{\mu} + \omega t\right) \tag{28}$$

Eq. (25) shows that the thermal diffusion length, μ, gives the distance at which an appreciable energy transfer takes place and that there is a phase lag between the excitation and the thermal response of the sample given by the term $x/\mu+\pi/4$ in the exponential term. Note that the thermal wave behaviour depends on the values of both, thermal effusivity, with determines the wave amplitude at x=0, and the thermal diffusivity, from which the attenuation and wave velocity depend.

Among other characteristics [Marín et al., 2002] a thermal wave described by Eq. (24) has a phase velocity, $v_f=\omega\mu=(2\alpha\omega)^{1/2}$. Because Eq. (21) is a linear ordinary differential equation describing the motion of a thermal wave, then the superposition of solutions will be also a solution of it (often, as doing above, the temperature distribution is approximated by just the first harmonic of that superposition because the higher harmonics damp out more quickly due to the damping coefficient increase with frequency). This superposition represents a group of waves with angular frequencies in the interval ω, ω+dω travelling in space as "packets" with a group velocity $v_g=2v_f$ [Marín et al., 2006]. It is worth to notice that both, phase and group velocities depend on the modulation frequency in such a way that if ω tends to infinite, they would approach infinite as well, what is physically inadmissible.

This apparent contradiction can be explained using the same arguments given in section 2. Starting from the hyperbolic heat diffusion equation (Eq. (17)) without internal heat sources, and making the separation of variables given by Eq. (20), the equation to be solve becomes

$$\frac{\partial^2\theta(x,t)}{\partial x^2} - q'^2\theta(x) = 0 \tag{29}$$

with the boundary condition at the surface [Salazar, 2006]

$$-k\left.\frac{\partial\theta(x,t)}{\partial x}\right|_{x=0} = Re\left[\frac{I_0}{2}(1+i\omega t)\right] \tag{30}$$

Eq. (29) is similar to Eq. (21) but with the complex wave number given by [Marín, 2007a]

$$q' = \omega\sqrt{\frac{\tau}{\alpha}}\sqrt{i\frac{\omega_l}{\omega} - 1} \tag{31}$$

where

$$\omega_l = \frac{1}{\tau} \tag{32}$$

Two limiting cases can be examined. First, for low modulation frequencies so that $\omega<<\omega_l$ the wave number becomes equal to q (Eq. (22)) and the solution becomes a thermal wave given by Eq. (25). But for high frequencies, i.e. for $\omega>>\omega_l$, the wave number becomes q=iω/u, and the solution of the problem has the form [Salazar, 2006.]

$$T(x,t) = \frac{I_0}{2Cu}exp\left(-\frac{xu}{2\alpha}\right)cos\left(\frac{\omega}{u}x - \omega t\right) \tag{33}$$

Thus according to the hyperbolic solution the amplitude of the surface temperature does not depend on the modulation frequency and depends on the specific heat capacity and the propagation velocity $u=(\alpha/\tau)^{1/2}$. There is not a phase lag, i.e. the excitation source and the surface temperature are in phase. Moreover, the penetration depth becomes also independent on the modulation frequency and depends on the wave propagation velocity. This case represents a form of heat transfer, which takes place through a direct coupling of vibrational modes (i.e. the acoustic phonon spectrum) of the material. At these high frequencies (short time scale) ballistic transport of heat can be dominant.

Measurement of the periodical temperature changes induced in a sample by harmonic heating is the basis of the working principle of the majority of the so-called photothermal (PT) techniques [Marín, 2009c]. These are methods widely used for thermal characterization because the thermal signal is dependent on thermal properties such as thermal diffusivity (see Eq. (29)). As mentioned above the time constant τ in condensed mater is related to the phonon relaxation time, which is in the picosecond range, so that the limiting frequency becomes about ω_1=10[12]Hz. Typical modulation frequencies used in PT experiments are between some Hz and several kHz, i.e. $\omega << \omega_1$, so that the more simpler parabolic approach is valid. This offers several advantages related with their use for thermal characterization of materials in situations where heat transport characteristic times are comparable to the relaxation time, τ [Marín, 2007b].

Following the above discussion in what follows the parabolic thermal wave approach will be used to explain a particular phenomenon observed in some experiments realized with PT techniques, which contradicts intuitively expectation. Suppose that a solid sample is subjected to periodical modulated heating at certain frequencies. Using different detection schemas some authors [Caerels et al., 1998) 2452-2458 ; Sahraoui et al., 2002; S. Longuemart et al., 2002; Depruester et al., 2005; Lima et al., 2006; Marin et al., 2010] have observed that when a sample is in contact with a liquid the resulting sample's temperature may be larger than that due to the bare sample, for certain values of the modulation frequencies. This contradicts the expected behavior that in the presence of a liquid the developed heat will always flow through the sample to the liquid, which acts as a heat sink.

In the PT techniques the periodical heating is mainly generated by impinging intensity modulated light (e.g. a laser beam) on a sample. When light energy is absorbed and subsequently totally or partially transformed into as heat, it results in sample heating, leading to temperature changes as well as changes in the thermodynamic parameters of the sample and its surroundings. Measurements of these changes are ultimately the basis for all photothermal methods. The temperature variations could be detected directly using a pyroelectric transducer in the so called Photopyroelectric (PPE) method [Mandelis & Zver, 1985]. The sample's temperature oscillations can be also the cause of periodical black body infrared electromagnetic waves that are radiated by the sample and that can be measured using an appropriate sensor in the PT radiometry [Chen et al., 1993] In the photoacoustic (PA) method the sample is enclosed in a gas (example air) tight cell. The temperature variations in the sample following the absorption of modulated radiation induce pressure fluctuations in the gas, i.e. acoustic waves, that can be detected by a sensitive microphone already coupled to the cell [Vargas & Miranda, 1980]. Other detection schemes have been devised too.

Consider the experimental setup schematically showed in Fig. 1. A glass sample covers one of the two openings of a PA cell, while the other is closed by a transparent glass window through which a modulated laser light beam impinges on the inner, metal coated (to warrant full absorption of the light) sample's surface, generating periodical heating (the so-called thermal waves) and hence a pressure fluctuation in a PA air chamber, which is detected with a microphone already enclosed in the PA cell. The microphone signal is fed into a Lock-in amplifier for measurement of its amplitude as a function of the modulation frequency, f.

Using this schema Lima *et al* [Lima et al., 2006);] and Marín *et al* [Marin et al., 2010)] have measured the PA signal as a function of the modulation frequency for a bare glass substrate, and then they have deposited about 100 μL of liquid and repeated the same measurement.
In Fig. 2 (a) the normalized signal amplitude (the ratio of the amplitude signal due to the substrate-liquid system and that due to the bare substrate) is showed as a function of f for the case of a distilled water liquid sample. One can see that in certain frequency intervals the normalized signal becomes greater than 1, a fact that, as discussed before, contradicts the intuitively awaited behavior.
In order to explain this apparent paradox the mentioned authors resorted to the thermal wave model supposing that, as other kind of waves do, they experiences reflection, refraction and interference. Consider two regions, 1 and 2, and a plane thermal wave (Eq. (24)) incident from region 1 that is partially reflected and transmitted at the interface.
One can show that for normal incidence the reflection and transmission coefficients can be written as [Bennett & Patty, 1982]:

$$R_{12} = \frac{1-b_{12}}{1+b_{12}} \tag{34}$$

and

$$T_{12} = \frac{2}{1+b_{12}} \tag{35}$$

where

$$b_{12} = \frac{\varepsilon_1}{\varepsilon_2} \tag{36}$$

is the ratio of the media thermal effusivities. Thermal effusivity can be also regarded, therefore, as a measure of the thermal mismatch between the two media.

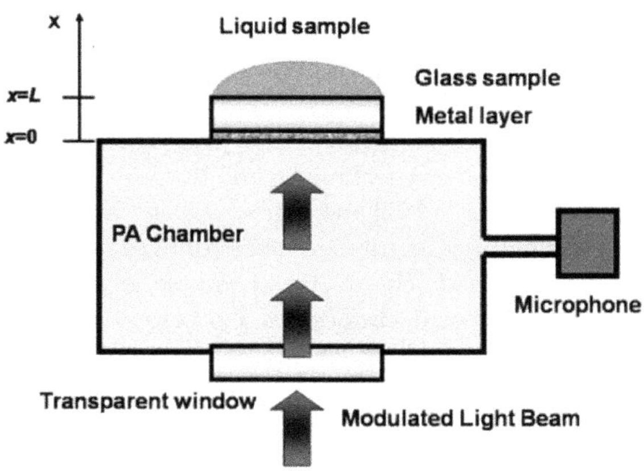

Fig. 1. Schema of a photothermal experimental setup with photoacoustic detection. In the experiment described here the glass substrate was 180 μm thick and it was coated with a 2 μm thick layer of Cu deposited by thermal evaporation in vacuum. The PA cell cylindrical cavity have a 5mm diameter and is 5 mm long. The light source was an Ar-ion laser beam of 50 mW modulated in intensity at 50% duty cycle with a mechanical chopper. An electret microphone was coupled to the cell through a 1mm diameter duct located at the cell wall.

Denoting with s the glass substrate of thickness L, which is sandwiched between two regions, namely 1 (the metallic coating) and 2 (the liquid sample or air). Supposing also that the surface of region 1 opposite to s is heated uniformly (so that a one dimensional analysis can be valid) by a light source of periodically modulated intensity, I_0. Because its thickness is much smaller than L it can be also supposed that region 1 acts only as a thin superficial light absorbing layer, where a thermal wave will be generated following the periodical heating and launched through the glass substrate. Consider the propagation of a thermal wave described by Eq. (25) through the substrate. The so-called thermal wave model shows that the thermal waves will propagate towards the interface between the sample and region 2 and back towards the sample's surface, 1. On striking the boundaries they will be partially reflected and transmitted, so that interference between the corresponding wave trains takes place. Because the PA signal will be proportional to the temperature at the glass-metal interface the interest is in the resulting temperature at x=0, which can be obtained by summing all the waves arriving at this point. The result is [Marín et al., 2010] (the time dependent second exponential term of Eq. (25) will be omitted from now on for sake of simplicity):

$$T(0) = T_0\left[1 + R_{s2}\frac{exp(-2q_sL)}{1-\gamma exp(-2q_sL)}\right] \tag{37}$$

where T_0 is a frequency dependent term, $\gamma=R_{s1}R_{s2}$, and R_{s1} and R_{s2} are the normal incidence thermal wave reflection coefficients at the s-1 and s-2 interfaces respectively.

The solid line in Fig. 2 represents the normalized signal as a function of f calculated using the above expression for the system composed of a glass substrate (L=180μm, ε_s=1480 $Ws^{1/2}m^{-2}K^{-1}$, α_s=3.5×10^{-6} m^2/s), a Copper (ε_{Cu}=37140 $Ws^{1/2}m^{-2}K^{-1}$) metallic layer as region 1, and water ε_w=1580 $Ws^{1/2}m^{-2}K^{-1}$) as a liquid sample (for air the value ε_a=5.5 $Ws^{1/2}m^{-2}K^{-1}$ was taking). The theoretical obtained results for higher frequencies [Marín et al., 2010] showed in part (b) of the same figure that the frequency intervals with amplitude ratios greater than unity are awaited to appear in a periodical manner, a typical result for wave phenomena.

A similar result has been reported by Depriester *et al* [Depriester et al., 2005] in the context of the photothermal infrared radiometry technique, and by Caerels *et al* [Caerels et al., 1998], Longuemart *et al* [Longuemart *et al.*, 2002] and Sahraoui *et al* [Sahraoui et al., 2002) using a photopyroelectric (PPE) technique. The measurement configuration is very similar as that described above for the PA method. The analyzed sample is placed in intimate thermal contact with one of the metal coated surfaces of the sensor (usually a polyvinyledene difluoride (PVDF) polymer film with metalized surfaces serving as electrodes or a pyroelectric ceramic crystal (e.g., $LiTaO_3$), while a periodical intensity modulated light beam impinges on its opposite metalized side, which acts as a light absorber. Following the absorption of light energy, the PE sensor temperature fluctuates periodically at the modulation frequency of the incident beam thereby generating a voltage, whose amplitude at a given frequency can be measured using a Lock-in amplifier. Recently Marin *et al* [Marín et al., 2011] used this last approach in order to explain the increase of the normalized PPE signal above unity for some frequencies.

The good agreement between experiment and theory shows that the described behavior can be explained as caused by a thermal wave interference phenomenon. The thermal wave

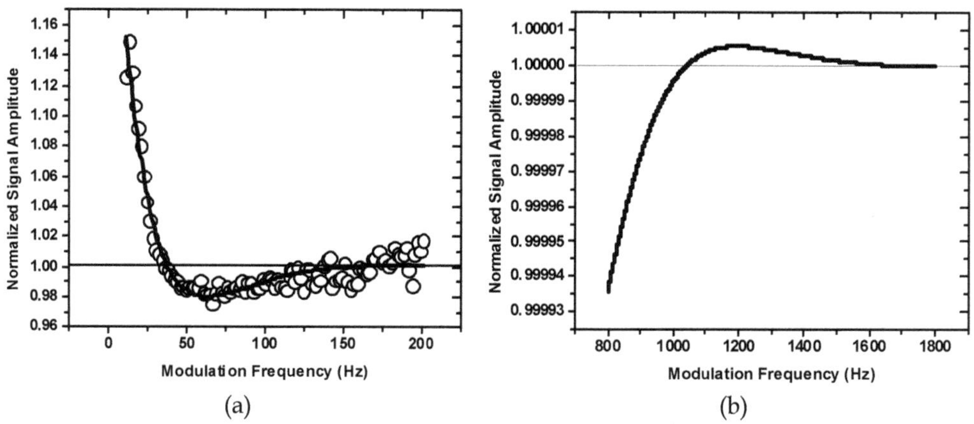

Fig. 2. (a): Normalized signal amplitude as a function of *f*. Circles: experimental points. Solid curve and (b): Result of theoretical simulation using Eq. (37). Reproduced from [Central Eur. J. Phys, 2010, 8, 4, 634-638].

approach could be helpful not only in the field of PA and PT techniques but it can be also used for the analysis of the phenomenon of heat transfer in the presence of modulated heat sources in multilayer structures, which appear frequently in men's made devices (for example semiconductor heterostructures lasers and LEDs driven by pulsed, periodical electrical current sources).

4.2 A finite sample exposed to a finite duration heat pulse

Considering a semi-infinite homogeneous medium exposed to a sudden temperature change at its surface at $x=0$ from T_0 to T_1. For the calculation of the temperature field created by a heat pulse at $t=0$ one has to solve the homogeneous heat diffusion equation (19) with the boundary conditions

$$T(x=0,\ t\geq 0) = T_1\,;\ \ T(x>0,\ t=0) = T_0. \qquad (38)$$

The solution for $t>0$ is [Carlslaw & Jaeger 1959]:

$$T(x,t) = T_1 + (T_0 - T_1)erf\left(\frac{x}{2\sqrt{\alpha t}}\right) \qquad (39)$$

where *erf* is the error function.

Using Fourier's law (Ec. (9)) one may obtain from the above equation for the heat flow

$$q(x,t) = \frac{\varepsilon(T_1-T_0)}{\sqrt{\pi t}}exp\left(-\frac{x^2}{\mu_t^2}\right) \qquad (40)$$

This expression describes a Gaussian spread of thermal energy with characteristic width

$$\mu_t = 2\sqrt{\alpha t} \qquad (41)$$

This characteristic distance is the thermal diffusion length (for pulsed excitation) and has a similar meaning as the thermal diffusion length defined by Eq. (23).

If Eq. (40) is scaled to three dimensions one can show that after a time t has elapsed the heat outspread over a sphere of radiusμ. Suppose that a spherical particle of radius R is heated in the form described above by a heat pulse at its surface. The particle requires for cooling a time similar to that the necessary for the heat to diffuse throughout its volume. The heat flux at the opposite surface of the particle could be expressed as

$$q(r = 2R) = q_0 exp\left[-\left(\frac{\tau_c}{t}\right)^2\right] \tag{42}$$

with q_0 as a time independent constant and a characteristic thermal time constant given by

$$\tau_c = \frac{R^2}{\alpha} \tag{43}$$

This time depend strongly on particle size and on its thermal diffusivity, α [Greffet, 2007; Wolf, 2004; Marín, 2010]. As for most condensed matter samples the order of magnitude of α is 10^{-6} m^2/s, for a sphere of diameter 1 cm one obtain τ_c=100 s and for a sphere with a radius of 6400 km, such as the Earth, this time is of around 10^{12} years, both values compatible with daily experience. But for spheres having diameters between 100 and 1 nm , these times values ranging from about 10 ns to 1 ps, i.e. they are very close to the relaxation times, τ, for which Fourier's Law of heat conduction is not more valid and the hyperbolic approach must be used as well. The above equations enclose the basic principle behind a well established method for thermal diffusivity measurement known as the Flash technique [Parker et al., 1961]. A sample with well known thickness is rapidly heated by a heat pulse while its temperature evolution with time is measured. From the thermal time constant the value of α can be determined straightforwardly. Care must be taken with the heat pulse duration if the parabolic approach will be used accurately. For time scales of the order of the relaxation time the solutions of the hyperbolic heat diffusion equation can differ strongly from those obtained with the parabolic one as has been shown elsewhere [Marín, et al.2005)].

Now, coming back to Eq. (40), one can see that the heat flux at the surface of the heated sample (x=0) is

$$q(x = 0, t) = \frac{\varepsilon(T_1 - T_0)}{\sqrt{\pi t}} \tag{44}$$

Thus the heat flow is not proportional to the thermal conductivity of the material, as under steady state conditions (see Eq. (23)), but to its thermal effusivity [Bein & Pelzl, 1989]. If two half infinite materials with temperatures T_1 and T_2 (T_1>T_2) touch with perfect thermal contact at t=0, the mutual contact interface acquires a contact temperature T_c in between. This temperature can be calculated from Eq. (44) supposing that heat flowing out from the hotter surface equals that flowing into the cooler one, i.e.

$$\frac{\varepsilon_1(T_1 - T_c)}{\sqrt{\pi t}} = \frac{\varepsilon_2(T_2 - T_c)}{\sqrt{\pi t}} \tag{45}$$

or

$$T_c = \frac{\varepsilon_1 T_1 + \varepsilon_2 T_2}{\varepsilon_1 + \varepsilon_2} \tag{46}$$

According to this result, if $\varepsilon_1 = \varepsilon_2$, T_c lies halfway between T_1 and T_2, while if $\varepsilon_1 > \varepsilon_2$, T_c will be closer to T_1 and if $\varepsilon_1 < \varepsilon_2$, T_c will be closer to T_2. The Eq. (46) shows that our perception of the temperature is often affected by several variables, such as the kind of material we touch, its absolute temperature and the time period of the "experiment", among others (note that the actual value of the contact temperature can be affected by factors such as objects surfaces roughness that have not taking into account in the above calculations). For example, at room temperature wooden objects feels warmer to the rapidly touch with our hands than those made of a metal, but when a sufficient time has elapsed both seem to be at the same temperature. Many people have the mistaken notion that the relevant thermophysical parameter for the described phenomena is the thermal conductivity instead of the thermal effusivity, as stated by Eq. (46). The source of this common mistake is the coincidence that in solids, a high effusivity material is also a good heat conductor. The reason arises from the almost constancy of the specific heat capacity of solids at room temperature explained at the beginning of this section. Using Eq. (13) the Eq. (18) can be written as $\varepsilon^2=Ck$. Then if ε^2 is plotted as a function of k for homogeneous solids one can see that all points are placed close to a straight line [Marín, 2007].

If we identify region 1 with our hand at T_1=37^{0}C and the other with a touched object at a different temperature, T_2, the contact temperature that our hand will reach upon contact can be calculated using Eq. (46) and tabulated values of the thermal effusivities. Calculation of the contact temperature between human skin at 37^{0}C and different bodies at 20^{0}C as a function of their thermal effusivities show [E Marín, 2007] that when touching a high thermal conductivity object such as a metal (e.g. Cu), as $\varepsilon_{metal} >> \varepsilon_{skin}$, the temperature of the skin drops suddenly to 20 ^{0}C and one sense the object as being "cold". On the other hand, when touching a body with a lower thermal conductivity, e.g. a wood's object ($\varepsilon_{wood} < \varepsilon_{skin}$) the skin temperature remains closest to 37 ^{0}C, and one sense the object as being "warm". This is the reason why a metal object feels colder than a wooden one to the touch, although they are both at the same, ambient equilibrium temperature. This is also the cause why human foot skin feels different the temperature of floors of different materials which are at the same room temperature and the explanation of why, when a person enters the cold water in a swimming pool, the temperature immediately felt by the swimmer is near its initial, higher, body temperature [Agrawal, 1999].

In Fig. 3 the calculated contact temperature between human skin at 37^{0}C and bodies of different materials at 1000^{0}C (circles) and 0^{0}C (squares) are represented as a function of their thermal effusivities. One can see that the contact temperature tends to be, in both cases, closer than that of the skin. This is one of the reasons why our skin is not burning when we make a suddenly (transient) contact to a hotter object or freezing when touching a very cold one (despite we fill that the object is hotter or colder, indeed).

Before concluding this subsection the following question merits further analysis. How long can be the contact time,τ_l, so that the transient analysis performed above becomes valid? The answer has to do with the very well known fact that when the skin touches very hot or cold objects a very thin layer of gas (with thickness L) is produced (e.g. water vapour exhaled when the outer layers of the skin are heated or evaporated from ice when it is heated by a warmer hand). This time can be calculated following a straightforward calculation starting from Eq. (44) and Fourier´s law in the form given by Eq. (5). It lauds [Marín ,2008]:

$$\tau_l = \frac{\varepsilon_2 L^2}{\pi k_{gas}^2}\left(\frac{T_2-T_c}{T_c-T_1}\right)^2 \tag{47}$$

It is represented in Fig. (4) for different thicknesses of the gas (supposed to be air) layer using for the skin temperature the value T_2=37⁰C.

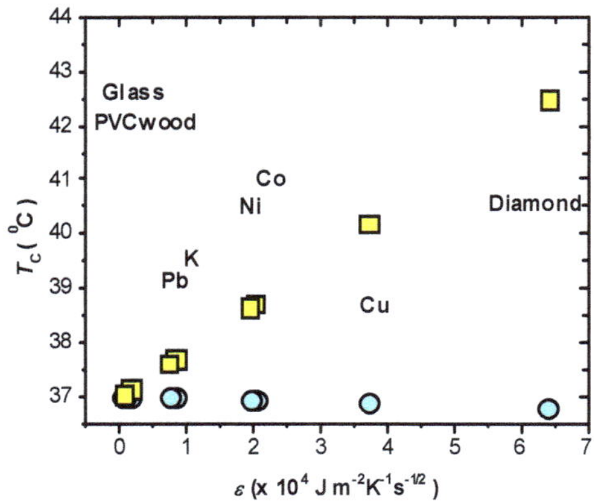

Fig. 3. Contact temperatures as a function of thermal effusivity calculated using Eq. (45) when touching with the hand at 37 ⁰C objects of different materials at 0 ⁰C (circles) and 1000⁰C (squares). Reproduced with permission from [Latin American Journal of Physics Education 2, 1, 15-17 (2008)]. Values of the thermal effusivities have been taken from [Salazar, 2003]

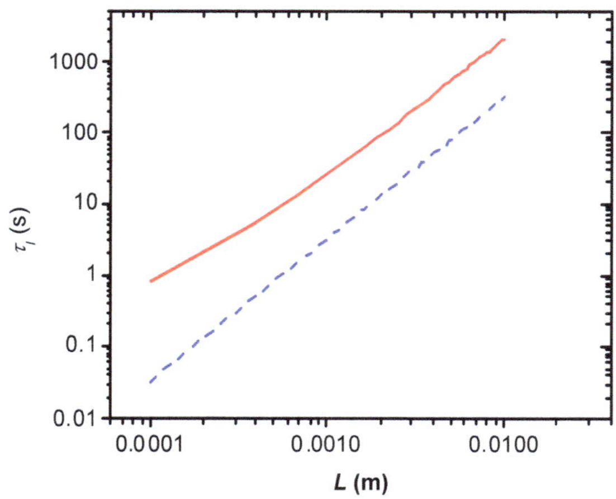

Fig. 4. The time required for the skin to reach values of the contact temperature of 0⁰C and 100⁰C without frostbitten or burning up respectively (see text), as a function of the hypothetical thickness of the gas layer evaporated at its surface. The solid and dotted curves correspond to the case of touching a cold (-196⁰C) and a hot (600⁰C) object, respectively Reproduced with permission from [Latin American Journal of Physics Education 2, 1, 15-17 (2008)].

The solid curve corresponds to the case of a cold touched object and the dotted line to that of the hotter ones. For the temperature of a colder object the value T_1=-196⁰C (e.g. liquid

Nitrogen) was taking. The corresponding limiting contact temperature will be T_c=0ºC (Eq. (46)). In the case of the hot object the value T_1=600ºC (T_c=100ºC) was taking. From the figure one can conclude that for gas layer thicknesses smaller than 1mm the time required to heat the skin to 100ºC by contact with an object at 600ºC is lower than 3s, a reasonable value. On the other hand, for the same layer thickness, liquid Nitrogen can be handled safely for a longer period of time which, in the figure, is about 25 s. These times are of course shorter, because the generated gas layers thicknesses are in reality much shorter than the here considered value.

The above examples try to clarify the role played by thermal effusivity in understanding thermal physics concepts. According to the definition of thermal conductivity, under steady-state conditions a good thermal conductor in contact with a thermal reservoir at a higher temperature extracts from it more energy per second than a poor conductor, but under transient conditions the density and the specific heat of the object also come into play through the thermal effusivity concept. Thermal effusivity is not a well known heat transport property, although it is the relevant parameter for surface heating or cooling processes.

4.3 A finite slab with superficial continuous uniform thermal excitation

The following phenomenon also contradicts common intuition of many people: As a result of superficial thermal excitation the front surface of a (thermally) thick sample reaches a higher equilibrium temperature than a (thermally) thin one [Salazar et al., 2010; Marín et al., 2011]. Consider a slab of a solid sample with thickness L at room temperature, T_0, is uniformly and continuously heated at its surface at x=0. The heating power density can be described by the function:

$$P = \begin{cases} 0 & t < 0 \\ P_0 & t > 0 \end{cases} \tag{48}$$

where P_0 is a constant.

The temperature field in a sample, $T(x,t)$, can be obtained by solving the one-dimensional heat diffusion problem (Eq. (19)) with surface energy losses, i.e., the third kind boundary condition:

During heating the initial condition lauds

$$\Delta T_{\uparrow}(x, t = 0) = T_{\uparrow}(x, t = 0) - T_{amb} = 0 \tag{49}$$

and the boundary conditions are:

$$H\Delta T_{\uparrow}(0, t) - k\left.\frac{\partial \Delta T_{\uparrow}(x,t)}{\partial x}\right|_{x=0} = P_0 \tag{50}$$

and

$$H\Delta T_{\uparrow}(L, t) - k\left.\frac{\partial \Delta T_{\uparrow}(x,t)}{\partial x}\right|_{x=L} = 0 \tag{51}$$

The heat transfer coefficients at the front (heated) and at the rear surface of the sample have been assumed to be the same and are represented by the variable H *(see Eq. (7))*.

When heating is interrupted, the equations (49) to (60) become

$$\Delta T_{\downarrow}(x, t = 0) = T_{\downarrow}(x, t = 0) - T_{amb} = T_{eq} \tag{52}$$

$$H\Delta T_{\downarrow}(0,t) - k\left.\frac{\partial \Delta T_{\downarrow}(x,t)}{\partial x}\right|_{x=0} = 0 \quad (53)$$

and

$$H\Delta T_{\downarrow}(L,t) - k\left.\frac{\partial \Delta T_{\downarrow}(x,t)}{\partial x}\right|_{x=L} = 0 \quad (54)$$

respectively, where T_{eq} is the equilibrium temperature that the sample becomes when thermal equilibrium is reached during illumination, being the initial sample temperature when illumination is stopped.
The solution of this problem is [Valiente et al., 2006]

$$\Delta T_{\downarrow}(x,t) = -\sum_{n=1}^{\infty} A_n e^{-\lambda_n a^2 t}\left[\frac{k\sqrt{\lambda_n}}{h_1}\cos\sqrt{\lambda_n}x + \sin\sqrt{\lambda_n}x\right] \quad (55)$$

and

$$\Delta T_{\uparrow}(x,t) = \frac{-P_0 x + (L + k/H)P_0}{2k + HL} + \sum_{n=1}^{\infty} A_n e^{-\lambda_n a^2 t}\left[\frac{k\sqrt{\lambda_n}}{h_1}\cos\sqrt{\lambda_n}x + \sin\sqrt{\lambda_n}x\right] \quad (56)$$

where $\alpha = a^2$,

$$\lambda_n = \left(\frac{\mu_n}{L}\right)^2 \quad (57)$$

$$\tan\mu = \frac{\frac{2L}{kH}\mu}{\frac{L}{k^2} - \left(\frac{\mu}{H}\right)^2} \quad (58)$$

and

$$A_n = -\frac{1}{\|X_n\|}\int_0^L w(\xi)X_n(\xi)\,d\xi \quad (59)$$

with

$$\|X_n\|^2 = \int_0^L \left[\frac{k\sqrt{\lambda_n}}{h_1}\cos\sqrt{\lambda_n}x + \sin\sqrt{\lambda_n}x\right]^2 dx \quad (60)$$

In order to examine under which condition a sample can be considered as a thermally thin and thick slab the thermodynamic equilibrium limit must be analyzed, i.e. the limit of infinitely long times.
Introducing the Biot Number defined in Eq. (8) and taking $t \to \infty$ after a straightforward calculation the following results are obtained:
At x=0:

$$\Delta T_{\uparrow}(0,\infty) = \frac{P_0}{H}\frac{B_i + 1}{B_i + 2} \quad (61)$$

and

$$\Delta T_{\uparrow}(L,\infty) = \frac{P_0}{H}\frac{1}{B_i + 2} \quad (62)$$

Two limiting cases can be analyzed:

a. Very large Biot number ($B_i >> 2$):

In this case Eq. (61) becomes

$$\Delta T_{\uparrow}(0,\infty) = \frac{P_0}{H} \tag{63}$$

while from Eq. (62) one has

$$\Delta T_{\uparrow}(L,\infty) = \frac{P_0}{H}\frac{1}{B_i} \tag{64}$$

For their quotient one can write

$$\frac{\Delta T_{\uparrow}(L,\infty)}{\Delta T_{\uparrow}(0,\infty)} = \frac{1}{B_i} \tag{65}$$

There is a thermal gradient across the sample so that the rear sample temperature becomes k/LH times lower than the front temperature. Note that the temperature difference will decrease as the heat losses do, as awaited looking at daily experience.

b. Very small Biot number (B_i<<1):

In this case both Eq. (61) and Eq. (62) lead to

$$\Delta T_{\uparrow}(0,\infty) = \Delta T_{\uparrow}(L,\infty) = \frac{P_0}{2H} \tag{66}$$

Thus, the equilibrium temperature becomes the same at both sample´s surfaces. The sample can be considered thin enough so that there is not a temperature gradient across it. Thus, the condition for a very thin sample is just:

$$B_i \ll 1 \tag{67}$$

With words, following the Biot´s number definition given in section 1, the temperature gradient across the sample can be neglected when the conduction heat transfer through its opposite surfaces of the samle is greater than convection and radiation losses.

The results presented above explain the phenomenon that the equilibrium temperature becomes greater for a thicker sample. Denoting the front (heated side) sample´s temperature of a thick sample (B_i >> 1) at $t \to \infty$ as $u_{\uparrow thick}$, and that of a thin ones (B_i << 1) as $u_{\uparrow thin}$. Their quotient is:

$$\frac{u_{\uparrow thick}(0,\infty)}{u_{\uparrow thin}(0,\infty)} = 2 \tag{68}$$

Here L_{thick} means that this is a thickness for which the sample is thermally thick. This means that after a sufficient long time the front surface temperature of a thick sample becomes two times higher than that for a thin sample. As discussed elsewhere [Marín et al., 2011]

The here presented results can have practical applications in the field of materials thermal characterization. When the thermally thin condition is achieved, the rise temperature becomes [Salazar et al., 2010; Valiente et al., 2006]

$$\Delta T_{\uparrow} = \frac{P_0}{2H}\left[1 - exp\left(-\frac{t}{\tau_r}\right)\right] \tag{69}$$

while when illumination is interrupted the temperature decreases as

$$\Delta T_{\downarrow} = \frac{P_0}{2H} exp\left(-\frac{t}{\tau_r}\right) \tag{70}$$

where

$$\tau_r = CL_{thin}/2H \tag{71}$$

and L_{thin} means that the sample thickness is such that it is thermally thin. If the front and/or rear temperatures (remember that both are the same for a thermally thin sample) are measured as a function of time during heating (and/or cooling) the value of τ_r can be determined by fitting to the Eq. (69) (and/or Eq. (70)) and then, using Eq. (71), the specific heat capacity can be calculated if the sample´s thickness is known. This is the basis of the so-called temperature relaxation method for measurement of C [Mansanares et al., 1990]. As we see from Eq. (71) precise knowledge of H is necessary.
On the other hand, from Eq. (65) follows that measurement of the asymptotic values of rear and front surface temperatures of a thermally thick sample leads to:

$$r = C\frac{u_{\uparrow}(L,\infty)}{u_{\uparrow}(0,\infty)} = \frac{1}{B_i} = \frac{k}{L_{thick}H} \tag{72}$$

from which thermal conductivity could be determined. Note that the knowledge of the H value is here necessary too.
From Eqs. (71) and (72) the thermal diffusivity value can be determined straightforwardly without the necessity of knowing H, i.e. it can calculated from the quotient [Marín et al., 2011]:

$$\frac{r}{\tau} = \frac{2k}{CL_{thin}L_{thick}} = 2\frac{\alpha}{L_{thin}L_{thick}} \tag{73}$$

Fig. 5 shows a kind of Heisler Plot [Heisler, 1947] of the percentile error associated to the thermally thick approximation as a function of the sample's thickness using a typical value of H=26 W/m^2 [Salazar et al., 2010] for a sample of plasticine (k=0.30 W/mK) and for a sample of cork (k=0.04 W/mK).

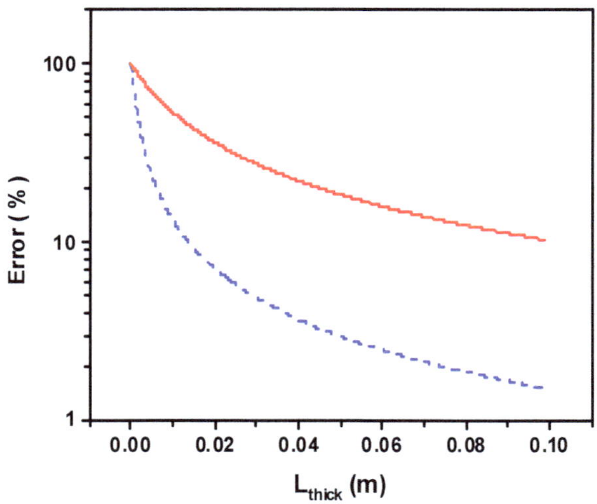

Fig. 5. Heisler Plots for Plasticine (solid line) and Cork (dashed line).

Note that for a 5 cm thick plasticine sample this error becomes about 20 %, while a considerable decrease is achieved for a low conductivity sample such as cork with the same thickness. These errors become lower for thicker samples, but rear surface temperature measurement can become difficult. Thus it can be concluded that practical applications of this method for thermal diffusivity measurement can be achieved better for samples with thermal conductivities ranging between 10^{-2} and 10^{-1} W/mK. Although limited, in this range of values are included an important class of materials such as woods, foams, porous materials, etc. For these the thermally thick approximation can be reached with accuracy lower than 10 % for thicknesses below about 2-3 cm.
Thermal diffusivity plays a very important role in non-stationary heat transfer problems because its value is very sensible to temperature and to structural and compositional changes in materials so that the development of techniques for its measurement is always impetuous. The above described method is simple and inexpensive, and renders reliable and precise results [Lara-Bernal et al., 2011]. The most important achievement of the method is that it cancels the influence of the heat losses by convection and radiation which is a handicap in other techniques because the difficulties for their experimental quantification.

5. Conclusion

Heat conduction in solids under time varying heating is a very interesting and important part of heat transfer from both, the phenomenological point of view and the practical applications in the field of thermal properties characterization. In this chapter a brief overview has been given for different kinds of thermal excitation. For each of them some interesting physical situations have been explained that are often misinterpreted by a general but also by specialized people. The incompatibility of the Fourier´s heat conduction model with the relativistic principle of the upper limit for the propagation velocity of signals imposed by the speed of light in vacuum was discussed, with emphasis of the limits of validity this approach and the corrections needed in situations where it is not applicable. Some applications of the thermal wave's analogy with truly wave fields have been described as well as the principal peculiarities of the heat transfer in the presence of pulsed and transient heating. It has been shown that although the four fundamental thermal parameters are related to one another by two equations, each of them has its own meaning. While static and stationary phenomena are governed by parameters like specific heat and thermal conductivity respectively, under non-stationary conditions the thermal effusivity and diffusivity are the more important magnitudes. While the former plays a fundamental role in the case of a body exposed to a finite duration short pulse of heat and in problems involving the propagation of oscillating wave fields at interfaces between dissimilar media, thermal diffusivity becomes the most important thermophysical parameter to describe the mathematical form of the thermal wave field inside a body heated by a non-stationary Source. It is worth to be noticed that the special cases discussed here are not the only of interest for thermal science scientists. There are several open questions that merit particular attention. For example, due to different reasons (e.g. the use of synchronous detection in PT techniquess and consideration of only the long-term temperature distribution once the system has forgotten its initial conditions in the transient methods), in the majority of the works the oscillatory part of the generated signal and the transient contribution have been analyzed separately, with no attention to the combined signal that appears due to the well known fact that when a thermal wave is switched on, it takes some time until phase and amplitude have reached their final values. Nevertheless, it is expected that this chapter will help scientists who wish to carry out theoretical or experimental research in the field of heat transfer by conduction and thermal characterization of materials, as well as students and teachers requiring a solid formation in this area.

6. Acknowledgment

This work was partially supported by SIP-IPN through projects 20090477 and 20100780, by SEP-CONACyT Grant 83289 and by the SIBE Program of COFAA-IPN. The standing support of J. A. I. Díaz Góngora and A. Calderón, from CICATA-Legaria, is greatly appreciated. Some subjects treated in this chapter have been developed with the collaboration of some colleagues and students. In particular the author is very grateful to A. García-Chéquer and O. Delgado-Vasallo.

7. References

Agrawal D. C. (1999) Work and heat expenditure during swimming. *Physics* Education. Vol. 34, No. 4, (July 1999), pp. 220-225, ISSN 0031-9120.

Ahmed, E. and Hassan, S.Z. (2000) On Diffusion in some Biological and Economic systems. *Zeitschrift für Naturforshung*. Vol. 55a, No. 8, (April 2000), pp. 669-672, ISSN 0932-0784.

Almond, D. P. and Patel, P. M. (1996). *Photothermal Science and Techniques* in *Physics and its Applications*, 10 Dobbsand, E. R. and Palmer, S. B. (Eds), ISBN 978-041-2578-80-9, Chapman and Hall, London, U.K..

Bein, B. K. and Pelzl, J. (1989). *Analysis of Surfaces Exposed to Plasmas by Nondestructive Photoacoustic and Photothermal Techniques*, in *Plasma Diagnostics,* Vol. 2, *Surface Analysis and Interactions,* pp. 211-326 Auciello, O. and Flamm D.L. (Eds.), ISBN 978-012-0676-36-1, Academic Press, New York, U.S.A.

Band, W. and Meyer, L. (1948) Second sound and the heat conductivity in helium II, *Physical Review,* Vol. 73, No. 3, (February 1948), pp. 226-229.

Bennett, C. A. and Patty R. R. (1982) Thermal wave interferometry: a potential application of the photoacoustic effect. Applied Optics Vol. 21, No. 1, (January 1982), pp. 49-54, ISSN 1559-128X

Boeker E and van Grondelle R (1999) *Environmental Physics* ISBN 978-047-1997-79-5, Wiley, New York, U.S.A.

Caerels J., Glorieux C. and Thoen J. (1998) Absolute values of specific heat capacity and thermal conductivity of liquids from different modes of operation of a simple photopyroelectric setup. *Review of Scientific Instruments*. Vol. 69 , No. 6, (June 1998) pp. 2452-2458, ISSN 0034-6748.

Carlslaw H. S. and Jaeger J. C. (1959) *Conduction of Heat in Solids.* ISBN 978-0198533030, Oxford University Press, London, U.K.

Cattaneo, C. (1948) *Sulla conduzione de calore,* Atti Semin. Mat. Fis. Univ. Modena Vol. 3, pp. 83-101.

Chen, Z. H., Bleiss, R. and Mandelis, A. (1993) Photothermal rate-window spectrometry for noncontact bulk lifetime measurements in semiconductors. *Journal of Applied Physics*, Vol. 73, No. 10, (May 1993), pp. 5043-5048, ISSN 0021-8979

Cahill, D. G., Ford, W. K., Goodson, K. E., Mahan, G., Majumdar, D. A., Maris, H. J., Merlin, R. and Phillpot, S. R. (2003) Nanoscale thermal transport. *Journal of Applied Physics.* Vol. 93, No. 2, (January 2003), pp. 793-818, ISSN 0021-8979

Depriester M., Hus P., Delenclos S. and Hadj Sahraoiui A. (2005) New methodology for thermal parameter measurements in solids using photothermal radiometry *Review of Scientific Instruments* Vol 76, No. 7, (July 2005), pp. 074902-075100, ISSN 0034-6748.

Fourier, J. (1878) *The analytical theory of heat*, Cambridge University Press, Cambridge, U.K., Translated by Alexander Freeman. Reprinted by Dover Publications, New York, 1955. French original: "Théorie analytique de la chaleur," Didot, Paris, 1822.

Govender, M., Maartens, R. and Maharaj, S. (2001). A causal model of radiating stellar collapse, Physics Letters A, Vol. 283, No. 1-2, pp. 71-79 (May 2001), ISSN: 0375-9601.

Greffet, J. (2007) *Laws of Macroscopic Heat Transfer and Their Limits in Topics in Applied Physics*, Voz, S. (Ed.) ISBN 978-354-0360-56-8, Springer, Paris, France, pp. 1-13

M P Heisler (1947) Temperature charts for induction and constant temperature heating. Transactions of ASME, Vol. 69, pp. 227-236.

Ivanov R., Marin E., Moreno I. and Araujo C. (2010) Electropyroelectric technique for measurement of the thermal effusivity of liquids. *Journal of Physics D: Applied Physics*, Vol. 43, No. 22, pp. 225501-225506 (June 2010), ISSN 0022-3727.

Joseph, D. D. and Preziosi, L. (1989) Heat Waves. *Reviews of Modern Physics*. Vol. 61, No. 1, (January-March), pp. 41-73, ISNN 0034-6861

Joseph, D. D. and Preziosi, L. (1990) Addendum to the paper "Heat Waves". *Reviews of Modern Physics*. Vol. 62, No. 2, (April-June 1990), pp. 375-391, ISNN 0034-6861

Landau, L. (1941) Theory of the Superfluidity of Helium II, Journal of Physics U.S.S.R. Vol. 5, No. 1, (January 1941), pp. 71-90, ISSN: 0368-3400.

Lara-Bernal et al. (2011) (submitted for publication).

Li, B., Xu, Y. and Choi, J. (1996). Applying Machine Learning Techniques, *Proceedings of ASME 2010 4th International Conference on Energy Sustainability*, pp. 14-17, ISBN 842-6508-23-3, Phoenix, Arizona, USA, May 17-22, 2010

C. A. S. Lima, L. C. M. Miranda and H. Vargas, (2006). Photoacoustics of Two-Layer Systems: Thermal Properties of Liquids and Thermal Wave Interference. *Instrumentation Science and Technology*. Vol. 34, No. 1-2.,(February 2006), pp. 191-209 ISSN 1073-9149

Longuemart S., Quiroz A. G., Dadarlat D., Hadj Sahraoui A., Kolinsky C., Buisine J. M., Correa da Silva E., Mansanares A. M., Filip X., Neamtu C. (2002). An application of the front photopyroelectric technique for measuring the thermal effusivity of some foods. *Instrumentation Science and Technology*. Vol. 30, No. 2, (June 2002), pp. 157-165, ISSN 1073-9149

Mandelis, A. (2000) Diffusion waves and their use. *Physics Today* Vol. 53, No. 8, (August 2000), pp. 29-36, ISSN 0031-9228.

Mandelis, A. and Zver, M. M. (1985) Theory of photopyroelectric spectroscopy of solids. Journal of Applied Physics, Vol. 57, No. 9, (May 1985), pp. 4421-4431, ISSN 0021-8979

Mansanares A.M., Bento A.C., Vargas H., Leite N.F., Miranda L.C.M. (1990) Phys Rev B Vol. 42, No. 7, (September 1990), pp. 4477-4486, ISSN 1098-0121.

Marín E., Marín-Antuña, J. and Díaz-Arencibia, P. (2002) On the wave treatment of the conduction of heat in experiments with solids. *European Journal of Physics*. Vol. 23, No. 5 (September 2002), pp. 523-532, ISSN 0143-0807

Marín, E., Marin, J. and Hechavarría, R., (2005) Hyperbolic heat diffusion in photothermal experiments with solids, *Journal de Physique IV*, Vol. 125, No. 6, (June 2005), pp. 365-368, ISSN 1155-4339.

Marín, E., Jean-Baptiste, E. and Hernández, M. (2006) Teaching thermal wave physics with soils. *Revista Mexicana de Física E* Vol. 52, No. 1, (June 2006), pp. 21–27, ISSN 1870-3542

Marin, E. (2007a) The role of thermal properties in periodic time-varying phenomena. *European Journal of Physics*. Vol. 28, No. 3, (May 2007), pp. 429-445, ISSN 0143-0807

Marín, E. (2007b) On the role of photothermal techniques for the thermal characterization of nanofluids. *Internet Electron. J. Nanoc. Moletrón*. Vol. 5, No. 2, (September 2007), pp 1007-1014, ISSN 0188-6150.

Marín, E. (2008) Teaching thermal physics by Touching. *Latin American Journal of Physics Education* Vol. 2, No. 1, (January 2008), pp. 15-17, ISSN 1870-9095

Marín, E. (2009a) Generalized treatment for diffusion waves. *Revista Mexicana de Física. E*. Vol. 55, No. 1, (June 2009), pp. 85–91, ISSN 1870-3542.

Marín, E. (2009b) Basic principles of thermal wave physics and related techniques. ChapterI in Thermal Wave Physics and Related Photothermal Techniques: Basic Principles and Recent Developments. Marin, E. (Ed.) ISBN 978-81-7895-401-1, pp. 1-28 . Transworld Research, Kerala, India.

Marín, E. (Ed.) (2009c) *Thermal Wave Physics and Related Photothermal Techniques: Basic Principles and Recent Developments*. ISBN978-81-7895-401-1, Transworld Research, Kerala, India.

Marín, E., Calderón, A. and Delgado-Vasallo, O. (2009) Similarity Theory and Dimensionless Numbers in Heat Transfer, *European Journal of Physics*. Vol. 30, No. 3, (May 2009), pp. 439-445, ISSN 0143-0807

Marin, E. (2010) Characteristic dimensions for heat transfer, *Latin American Journal of Physics Education*. Vol. 4, No. 1, (January 2010), pp. 56-60, ISSN 1870-9095

Marin, E., García, A., Vera-Medina, G. and Calderón, A., (2010) On the modulation frequency dependence of the photoacoustic signal for a metal coated glass-liquid system. *Central European Journal of Physics*. Vol. 8, No. 4, (August 2010), pp. 634-638, ISSN 1895-1082.

Marín, E., García, A., Juárez, G., Bermejo-Arenas, J. A. and Calderón, A., (2011) On the heating modulation frequency dependence of the photopyroelectric signal in experiments for liquid thermal characterization, Infrared Physics & Technology (in press).

Marín, E., Lara-Bernal, A., Calderón, A. and Delgado-Vasallo O. (2011) On the heat transfer through a solid slab heated uniformly and continuously on one of its surfaces. *European Journal of Physics*. Vol. 32, No. 4, (May 2011), pp. 783–791, ISSN 0143-0807.

Narasimhan, T. N. (1999) Fourier´s heat conduction equation: History, influence, and connections. *Reviews of Geophysics*. Vol. 37, No. 1, (February 199), pp. 151-172, ISSN 8755–1209.

Parker W. J., Jenkins, W. J., Butler, C. P., Abbott, G. L. (1961) Flash Method of Determining Thermal Diffusivity, Heat Capacity and Thermal Conductivity. *Journal of Applied Physics Vol.* 32, No. 9, (September 1961), pp. 1679-1684, ISSN 0021-8979

Peshkov, V. (1944) Second Sound in Helium II, *Journal of Physics U.S.S.R.* Vol. 8, No. 2, (February 1944), pp. 381-383, ISSN: 0368-3400

Sahraoui, H. Longuemart, S., Dadarlat, D., Delenclos, S., Kolinsky C. and Buisine, J. M. (2002) Review of Scientific Instruments Vol. 73, No. 7, (July 2002), pp. 2766-2771, ISSN 0034-6748

Salazar, A. (2003) On thermal diffusivity *European Journal of Physics.* Vol. 24, No. 4, (July 2003), pp. 351-358, ISSN 0143-0807

Salazar, A. (2006), Energy propagation of thermal waves. *European Journal of Physics*. Vol. 27, No. 6, (November 2006). pp. 1349-1356, ISSN 0143-0807

Salazar, A., Apiñaniz, E., Mendioroz, A., Aleaga, A. (2010) A thermal paradox: which gets warmer? *European Journal of Physics.* Vol. 31 No. 5, (September 2010), pp. 1053-1060, ISSN 0143-0807

Tisza, L. (1938) Sur la Supraconductibilit e thermique de l'helium II liquide et la statistique de Bose-Einstein, Comptes Rendus de l'Académie des Sciences., Paris *Vol.* 207, pp. 1035-1038.

Tzou, D. Y. (1989) Schock wave formation around a moving heat source in a solid with finite speed of heat propagation. *International Journal of Mass and Heat Transfer*. Vol. 32, No. 10, (October 1989), pp. 1979-1987, ISSN 0017-9310

Tzou, D. Y. (1991) The resonance phenomenon in thermal waves. *International Journal of Engineering Science and Technology* Vol. 29, No. 5, (May 1991), pp. 1167-1177 ISSN: 0975–5462

Valiente, H., Delgado-Vasallo, O., Abdelarrague, R., Calderón, A., Marín, E. (2006), Specific Heat Measurements by a Thermal Relaxation Method: Influence of Convection and Conduction. *International Journal of Thermophysics Vol.* 27, No. 6, (November 2006), pp. 1859-1872, ISSN 0195-928X

Vargas, H. and Miranda, L.C.M. (1988) Photoacoustic and related photothermal techniques. *Physics Reports*, Vol. 161, No. 2, (April 1988) pp. 43-101, ISSN: 0370-1573

Vernotte, P. (1958) La véritable équation de la chaleur, *Comptes Rendus de l'Académie des Sciences, Paris Vol.* 247, pp. 2103-2105.

Wautelet, M. and Duvivier, D. (2007) The characteristic dimensions of the nanoworld. *European Journal of Physics*. Vol. 28, No. 5, (September 2007), pp. 953-960, , ISSN 0143-0807

Wolf, E. L. (2004) *Nanophysics and Nanotechnology: An Introduction to Modern Concepts in Nanoscience,* ISBN 978-352-7406-51-7, Wiley-VCH, Weinheim, Germany.

6

Exact Travelling Wave Solutions for Generalized Forms of the Nonlinear Heat Conduction Equation

Mohammad Mehdi Kabir Najafi
Department of Engineering, Aliabad Katoul branch, Islamic Azad University, Golestan Province, Iran

1. Introduction

"*The most incomprehensible thing about the world is that it is at all comprehensible*" (Albert Einstein), but the question is how do we fully understand incomprehensible things? Nonlinear science provides some clues in this regard (He, 2009).

The world around us is inherently nonlinear. For instance, nonlinear evolution equations (NLEEs) are widely used as models to describe complex physical phenomena in various fields of sciences, especially in fluid mechanics, solid-state physics, plasma physics, plasma waves, and biology. One of the basic physical problems for these models is to obtain their travelling wave solutions. In particular, various methods have been utilized to explore different kinds of solutions of physical models described by nonlinear partial differential equations (PDEs). For instance, in the numerical methods, stability and convergence should be considered, so as to avoid divergent or inappropriate results. However, in recent years, a variety of effective analytical and semi-analytical methods have been developed to be used for solving nonlinear PDEs, such as the variational iteration method (VIM) (He, 1998; He et al., 2010), the homotopy perturbation method (HPM) (He, 2000, 2006), the homotopy analysis method (HAM) (Abbasbandy, 2010), the tanh-method (Fan, 2002; Wazwaz, 2005, 2006), the sine-cosine method (Wazwaz, 2004), and others. Likewise, He and Wu (2006) proposed a straightforward and concise method called the Exp-function method to obtain the exact solutions of NLEEs. The method, with the aid of Maple or Matlab, has been successfully applied to many kinds of NLEE (He & Zhang, 2008; Kabir & Khajeh, 2009; Borhanifar & Kabir, 2009, 2010; Borhanifar et al., 2009; Kabir et al., 2011). Lately, the (G'/G)-expansion method, first introduced by Wang et al. (2008), has become widely used to search for various exact solutions of NLEEs (Bekir & Cevikel, 2009; Zhang et al., 2009; Zedan, 2010; Kabir et al., 2011). The results reveal that the two recent methods are powerful techniques for solving nonlinear partial differential equations (NPDEs) in terms of accuracy and efficiency. This is important, since systems of NPDEs have many applications in engineering.

The generalized forms of the nonlinear heat conduction equation can be given as

$$u_t - a(u^n)_{xx} - u + u^n = 0, \qquad a > 0\,, n > 1 \tag{1.1}$$

and in (2 + 1)-dimensional space

$$u_t - a(u^n)_{xx} - a(u^n)_{yy} - u + u^n = 0. \tag{1.2}$$

The heat equation is an important partial differential equation which describes the distribution of heat (or variation in temperature) in a given region over time. The heat equation is a consequence of Fourier's law of cooling. In this chapter, we consider the heat equation with a nonlinear power-law source term. The equations (1.1) and (1.2) describe one-dimensional and two-dimensional unsteady thermal processes in quiescent media or solids with the nonlinear temperature dependence of heat conductivity. In the above equations, u= u(x,y,t) is temperature as a function of space and time; u_t is the rate of change of temperature at a point over time; u_{xx} and u_{yy} are the second spatial derivatives (thermal conductions) of temperature in the x and y directions, respectively; also u_x and u_y are the temperature gradient.

Many authors have studied some types of solutions of these equations. Wazwaz (2005) used the tanh-method to find solitary solutions of these equations and a standard form of the nonlinear heat conduction equation (when $n=3$ in Eq. (1.1)). Also, Fan (2002) applied the solutions of Riccati equation in the tanh-method to obtain the travelling wave solution when $n=2$ in Eq. (1.1). More recently, Kabir et al. (2009) implemented the Exp-function method to find exact solutions of Eq. (1.1), and obtained more general solutions in comparison with Wazwaz's results.

Considering all the indispensably significant issues mentioned above, the objective of this paper is to investigate the travelling wave solutions of Eqs. (1.1) and (1.2) systematically, by applying the (G'/G)-expansion and the Exp-function methods. Some previously known solutions are recovered as well, and, simultaneously, some new ones are also proposed.

2. Description of the two methods

2.1 The (G'/G)-expansion method

Suppose that a nonlinear PDE, say in two independent variables x and t, is given by

$$P(u, u_t, u_x, u_{xx}, u_{tt}, u_{tx}, \cdots) = 0, \tag{2.1}$$

or in three independent variables x, y and t, is given by

$$P(u, u_t, u_x, u_y, u_{xx}, u_{yy}, u_{tt}, u_{tx}, u_{ty}, \cdots) = 0, \tag{2.2}$$

where P is a polynomial in its arguments, which include nonlinear terms and the highest order derivatives.

Introducing a complex variable η defined as

$$u(x,t)=U(\eta) \quad , \quad \eta=k(x-ct) \tag{2.3}$$

or

$$u(x,y,t)=U(\eta) \quad , \quad \eta=k(x+y-ct) \tag{2.4}$$

Eq. (2.1) and (2.2) reduce to the ordinary differential equations (ODEs)

$$P(U,-kcU',kU',k^2U'',k^2c^2U'',-k^2cU'',\cdots)=0, \tag{2.5}$$

and

$$P(U,-kcU',kU',kU',k^2U'',k^2U'',k^2c^2U'',-k^2cU'',-k^2cU'',\cdots)=0, \tag{2.6}$$

respectively, where k and c are constants to be determined later. According to the (G'/G)-expansion method, it is assumed that the travelling wave solution of Eq. (2.5) or (2.6) can be expressed by a polynomial in $\left(\frac{G'}{G}\right)$ as follows:

$$U(\eta)=\sum_{i=1}^{m}\alpha_i\left(\frac{G'}{G}\right)^i+\alpha_0\,, \qquad \alpha_m\neq 0 \tag{2.7}$$

where α_0, and α_i, for $i=1,2,\dots,m$, are constants to be determined later, and $G(\eta)$ satisfies a second-order linear ordinary differential equation (LODE):

$$\frac{d^2G(\eta)}{d\eta^2}+\lambda\frac{dG(\eta)}{d\eta}+\mu\,G(\eta)=0 \tag{2.8}$$

where λ and μ are arbitrary constants. Using the general solutions of Eq. (2.8), we have

$$\frac{G'(\eta)}{G(\eta)}=\begin{cases}\dfrac{\sqrt{\lambda^2-4\mu}}{2}\left(\dfrac{C_1\sinh\left(\dfrac{\sqrt{\lambda^2-4\mu}}{2}\eta\right)+C_2\cosh\left(\dfrac{\sqrt{\lambda^2-4\mu}}{2}\eta\right)}{C_1\cosh\left(\dfrac{\sqrt{\lambda^2-4\mu}}{2}\eta\right)+C_2\sinh\left(\dfrac{\sqrt{\lambda^2-4\mu}}{2}\eta\right)}\right)-\dfrac{\lambda}{2}, & \lambda^2-4\mu>0,\\[2ex] \dfrac{\sqrt{4\mu-\lambda^2}}{2}\left(\dfrac{-C_1\sin\left(\dfrac{\sqrt{4\mu-\lambda^2}}{2}\eta\right)+C_2\cos\left(\dfrac{\sqrt{4\mu-\lambda^2}}{2}\eta\right)}{C_1\cos\left(\dfrac{\sqrt{4\mu-\lambda^2}}{2}\eta\right)+C_2\sin\left(\dfrac{\sqrt{4\mu-\lambda^2}}{2}\eta\right)}\right)-\dfrac{\lambda}{2}, & \lambda^2-4\mu<0,\end{cases} \tag{2.9}$$

and it follows from (2.7) and (2.8), that

$$U' = -\sum_{i=1}^{m} i\alpha_i \left[\left(\frac{G'}{G}\right)^{i+1} + \lambda \left(\frac{G'}{G}\right)^{i} + \mu \left(\frac{G'}{G}\right)^{i-1} \right],$$

$$U'' = \sum_{i=1}^{m} i\alpha_i \left[\begin{array}{l} (i+1)\left(\frac{G'}{G}\right)^{i+2} + \lambda\,(2i+1)\left(\frac{G'}{G}\right)^{i+1} + i\,(\lambda^2+2\mu)\left(\frac{G'}{G}\right)^{i} \\ +\mu\lambda\,(2i-1)\left(\frac{G'}{G}\right)^{i-1} + \mu^2(i-1)\left(\frac{G'}{G}\right)^{i-2} \end{array} \right] \tag{2.10}$$

and so on. Here, the prime denotes the derivative with respective to η.
To determine u explicitly, we take the following four steps:
Step 1. Determine the integer m by substituting Eq. (2.7) along with Eq. (2.8) into Eq. (2.5) or (2.6), and balancing the highest-order nonlinear term(s) and the highest-order partial derivative.
Step 2. Substitute Eq. (2.7) with the value of m determined in *Step 1*, along with Eq. (2.8) into Eq. (2.5) or (2.6) and collect all terms with the same order of $\left(\frac{G'}{G}\right)$ together; the left-hand side of Eq. (2.5) or (2.6) is converted into a polynomial in $\left(\frac{G'}{G}\right)$. Then set each coefficient of this polynomial to zero to derive a set of algebraic equations for k, c, α_0 and α_i, for $i = 1, 2, \ldots, m$.
Step 3. Solve the system of algebraic equations obtained in *Step 2*, for k, c, α_0 and α_i, for $i = 1, 2, \ldots, m$, by use of Maple.
Step 4. Use the results obtained in the above steps to derive a series of fundamental solutions $u(\eta)$ of Eq. (2.5) or (2.6) depending on $\left(\frac{G'}{G}\right)$; since the solutions of Eq. (2.8) have been well known for us, we can obtain exact solutions of Eqs. (2.1) and (2.2).

2.2 The Exp-function method

According to the classic Exp-function method, it is assumed that the solution of ODEs (2.5) or (2.6) can be written as

$$u(\eta) = \frac{\sum_{n=-f}^{g} a_n \exp(n\eta)}{\sum_{m=-p}^{q} b_m \exp(m\eta)} = \frac{a_f \exp(f\eta) + \cdots + a_{-g}\exp(-g\eta)}{b_p \exp(p\eta) + \cdots + b_{-q}\exp(-q\eta)}, \tag{2.11}$$

where f, g, p and q are positive integers which are unknown, to be further determined, and a_n and b_m are unknown constants.

3. A generalized form of the nonlinear heat conduction equation

3.1 Application of the (G'/G)-expansion method

Introducing a complex variable η defined as Eq. (2.3), Eq. (1.1) becomes an ordinary differential equation, which can be written as

$$-kcU' - ak^2(U^n)'' - U + U^n = 0, \qquad a > 0 \tag{3.1}$$

or, equivalently,

$$-kcU' - ak^2n(n-1)U^{n-2}U'^2 - ak^2nU^{n-1}U'' - U + U^n = 0, \tag{3.2}$$

To get a closed-form analytic solution, we use the transformation (Kabir & Khajeh, 2009; Wazwaz, 2005)

$$U(\eta) = V^{\frac{-1}{n-1}}(\eta), \tag{3.3}$$

which will convert Eq. (3.2) into

$$kc(n-1)V'V^2 + ak^2n(1-2n)V'^2 + ak^2n(n-1)VV'' - (n-1)^2V^3 + (n-1)^2V^2 = 0, \tag{3.4}$$

According to Step 1, considering the homogeneous balance between VV'' and $V'V^2$ in Eq. (3.4) gives

$$2m + 2 = 3m + 1, \tag{3.5}$$

so that

$$m = 1. \tag{3.6}$$

Suppose that the solutions of (3.4) can be expressed by a polynomial in $\left(\frac{G'}{G}\right)$ as follows:

$$V(\eta) = \alpha_0 + \alpha_1\left(\frac{G'}{G}\right), \qquad \alpha_1 \neq 0. \tag{3.7}$$

where α_0 and α_1, are constants which are unknown, to be determined later.

Substituting Eq. (3.7) along with Eq. (2.8) into Eq. (3.4) and collecting all terms with the same power of $\left(\frac{G'}{G}\right)$ together, the left-hand side of Eq. (3.4) is converted into a polynomial in $\left(\frac{G'}{G}\right)$. Equating each coefficient of this polynomial to zero yields a set of simultaneous algebraic equations for $\alpha_0, \alpha_1, k, c, \lambda$ and μ. Solving the system of algebraic equations with the aid of Maple 12, we obtain the following.

Case A: When $\lambda^2 - 4\mu > 0$

Case A-1.

$$\alpha_0 = \frac{1}{2} + \frac{\lambda}{2\sqrt{\lambda^2 - 4\mu}},\ \alpha_1 = \frac{1}{\sqrt{\lambda^2 - 4\mu}},\ k = \pm\frac{n-1}{n\sqrt{a}}.\frac{1}{\sqrt{\lambda^2 - 4\mu}},\ c = \mp\sqrt{a} \tag{3.8}$$

where λ and μ are arbitrary constants.
By using Eq. (3.8), expression (3.7) can be written as

$$V(\eta) = \frac{1}{2} + \frac{\lambda}{2\sqrt{\lambda^2 - 4\mu}} + \frac{1}{\sqrt{\lambda^2 - 4\mu}}\left(\frac{G'}{G}\right), \tag{3.9}$$

Substituting the general solution of (2.9) into Eq. (3.9), we get the generalized travelling wave solution as follows:

$$V(\eta) = \frac{1}{2}\left(1 + \left[\frac{C_1 \sinh\left(\frac{\sqrt{\lambda^2 - 4\mu}}{2}\eta\right) + C_2 \cosh\left(\frac{\sqrt{\lambda^2 - 4\mu}}{2}\eta\right)}{C_1 \cosh\left(\frac{\sqrt{\lambda^2 - 4\mu}}{2}\eta\right) + C_2 \sinh\left(\frac{\sqrt{\lambda^2 - 4\mu}}{2}\eta\right)}\right]\right), \tag{3.10}$$

where

$$\eta = \pm\frac{n-1}{n\sqrt{a}}.\frac{1}{\sqrt{\lambda^2 - 4\mu}}\left(x \pm \sqrt{a}t\right).$$

inserting Eq. (3.10) into Eq. (3.3), it yields the following exact solution of Eq. (1.1):

$$u(x,t) = \left[\frac{1}{2}\left(1 + \left[\frac{C_1 \sinh\left(\pm\frac{n-1}{2n\sqrt{a}}\left(x \pm \sqrt{a}t\right)\right) + C_2 \cosh\left(\pm\frac{n-1}{2n\sqrt{a}}\left(x \pm \sqrt{a}t\right)\right)}{C_1 \cosh\left(\pm\frac{n-1}{2n\sqrt{a}}\left(x \pm \sqrt{a}t\right)\right) + C_2 \sinh\left(\pm\frac{n-1}{2n\sqrt{a}}\left(x \pm \sqrt{a}t\right)\right)}\right]\right)\right]^{\frac{-1}{n-1}} \tag{3.11}$$

in which C_1 and C_2 are arbitrary parameters that can be determined by the related initial and boundary conditions.
Now, to obtain some special cases of the above general solution, we set $C_2 = 0$; then (3.11) leads to the formal solitary wave solution to (1.1) as follows:

$$u(x,t) = \left[\frac{1}{2}\left(1 + \tanh\left(\pm\frac{n-1}{2n\sqrt{a}}\left(x \pm \sqrt{a}t\right)\right)\right)\right]^{\frac{-1}{n-1}}, \tag{3.12}$$

and, when $C_1 = 0$, the general solution (3.11) reduces to

$$u(x,t)=\left[\frac{1}{2}\left(1+\coth\left(\pm\frac{n-1}{2n\sqrt{a}}\left(x\pm\sqrt{a}t\right)\right)\right)\right]^{\frac{-1}{n-1}}, \tag{3.13}$$

Comparing the particular cases of our general solution, Eqs. (3.12) and (3.13), with Wazwaz's results (2005), Eqs. (73) and (74), it can be seen that the results are exactly the same.

Case A-2.

$$\alpha_0=\frac{1}{2}-\frac{\lambda}{2\sqrt{\lambda^2-4\mu}},\ \alpha_1=\frac{-1}{\sqrt{\lambda^2-4\mu}},\ k=\pm\frac{n-1}{n\sqrt{a}}.\frac{1}{\sqrt{\lambda^2-4\mu}},\ c=\pm\sqrt{a} \tag{3.14}$$

Inserting Eq. (3.14) into (3.7) yields

$$V(\eta)=\frac{1}{2}-\frac{\lambda}{2\sqrt{\lambda^2-4\mu}}-\frac{1}{\sqrt{\lambda^2-4\mu}}\left(\frac{G'}{G}\right), \tag{3.15}$$

Substituting the general solution of (2.9) into Eq. (3.15), we obtain

$$V(\eta)=\frac{1}{2}\left(1-\left[\frac{C_1\sinh\left(\frac{\sqrt{\lambda^2-4\mu}}{2}\eta\right)+C_2\cosh\left(\frac{\sqrt{\lambda^2-4\mu}}{2}\eta\right)}{C_1\cosh\left(\frac{\sqrt{\lambda^2-4\mu}}{2}\eta\right)+C_2\sinh\left(\frac{\sqrt{\lambda^2-4\mu}}{2}\eta\right)}\right]\right), \tag{3.16}$$

where $\eta=\pm\frac{n-1}{n\sqrt{a}}.\frac{1}{\sqrt{\lambda^2-4\mu}}\left(x\mp\sqrt{a}t\right)$.

Substituting Eq. (3.16) into the transformation (3.3) leads to the generalized solitary wave solution of Eq. (1.1) as follows:

$$u(x,t)=\left[\frac{1}{2}\left(1-\left[\frac{C_1\sinh\left(\pm\frac{n-1}{2n\sqrt{a}}\left(x\mp\sqrt{a}t\right)\right)+C_2\cosh\left(\pm\frac{n-1}{2n\sqrt{a}}\left(x\mp\sqrt{a}t\right)\right)}{C_1\cosh\left(\pm\frac{n-1}{2n\sqrt{a}}\left(x\mp\sqrt{a}t\right)\right)+C_2\sinh\left(\pm\frac{n-1}{2n\sqrt{a}}\left(x\mp\sqrt{a}t\right)\right)}\right]\right)\right]^{\frac{-1}{n-1}} \tag{3.17}$$

Similarly, to derive some special cases of the above general solution (3.17), we choose $C_2 = 0$; then (3.17) leads to

$$u(x,t)=\left[\frac{1}{2}\left(1-\tanh\left(\pm\frac{n-1}{2n\sqrt{a}}\left(x\mp\sqrt{a}t\right)\right)\right)\right]^{\frac{-1}{n-1}}, \tag{3.18}$$

and, when $C_1 = 0$, the general solution (3.17) reduces to

$$u(x,t) = \left[\frac{1}{2}\left(1 - \coth\left(\pm\frac{n-1}{2n\sqrt{a}}\left(x \mp \sqrt{a}t\right)\right)\right)\right]^{\frac{-1}{n-1}}, \quad (3.19)$$

Validating our results, Eqs. (3.18) and (3.19), with Wazwaz's solutions (2005), Eqs. (71) and (72), we can conclude that the results are exactly the same.

Case B: When $\lambda^2 - 4\mu < 0$

Case B-1.

$$\alpha_0 = \frac{1}{2} + \frac{\lambda i}{2\sqrt{4\mu - \lambda^2}}, \; \alpha_1 = \frac{i}{\sqrt{4\mu - \lambda^2}}, \; k = \pm\frac{n-1}{n\sqrt{a}} \cdot \frac{i}{\sqrt{4\mu - \lambda^2}}, \; c = \mp\sqrt{a} \quad (3.20)$$

Inserting Eq. (3.20) into (3.7) results

$$V(\eta) = \frac{1}{2} + \frac{\lambda i}{2\sqrt{4\mu - \lambda^2}} + \frac{i}{\sqrt{4\mu - \lambda^2}}\left(\frac{G'}{G}\right), \quad (3.21)$$

Substituting the general solution of (2.9) for $\lambda^2 - 4\mu < 0$ into Eq. (3.21), we get

$$V(\eta) = \frac{1}{2}\left(1 + i\left[\frac{-C_1 \sin\left(\frac{\sqrt{4\mu - \lambda^2}}{2}\eta\right) + C_2 \cos\left(\frac{\sqrt{4\mu - \lambda^2}}{2}\eta\right)}{C_1 \cos\left(\frac{\sqrt{4\mu - \lambda^2}}{2}\eta\right) + C_2 \sin\left(\frac{\sqrt{4\mu - \lambda^2}}{2}\eta\right)}\right]\right), \quad (3.22)$$

where

$$\eta = \pm\frac{n-1}{n\sqrt{a}} \cdot \frac{i}{\sqrt{4\mu - \lambda^2}}\left(x \pm \sqrt{a}t\right).$$

Using the following transformation,

$$\begin{aligned} &\eta = i\zeta, \\ &\sinh\left(\frac{\sqrt{4\mu - \lambda^2}}{2}\zeta\right) = -i\sin\left(\frac{\sqrt{4\mu - \lambda^2}}{2}\eta\right), \\ &\cosh\left(\frac{\sqrt{4\mu - \lambda^2}}{2}\zeta\right) = \cos\left(\frac{\sqrt{4\mu - \lambda^2}}{2}\eta\right). \end{aligned} \quad (3.23)$$

in Eq. (3.22) and substituting the result into (3.3), we obtain the following exact solution of Eq. (1.1):

$$u(x,t)=\left[\frac{1}{2}\left(1+\left[\frac{C_1\sinh\left(\pm\frac{n-1}{2n\sqrt{a}}\left(x\pm\sqrt{a}t\right)\right)+C_2 i\cosh\left(\pm\frac{n-1}{2n\sqrt{a}}\left(x\pm\sqrt{a}t\right)\right)}{C_1\cosh\left(\pm\frac{n-1}{2n\sqrt{a}}\left(x\pm\sqrt{a}t\right)\right)+C_2 i\sinh\left(\pm\frac{n-1}{2n\sqrt{a}}\left(x\pm\sqrt{a}t\right)\right)}\right]\right)\right]^{\frac{-1}{n-1}} \quad (3.24)$$

We note that if we set $C_2=0$ and $C_1=0$ in the general solution (3.24), we can recover the solutions (3.12) and (3.13), respectively.

Case B-2.

$$\alpha_0=\frac{1}{2}-\frac{\lambda i}{2\sqrt{4\mu-\lambda^2}},\ \alpha_1=\frac{-i}{\sqrt{4\mu-\lambda^2}},\ k=\pm\frac{n-1}{n\sqrt{a}}\cdot\frac{i}{\sqrt{4\mu-\lambda^2}},\ c=\pm\sqrt{a} \quad (3.25)$$

Inserting Eq. (3.25) into (3.7) leads to

$$V(\eta)=\frac{1}{2}-\frac{\lambda i}{2\sqrt{4\mu-\lambda^2}}-\frac{i}{\sqrt{4\mu-\lambda^2}}\left(\frac{G'}{G}\right), \quad (3.26)$$

Substituting the general solution of (2.9) for $\lambda^2-4\mu<0$ into Eq. (3.26), we have

$$V(\eta)=\frac{1}{2}\left(1-i\left[\frac{-C_1\sin\left(\frac{\sqrt{4\mu-\lambda^2}}{2}\eta\right)+C_2\cos\left(\frac{\sqrt{4\mu-\lambda^2}}{2}\eta\right)}{C_1\cos\left(\frac{\sqrt{4\mu-\lambda^2}}{2}\eta\right)+C_2\sin\left(\frac{\sqrt{4\mu-\lambda^2}}{2}\eta\right)}\right]\right), \quad (3.27)$$

in which $\eta=\pm\frac{n-1}{n\sqrt{a}}\cdot\frac{i}{\sqrt{4\mu-\lambda^2}}\left(x\mp\sqrt{a}t\right)$.

Using the transformation (3.23) into Eq. (3.27), and substituting the result into (3.3) yields the following exact solution:

$$u(x,t)=\left[\frac{1}{2}\left(1-\left[\frac{C_1\sinh\left(\pm\frac{n-1}{2n\sqrt{a}}\left(x\mp\sqrt{a}t\right)\right)+C_2 i\cosh\left(\pm\frac{n-1}{2n\sqrt{a}}\left(x\mp\sqrt{a}t\right)\right)}{C_1\cosh\left(\pm\frac{n-1}{2n\sqrt{a}}\left(x\mp\sqrt{a}t\right)\right)+C_2 i\sinh\left(\pm\frac{n-1}{2n\sqrt{a}}\left(x\mp\sqrt{a}t\right)\right)}\right]\right)\right]^{\frac{-1}{n-1}} \quad (3.28)$$

Similarly, if we set $C_2=0$ and $C_1=0$ in the general solution (3.28), we arrive at the same solutions (3.18) and (3.19), respectively.

3.2 Application of the Exp-function method

In order to determine values of f and p, we balance the term v^3 with vv'' in Eq. (3.4); we have

$$v^3=\frac{c_1\exp(3f\eta)+\cdots}{c_2\exp(3p\eta)+\cdots}, \quad (3.29)$$

$$vv'' = \frac{c_3 \exp([2f + 3p]\eta) + \cdots}{c_4 \exp(5p\eta) + \cdots}, \tag{3.30}$$

where c_i are determined coefficients only for simplicity. Balancing the highest order of the Exp-function in Eqs. (3.29) and (3.30), we have

$$3f + 2p = 2f + 3p, \tag{3.31}$$

which leads to the result

$$p = f, \tag{3.32}$$

Similarly, to determine values of g and q, we have

$$v^3 = \frac{\cdots + d_1 \exp(-3g\eta)}{\cdots + d_2 \exp(-3q\eta)}, \tag{3.33}$$

$$vv'' = \frac{\cdots + d_3 \exp(-[2g + 3q]\eta)}{\cdots + d_4 \exp(-5p\eta)}, \tag{3.34}$$

where d_i are determined coefficients for simplicity. Balancing the lowest order of the Exp-function in Eqs. (3.33) and (3.34), we have

$$3g + 2q = 2g + 3q, \tag{3.35}$$

which leads to the result

$$q = g. \tag{3.36}$$

Case A: $p = f = 1, q = g = 1$

We can freely choose the values of p and q. For simplicity, we set $p = f = 1$ and $q = g = 1$, so Eq. (2.11) reduces to

$$v(\eta) = \frac{a_1 \exp(\eta) + a_0 + a_{-1} \exp(-\eta)}{\exp(\eta) + b_0 + b_{-1} \exp(-\eta)}, \tag{3.37}$$

Substituting Eq. (3.37) into Eq. (3.4), and making use of Maple, we arrive at

$$\begin{aligned} &\frac{1}{A}[c_4 \exp(4\eta) + c_3 \exp(3\eta) + c_2 \exp(2\eta) + c_1 \exp(\eta) + c_0 + c_{-1} \exp(-\eta) \\ &+ c_{-2} \exp(-2\eta) + c_{-3} \exp(-3\eta) + c_{-4} \exp(-4\eta)] = 0, \end{aligned} \tag{3.38}$$

in which

$$A = [\exp(\eta) + b_0 + b_{-1} \exp(-\eta)]^4, \tag{3.39}$$

And the c_n are coefficients of $\exp(n\eta)$. Equating to zero the coefficients of all powers of $\exp(n\eta)$ yields a set of algebraic equations for $a_0, b_0, a_1, a_{-1}, b_{-1}, k$, and c. Solving the system of algebraic equations with the aid of Maple 12, we obtain:

Case 1.

$$a_0 = 0,\ b_0 = 0,\ a_1 = 0,\ a_{-1} = b_{-1},\ b_{-1} = b_{-1},\ k = \pm\frac{n-1}{2n\sqrt{a}},\ c = \pm\sqrt{a} \tag{3.40}$$

Substituting Eq. (3.40) into (3.37) and inserting the result into the transformation (3.3), we get the generalized solitary wave solution of Eq. (1.1) as follows:

$$u(x,t) = \left[\frac{b_{-1}\exp(-\eta)}{\exp(\eta) + b_{-1}\exp(-\eta)}\right]^{\frac{-1}{n-1}}, \tag{3.41}$$

where $\eta = \pm\frac{n-1}{2n\sqrt{a}}\left(x \mp \sqrt{a}t\right)$ and b_{-1} is an arbitrary parameter which can be determined by the initial and boundary conditions.

If we set $b_{-1} = 1$ and $b_{-1} = -1$ in (3.41), the solutions (3.18) and (3.19) can be recovered, respectively.

Case 2.

$$a_0 = 0,\ b_0 = 0,\ a_1 = 1,\ a_{-1} = 0,\ b_{-1} = b_{-1},\ k = \pm\frac{n-1}{2n\sqrt{a}},\ c = \mp\sqrt{a} \tag{3.42}$$

By the same procedure as illustrated above, we obtain

$$u(x,t) = \left[\frac{\exp(\eta)}{\exp(\eta) + b_{-1}\exp(-\eta)}\right]^{\frac{-1}{n-1}}, \tag{3.43}$$

in which $\eta = \pm\frac{n-1}{2n\sqrt{a}}\left(x \pm \sqrt{a}t\right)$ and b_{-1} is a free parameter.

If we set $b_{-1} = 1$ and $b_{-1} = -1$ in (3.43), then it can be easily converted to the same solutions (3.12) and (3.13), respectively.

Case 3.

$$a_1 = 0,\ b_{-1} = 0,\ a_0 = a_0,\ b_0 = b_0,\ a_{-1} = a_0 b_0,\ k = \pm\frac{n-1}{n\sqrt{a}},\ c = \mp n\sqrt{a} \tag{3.44}$$

and consequently we get

$$u(x,t)=\left[\frac{a_0+a_0b_0\exp(-\eta)}{\exp(\eta)+b_0}\right]^{\frac{-1}{n-1}}, \tag{3.45}$$

where $\eta=\pm\frac{n-1}{n\sqrt{a}}\left(x\pm n\sqrt{a}t\right)$ and a_0, b_0, are arbitrary parameters; for example, if we put $b_0=0$, solution (3.45) reduces to

$$u(x,t)=\left[a_0\left(\cosh\eta-\sinh\eta\right)\right]^{\frac{-1}{n-1}}, \tag{3.46}$$

Case 4.

$$\begin{aligned}&a_1=0,\ a_0=a_0,\ b_0=b_0,\ b_{-1}=-a_0(a_0-b_0),\ a_{-1}=-a_0(a_0-b_0),\\&k=\pm\frac{n-1}{n\sqrt{a}},\ c=\pm\sqrt{a}\end{aligned} \tag{3.47}$$

and

$$u(x,t)=\left[\frac{a_0-a_0(a_0-b_0)\exp(-\eta)}{\exp(\eta)+b_0-a_0(a_0-b_0)\exp(-\eta)}\right]^{\frac{-1}{n-1}}, \tag{3.48}$$

where $\eta=\pm\frac{n-1}{n\sqrt{a}}\left(x\mp\sqrt{a}t\right)$ and a_0, b_0 are free parameters; for example, if we set $a_0=1,\ b_0=0$ in Eq. (3.48), it can be easily converted to

$$u(x,t)=\left[\frac{1}{2}(1-\coth\eta+\operatorname{csc}h\eta)\right]^{\frac{-1}{n-1}}, \tag{3.49}$$

Case 5.

$$a_1=1,\ a_0=0,\ b_0=b_0,\ b_{-1}=0,\ a_{-1}=0,\ k=\pm\frac{n-1}{n\sqrt{a}},\ c=\mp\sqrt{a} \tag{3.50}$$

and finally we obtain

$$u(x,t)=\left[\frac{\exp(\eta)}{\exp(\eta)+b_0}\right]^{\frac{-1}{n-1}}. \tag{3.51}$$

in which $\eta=\pm\frac{n-1}{n\sqrt{a}}\left(x\pm\sqrt{a}t\right)$ and b_0 is a free parameter.

Case B: $p=f=2,\ q=g=1$

Since the values of g and f can be freely chosen, we can put $p = f = 2$ and $q = g = 1$, the trial function, Eq. (2.11) becomes

$$v(\eta) = \frac{a_2 \exp(2\eta) + a_1 \exp(\eta) + a_0 + a_{-1} \exp(-\eta)}{\exp(2\eta) + b_1 \exp(\eta) + b_0 + b_{-1} \exp(-\eta)}, \tag{3.52}$$

By the same manipulation as illustrated above, we have the following sets of solutions:

Case 1.

$$a_1 = 0,\ a_0 = a_0,\ b_0 = 0,\ b_{-1} = 0,\ a_{-1} = 0,\ a_2 = 0,\ b_1 = 0,\ k = \pm\frac{n-1}{2n\sqrt{a}},\ c = \mp n\sqrt{a} \tag{3.53}$$

Substituting Eq. (3.53) into (3.52), we have

$$v(\eta) = a_0 \exp(-2\eta), \tag{3.54}$$

Substituting Eq. (3.54) into Eq. (3.3), we get the generalized solitary wave solution of Eq. (1.1) as

$$u(x,t) = [a_0 \exp(-2\eta)]^{\frac{-1}{n-1}}, \tag{3.55}$$

where $\eta = \pm\frac{n-1}{2n\sqrt{a}}(x \pm n\sqrt{a}t)$ and a_0 is an arbitrary parameter. Using the transformation $\begin{cases}\exp(\eta) = \cosh\eta + \sinh\eta \\ \exp(-\eta) = \cosh\eta - \sinh\eta\end{cases}$, Eq. (3.55) yields the same solution (3.46).

Case 2.

$$a_1 = 0,\ a_0 = b_0,\ b_0 = b_0,\ b_{-1} = 0,\ a_{-1} = 0,\ a_2 = 0,\ b_1 = 0,\ k = \pm\frac{n-1}{2n\sqrt{a}},\ c = \pm\sqrt{a} \tag{3.56}$$

Substituting Eq. (3.56) into (3.52), we have

$$v(\eta) = \frac{b_0}{\exp(2\eta) + b_0}, \tag{3.57}$$

Inserting Eq. (3.57) into (3.3), it admits to the generalized solitary wave solution of Eq. (1.1) as follows:

$$u(x,t) = \left[\frac{b_0}{\exp(2\eta) + b_0}\right]^{\frac{-1}{n-1}}, \tag{3.58}$$

where $\eta = \pm\frac{n-1}{2n\sqrt{a}}(x \mp \sqrt{a}t)$ and b_0 is a free parameter.

We note that if we set $a_0 = b_0$ in Eq. (3.48), we can recover the solution (3.58).

Case 3.

$$a_1 = 0,\ a_0 = 0,\ b_0 = 0,\ b_{-1} = b_{-1},\ a_{-1} = b_{-1},\ a_2 = 0,\ b_1 = 0,\ k = \pm\frac{n-1}{3n\sqrt{a}},\ c = \pm\sqrt{a} \tag{3.59}$$

Substituting Eq. (3.59) into (3.52) we obtain

$$v(\eta) = \frac{b_{-1}\exp(-\eta)}{\exp(2\eta) + b_{-1}\exp(-\eta)}, \tag{3.60}$$

and by inserting Eq. (3.60) into (3.3), we get the generalized solitary wave solution of (1.1) as

$$u(x,t) = \left[\frac{b_{-1}\exp(-\eta)}{\exp(2\eta) + b_{-1}\exp(-\eta)}\right]^{\frac{-1}{n-1}}, \tag{3.61}$$

in which $\eta = \pm\frac{n-1}{3n\sqrt{a}}(x \mp \sqrt{a}t)$ and b_{-1} is a free parameter that can be determined by the initial and boundary conditions.

4. The generalized nonlinear heat conduction equation in two dimensions

4.1 Application of the (G'/G)-expansion method

Using the wave variable (2.4) transforms Eq. (1.2) to the ODE

$$-kcU' - 2ak^2(U^n)'' - U + U^n = 0, \qquad a > 0 \tag{4.1}$$

or, equivalently,

$$-kcU' - 2ak^2 n(n-1)U^{n-2}U'^2 - 2ak^2 nU^{n-1}U'' - U + U^n = 0, \tag{4.2}$$

Then we use the transformation (3.3), which will convert Eq. (4.2) into

$$kc(n-1)V'V^2 + 2ak^2 n(1-2n)V'^2 + 2ak^2 n(n-1)VV'' - (n-1)^2V^3 + (n-1)^2V^2 = 0, \tag{4.3}$$

By the same manipulation as illustrated in Section 3.1, we obtain the following sets of solutions.

Case A: When $\lambda^2 - 4\mu > 0$

Case A-1.

$$\alpha_0 = \frac{1}{2} + \frac{\lambda}{2\sqrt{\lambda^2 - 4\mu}},\ \alpha_1 = \frac{1}{\sqrt{\lambda^2 - 4\mu}},\ k = \pm\frac{n-1}{n\sqrt{2a}}.\frac{1}{\sqrt{\lambda^2 - 4\mu}},\ c = \mp\sqrt{2a} \tag{4.4}$$

By the same procedure as illustrated in Case A-1 of Section 3.1, Eqs. (3.9) and (3.10), we can finally find the generalized solitary wave solution of Eq. (1.2) as

$$u(x,y,t)=\left[\frac{1}{2}\left(1+\left[\frac{C_1\sinh\left(\pm\frac{n-1}{2n\sqrt{2a}}\left(x+y\pm\sqrt{2a}t\right)\right)+C_2\cosh\left(\pm\frac{n-1}{2n\sqrt{2a}}\left(x+y\pm\sqrt{2a}t\right)\right)}{C_1\cosh\left(\pm\frac{n-1}{2n\sqrt{2a}}\left(x+y\pm\sqrt{2a}t\right)\right)+C_2\sinh\left(\pm\frac{n-1}{2n\sqrt{2a}}\left(x+y\pm\sqrt{2a}t\right)\right)}\right]\right)\right]^{\frac{-1}{n-1}} \tag{4.5}$$

in which C_1 and C_2 are arbitrary parameters that can be determined by the related initial and boundary conditions.

Now, to obtain some special cases of the above general solution, we set $C_2=0$; then (4.5) leads to

$$u(x,y,t)=\left[\frac{1}{2}\left(1+\tanh\left(\pm\frac{n-1}{2n\sqrt{2a}}\left(x+y\pm\sqrt{2a}t\right)\right)\right)\right]^{\frac{-1}{n-1}}, \tag{4.6}$$

and, when $C_1=0$, the exact solution (4.5) reduces to

$$u(x,y,t)=\left[\frac{1}{2}\left(1+\coth\left(\pm\frac{n-1}{2n\sqrt{2a}}\left(x+y\pm\sqrt{2a}t\right)\right)\right)\right]^{\frac{-1}{n-1}}, \tag{4.7}$$

Comparing the particular cases of our general solution, Eqs. (4.6) and (4.7), with Wazwaz's results (2005), Eqs. (87) and (88), it can be seen that the results are exactly the same.

Case A-2.

$$\alpha_0=\frac{1}{2}-\frac{\lambda}{2\sqrt{\lambda^2-4\mu}},\quad \alpha_1=\frac{-1}{\sqrt{\lambda^2-4\mu}},\quad k=\pm\frac{n-1}{n\sqrt{2a}}\cdot\frac{1}{\sqrt{\lambda^2-4\mu}},\quad c=\pm\sqrt{2a} \tag{4.8}$$

By the similar process as illustrated in Case A-2 of Section 3.1, Eqs. (3.15) and (3.16), we can easily gain the following exact solution of Eq. (1.2):

$$u(x,y,t)=\left[\frac{1}{2}\left(1-\left[\frac{C_1\sinh\left(\pm\frac{n-1}{2n\sqrt{2a}}\left(x+y\mp\sqrt{2a}t\right)\right)+C_2\cosh\left(\pm\frac{n-1}{2n\sqrt{2a}}\left(x+y\mp\sqrt{2a}t\right)\right)}{C_1\cosh\left(\pm\frac{n-1}{2n\sqrt{2a}}\left(x+y\mp\sqrt{2a}t\right)\right)+C_2\sinh\left(\pm\frac{n-1}{2n\sqrt{2a}}\left(x+y\mp\sqrt{2a}t\right)\right)}\right]\right)\right]^{\frac{-1}{n-1}} \tag{4.9}$$

Similarly, to derive some special cases of the above general solution, we choose $C_2=0$; then (4.9) leads to the formal solitary wave solution as follows:

$$u(x,y,t)=\left[\frac{1}{2}\left(1-\tanh\left(\pm\frac{n-1}{2n\sqrt{2a}}\left(x+y\mp\sqrt{2a}t\right)\right)\right)\right]^{\frac{-1}{n-1}}, \tag{4.10}$$

and, when $C_1 = 0$, the general solution (4.9) reduces to

$$u(x,y,t) = \left[\frac{1}{2}\left(1 - \coth\left(\pm\frac{n-1}{2n\sqrt{2a}}\left(x + y \mp \sqrt{2a}t\right)\right)\right)\right]^{\frac{-1}{n-1}}, \tag{4.11}$$

Validating our results, Eqs. (4.10) and (4.11), with Wazwaz's solutions (2005), Eqs. (85) and (86), it can be seen that the results are exactly the same.

Case B: When $\lambda^2 - 4\mu < 0$

Case B-1.

$$\alpha_0 = \frac{1}{2} + \frac{\lambda i}{2\sqrt{4\mu - \lambda^2}},\ \alpha_1 = \frac{i}{\sqrt{4\mu - \lambda^2}},\ k = \pm\frac{n-1}{n\sqrt{2a}}.\frac{i}{\sqrt{4\mu - \lambda^2}},\ c = \mp\sqrt{2a} \tag{4.12}$$

By the same manipulation as illustrated in Case B-1 of Section 3.1, Eqs. (3.21)-(3.23), we can finally obtain the following exact solution:

$$u(x,y,t) = \left[\frac{1}{2}\left(1 + \left[\frac{C_1 \sinh\left(\pm\frac{n-1}{2n\sqrt{2a}}\left(x + y \pm \sqrt{2a}t\right)\right) + C_2 i \cosh\left(\pm\frac{n-1}{2n\sqrt{2a}}\left(x + y \pm \sqrt{2a}t\right)\right)}{C_1 \cosh\left(\pm\frac{n-1}{2n\sqrt{2a}}\left(x + y \pm \sqrt{2a}t\right)\right) + C_2 i \sinh\left(\pm\frac{n-1}{2n\sqrt{2a}}\left(x + y \pm \sqrt{2a}t\right)\right)}\right]\right)\right]^{\frac{-1}{n-1}} \tag{4.13}$$

We note that, if we set $C_2 = 0$ and $C_1 = 0$ in the general solution (4.13), we can recover the solutions (4.6) and (4.7), respectively.

Case B-2.

$$\alpha_0 = \frac{1}{2} - \frac{\lambda i}{2\sqrt{4\mu - \lambda^2}},\ \alpha_1 = \frac{-i}{\sqrt{4\mu - \lambda^2}},\ k = \pm\frac{n-1}{n\sqrt{2a}}.\frac{i}{\sqrt{4\mu - \lambda^2}},\ c = \pm\sqrt{2a} \tag{4.14}$$

Similar to Case B-2 of Section 3.1, we can find the following result:

$$u(x,y,t) = \left[\frac{1}{2}\left(1 - \left[\frac{C_1 \sinh\left(\pm\frac{n-1}{2n\sqrt{2a}}\left(x + y \mp \sqrt{2a}t\right)\right) + C_2 i \cosh\left(\pm\frac{n-1}{2n\sqrt{2a}}\left(x + y \mp \sqrt{2a}t\right)\right)}{C_1 \cosh\left(\pm\frac{n-1}{2n\sqrt{2a}}\left(x + y \mp \sqrt{2a}t\right)\right) + C_2 i \sinh\left(\pm\frac{n-1}{2n\sqrt{2a}}\left(x + y \mp \sqrt{2a}t\right)\right)}\right]\right)\right]^{\frac{-1}{n-1}} \tag{4.15}$$

In particular, if we take $C_2 = 0$ and $C_1 = 0$ in the general solution (4.15), we arrive at the same solutions (4.10) and (4.11), respectively.

4.2 Application of the Exp-function method

By the same manipulation as illustrated in Section 3.2, we obtain the following sets of solutions.

Case 1.

$$a_{-1}=a_{-1},\ a_1=0,\ a_0=0,\ b_0=0,\ b_{-1}=a_{-1},\ k=\pm\frac{n-1}{2n\sqrt{2a}},\ c=\pm\sqrt{2a} \tag{4.16}$$

Substituting Eq. (4.16) into (3.37) and inserting the result into the transformation (3.3), we get the generalized solitary wave solution of Eq. (1.2) as follows:

$$u(x,y,t)=\left[\frac{a_{-1}\exp(-\eta)}{\exp(\eta)+a_{-1}\exp(-\eta)}\right]^{\frac{-1}{n-1}}, \tag{4.17}$$

where $\eta=\pm\frac{n-1}{2n\sqrt{2a}}\left(x+y\mp\sqrt{2a}t\right)$ and a_{-1} is an arbitrary parameter which can be determined by the initial and boundary conditions.

If we set $a_{-1}=1$ and $a_{-1}=-1$ in (4.17), the solutions (4.10) and (4.11) can be recovered, respectively.

Case 2.

$$a_0=0,\ b_0=0,\ a_1=1,\ a_{-1}=0,\ b_{-1}=b_{-1},\ k=\pm\frac{n-1}{2n\sqrt{2a}},\ c=\mp\sqrt{2a} \tag{4.18}$$

By the same process as illustrated in the previous case, we obtain

$$u(x,y,t)=\left[\frac{\exp(\eta)}{\exp(\eta)+b_{-1}\exp(-\eta)}\right]^{\frac{-1}{n-1}}, \tag{4.19}$$

in which $\eta=\pm\frac{n-1}{2n\sqrt{2a}}\left(x+y\pm\sqrt{2a}t\right)$ and b_{-1} is a free parameter.

If we set $b_{-1}=1$ and $b_{-1}=-1$ in (4.19), then it can be easily converted to the same solutions (4.6) and (4.7), respectively.

Case 3.

$$a_{-1}=a_{-1},\ a_1=0,\ a_0=0,\ b_{-1}=0,\ b_0=0,\ k=\pm\frac{n-1}{2n\sqrt{2a}},\ c=\mp n\sqrt{2a} \tag{4.20}$$

and consequently we get

$$u(x,y,t)=\left[a_{-1}\exp(-2\eta)\right]^{\frac{-1}{n-1}}=\left[a_{-1}\left(\cosh 2\eta-\sinh 2\eta\right)\right]^{\frac{-1}{n-1}}, \tag{4.21}$$

where $\eta=\pm\frac{n-1}{2n\sqrt{2a}}\left(x+y\pm n\sqrt{2a}t\right)$ and a_{-1} is an arbitrary parameter.

Case 4.

$$a_1 = 1,\ a_0 = a_0,\ a_{-1} = 0, b_{-1} = b_{-1},\ b_0 = \frac{b_{-1} + a_0{}^2}{a_0},\ k = \pm\frac{n-1}{n\sqrt{2a}},\ c = \mp\sqrt{2a} \tag{4.22}$$

and

$$u(x,y,t) = \left[\frac{\exp(\eta) + a_0}{\exp(\eta) + \frac{b_{-1} + a_0{}^2}{a_0} + b_{-1}\exp(-\eta)}\right]^{\frac{-1}{n-1}}, \tag{4.23}$$

where $\eta = \pm\frac{n-1}{n\sqrt{2a}}\left(x + y \pm \sqrt{2a}t\right)$ and a_0, b_{-1} are free parameters.

Case 5.

$$a_1 = 0, a_{-1} = a_{-1},\ a_0 = a_0, b_{-1} = a_{-1}, b_0 = \frac{a_{-1} + a_0^2}{a_0},\ k = \pm\frac{n-1}{n\sqrt{2a}},\ c = \pm\sqrt{2a} \tag{4.24}$$

and finally we obtain

$$u(x,y,t) = \left[\frac{a_0 + a_{-1}\exp(-\eta)}{\exp(\eta) + \frac{a_{-1} + a_0^2}{a_0} + a_{-1}\exp(-\eta)}\right]^{\frac{-1}{n-1}} \tag{4.25}$$

in which $\eta = \pm\frac{n-1}{n\sqrt{2a}}(x + y \mp \sqrt{2a}t)$ and a_0, a_{-1} are free parameters.

Remark 1. We have verified all the obtained solutions by putting them back into the original equations (1.1) and (1.2) with the aid of Maple 12.

Remark 2. The solutions (3.12), (3.13), (3.18), (3.19), (4.6), (4.7), (4.10), (4.11) have been obtained by the tanh method (Wazwaz, 2005); the other solutions are new and more general solutions for the generalized forms of the nonlinear heat conduction equation.

5. Conclusions

To sum up, the purpose of the study is to show that exact solutions of two generalized forms of the nonlinear heat conduction equation can be obtained by the (G'/G)-expansion and the Exp-function methods. The final results from the proposed methods have been compared and verified with those obtained by the tanh method. New exact solutions, not obtained by the previously available methods, are also found. It can be seen that the Exp-function method yields more general solutions in comparison with the other method. Overall, the results reveal that the (G'/G)-expansion and the Exp-function methods are powerful mathematical tools to solve the nonlinear partial differential equations (NPDEs) in the terms of accuracy and efficiency. This is important, since systems of NPDEs have many applications in engineering.

6. References

Abbasbandy, S. (2010). Homotopy analysis method for the Kawahara equation. *Nonlinear Analysis: Real World Applications*, 11, 1, 307-312.

Bekir, A., Cevikel, C. (2009). New exact travelling wave solutions of nonlinear physical models. Chaos, Solitons and Fractals, 41, 1733–1739.

Borhanifar, A., Kabir, MM. (2009). New periodic and soliton solutions by application of Exp-function method for nonlinear evolution equations. *Journal of Computational & Applied Mathematics*, 229, 158-167.

Borhanifar, A., Kabir, MM., Vahdat Lasemi, M. (2009). New periodic and soliton wave solutions for the generalized Zakharov system and (2+1)-dimensional Nizhnik-Novikov–Veselov system. *Chaos, Solitons & Fractals*, 42, 1646–1654.

Borhanifar, A., Kabir, MM. (2010). Soliton and Periodic solutions for (3+1)-dimensional nonlinear evolution equations by Exp-function method. *Applications and Applied Mathematics: International Journal (AAM)*, 5, 1, 59-69.

Fan, E. (2002). Traveling wave solutions for nonlinear equations using symbolic computation. *Comput. Math. Appl.*, 43, 671–680.

He, JH. (1998). Approximate analytical solution for seepage flow with fractional derivatives in porous media. *Comput. Methods Appl. Mech. Eng.*, 167, 57-68.

He, JH. (2000). A coupling method of a homotopy technique and a perturbation technique for non-linear problems. *Int. J. Non-Linear Mechanics*, 35, 37-43.

He, JH. (2006). New interpretation of homotopy perturbation method. Int. J. Mod. Phys. B, 20, 18, 2561–2568.

He, JH., Wu, XH. (2006). Exp-function method for nonlinear wave equations. *Chaos, Solitons & Fractals*, 30, 3, 700–708.

He, JH., Zhang, LN. (2008). Generalized solitary solution and compacton-like solution of the Jaulent–Miodek equations using the Exp-function method. Physics Letters A, 372, 1044–1047.

He, JH. (2009). Nonlinear science as a fluctuating research frontier. *Chaos, Solitons and Fractals*, 41, 2533–2537.

He, JH., Wu, GC., Austin, F. (2010). The variational iteration method which should be followed. Nonlinear Science Letters A, 1, 1, 1-30.

Kabir, MM., Khajeh, A. (2009). New explicit solutions for the Vakhnenko and a generalized form of the nonlinear heat conduction equations via Exp-function method. *International Journal of Nonlinear Sciences & Numerical Simulation*, 10, 10, 1307-1318.

Kabir, MM., Khajeh, A., Abdi Aghdam, E., Yousefi Koma, A. (2011). Modified Kudryashov method for finding exact solitary wave solutions of higher-order nonlinear equations. Mathematical Methods in the Applied Sciences, 34, 213-219.

Kabir, MM., Borhanifar, A., Abazari, R. (2011). Application of (G'/G)-expansion method to Regularized Long Wave (RLW) equation. Computers and Mathematics with Applications, 61, 8, 2044–2047.

Wang, M., Li, X., Zhang, J. (2008). The (G'/G)-expansion method and traveling wave solutions of nonlinear evolution equations in mathematical physics. *Phys. Lett. A*, 372, 417–423.

Wazwaz, AM. (2004). A sine-cosine method for handling nonlinear wave equations. *Math. Comput. Model*, 40, 499–508.

Wazwaz, AM. (2005). The tanh method for generalized forms of nonlinear heat conduction and Burgers–Fisher equations. *Applied Mathematics and Computation*, 169, 321–338.

Wazwaz, AM. (2006). New solitary wave solutions to the Kuramoto–Sivashinsky and the Kawahara equations. Appl Math. Comput., 182, 1642-1650.

Zedan, HA. (2010). New classes of solutions for a system of partial differential equations by G'/G)-expansion method. Nonlinear Science Letters A, 1, 3, 219–238.

Zhang, S., Wang, W., Tong, J. (2009). A generalized (G'/G)-expansion method and its application to the (2+1)-dimensional Broer-Kaup equations. *Appl. Math. Comput.*, 209, 399-404.

7

Heat Conduction Problems of Thermosensitive Solids under Complex Heat Exchange

Roman M. Kushnir and Vasyl S. Popovych
*Pidstryhach Institute for Applied Problems of Mechanics and Mathematics,
Ukrainian National Academy of Sciences
Ukraine*

1. Introduction

To provide efficient investigations for engineering problems related to heating/cooling process in solids, the effect of thermosensitivity (the material characteristics depend on the temperature) should be taken into consideration when solving the heat conductivity problems (Carslaw & Jaeger, 1959; Noda, 1986; Nowinski, 1962; Podstrihach & Kolyano, 1972). It is important to construct the solutions to the aforementioned heat conduction problems in analytical form. This requirement is motivated, for instance, by the need to solve the thermoelasticity problems for thermosensitive bodies, for which the determined temperature is a kind of input data, and thus, is desired in analytical form.
In general, the model of a thermosensitive body leads to a nonlinear heat conductivity problem. It is mentioned in (Carslaw & Jaeger, 1959) that the exact solutions of such problems can be determined when the temperature or heat flux is given on the surface by assuming the material to be "simply nonlinear" (thermal conductivity λ_t and volumetric volumetric heat capacity c_v depend on the temperature, but the relation, called thermal diffusivity $a = \lambda_t / c_v$, is assumed to be constant). For construction of the solution in this case, it is sufficient to use the Kirchhoff's transformation to obtain the corresponding linear problem for the Kirchhoff's variable. This problem can be solved (Ditkin & Prudnikov, 1975; Galitsyn & Zhukovskii, 1976; Sneddon, 1951) by application of classical methods (separation of variables, integral transformations, etc.). The solutions to the heat conductivity problems for crystal bodies, whose thermal characteristics are proportional to the third power of the absolute temperature, can be constructed in a similar manner for the case of radiation heat exchange with environment.
In the case of complex heat exchange, the Kirchhoff transform makes the heat conductivity problem to be linear only in part. In the heat conductivity problem for the Kirchhoff's variable, the heat conduction equation is nonlinear due to dependence of the thermal diffusivity on the Kirchhoff's variable. The boundary condition of the complex heat exchange is also nonlinear due to a nonlinear expression of the temperature on the surface. Herein we discuss several approaches, developed by the authors for determining temperature distribution in thermosensitive bodies of classical shape under complex (convective, radiation or convective-radiation) heat exchange on the surface (Kushnir & Popovych, 2006, 2007, 2009; Kushnir & Protsiuk, 2009; Kushnir et al., 2001, 2008; Popovych,

1993a, 1993b; Popovych & Harmatiy, 1996, 1998; Popovych & Sulym, 2004; Popovych et al. 2006). Note that the necessity of these investigations is emphasized in (Carslaw & Jaeger, 1959).

2. The step-by-step linearization method for solving the one-dimensional transient heat conductivity problems with simple thermal non-linearity

Let us consider the step-by-step method for determining one-dimensional transient temperature field $t(x,\tau)$, which can be found from the following non-linear heat conduction equation:

$$\frac{1}{x^m}\frac{\partial}{\partial x}\left(x^m \lambda_t(t)\frac{\partial t}{\partial x}\right) = c_v(t)\frac{\partial t}{\partial \tau} - W, \qquad (1)$$

where $\lambda_t(t)$ is the thermal conductivity; $c_v(t)$ is the volumetric heat capacity; $m = 0; 1; 2$ corresponds to Cartesian, cylindrical and spherical coordinate systems, respectively; $a \le x \le b,\ a \ge 0,\ a < b \le \infty$. The thermosensitive body of consideration is made of a material with simple nonlinearity. The density of heat sources W is a function of coordinate x and time τ. Let the surface $x = a$, for instance, is exposed to convective-radiation heat exchange

$$\left[\lambda_t(t)\frac{\partial t}{\partial x} - \alpha_a(t)(t - t_a) - \sigma\varepsilon_a(t)(t^4 - t_a^4)\right]_{x=a} = 0 \qquad (2)$$

with the environment of constant temperature t_a, where $\alpha_a(t)$ is the temperature dependent coefficient of heat exchange between the surface and the environment; $\varepsilon_a(t)$ is the temperature dependent emittance; σ is the Stefan-Boltzmann constant. The surface $x = b$ is heated with constant temperature t_b or constant heat flux q_b:

$$t\big|_{x=b} = t_b \quad \text{or} \quad \lambda_t(t)\frac{\partial t}{\partial x}\bigg|_{x=b} = q_b. \qquad (3)$$

At the initial moment of time, the temperature is uniformly distributed within the body:

$$t\big|_{\tau=0} = t_p \ . \qquad (4)$$

The key point of the solution method for the formulated non-linear heat conductivity problem (1)-(4), which is presented below, consists in the step-by-step linearization involving the Kirchhoff transformation along with linearization of the nonlinear term in the boundary conditions by means of the spline approximation.

By introducing the dimensionless coordinates $\overline{x} = x/l_0$, temperature $T = t/t_0$, and time $\text{Fo} = a\tau/l_0^2$ (the Fourier number), we can present the functional parameters $\lambda_t(t)$, $c_v(t)$, $\alpha_a(t)$, and $\varepsilon_a(t)$ in the form $\chi(t) = \chi_0 \chi^*(T)$, where χ_0 is a reference value and $\chi^*(T)$ stands for the dimensionless function; t_0 is a reference temperature and l_0 is a characteristic dimension. The density of heat sources can be presented as $W = q_0\ q(\overline{x},\text{Fo})$, where q_0 is the dimensional constants, $q(\overline{x},\text{Fo})$ is the dimensionless function describing the time variation of the heat sources. As a result, the problem (1)-(4) takes the form

$$\frac{1}{\overline{x}^m}\frac{\partial}{\partial\overline{x}}\left(\overline{x}^m\lambda_t^*(T)\frac{\partial T}{\partial\overline{x}}\right)=c_v^*(T)\frac{\partial T}{\partial\text{Fo}}-\text{Po}\,q(\overline{x},\text{Fo})\,, \tag{5}$$

$$\left[\lambda_t^*(T)\frac{\partial T}{\partial\overline{x}}-\text{Bi}_a\alpha_a^*(T)(T-T_a)-\text{Sk}_a\varepsilon_a^*(T)(T^4-T_a^4)\right]_{\overline{x}=\overline{a}}=0\,, \tag{6}$$

$$T\big|_{\overline{x}=\overline{b}}=T_b \quad\text{or}\quad \lambda_t^*(T)\frac{\partial T}{\partial\overline{x}}\bigg|_{\overline{x}=\overline{b}}=\text{Ki}_b\,, \tag{7}$$

$$T\big|_{\text{Fo}=0}=T_p\,. \tag{8}$$

Here $P_0=q_0l_0^2/(t_0\lambda_{t0})$ (the Pomerantsev number), $\text{Bi}_a=\alpha_a^{(0)}r_0/\lambda_{t0}$ (the Biot number), $\text{Sk}_a=\sigma\varepsilon_a^{(0)}l_0t_0^3/\lambda_{t0}$ (the Starc number), $\text{Ki}_b=q_bl_0/(t_0\lambda_{t0})$ (the Kirpichev number), $T_b=t_b/t_0$, $T_p=t_p/t_0$.

Let us apply the Kirchhoff's integral transformation (Carslaw & Jaeger, 1959; Noda, 1986; Podstrihach & Kolyano, 1972)

$$\theta=\int_{T_p}^{T}\lambda_t^*(t)dt \tag{9}$$

to the problem (5)-(8). By taking into account the feature of simple nonlinearity ($\lambda_t^*(T)\approx c_v^*(T)$) and expressions $\frac{\partial\theta}{\partial\overline{x}}=\lambda_t^*(T)\frac{\partial T}{\partial\overline{x}}$, $\frac{\partial\theta}{\partial\text{Fo}}=\lambda_t^*(T)\frac{\partial T}{\partial\text{Fo}}\approx c_v^*(T)\frac{\partial T}{\partial\text{Fo}}$, the equation

$$\frac{1}{\overline{x}^m}\frac{\partial}{\partial\overline{x}}\left(\overline{x}^m\frac{\partial\theta}{\partial\overline{x}}\right)=\frac{\partial\theta}{\partial\text{Fo}}-\text{Po}\,q(\overline{x},\text{Fo}) \tag{10}$$

follows from the nonlinear heat conductivity equation (5). The boundary condition of convective-radiation heat exchange (6) can be partially linearized and presented as

$$\left[\frac{\partial\theta}{\partial\overline{x}}-Q_a(T(\theta))\right]_{\overline{x}=\overline{a}}=0\,, \tag{11}$$

where $Q_a(T(\theta))=\text{Bi}_a\alpha_a^*(T(\theta))(T(\theta)-T_a)+\text{Sk}_a\varepsilon_a^*(T(\theta))(T(\theta)^4-T_a^4)$. The boundary conditions (7) and initial condition (8) yield

$$\theta\big|_{\overline{x}=\overline{b}}=\theta_b \quad\text{or}\quad \frac{\partial\theta}{\partial\overline{x}}\bigg|_{\overline{x}=\overline{b}}=\text{Ki}_b\,, \tag{12}$$

$$\theta\big|_{\text{Fo}=0}=0\,, \tag{13}$$

where $\theta_b = \int_{T_p}^{T_b} \lambda_t^*(T)dT$, $T(\theta)$ denotes the temperature expressed through the Kirchhoff's variable and determined for certain $\lambda_t^*(T)$ by means of the integral equation (9).

Application of the Kirchhoff's variable allows us to linearize the nonlinear heat conductivity equation (5) and the second boundary condition (7) completely, whereas the convective-radiation heat exchange condition is linearized in a part. Due to the nonlinear expression $Q_a(T(\theta))$, it is impossible to apply any classical method to solve the boundary problem (10)-(13). Therefore, it is necessary to linearize the boundary condition (11). In (Nedoseka, 1988; Podstrihach & Kolyano, 1972), the convective heat exchange condition has been considered. Therefore, the nonlinear expression $T(\theta)$ is simply replaced by θ. As a result, the nonlinear convective heat exchange condition on θ becomes linear. However, it has been shown in (Kushnir & Popovych, 2009; Popovych, 1993b; Popovych & Harmatiy, 1996) that this unsubstantiated linearization leads to the numerically or physically incorrect results. In our case, when we take into account the radiation constituent (which is nonlinear even for a non-thermosensitive material) and dependence of the heat transfer coefficient and emittance on the temperature, the considered substitution does not provide the complete linearization of the condition (11). Instead, the boundary condition (11) can be linearized by means of interpolation of the nonlinear expression $Q_a(T(\theta))$ by special splines with order 0 or 1. For $\overline{x} = \overline{a}$, the expression $Q_a(T(\theta))$ is a function of Fo only. Let us select a finite set of points Fo_i $(i = \overline{1,n};\ 0 = \text{Fo}_0 < \text{Fo}_1 < \text{Fo}_2 < \ldots < \text{Fo}_n)$, which divides the region of time variation into $n+1$ intervals. Let us construct the spline $S_a^{(0)}(\text{Fo})$ with order 0, whose values coincide with the values of expression $Q_a(\text{Fo}) = Q_a(T(\theta))\big|_{\overline{x}=\overline{a}}$ at $\text{Fo} = \text{Fo}_i$ and

$$S_a^{(0)}(\text{Fo}) = Q_1^{(a)} + \sum_{i=1}^{n-1} (Q_{i+1}^{(a)} - Q_i^{(a)}) S_+(\text{Fo} - \text{Fo}_i)\,; \tag{14}$$

$$Q_i^{(a)} = \text{Bi}_a \alpha_a^*(T_i^{(a)})(T_i^{(a)} - T_a) + \text{Sk}_a \varepsilon_a^*(T_i^{(a)})\left((T_i^{(a)})^4 - T_a^4\right) \tag{15}$$

on the every interval of interpolation. Here $T_i^{(a)}$ $(i \equiv \overline{1,n})$ are the values of temperature $T(\overline{x},\text{Fo})$, which are to be found on the surface $\overline{x} = \overline{a}$ at the moments of time Fo_i (the unknown parameters of spline approximation), $S_+(\cdot)$ denotes the asymmetric unit Heaviside function (H. Korn & T. Korn, 1977).

Having presented the nonlinear expression $Q_a(T(\theta))\big|_{\overline{x}=\overline{a}}$ by spline (14), the boundary condition (11) becomes linear

$$\left.\frac{\partial\theta}{\partial\overline{x}}\right|_{\overline{x}=\overline{a}} - S_a^{(0)}(\text{Fo}) = 0\ . \tag{16}$$

Similarly, the first-order spline $S_a^{(0)}(\text{Fo})$, whose values coincide with values of expression $Q_a(\text{Fo})$ at the points Fo_i and on every segment of decomposition approximates $Q_a(\text{Fo})$ by the linear polynom $P_i^{(a)}(\mathrm{Fo}) = k_i^{(a)}\mathrm{Fo} + b_i^{(a)}$, can be constructed by the abovementioned decomposition. This spline can be written as

$$S_a^{(1)}(\mathrm{Fo}) = P_1^{(a)}(\mathrm{Fo}) + \sum_{i=1}^{n-1}\left(P_{i+1}^{(a)}(\mathrm{Fo}) - P_i^{(a)}(\mathrm{Fo})\right)S_+(\mathrm{Fo} - \mathrm{Fo}_i). \tag{17}$$

Here the coefficients $k_i^{(a)}$, $b_i^{(a)}$ of polynom $P_i^{(a)}(\mathrm{Fo})$ are calculated by formulae

$$k_i^{(a)} = \frac{Q_i^{(a)} - Q_{i-1}^{(a)}}{\mathrm{Fo}_i - \mathrm{Fo}_{i-1}}, \quad b_i^{(a)} = Q_{i-1}^{(a)} - k_i^{(a)}\mathrm{Fo}_{i-1}, \tag{18}$$

where $Q_i^{(a)}$ is expressed through $T_i^{(a)}$ by means of formula (15).

If $Q_a(T(\theta))\big|_{\bar{x}=\bar{a}}$ is expressed as the first-order spline (17), then boundary condition (11) becomes linear

$$\left.\frac{\partial\theta}{\partial\bar{x}}\right|_{\bar{x}=\bar{a}} - S_a^{(1)}(\mathrm{Fo}) = 0. \tag{19}$$

Having solved the obtained linear problem (10), (12), (13), (16) or (10), (12), (13), (19) by means of the classical methods, the Kirchhoff's variable is found as a function of $\bar{x}$ and Fo. Besides the input data of the problem, this variable contains $\mathrm{Fo_i}$ and unknown values $T_i^{(a)} = T(\bar{a}, \mathrm{Fo}_i)$:

$$\theta = \theta(\bar{x}, \mathrm{Fo}, \mathrm{Fo}_1, \ldots, \mathrm{Fo}_n, T_1^{(a)}, \ldots, T_n^{(a)}). \tag{20}$$

By substitution θ into the expression for $T(\theta)$ (for specific dependence $\lambda_t^*(T)$), the formula for determination of the temperature

$$T = f(\bar{x}, \mathrm{Fo}, \mathrm{Fo}_1, \ldots, \mathrm{Fo}_n, T_1^{(a)}, \ldots, T_n^{(a)}) \tag{21}$$

can be obtained at arbitrary point $\bar{x}$ and arbitrary moment of time Fo. For determination of unknown values $T_i^{(a)}$ in the expressions for temperature (21), the collocation method is used. Assuming $\mathrm{Fo} = \mathrm{Fo}_i$ ($i = \overline{1,n}$) in (21), the system of equation for determination $T_i^{(a)}$

$$\begin{cases} T_1 = f(\bar{a}, \mathrm{Fo}_1, T_1^{(a)}), \\ T_2 = f(\bar{a}, \mathrm{Fo}_1, \mathrm{Fo}_2, T_1^{(a)}, T_2^{(a)}), \\ \quad\cdots \\ T_n = f(\bar{a}, \mathrm{Fo}_1, \ldots, \mathrm{Fo}_n, T_1^{(a)}, \ldots, T_n^{(a)}) \end{cases} \tag{22}$$

is obtained. The structure of system (22) makes it possible to determine all unknown values $T_i^{(a)}$, starting from $T_1^{(a)}$. Substitution of values, determined from (22), into the formula (21) completes the solution procedure.

The temperature at given point $\bar{x}$ and moment of time can be calculated in accordance to the following scheme:

a. to divide the time axis by Fo_i and then to determine the approximation parameters $T_i^{(a)}$ from the system (22); as a result, the value of temperature (21) $T_n^{(a)}$ is obtained;

b. to divide every interval in two; to compute the values of parameters $T_i^{(a)}$ for this new time-segmentation and then to obtain the values of temperature $T_{n+1}^{(a)}$;

c. to calculate the difference $T_{n+1}^{(a)} - T_n^{(a)}$. If $|T_{n+1} - T_n| < \varepsilon$, where ε is the accuracy, then the calculation is over. Otherwise, we shall return to the stage *b*.

The temperature can be computed with any given accuracy ε for arbitrary segmentation of the time axis. However, the increasing of number of time-segments decreases the convergence of the proposed scheme. An appropriate choice of the initial moment of time can be done by means of the estimated 'a priory' time-dependence of the temperature on the surface $\bar{x} = \bar{a}$. We can also use the solution of corresponding boundary value problem for the body of the same shape with constant characteristics. Then the initial choice for values Fo_i can be used as the appropriate one for the thermosensitive body.

The method of step-by-step linearization is applicable for determination of the temperature fields in thermosensitive plates, half-space, solid and hollow cylinders or spheres, space with cylindrical or spherical cavities, on the surfaces of which, the conditions of convective, radiation or convective-radiation heat exchange may be given. This method has been efficiently used for solving the two-dimensional steady problem in thermosensitive body.

3. Method of linearizing parameters

The method of step-by-step linearization makes it possible to determine the solutions to the two-dimensional heat conductivity problems in thermosensitive bodies with simple nonlinearity, when the nonlinear term in the condition of complex heat exchange for the Kirchhoff's variable depends on one (spatial or time) variable only. In this section, we consider an efficient method for solving the steady-state and transient heat conductivity problems of arbitrary dimension those describe the propagation of heat in thermosensitive bodies with simple nonlinearity under the convective heat exchange with environment.

Let the body occupies region D with surface S. The surface (whole or a part) is subjected to the convective heat exchange with the environment of temperature t_p. From the moment of time $\tau > 0$, the heat sources $W(x,y,z,\tau)$ are acting in the body. The temperature in the body shall be determined from the following heat conduction equation:

$$\operatorname{div}\left(\lambda_t(t)\operatorname{grad} t\right) = c_v(t)\frac{\partial t}{\partial \tau} - W \tag{23}$$

and the boundary

$$\left[\lambda_t(t)\frac{\partial t}{\partial n} + \alpha(t - t_c)\right]_S = 0 \tag{24}$$

and initial

$$t\,\big|_{\tau=0} = t_p \tag{25}$$

conditions, where α is the constant heat transfer coefficient; n is the external normal to surface S.

By making use of the above-introduced presentation for the material characteristics, heat sources, and dimensionless variables, the boundary value problem (23)-(25) can be reduced to the dimensionless form. After application of the Kirchhoff's transformation, the following boundary value problem for variable θ

$$\operatorname{div}\operatorname{grad}\theta = \frac{\partial\theta}{\partial \text{Fo}} - \text{Po}\, q(X,Y,Z,\text{Fo}), \tag{26}$$

$$\left[\frac{\partial\theta}{\partial\overline{n}} + \text{Bi}\left(T(\theta) - T_c\right)\right]_S = 0, \tag{27}$$

$$\theta\big|_{\text{Fo}=0} = 0 \tag{28}$$

is obtained, where $X = x/l_0$, $Y = y/l_0$, $Z = z/l_0$ are dimensionless coordinates; $\overline{n} = n/l_0$, $q(X,Y,Z,\text{Fo})$ is the dimensionless function of heat sources. As a result, the initial problem is partially linearized, meanwhile the condition (27) remains nonlinear. The latter conditions have been obtained from the conditions of convective heat exchange due to nonlinear expression $T(\theta)$ on the surface S. For solving the problem (26)-(28) by using an analytical method, it is necessary to linearize this condition. Let us prove the possibility of such linearization.

Consider the simplest case of linear dependence of heat conductivity coefficient on the temperature:

$$\lambda_t(t) = \lambda_{to}\ \lambda_t^*(T) = \lambda_{to}\left[1 + k(T - T_p)\right], \tag{29}$$

where k is a constant. From the equation (9), the formula

$$\theta = (T - T_p) + \frac{k}{2}(T - T_p)^2 \tag{30}$$

follows, where

$$T(\theta) = k^{-1}(\sqrt{1 + 2k\theta} - 1) + T_p\,. \tag{31}$$

From the physical standpoint, the square root is chosen to be positive. After substitution of the equation (31) into the boundary condition (27), the last one takes the form

$$\left[\frac{\partial\theta}{\partial\overline{n}} + \text{Bi}\left(\frac{\sqrt{1+2k\theta} - 1}{k} + T_p - T_c\right)\right]_S = 0\,. \tag{32}$$

Be decomposing the square root in (32) into the series and restricting this series with two terms, the boundary condition

$$\left[\frac{\partial\theta}{\partial\overline{n}} + \text{Bi}\left(\theta - (T_c - T_p)\right)\right]_S = 0 \tag{33}$$

is obtained. The solution of equation (26) with boundary conditions (28), (33) is an approximate solution to the boundary value problem (26), (28), (32). To determine the exact solution, the equation (26) is to be solved under initial condition (28) and the following linear boundary condition

$$\left[\frac{\partial\theta}{\partial\overline{n}} + \mathrm{Bi}\Big((1+\kappa)\theta - (T_c - T_p)\Big)\right]_S = 0 \tag{34}$$

instead of the nonlinear condition (32), where κ is an unknown constant (linearized parameter). Note that the boundary condition (34) coincides at $\kappa = 0$ with the condition (33). Since the problem (26), (28), (34) is linear, the appropriate classical analytical method can be used for its solution. In addition to the original parameters of the problem ($\mathrm{Po}, \mathrm{Bi}, T_c, T_p$, dimensions of the body, coordinates and time), the solution involves the unknown linearized parameter κ:

$$\theta = \theta(X, Y, Z, \mathrm{Fo}, \kappa)\,. \tag{35}$$

For an arbitrary value of κ, the solution (35) meets the equation (26) and the initial condition (28). In order the solution (35) to satisfy the nonlinear conditions (32) and (34), the parameter κ is to be the solution of the equation

$$\left[\frac{\sqrt{1+2k\theta}-1}{k} - (1+\kappa)\theta\right]_S = 0\,.$$

After some transformations, this equation can be given as

$$\theta\big|_S = -\frac{2\kappa}{k(1+\kappa)^2}\,. \tag{36}$$

This equation holds for every moment of time Fo. After the paramenter κ is found, we substitute it into (35). In such manner, the expression for Kirchhoff's variable is obtained. The temperature in the body is then calculated by means of the relation (31).

Note that the boundary condition (34) can be represented as

$$\left[\frac{\partial\theta}{\partial\overline{n}} + \mathrm{Bi}^*(\theta - T_c^*)\right]_S = 0\,, \tag{37}$$

where $\mathrm{Bi}^* = \mathrm{Bi}(1+\kappa)$; $T_c^* = (T_c - T_p)/(1+\kappa)$. This condition can be interpreted as a condition of convective heat exchange with certain parameters (the Biot number Bi^* and the temperature T_c^* of external environment) depending on the unknown parameter κ.

The equation (36) is nonlinear. It provides analytical solutions only for some cases of steady-state problems with substantional use of the numerical methods. Therefore, these solutions can be regarded as analytico-numerical solutions.

Let us consider the non-linear dependence of the heat conductivity coefficient on the temperature. For linearization of the boundary condition (27), we shall find the Kirhoff's variable for the case when the surface temperature of the thermosensitive body is equal to the surface temperature of the body with constant characteristics. The latter temperature is to be found from the problem:

$$\operatorname{div}\operatorname{grad} T_H = \frac{\partial T_H}{\partial \mathrm{Fo}} - \mathrm{Po}\, q(X,Y,Z,\mathrm{Fo}), \tag{38}$$

$$\left[\frac{\partial T_H}{\partial \bar{n}} + \mathrm{Bi}((T_H - T_c)\right]_S = 0 , \tag{39}$$

$$T_H\big|_{\mathrm{Fo}=0} = T_p , \tag{40}$$

where $T_H = t_H/t_o$, t_H is temperature of the body with constant characteristics.
By subtraction equations of the problem (26)-(28) from corresponding equation of the problem (38)-(40) and taking into account that $T(\theta)\big|_S = T_H\big|_S$, we obtain:

$$\operatorname{div}\operatorname{grad}(T_H - \theta) = \frac{\partial (T_H - \theta)}{\partial \mathrm{Fo}}, \tag{41}$$

$$\left.\frac{\partial (T_H - \theta)}{\partial \bar{n}}\right|_S = 0, \tag{42}$$

$$(T_H - \theta)\big|_{\mathrm{Fo}=0} = T_p . \tag{43}$$

The boundary value problem (41)-(43) is a problem of heat conductivity in the body with the surface S and uniform initial temperature T_p. The heat sources are absent and the boundary of the body is thermoinsulated. The evident solution of this problem is $T_H - \theta = T_p$. Consequently, if in the problem (26)-(28) for the Kirchhoff's variable the surface temperature for the thermosensitive body is replaced with the surface temperature for the body with constant characteristics (whose thermal diffusivity is equal to the thermal diffusivity of thermosensitive body and the heat conductivity coefficient is equal to the reference value of the heat conductivity coefficient λ_{t0}), then $\theta = T_H - T_p$.

Thus, if the surface temperature $T(\theta)\big|_S$ of the thermosensitive body in the condition (27) is equal to the corresponding temperature of the body with constant characteristics, then the boundary value problem for the Kirchhoff's variable θ should be solved with the condition (33). Then the solution of this problem presents the difference of the temperature in the same-shape body with constant characteristics and the initial temperature:

$$\theta = T_H - T_p . \tag{44}$$

As it was mentioned above, the substitution of $T(\theta)$ for $\theta + T_p$ in the case of linear dependence of the heat conductivity coefficient on the temperature is equivalent to keeping only two terms in the series, into which the square root in expression for the temperature through the Kirhoff's variable has been decomposed. This linearization does not guarantee a sufficient solution approximation. To overcome this difficulty, we consider the boundary value problem for the variable θ with the linear condition (37) instead of the nonlinear condition (27), which involves an additional parameter κ. Having solved the obtained linear problem, the Kirhoff's variable θ is found as a function of the coordinates and parameter κ. The parameter κ should be chosen in the way to satisfy the nonlinear condition (27) with any given accuracy. Thus for determination of the temperature field in the body with simple nonlinearity for arbitrary temperature dependence of heat conductivity coefficient under convective heat exchange between the surface and environment, the corresponding solution of the nonlinear heat conductivity problem can be determined by following the proposed algorithm of the method of linearized parameters:

- to present the problem in dimensionless form;
- to linearize the problem in part by using integral Kihhoff transformation;
- to linearize the problem completely by linearizing the nonlinear condition on Kirchhoff's variable θ obtained from condition of convective heat exchange due to replacement of nonlinear expression $T(\theta)$ by $(1+\kappa)\theta+T_p$ with unknown parameter κ;
- to solve the obtained linear boundary value problem for variable θ by means of an appropriate classical method;
- to satisfy with given accuracy the nonlinear condition for variable θ by using the parameter κ;
- to determine the temperature using the obtained Kirchhoff's variable.

The main feature of the method of linearizing parameters consists in a possibility to obtain the solution of linearized boundary value problem for the Kirchhoff's variable in a thermosensitive body by solving the heat conductivity problem in the body with constant characteristics under convective heat exchange. This solution is obtained from (44) by setting $\mathrm{Bi}^* = \mathrm{Bi}(1+\kappa)$ and $T_c^* = (T_c - T_p)/(1+\kappa)$ instead of T_H Bi and T_c, respectively.

4. The method of linearizing parameters for the steady-state heat conduction problems in piecewise-homogeneous thermosensitive bodies

Determination of the temperature fields in piecewise-homogeneous bodies subjected to intensive thermal loadings is an initial stage that precedes the determination of steady-state or transient thermal stresses in the mentioned bodies. Let us assume that the elements of piecewise-homogeneous body are in the ideal thermal contact and the limiting surface is under the condition of complex heat exchange with environment. Mathematical model for determination of the temperature fields in such structures leads to the coupled problem for a set of nonlinear heat conduction equations with temperature-dependent material characteristics in the coupled elements. By making use of the Kirhoff's integral transformation for each element by assuming the thermal conductivity to be constants, the problem can be partially linearized. The nonlinearities remain due to the thermal contact conditions on the interfaces and the conditions of complex heat exchange on the surfaces. To obtain an analytical solution to the coupled problem for the Kirchhoff's variable, it is necessary to linearize this problem. The possible ways of such a linearization and, thus, determination of the general solution to the heat conduction problems in piecewise-homogeneous bodies are considered below in this section.

Let us adopt the method of linearizing parameters to solution of the steady-state heat conduction problems for coupled bodies of simple shape, for instance, n-layer thermosensitive cylindrical pipe. The pipe is of inner and outer radii $r=r_0$ and $r=r_n$, respectively, with constant temperatures t_b and t_H on the inner and outer surfaces. The layers of different temperature-dependent heat conduction coefficients are in the ideal thermal contact. The cylindrical coordinate system r,ϕ,z is chosen with z-axis coinciding with the axis of pipe. The temperature field in this pipe can be determined from the set of heat conduction equations

$$\frac{1}{r}\frac{d}{dr}\left[r\lambda_t^{(i)}(t_i)\frac{dt_i}{dr}\right]=0,\ i=\overline{1,\ n}\ , \tag{45}$$

with the boundary conditions

$$t_1\big|_{r=r_0}=t_b,\ \ t_n\big|_{r=r_n}=t_H\ , \tag{46}$$

$$t_i=t_{i+1},\ \lambda_t^{(i)}(t_i)\frac{dt_i}{dr}=\lambda_t^{(i+1)}(t_{i+1})\frac{dt_{i+1}}{dr}\ ,\ r=r_i,\ i=\overline{1,\ n-1}\ , \tag{47}$$

where $\lambda_t^{(i)}(t_i)$ denotes the heat conduction coefficient of the layers. We introduce the dimensionless values $T_i=t_i/t_0$, $\rho=r/r_0$ and $\lambda_t^{(i)}(t_i)=\lambda_{t0}^{(i)}\lambda_t^{*(i)}(T_i)$, where the constituents with the indices "0" are dimensional constants and the asterisked terms are dimensionless functions, t_0 is the reference temperature. In the dimensionless form, the problem (45)-(47) appears as

$$\frac{1}{\rho}\frac{d}{d\rho}\left[\rho\lambda_t^{*(i)}(T_i)\frac{dT_i}{d\rho}\right]=0,\ i=\overline{1,\ n}\ , \tag{48}$$

$$T_1\big|_{\rho=1}=T_b,\ \ T_n\big|_{\rho=\rho_n}=T_H\ , \tag{49}$$

$$T_i=T_{i+1},\ \lambda_{t0}^{(i)}\lambda_t^{*(i)}(T_i)\frac{dT_i}{d\rho}=\lambda_{t0}^{(i+1)}\lambda_t^{*(i+1)}(T_{i+1})\frac{dT_{i+1}}{d\rho}\ ,\ \rho=\rho_i,\ i=\overline{1,\ n-1}. \tag{50}$$

Consider the heat conduction coefficients in the form of linear dependence on the temperature $\lambda_t^{(i)}(t_i)=\lambda_{t0}^{(i)}(1+k_iT_i)$, where k_i are constants. By introducing the Kirchhoff's variable

$$\theta_i=\int_0^{T_i}\lambda_t^{*(i)}(T)dT \tag{51}$$

in each layer, the following problem on Kirchhoff's variable

$$\frac{1}{\rho}\frac{d}{d\rho}\left(\rho\frac{d\theta_i}{d\rho}\right)=0,\ i=\overline{1,\ n}\ , \tag{52}$$

$$\theta_1\big|_{\rho=1} = \theta_b,\ \theta_n\big|_{\rho=\rho_n} = \theta_{\text{н}}, \tag{53}$$

$$\left.\begin{aligned} &\left(\sqrt{1+2k_i\theta_i}-1\right)/k_i = \left(\sqrt{1+2k_{i+1}\theta_{i+1}}-1\right)/k_{i+1}, \\ &\lambda_{t0}^{(i)}\frac{d\theta_i}{d\rho} = \lambda_{t0}^{(i+1)}\frac{d\theta_{i+1}}{d\rho} \end{aligned}\right\} \text{ at } \rho=\rho_i,\, i=\overline{1,n-1}, \tag{54}$$

is obtained from the problem (48)-(50). Here $\theta_b = \int\limits_0^{T_b} \lambda_t^{*(1)}(T)dT$; $\theta_{\text{н}} = \int\limits_0^{T_{\text{н}}} \lambda_t^{*(n)}(T)dT$.

The initially nonlinear heat conduction problem is partially linearized due to application of the Kirchhoff's variables. However, the conditions for temperature, that reflects the temperature equalities of the neighbouring layers, remain nonlinear (the first group of conditions (54)). By integrating the set of equations (52) with boundary conditions (53) and contact conditions (54), the set of transcendent equation can be obtained for determination of constant of integration. This set can be solved numerically. The efficiency of numerical methods depends on the appropriate initial approximation. Unfortunately, it is very complicated to determine the definition domain for the solution of this set of equations and thus to present a constructive algorithm for determination of the initial approximation.

The possible way around this problem is to decompose the square root in the first conditions (54) into series by holding only two terms. Then, instead of mentioned conditions, the following approximated conditions are obtained:

$$\theta_i = \theta_{i+1} \text{ at } \rho=\rho_i,\, i=\overline{1,n-1}. \tag{55}$$

Application of the conditions (55), instead of exact ones, separates the interfacial conditions. This fact allows us to consider the boundary problem (52)–(54) replacing the conditions (54) by the following ones:

$$(1+\kappa_i)\theta_i = (1+\kappa_{i+1})\theta_{i+1} \text{ at } \rho=\rho_i,\, i=\overline{1,\ n-1}, \tag{56}$$

where κ_i are unknown constants (linearizing parameters). By substitution

$$\theta_i^* = (1+\kappa_i)\theta_i, \tag{57}$$

we obtain

$$\frac{1}{\rho}\frac{d}{d\rho}\left(\rho\frac{d\theta_i^*}{d\rho}\right) = 0, \tag{58}$$

$$\theta_1^*\big|_{\rho=1} = \theta_b^*,\ \theta_n^*\big|_{\rho=\rho_n} = \theta_n^*, \tag{59}$$

$$\theta_i^* = \theta_{i+1}^*,\ \gamma_i\frac{d\theta_i^*}{d\rho} = \gamma_{i+1}\frac{d\theta_{i+1}^*}{d\rho} \text{ at } \rho=\rho_i,\, i=\overline{1,\ n-1}, \tag{60}$$

where $\theta_b^* = (1+\kappa_i)\theta_b$; $\theta_{\text{н}}^* = (1+\kappa_n)\theta_{\text{н}}$; $\gamma_i = \lambda_{t0}^{(i)}/(1+\kappa_i)$, $i=\overline{1,n-1}$.

It can be shown (Podsdrihach et al., 1984) that the boundary value problem (58)-(60) is equivalent to the problem

$$\frac{1}{\rho}\frac{d}{d\rho}\left(\rho\gamma(\rho)\frac{d\theta^*}{d\rho}\right)=0, \tag{61}$$

$$\theta_1^*\Big|_{\rho=1}=\theta_b^*,\quad \theta_n^*\Big|_{\rho=\rho_n}=\theta_n^*, \tag{62}$$

where $\gamma(\rho)=\gamma_1+\sum_{j=1}^{n-1}(\gamma_{j+1})-\gamma_j)S_-(\rho-\rho_j)$. After integration of the equation (61), we obtain

$$\theta^*=C_1\left[\frac{\ln\rho}{\gamma(\rho)}-\sum_{j=1}^{n-1}\left(\frac{1}{\gamma_{j+1}}-\frac{1}{\gamma_j}\right)\ln\rho_j S_-(\rho-\rho_j)\right]+C_2. \tag{63}$$

Substitution of (63) into (62) yields

$$\theta^*=\frac{\theta_{\text{H}}^*-\theta_b^*}{\dfrac{\ln\rho_n}{\gamma_n}-\sum_{j=1}^{n-1}\left(\dfrac{1}{\gamma_{j+1}}-\dfrac{1}{\gamma_j}\right)\ln\rho_j}\left[\frac{\ln\rho}{\gamma(\rho)}-\sum_{j=1}^{n-1}\left(\frac{1}{\gamma_{j+1}}-\frac{1}{\gamma_j}\right)\ln\rho_j S_-(\rho-\rho_j)\right]+\theta_b^*, \tag{64}$$

or

$$\theta_i^*=A_i^*\ln\rho+B_i^*, \tag{65}$$

where

$$A_i^*=\left(\theta_{\text{H}}^*-\theta_b^*\right)\left(\gamma_j\left[\frac{\ln\rho_n}{\gamma_n}-\sum_{j=1}^{n-1}\left(\frac{1}{\gamma_{j+1}}-\frac{1}{\gamma_j}\right)\ln\rho_j\right]\right)^{-1};\ B_i^*=\theta_b^*-A_i^*\gamma_i\sum_{j=1}^{i-1}\left(\frac{1}{\gamma_{j+1}}-\frac{1}{\gamma_j}\right)\ln\rho_j.$$

For the Kirchhoff's variables, we have

$$\theta_i=A_i\ln\rho+B_i, \tag{66}$$

where

$$A_i=\left((1+\kappa_n)\theta_{\text{H}}-(1+\kappa_1)\theta_b\right)\left(\lambda_{t0}^{(i)}\left[\frac{1+\kappa_n}{\lambda_{t0}^{(n)}}\ln\rho_n-\sum_{j=1}^{n-1}\left(\frac{1+\kappa_{j+1}}{\lambda_{t0}^{(j+1)}}-\frac{1+\kappa_j}{\lambda_{t0}^{(j)}}\right)\ln\rho_j\right]\right)^{-1};$$

$$B_i=\frac{1}{1+\kappa_i}\left[(1+\kappa_1)\theta_b-A_i\lambda_{t0}^{(i)}\sum_{j=1}^{i-1}\left(\frac{1+\kappa_{j+1}}{\lambda_{t0}^{(j+1)}}-\frac{1+\kappa_j}{\lambda_{t0}^{(j)}}\right)\ln\rho_j\right].$$

Besides the initial data, the solution (66) contains n arbitrary constants κ_i and satisfies the equation (52), boundary conditions (53) and the second group of the contact conditions (54).

The linearized parameters κ_i will be selected to satisfy the first group of the conditions (54). By assuming that one of the linearizing parameters κ_i, for instance, is equal to zero, the following set of $n-1$ equations can be obtained

$$\left(\sqrt{1+2k_i\,\theta_i\big|_{\rho=\rho_i}}-1\right)/\,k_i=\left(\sqrt{1+2k_{i+1}\,\theta_{i+1}\big|_{\rho=\rho_i}}-1\right)/\,k_{i+1},\quad i=\overline{1,n-1} \tag{67}$$

for determination of the rest $n-1$ linearizing parameters. The solution should be found in a neighborhood of zero. From the set (67), we determine the values of linearization parameters and thus the Kirchhoff's variables. Then the temperature in layers is

$$T_i=k_i^{-1}(\sqrt{1+2k_i\theta_i}-1)\,. \tag{68}$$

For example, we consider the two-layer pipe $(n=2)$. The Kirchhoff's variables for this case are expressed as

$$\theta_1=K_\lambda\frac{(1+\kappa)\theta_{\text{н}}-\theta_b}{(1+\kappa)\ln\dfrac{\rho_2}{\rho_1}+K_\lambda\ln\rho_1}\ln\rho+\theta_b\,,\;\theta_2=\frac{(1+\kappa)\theta_{\text{н}}-\theta_b}{(1+\kappa)\ln\dfrac{\rho_2}{\rho_1}+K_\lambda\ln\rho_1}\ln\frac{\rho}{\rho_2}+\theta_{\text{н}}\,, \tag{69}$$

where $K_\lambda=\lambda_{t0}^{(2)}/\lambda_{t0}^{(1)}$; κ_1 is equal to zero, and κ_2 is denoted as κ. The value of κ shall be obtained from the equation

$$\frac{1}{k_1}\left[\sqrt{1+2k_1\left(K_\lambda\frac{(1+\kappa)\theta_{\text{н}}-\theta_b}{(1+\kappa)\ln\rho_2/\rho_1+K_\lambda\ln\rho_1}\ln\rho_1+\theta_b\right)}-1\right]=$$

$$=\frac{1}{k_2}\left[\sqrt{1+2k_2\left(\frac{(1+\kappa)\theta_{\text{н}}-\theta_b}{(1+\kappa)\ln\rho_2/\rho_1+K_\lambda\ln\rho_1}\ln\frac{\rho_1}{\rho_2}+\theta_{\text{н}}\right)}-1\right]. \tag{70}$$

If the heat conduction coefficients of the layers $\lambda_t^{(i)}$ $(i=1,2)$ are constants, then the temperature in each layer is determined by formula

$$T_1=NK_\lambda^*\ln\rho+T_b,\;T_2=N\ln\rho+T_{\text{н}}\,, \tag{71}$$

where $N=\left(T_{\text{н}}-T_b\right)\big/\left((K_\lambda^*-1)\;\ln\rho_1+\ln\rho_2\right),\,K_\lambda^*=\lambda_t^{(2)}/\lambda_t^{(1)}$.

Let the first layer of thickness $e-1\,(\rho_1=e)$ is made of steel C12 and the second layer of thickness $e^2-e(\rho_2=e^2)$ is made of steel C8 (Sorokin et al., 1989). Let $t_b=700^\circ\text{C}$, $t_{\text{н}}=0^\circ\text{C}$, and $t_0=t_b$. The heat conduction coefficients in the temperature range $0...700^\circ\text{C}$ are given in the form of linear relations: $\lambda_t^{(1)}=47.5(1-0.37T)$ $[W/(m\cdot K)]$, $\lambda_t^{(2)}=$ $64.5(1-0.49\,\text{T})$ $[W/(m\cdot K)]$. Then $k_1=-0.37$, $\lambda_{t0}^{(1)}=47.5$, $k_2=-0.49$, $\lambda_{t0}^{(2)}=64.5$, $K_\lambda=1.36$, $T_b=1$, $T_{\text{н}}=0$, $\theta_b=0.815$, $\theta_{\text{н}}=0$. At reference values, the linearized parameter κ (determined from equation (70)), is equal to 0.0249.

ρ	Thermosensitive layers				Layers with constant characteristics			
	$\kappa = 0{,}0249$		$\kappa = 0$		$\lambda_{tc}^{(1)}\ \lambda_{tc}^{(2)}$		$\lambda_{t0}^{(1)}\ \lambda_{t0}^{(2)}$	
	T	t°C	T	t°C	T	t°C	T	t°C
1	1	700	1	700	1	700	1	700
1,34	0,7945	556,1	0,7924	554,7	0,8369	585,9	0,8314	582,0
1,69	0,6500	455,0	0,6466	452,6	0,7077	495,4	0,6978	488,5
2,03	0,5395	377,7	0,5352	374,6	0,6055	423,9	0,5922	414,6
2,37	0,4506	315,4	0,4455	311,9	0,5193	363,5	0,5031	352,1
$e-0$	0,3764	263,5	0,3707	259,5	0,4429	310,0	0,4241	296,9
$e+0$	0,3765	263,6	0,3810	266,7	0,4429	310,0	0,4241	296,9
3,65	0,2570	179,9	0,2600	182,0	0,3124	218,6	0,2991	209,4
4,59	0,1701	119,1	0,1720	120,4	0,2109	147,6	0,2019	141,3
5,52	0,1023	71,6	0,1037	72,4	0,1292	90,4	0,1237	86,6
6,49	0,0468	32,8	0,0473	33,1	0,0602	42,1	0,0576	40,4
e^2	0	0	0	0	0	0	0	0

Table 1. Distribution of temperature in a two layer pipe along its radius

Table 1 presents the temperature values in two-layer pipe versus its radius. In the first four columns, the values of dimensionless and real temperature T and t, respectively, are given; the first and second columns present the temperature values, obtained by method of linearizing parameters (formulae (68)-(70)); the third and fourth columns present the approximate values of the temperature, obtained by holding only two terms in the series into which the square roots in the first group of the conditions (54) were decomposed (formulae (68), (69) at $\kappa = 0$). The maximum difference between the exact and approximate values of temperature falls within 1.5%. But the approximate solution has a gap $7.2°\mathrm{C}$ on the interface. This fact shows that the condition of the ideal thermal contact is not satisfied, which is physically improper result. In the last four columns, the values of dimensionless and real temperature in the pipe with constant thermal characteristics are presented. The values in the fifth and sixth columns describe the case when the heat conduction coefficients have the mean value in the temperature region 0...700°C i.e. $\lambda_{tc}^{(1)} = \dfrac{1}{700}\int_0^{700} \lambda_t^{(1)}(t)dt = 38.7$ $[W/(m\cdot K)]$, $\lambda_{tc}^{(2)} = \dfrac{1}{700}\int_0^{700} \lambda_t^{(2)}(t)dt = 48.7$ $[W/(m\cdot K)]$; the seventh and eighth columns present the maximum values of the heat conduction coefficients in the considered temperature range $\lambda_t^{(1)} = \lambda_{t0}^{(1)}$, $\lambda_t^{(2)} = \lambda_{t0}^{(2)}$. Thus, the maximum difference between the values of the temperature computed for the mean values of the heat conduction coefficients is about $15\%\ (\approx 48°\mathrm{C})$. If the temperature is computed for the maximum values of the heat conduction coefficients, this difference is about $10\%\ (\approx 37°\mathrm{C})$.

To simplify the explanation of the linearized parameters method for solving the heat conductivity problem in the coupling thermal sensitive bodies, the constant temperatures on bounded surfaces of piecewise-homogeneous bodies were considered. If the conditions of convective heat exchange are given, then the final linearization of the obtained nonlinear conditions on Kirchhoff's variables may be fulfilled using the method of linearizing parameters.

The method of linearizing parameters can be successfully used for solution of the transient heat conduction problems.

5. Determination of the temperature fields by means of the step-by-step linearization method

To illustrate the step-by-step linearization method, consider the solution of the centro-symmetrical transient heat conduction problem. Let us consider the thermosensitive hollow sphere of inner radius r_1 and outer radius r_2. The sphere is subjected to the uniform temperature distribution t_p and, from the moment of time $\tau = 0$, to the convective-radiation heat exchange trough the surfaces $r = r_1$ and $r = r_2$ with environments of constant temperatures t_{c1} and t_{c2}, respectively. The transient temperature field in the sphere shall be determined from nonlinear heat conduction equation

$$\frac{1}{r^2}\frac{\partial}{\partial r}\left(r^2 \lambda_t(t)\frac{\partial t}{\partial r}\right) = c_v(t)\frac{\partial t}{\partial \tau}, \tag{72}$$

with boundary and initial conditions

$$\left[\lambda_t(t)\frac{\partial t}{\partial r} + (-1)^j\left(\alpha_j(t)(t - t_{cj}) + \sigma\varepsilon_j(t)(t^4 - t_{cj}^4)\right)\right]_{r=r_j} = 0 \quad (j = 1,2), \tag{73}$$

$$t\big|_{\tau=0} = t_p \ . \tag{74}$$

Let us construct the solution to the problem (72)–(74) for the material with simple nonlinearity $(a = \lambda_t(t)/c_v(t) \approx \text{const})$. The temperature-dependent characteristics of the material are given as $\chi(t) = \chi_0 \chi^*(T)$, where the values with indices zero are dimensional and the asterisked terms are dimensionless functions of the dimensionless temperature $T = t/t_0$ (t_0 denotes the reference temperature). Let the thickness of spherical wall $r_0 = r_2 - r_1$ be the characteristic dimension, and $\rho = r/r_0$, $\text{Fo} = a\tau/r_0^2$, $\text{Bi}_j = \alpha_a^{(j)} r_0/\lambda_{t0}$ (Biot number), and $\text{Sk}_j = \sigma\varepsilon_a^{(j)} r_0 t_0^3/\lambda_{t0}$ (Starc number). Then the problem (72)–(74) takes the dimensionless form

$$\frac{1}{\rho^2}\frac{\partial}{\partial \rho}\left(\rho^2 \lambda_t^*(T)\frac{\partial T}{\partial \rho}\right) = c_v^*(T)\frac{\partial T}{\partial \text{Fo}}, \tag{75}$$

$$\left[\lambda_t^*(T)\frac{\partial T}{\partial \rho} + (-1)^j\left(\text{Bi}_j \alpha^*(T)(T - T_{cj}) + \text{Sk}_j \varepsilon^*(T)(T^4 - T_{cj}^4)\right)\right]_{\rho=\rho_j} = 0 \ (j = 1,2), \tag{76}$$

$$T\big|_{\text{Fo}=0} = T_p \ , \tag{77}$$

where $T_{cj} = t_{cj}/t_0$. By application of the Kirchhoff transformation (9) to the nonlinear problem (75)-(77), the following problem for θ

$$\frac{\partial^2(\rho\theta)}{\partial\rho^2} = \frac{\partial(\rho\theta)}{\partial \mathrm{Fo}}, \tag{78}$$

$$\left[\frac{\partial\theta}{\partial\rho} + (-1)^j Q^{(j)}(T(\theta))\right]_{\rho=\rho_j} = 0 \quad (j=1,2), \tag{79}$$

$$\theta\big|_{\mathrm{Fo}=0} = 0 \tag{80}$$

is obtained, where

$$Q^{(j)}(T(\theta)) = \mathrm{Bi}_j\alpha_j^*(T(\theta))(T(\theta)-T_{cj}) + \mathrm{Sk}_j\varepsilon_j^*(T(\theta))\left((T(\theta))^4 - T_{cj}^4\right). \tag{81}$$

The heat conduction equation for the Kirchhoff's variable θ is linear, meanwhile the conditions of convective-radiation heat exchange are partially linearized with the nonlinearities in the expressions $Q^{(j)}(T(\theta))$. These expressions depend on the temperature which is to be determined on the surfaces $\rho=\rho_j$. The temperature of the sphere $T(\rho,\mathrm{Fo})$ on each surface $\rho=\rho_j$ is continuous and monotonic function of time. Because every continuous and monotonic function is an uniform limit of a linear combination of unit functions, these functions can be interpolated by means of the splines of order 0 as

$$Q^{(j)}(\mathrm{Fo}) = Q_i^{(j)} + \sum_{i=1}^{m_j-1}(Q_{i+1}^{(j)} - Q_i^{(j)})S_-(\mathrm{Fo}-\mathrm{Fo}_i^{(j)}), \tag{82}$$

$$Q_i^{(j)}(T(\theta)) = \mathrm{Bi}_j\alpha_j^*(T_i^{(j)})(T_i^{(j)} - T_{cj}) + \mathrm{Sk}_j\varepsilon_j^*(T_i^{(j)})((T_i^{(j)})^4 - T_{cj}^4), \tag{83}$$

where $T_1^{(j)} = T_p, T_i^{(j)}$ $(i=\overline{2,m_j})$ are unknown parameters of spline interpolation for the temperature which is to be determined on the surfaces $\rho=\rho_j$ at $\mathrm{Fo}_{i-1}^{(j)} \le \mathrm{Fo} < \mathrm{Fo}_i^{(j)}$ and $\mathrm{Fo}_{m_j}^{(j)} = \infty$, $S_-(\eta) = \begin{cases} 0, \eta<0, \\ 1, \eta\ge 0\end{cases}$ is asymmetric unit function (H. Korn & T. Korn, 1977; Podstrihach et al., 1984), $\mathrm{Fo}_i^{(j)}$ are the points of segmentation of the time axis $(0;\mathrm{Fo})$. After substitution of the expression (82) into the boundary conditions (79), the boundary value problem (78)-(80) becomes linear. For its solving, the Laplace integral transformation can be used (Ditkin & Prudnikov, 1975). As a result, the Laplace transforms of the Kirhoff's variables are determined as

$$\tilde{\theta} = -\frac{1}{\rho}\left[\rho_1^2\left(Q_1^{(1)} + \sum_{i=1}^{m_1-1}(Q_{i+1}^{(1)} - Q_i^{(1)})e^{-s\mathrm{Fo}_i^{(1)}}\right)\frac{\Phi_2(s)}{\psi(s)} + \right.$$

$$+\rho_2^2\left(Q_1^{(2)}+\sum_{i=1}^{m_2-1}(Q_{i+1}^{(2)}-Q_i^{(2)})e^{-s\mathrm{Fo}_i^{(2)}}\right)\frac{\Phi_1(s)}{\psi(s)}\Bigg], \tag{84}$$

where $\Phi_j(s)=\rho_j\, ch(\rho-\rho_j)\sqrt{s}+\dfrac{sh(\rho-\rho_j)\sqrt{s}}{\sqrt{s}}$; $\psi(s)=s\left[(\rho_1\rho_2 s-1)\dfrac{sh\sqrt{s}}{\sqrt{s}}+ch\sqrt{s}\right]$; s is the parameter of Laplace transformation; $\tilde{\theta}=\int_0^\infty \theta e^{-s\mathrm{Fo}}d\mathrm{Fo}$.

The inverse Laplace transformation can be found by means of the Vashchenko-Zakharchenko expansion theorem of and shift theorem (Lykov, 1967). As a result, the following expression for Kirchhoff's variable

$$\theta=-\frac{1}{\rho}\Bigg[\rho_1^2\left(Q_1^{(1)}\Psi_2(\rho,\mathrm{Fo})+\sum_{i=1}^{m_1-1}(Q_{i+1}^{(1)}-Q_i^{(1)})\Psi_2(\rho,\mathrm{Fo}-\mathrm{Fo}_i^{(1)})S_-(\mathrm{Fo}-\mathrm{Fo}_i^{(1)})\right)+$$

$$+\rho_2^2\left(Q_1^{(2)}\Psi_1(\rho,\mathrm{Fo})+\sum_{i=1}^{m_2-1}(Q_{i+1}^{(2)}-Q_i^{(2)})\Psi_1(\rho,\mathrm{Fo}-\mathrm{Fo}_i^{(2)})S_-(\mathrm{Fo}-\mathrm{Fo}_i^{(2)})\right) \tag{85}$$

is obtained, where

$$\Psi_j(\rho,\mathrm{Fo})=\frac{1}{1+3\rho_1\rho_2}\left(3\rho\mathrm{Fo}+\frac{1}{2}(\rho-\rho_j)^2(\rho-2\rho_j)-\frac{3\rho(1+5\rho_1\rho_2)}{10(1+3\rho_1\rho_2)}\right)\sum_{n=1}^{\infty}A_n^{(j)}e^{-\mu_n^2\mathrm{Fo}};$$

$$A_n^{(j)}=\frac{2(1+\rho_1\rho_2\mu_n^2)}{\mu_n^2(1+3\rho_1\rho_2+\rho_1^2\rho_2^2\mu_n^2)\cos\mu_n}\left(\rho_j\cos(\rho-\rho_j)\mu_n+\frac{\sin(\rho-\rho_j)\mu_n}{\mu_n}\right); \tag{86}$$

μ_n are roots of characteristic equation

$$(1+\rho_1\rho_2\mu^2)\,\mathrm{tg}\,\mu=\mu. \tag{87}$$

For example, let the heat conduction coefficient be a linear function of the temperature $\lambda_t^*(T)=1-kT$. Then on the basis of formula (9),

$$T=k^{-1}\left(1-\sqrt{(1-kT_p)^2-2k\theta}\right). \tag{88}$$

The determined temperature is a function of coordinate ρ and time Fo; it contains $2(m_1+m_2)$ approximation parameters: m_1 values of the temperature $T_i^{(1)}$ on the surface $\rho=\rho_1$ (due to the expressions of $Q_i^{(1)}$) and $\mathrm{Fo}_i^{(1)}$ and m_2 values of the temperature $T_i^{(2)}$ on the surface $\rho=\rho_2$ (due to the expressions of $Q_i^{(2)}$) and $\mathrm{Fo}_i^{(2)}$. The collocation method has been used to determine the approximation parameters. If $\rho=\rho_j$ in (88), the expression of the temperature on the surface $\rho=\rho_j$ are determined as

$$T\big|_{\rho=\rho_j} = k^{-1}\left(1-\sqrt{(1-kT_p)^2 - 2k\theta\big|_{\rho=\rho_j}}\right) \quad (j=1,2). \tag{89}$$

If the values $\mathrm{Fo}_i^{(1)}$ and $\mathrm{Fo}_i^{(2)}$ are given ($\mathrm{Fo}=\mathrm{Fo}_1^{(j)}$, $\mathrm{Fo}=\mathrm{Fo}_2^{(j)}$, etc.) and $T\big|_{\substack{\rho=\rho_j \\ \mathrm{Fo}=\mathrm{Fo}_i^{(j)}}} = T_i^{(j)}$, then the set of m_1+m_2 algebraic equations will be obtained to determine m_1 values of $T_i^{(1)}$ and m_2 values of $T_i^{(2)}$:

$$\begin{cases} T_{i+1}^{(1)} = k^{-1}\left(1-\sqrt{(1-kT_p)^2 - 2k\theta\big|_{\substack{\rho=\rho_1 \\ \mathrm{Fo}=\mathrm{Fo}_i^{(1)}}}}\right) & (i=\overline{1,m_1-1}), \\ T_{i+1}^{(2)} = k^{-1}\left(1-\sqrt{(1-kT_p)^2 - 2k\theta\big|_{\substack{\rho=\rho_2 \\ \mathrm{Fo}=\mathrm{Fo}_i^{(2)}}}}\right) & (i=\overline{1,m_2-1}). \end{cases} \tag{90}$$

After solving this set of equations and substituting the values $T_i^{(j)}$ $(j=1,2)$ into (88), the expression for the temperature can be obtained.

For approximation of the nonlinear expressions $Q^{(j)}(T(\theta))$, we use the same segmentation of the time axis ($m_1=m_2=m$, $\mathrm{Fo}_i^{(1)}=\mathrm{Fo}_i^{(2)}=\mathrm{Fo}_i$) on the sphere surfaces $\rho=\rho_j$. In this case, the set of equations for determination of unknown values $T_i^{(1)}, T_i^{(2)}$ $(i=\overline{1,m})$ takes the following form: the first and second equations (obtained from (90) at $\mathrm{Fo}=\mathrm{Fo}_1$) contain only $T_2^{(1)}$ and $T_2^{(2)}$; the third and fourth equations (obtained from (90) at $\mathrm{Fo}=\mathrm{Fo}_2$) contain four values $T_i^{(1)}$ and $T_i^{(2)}$ $(i=2,3)$, etc.; in the last two equations (obtained from (90) at $\mathrm{Fo}=\mathrm{Fo}_{m-1}$), all $2(m-1)$ unknown values $T_i^{(1)}$ and $T_i^{(2)}$ $(i=\overline{2,m})$ are presented. After solving the first and second equations, the values $T_2^{(1)}$ and $T_2^{(2)}$ are determined. After substitution of these values into the third and fourth equations, the following two unknown values can be determined. The same procedure shall be repeated until all $T_i^{(j)}$ $(j=1,2)$ are determined.

Consider the transient temperature field in a solid thermosensitive sphere with simple nonlinearity under convective-radiation heat exchange between surface and environment of constant temperature t_c. The solution of such heat conduction problem can be obtained from solution of the problem for a hollow sphere. Putting $\rho_1=0$ and $\rho_2=1$ in (85) and denoting $\mathrm{Bi}_2=\mathrm{Bi}$, $Q_i^{(2)}=Q_i$, $T_{c2}=T_c$, the following expression for the Kirchhoff's variable

$$\theta = Q_1\Psi(\rho,\mathrm{Fo}) + \sum_{i=1}^{m-1}(Q_{i+1}-Q_i)\Psi(\rho_1,\mathrm{Fo}-\mathrm{Fo}_i)S_-(\mathrm{Fo}-\mathrm{Fo}_i) \tag{91}$$

can be obtained for the solid sphere, where $\Psi(\rho,\mathrm{Fo})=\frac{3}{10}-\frac{\rho^2}{2}-3\mathrm{Fo}-\frac{2}{\rho}\sum_{n=1}^{\infty}\frac{\sin\rho\mu_n}{\mu_n^3\cos\mu_n}e^{-\mu_n^2\mathrm{Fo}}$, $Q_i=\mathrm{Bi}\alpha^*(T_i)(T_i-T_c)+\mathrm{Sk}\varepsilon^*(T_i)(T_i^4-T_c^4)$, $T_1=T_p$, and μ_n are roots of the equation

$$tg\,\mu=\mu\,. \tag{92}$$

The unknown parameters $T_i=T(\rho,\mathrm{Fo})\big|_{\substack{\rho=1\\ \mathrm{Fo}=\mathrm{Fo}_i}}$ are determined from the equations

$$T_{i+1}=k^{-1}\left(1-\sqrt{(1-kT_p)^2-2k\theta\big|_{\substack{\rho=1\\ \mathrm{Fo}=\mathrm{Fo}_i}}}\right)\quad (i=\overline{1,m-1})\,. \tag{93}$$

If the Kirchhoff's variable is obtained, then the temperature in the sphere can be calculated by means of the formula (88).
For the case when $\mathrm{Sk}=0$ and the heat exchange coefficient is independent of the temperature $(\alpha^*(T)=1)$, then formula (91) yields

$$\theta=\mathrm{Bi}\left[(T_p-T_c)\Psi(\rho,\mathrm{Fo})+\sum_{i=1}^{m-1}\Big(T_{i+1}-T_i\ \Psi(\rho,\mathrm{Fo}-\mathrm{Fo}_i)S_-(\mathrm{Fo}-\mathrm{Fo}_i)\Big)\right]\,. \tag{94}$$

The unknown parameters of spline approximation T_i $(i=\overline{2,m})$ are determined from the set of equations (93) in the following manner. From the first equation of this set, T_2 can be found as

$$T_2=\frac{1}{k}\left[L-\sqrt{L^2-2k\Big(\theta_0+\mathrm{Bi}\big[(T_p-T_c)\Psi(1,\mathrm{Fo}_1)-T_p\Psi(1,0)\big]\Big)}\right], \tag{95}$$

where $L=1-\mathrm{Bi}\Psi(1,0)$; $\theta_0=T_p-kT_p^2/2$. Then the solutions of second, third, and all the following equations can be written as

$$T_i=\frac{1}{k}[L-\{L^2-2k\Bigg(\theta_0+\mathrm{Bi}\Bigg[(T_p-T_c)\Psi(1,\mathrm{Fo}_{i-1})+\sum_{j=1}^{i-2}(T_{j+1}-$$

$$-T_j)\Psi(1,\mathrm{Fo}_{i-1}-\mathrm{Fo}_j)-T_{i-1}\Psi(1,0)\Bigg]\Bigg)\Bigg\}^{\frac{1}{2}}]\ (i=\overline{3,m})\,. \tag{96}$$

To linearize the nonlinear boundary condition

$$\left[\frac{\partial\theta}{\partial\rho}+\mathrm{Bi}\big(T(\theta)-T_c\big)\right]_{\rho=1}=0\,, \tag{97}$$

the substitution of the nonlinear expression $T(\theta)$ by θ (Nedoseka, 1988; Podstrihach & Kolyano, 1972) can be employed. Then the Kirchhoff's variable can be given as

$$\theta = T_c \left(1 - \frac{2}{\rho} \sum_{n=1}^{\infty} \frac{\sin\gamma_n - \gamma_n \cos\gamma_n}{\gamma_n(\gamma_n - \sin\gamma_n \cos\gamma_n)} \sin\rho\gamma_n e^{-\gamma_n^2 \mathrm{Fo}} \right), \tag{98}$$

where γ_n are roots of the characteristic equation

$$(1 + \mathrm{Bi})\mathrm{tg}\,\gamma = \gamma \,. \tag{99}$$

Let us provide the numerical implementation of the proposed solution method to determine the time-variation of the temperature on the surface $\rho = 1$ of solid sphere exposed to the condition of convective heat exchange. We assume t_c = 300°C (573 K) and this value is also chosen to be the reference temperature; the initial temperature is t_p = 20°C (293 K); the Biot number is $\mathrm{Bi} = 10$. In the expression $\lambda_t(t) = \lambda_{to}(1 - kT)$ we set $\lambda_{to} = 50{,}2$ W/(m°K) and $k = 0{,}018$. The results of computation are shown in Figure 1.

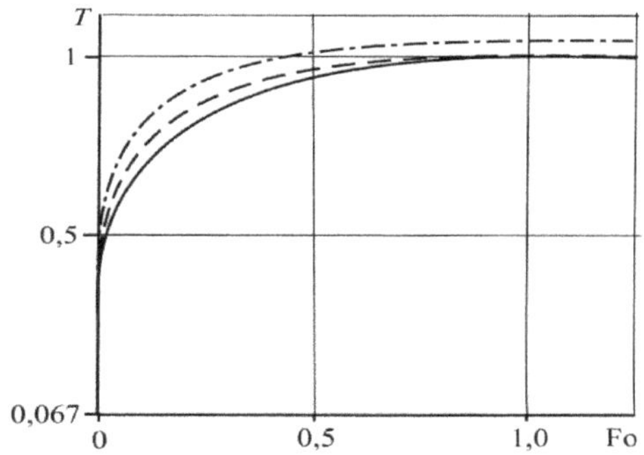

Fig. 1. Dependence of $T(\theta)$ on Fo

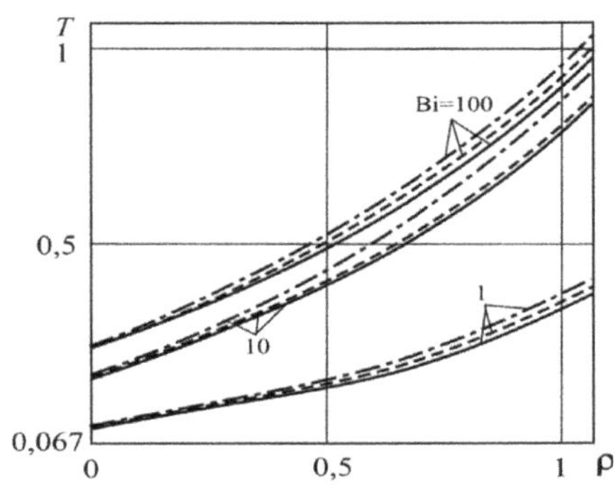

Fig. 2. Dependence of $T(\theta)$ on ρ

In Figure 2, the dependence of the temperature on the radial coordinate at the moment of time $\mathrm{Fo} = 0{,}1$ is shown for some values of the Biot number. The solid lines correspond to the solution of the heat conduction problem, obtained by using the step-by-step method, i.e., when the Kirchhoff's variable is computed by the formula (94). The dash-dot line corresponds to the solution of the problem when the boundary condition is linearized by changing $T(\theta)$ for θ. In this case, the Kirchhoff's variable is calculated by formula (98). The dashed line presents the solution of corresponding linear problem when thermal characteristics are constant. In the considered case, neglecting the temperature dependence in thermal properties leads to the increasing of the temperature values. In the same time, the unsubstantiated linearization of boundary condition increases the temperature and leads to physically improper results. As it follows from the figures, at some moments of time, the temperature on surface of sphere is greater than the temperature of heating environment. The authors (Nedoseka, 1988; Podstrihach & Kolyano, 1972) did not give much attention to this matter because mainly they considered the temperature fields in thermosensitive bodies due to the internal heat sources. In this case increasing of the temperature is unbounded.

6. Conclusion

In this chapter, the formulations of non-linear heat conduction problems for the bodies with temperature-dependent characteristics (thermosensitive bodies) are given. The efficient analytico-numerical methods for solution of the formulated problems are developed. Particularly, the step-by-step linearization method is proposed for solution of one-dimensional transient problems of heat conduction, which describe the temperature fields in thermosensitive structure members of simple nonlinearity under complex (convective, radiation or convective-radiation) heat exchange boundary conditions. The coefficient of heat exchange and emissivity of the surface, that is under heat exchange with environment, are also dependent on the temperature. The method provides:

- reduction of the heat conduction problem to the corresponding dimensionless problem;
- partial linearization of the obtained problem by means of the Kirchhoff's transform;
- complete linearization of the nonlinear condition on the Kirchhoff's variable θ, that has been obtained from the condition of complex heat exchange due to approximation of the nonlinear term by specially constructed spline of zero or first order;
- construction of the solution to the linearized boundary value problem for θ by means of the appropriate analytical method;
- determination of the temperature in question by means of the inverse Kirchhoff's transform;
- determination of the unknown parameters of spline-approximation, those remain in the expression for the temperature, by means of the collocation method.

The method is verified by the solutions of transient heat conduction problems for thermosensitive solid and hollow spheres subjected to heating (cooling) due to the heat exchange over the limiting surface. This method can be efficiently used fro solution of two-dimensional steady-state heat conduction problems.

The efficient method of linearizing parameters is proposed for determination of the temperature fields in structure members with simple nonlinearity due to convective heat exchange through the limiting surfaces for an arbitrary dependence of the heat conduction coefficient on the temperature. The main feature of this method consists in the fact that the complete linearization of the nonlinear condition for the Kirchhoff's variable θ (obtained form the condition of convective heat exchange) is achieved by substitution of the nonlinear term $T(\theta)$ by $(1+\kappa)\theta+T_p$ with unknown parameter κ. This parameter can be found by satisfaction of the nonlinear condition for θ with required accuracy.

The method of linearizing parameters is adopted to solution of the nonlinear steady-state and transient heat conduction problems for contacting thermosensitive bodies of simple geometrical shape under conditions of the ideal thermal contact at the interfaces and complex heat exchange on the limiting surfaces. Its approbation is provided for the n-layer cylindrical pipe under given temperatures on its inner and outer surfaces. It these surfaces are subjected to the convective heat exchange, then the complete linearization of the obtained nonlinear conditions for the Kirchhoff's variable θ can be done by means of the method of linearizing parameters.

7. Acknowledgment

This research is provided under particular support of the project within the joint program of scientific research between the Ukrainian National Academy of Sciences and Russian Foundation of Basic Research (2010-1011).

8. References

Carslaw, H.S. & Jaeger, J.C. (1959). *Conduction of Heat in Solids,* Clarendon, ISBN: 978-0-19-853368-9, Oxford, UK

Ditkin, V.A. & Prudnikov, A.P. (1975). *Operational Calculus,* Vysshaja shkola, Moscow, Rusia (in Russian)

Galitsyn, A.S. & Zhukovskii, A.N. (1976). *Integral Transforms and Special Functions in Problems of Heat Conduction,* Naukova Dumka, Kyiv, Ukraine (in Russian)

Korn, H. & Korn, T. (1977). *Handbook on Mathematics for Scientifik Researchers,* Nauka, Moscow, Rusia (in Russian)

Kushnir, R.M. & Popovych, V.S. (2006). Stressed State of Thermosensitive Plate in a Central-Symmetric Temperature Field, *Materials Science,* Vol.42, No.2, pp. 145–154, ISSN 1068-820X

Kushnir, R.M. & Popovych, V.S. (2007). Thermostressed State of Thermal Sensitive Sphere Under Complex Heat Exchange with the Surroundings, *Proceedings of 7th International Congress on Thermal Stresses,* Vol.1, pp. 369-372, ISBN: 978-986-00-9556-2, Taipei, Taiwan, June 4-7, 2007

Kushnir, R.M. & Popovych, V.S. (2009). *Thermoelasticity of Thermosensitive Solids,* SPOLOM, ISBN 978-966-665-529-8, L'viv, Ukraine (in Ukrainian)

Kushnir, R.M.; Popovych, V.S. & Harmatiy, H.Yu. (2001). Analytic-Numerical Solution of Contact Problems of Thermoelasticity for Thermal Sensitive Bodies, *Materials Science,* Vol.37, No.6, pp. 893–901, ISSN 1068-820X

Kushnir, R.M.; Popovych, V.S. & Vovk, O.M. (2008). The Thermoelastic State of a Thermosensitive Sphere and Space with a Spherical Cavity Subject to Complex Heat Exchange. *Journal of Engineering Mathematics,* Vol.61, No.2-4, (August 2008), pp. 357-369, ISSN 0022-0833

Kushnir, R. & Protsiuk, B. (2009). A Method of the Green's Functions for Quasistatic Thermoelasticity Problems in Layered Thermosensitive Bodies under Complex Heat Exchang. In: *Operator Theory: Advances and Applications, Vol.191,* V.Adamyan,

Yu.Berezansky, I.Gohberg & G.Popov, (Eds.), 143-154, Birkhauser Verlag, ISBN 978-3-7643-9920-7, Basel, Switzerland

Lykov, A.V. (1967). *Heat Conduction Theory*, Vysshaja shkola, Moscow, Rusia (in Russian)

Nedoseka, A.Ya. (1988). *Fundamentals of Design Computation for Welded Structures*, Vyscha shkola, Kyiv, Ukraine (in Russian)

Noda, N. (1986). Thermal Stresses in Materials with Temperature-Dependent Properties, In: *Thermal Stresses I*, R.B. Hetnarski, (Ed.), 391-483, North-Holland, Elsevier, ISBN 0444877282 , Amsterdam, Netherland

Nowinski, J. (1962). Transient Thermoelastic Problem for an Infinite Medium with a Spherical Cavity Exhibiting Temperature-Dependent Properties. *Journal of Applied Mechanics*, Vol.29, pp. 399-407, ISSN: 0021-8936

Podstrihach, Ya.S. & Kolyano, Yu.M. (1972). *Nonstationary Temperature Fields and Stresses in Thin Plates*, Naukova Dumka, Kyiv, Ukraine (in Russian)

Podstrihach, Ya.S.; Lomakin, V.A. & Kolyano, Yu.M. (1984). *Thermoelasticity of Bodies of Inhomogeneous Structure*, Nauka, Moscow, Rusia (in Russian)

Popovych, V.S. (1993a). On the Solution of Heat Conduction for Thermo-Sensitive Bodies, Heated by Convective Heat Exchange, *Journal of Soviet Mathematics*, Vol.63, No.1, pp. 94–97, ISSN 1072-3374

Popovych, V.S. (1993b). On the Solution of Stationary Problem for the Thermal Conductivity of Heat-Sensitive Bodies in Contact, *Journal of Soviet Mathematics*, Vol.65, No.4, pp. 1762–1766, ISSN 1072-3374

Popovych, V.S. & Harmatiy, H.Yu. (1996). The Nonstationary Heat-Conduction Problem for Heat-Sensitive Space with a Spherical Cavity, *Journal of Mathematical Sciences*, Vol.79, No.6, pp. 1478–1482, ISSN 1072-3374

Popovych, V.S. & Harmatiy, H.Yu. (1998). Solution of Nonstationary Heat-Conduction Problems for Thermosensitive Bodies Under Convective Heat Exchange, *Journal of Mathematical Sciences*, Vol.90, No.2, pp. 2037–2040, ISSN 1072-3374

Popovych, V.S.; Harmatiy, H.Yu. & Vovk, O.M. (2006). Thermoelastic State of a Thermosensitive Hollow Sphere Under Convective-Radiant Heat Exchange with the Environment. *Materials Science*, Vol.42, No.6, pp. 756–770, ISSN 1068-820X

Popovych, V.S. & Sulym, G.T. (2004). Centrally-Symmetric Quasi-Static Thermoelasticity Problem for Thermosensitive Body. *Materials Science*, Vol.40, No.3, pp. 365-375, ISSN 1068-820X

Sneddon, I.N. (1951). *Fourier Transforms*, McGraw-Hill Book Company, ISBN 0-486-68522-5, New York, USA

Sorokin, V.G.; Volosnikova A.V. & Vjatkin S.A. (1989). *Grades of Steels and Alloys*, Mashinostroenije, Moscow, Rusia (in Russian)

8

Can a Lorentz Invariant Equation Describe Thermal Energy Propagation Problems?

Ferenc Márkus
Department of Physics, Budapest University of Technology and Economics
Hungary

1. Introduction

In the new technologies the development towards the small scales initiates and encourages the reformulation of those well-known transport equations, like heat and electric conduction, that were applied for bulk materials. The reason of it is that there are several physical evidences for the changes of the behavior of the signal propagation as the sample size is decreasing (Anderson & Tamma, 2006; Cahill et al., 2003; Chen, 2001; Liu & Asheghi, 2004; Schwab et al., 2000; Vázquez et al., 2009). The constructed different mathematical models clearly belong to the phenomena of the considered systems. However, presently, there is no a well-trodden way how to establish the required formulations in general. A great challenge is to establish and exploit the Lagrangian and Lorentz invariant formulation of the thermal energy propagation, since, on the one hand, the connection with other field theories including the interactions of fields can be done on this level, on the other hand, these provide the finite physical action and signal propagation. The results of the presented theory ensures a deeper insight into the phenomena, thus hopefully it will contribute to the technical progress in the near future.

It is an old and toughish question how to introduce the finite speed propagation of action in such physical processes like the thermal energy propagation (Eckart, 1940; Joseph & Preziosi, 1989; Jou et al., 2010; Márkus & Gambár, 2005; Sandoval-Villalbazo & García-Colín, 2000; Sieniutycz, 1994; Sieniutycz & Berry, 2002). There is no doubt that the solution must exist somehow and the suitable description should be Lorentz invariant. Moreover, this Lorentz invariant formulation needs to involve anyway the Fourier heat conduction as the classical limit. The elaborated theory ensures that in the case of Lorentz invariant formulation both the speed of the signal and the action propagation is finite. Furthermore, for the Fourier heat conduction the temperature propagation is finite, however, the speed of action is infinite.

This chapter treats the consequent mathematical formulation of a suitable relativistic invariant description of the above problem and its consequences, connections with other topics are also treated. As the author hopes it will be noticeable step-by-step that this synthesis theory may have a prominent role in the phenomena of nature. The construction of the Lorentz invariant thermal energy propagation, the Klein-Gordon equation with negative "mass term" providing the expected propagation modes, the limit to the classical heat conduction and the related dynamic phase transition between the dissipative – non-dissipative dynamic phase transitions are discussed in a coherent frame within Sec. 2. Two mechanical analogies are shown in Sec. 3 for the two kinds of Klein-Gordon type equations to see the distinct behavior due to the opposite sign of the mass term. On the one hand, it will be convincing to see how the negative "mass term" can govern the above mentioned change in the dynamics, and, on the other hand, it clarifies the physical role of the similar term in the Lorentz invariant propagation studied in Sec. 2. It is assumable that the efficiency of the relativistic invariant theory can be demonstrated via other physical phenomena. The spectacular description of the inflationary cosmology with the inflaton-thermal field coupling, the resulted time evolution of the inflaton field and the dynamic temperature show this fact clearly in Sec. 4. Finally, to achieve a deeper insight into the soul of this new theory and to be sure that the causality principle is completed, for this reason the Wheeler propagator is calculated in Sec. 5 as well. The main ideas, results of the chapter and some concluding remarks are summarized in Sec. 6. Finally, Sec. 7 is for the acknowledgment.

2. Lorentz invariant thermal energy propagation

The mathematical description is based on the least action principle (Hamilton's principle)

$$S = \int L\, d^3x dt = extremum, \tag{1}$$

i.e., there exists a Lagrange density function L by which the calculated action S is extremal for the real physical processes. The Hamiltonian formulation can be also achieved for certain differential equations involving non-selfadjoint operators like the first time derivative in the classical Fourier heat conduction. Then such potential functions are required to introduce by which the Lagrange functions can be expressed and the whole Hamiltionian theory can be constructed (Gambár & Márkus, 1994; Gambár, 2005; Márkus, 2005). The long scientific experience on this topic showed that the theories are comparable and connectable on this — Lagrangian-Hamiltonian — level, thus in the further development of the theory it is useful to apply this idea and scheme. In order to generate a dynamic temperature and the related covariant Klein-Gordon type field equation, to describe the heat propagation with finite speed — less than the speed of light — of action an abstract scalar potential field has been introduced (Gambár & Márkus, 2007). In this case the thermal energy propagation has wave-like modes. It is important to emphasize that, on the other hand, this scalar field can be connected to the usual (local equilibrium) temperature and the Fourier's heat conduction in the classical limit. This treating is an attempt to point out that the dynamic phase transition (Ma, 1982) between the two kinds of propagation, between a wave and a non-wave, or with another context it is better to say — between a non-dissipative and a dissipative thermal process — has a more general role and manifestation in the processes.

As a starting point the Lagrange functions are given for both the Lorentz invariant heat propagation (Márkus & Gambár, 2005) and for the classical heat conduction (Fourier's heat conduction) (Gambár & Márkus, 1994). The first description is based on a Klein-Gordon type equation formulated by a negative "mass term". It will be shown that this pertains to a repulsive potential, which repulsive interaction produces a tachyon solution leading to the so-called spinodal instability which effect is often applied in modern field theories (Borsányi et al., 2000; 2002; 2003). Now, the Hamiltonian descriptions are written side by side — to prepare the later comparison — showing how the Lorentz invariant solution provides the classical solution in the limit of speed of light. The relevant Lagrangians, L_w for the wave-like solution (Márkus & Gambár, 2005) and L_c for the classical heat conduction (Gambár & Márkus, 1994) restricting our examination for the one dimensional case, are

$$L_w = \frac{1}{2}\left(\frac{\partial^2 \varphi}{\partial x^2}\right)^2 + \frac{1}{2c^4}\left(\frac{\partial^2 \varphi}{\partial t^2}\right)^2 - \frac{1}{c^2}\frac{\partial^2 \varphi}{\partial x^2}\frac{\partial^2 \varphi}{\partial t^2} - \frac{1}{2}\frac{c^4 c_v^4}{16\lambda^4}\varphi^2, \tag{2a}$$

$$L_c = \frac{1}{2}\left(\frac{\partial \varphi}{\partial t}\right)^2 + \frac{1}{2}\frac{\lambda^2}{c_v^2}\left(\frac{\partial^2 \varphi}{\partial x^2}\right)^2, \tag{2b}$$

where φ is a four times differentiable and Lorentz invariant scalar field that generates the measurable thermal field, and c denotes the speed of light, λ is the heat conductivity, c_v is the specific heat. Applying the calculus of variation the corresponding Euler-Lagrange equations as equations of motion for the field φ can be obtained

$$0 = \frac{1}{c^4}\frac{\partial^4 \varphi}{\partial t^4} + \frac{\partial^4 \varphi}{\partial x^4} - \frac{2}{c^2}\frac{\partial^4 \varphi}{\partial t^2 \partial x^2} - \frac{c^4 c_v^4}{16\lambda^4}\varphi, \tag{3a}$$

$$0 = -\frac{\partial^2 \varphi}{\partial t^2} + \frac{\lambda^2}{c_v^2}\frac{\partial^4 \varphi}{\partial x^4}. \tag{3b}$$

It is expected that the above scalar field is able to define the measurable physical quantities, namely, in the present case, the temperature. Let the temperature T be a Lorentz invariant temperature, which is defined from a dynamical point of view, thus it can be considered as the dynamic temperature. Furthermore, temperature $\mathcal{T}$ denotes the usual local equilibrium temperature

$$T = \frac{1}{c^2}\frac{\partial^2 \varphi}{\partial t^2} - \frac{\partial^2 \varphi}{\partial x^2} + \frac{c^2 c_v^2}{4\lambda^2}\varphi, \tag{4a}$$

$$\mathcal{T} = -\frac{\partial \varphi}{\partial t} - \frac{\lambda}{c_v}\frac{\partial^2 \varphi}{\partial x^2}. \tag{4b}$$

Eliminating the potentials in Equations (3a) and (3b) by the help of the corresponding Equations (4a) and (4b), for the relevant case, a differential equation for the time evolution of the temperature can be obtained

$$\frac{1}{c^2}\frac{\partial^2 T}{\partial t^2} - \frac{\partial^2 T}{\partial x^2} - \frac{c^2 c_v^2}{4\lambda^2}T = 0, \tag{5a}$$

$$\frac{\partial \mathcal{T}}{\partial t} - \frac{\lambda}{c_v}\frac{\partial^2 \mathcal{T}}{\partial x^2} = 0. \tag{5b}$$

Here, Equation (5a) — the hyperbolic one — is a Klein-Gordon type equation with a negative "mass term" $-(c^2 c_v^2/4\lambda^2)T$ which means a kind of repulsive interaction. This term is responsible for the tachyon solution leading to a spinodal instability as it will be also seen in Sec. 3 in the case of classical Klein-Gordon equation of the mechanical analogy. On the other hand, Equation (5b) — the parabolic one — pertains to the Fourier's heat equation. The signal propagation mechanism can be examined by the calculation of the dispersion relations for both cases

$$\omega(k) = \sqrt{c^2 k^2 - \frac{c^4 c_v^2}{4\lambda^2}} = ck\sqrt{1 - \frac{c^2}{4D^2 k^2}}, \tag{6a}$$

$$\omega(k) = -i\frac{\lambda}{c_v}k^2 = -iDk^2. \tag{6b}$$

Here, the diffusivity parameter $D = \lambda / c_v$ is introduced to simplify the forms. The dispersion relation in Equation (6a) pertains to the Klein-Gordon wave equation in Equation (5a) from which we obtain the phase velocity w_f

$$w_f(k) = \frac{\omega}{k} = c\sqrt{1 - \frac{c^2}{4D^2k^2}}. \tag{7}$$

The dispersion relation in Equation (6b) belongs to the classical (non-wave) Fourier's heat conduction. The models can be compared by the calculation of the group velocities since these pertain to the signal propagations. Thus, from Equation (6a) the group velocity $v_g = d\omega/dk$ of the wave-like propagation can be directly calculated. Then, tending to the infinity with the speed of light, the group velocity v_T of the classical heat conduction can be obtained, as it is expected

$$v_g = \frac{d\omega}{dk} = \frac{c}{\sqrt{1 - \frac{c^2}{4D^2k^2}}} \qquad \longrightarrow \qquad \left.\frac{d\omega}{dk}\right|_{c\to\infty} = -i2Dk; \quad v_T = 2Dk \ll c. \tag{8}$$

This limit shows clearly that the Lorentz invariant description covers both cases, and the wave-like and the non-wave heat propagation can be discussed in the same frame.

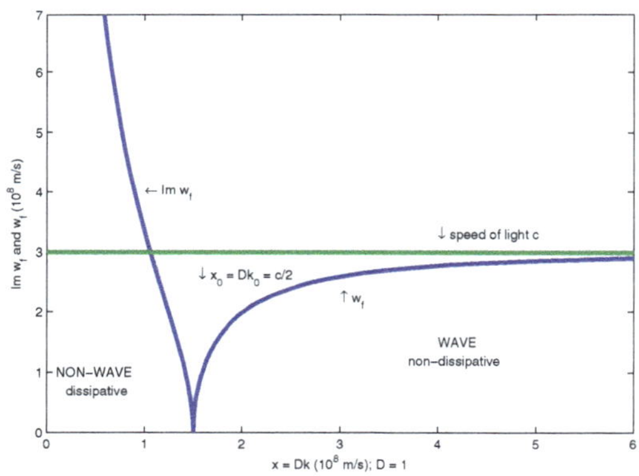

Fig. 1. Phase transition between the non-wave (dissipative) [left] and the wave (non-dissipative) solution [right]. The critical transition point is at $x_0 = Dk_0 = c/2$. The value of diffusity is taken $D = 1$. The phase velocity w_f of the wave-like propagation is always smaller than the speed of light.

It can be recognized that there is a value of the wave number k when the discriminant changes its sign in Equation (7) at the value $k_0 = c/2D$. Now, the solutions can be split into two parts. On the one hand, we can consider the case $k > k_0$, when the solution is real and wave-like (non-dissipative), and on the other hand, we take the case $k < k_0$, when the solution is imaginary and non-wave (dissipative). The real and the imaginary part of the phase velocity w_f can be written for both cases

$$w_f = \frac{\omega}{k} = c\sqrt{1 - \frac{c^2}{4D^2k^2}} < c \quad k > k_0, \tag{9a}$$

$$Im\, w_f = \frac{\omega}{k} = c\sqrt{\frac{c^2}{4D^2k^2} - 1} \quad k < k_0. \tag{9b}$$

The above physical discussion can be easily followed in Fig. 1.
In order to couple the thermal field given in Equation (2a) with other fields (like the inflaton field in the cosmology shown in Sec. 4) it is worthy to reformulate it for this later use. It has been shown in the literature (Márkus & Gambár, 2005) that the quantization of the thermal field generates quasi particles and these particles may have a mass

$$M_0 = \frac{\hbar}{2D}, \tag{10}$$

where $\hbar$ is the Planck constant. Moreover, the Planck units are applied for the present case ($c = 1; \hbar = 1$). Then the 3D Lagrangian given by Eq. (2a) should be rewritten

$$L_w = \frac{1}{2}(\Delta\varphi)^2 + \frac{1}{2}\left(\frac{\partial^2\varphi}{\partial t^2}\right)^2 - \frac{\partial^2\varphi}{\partial t^2}\Delta\varphi - \frac{1}{2}M_0^4\varphi^2, \tag{11}$$

where Δ is the Laplace operator.

3. Mechanical analogies for the two kinds of Klein-Gordon equations

It is instructive to study the set-up of the classical model of the Klein-Gordon equation (Morse & Feshbach, 1953) to make comparisons and conclusions on the physical meaning of the relevant terms that may appear similarly in a more general and abstract theory. The mechanical model is a stretched string with little vertically oriented springs along the string which pull back the spring to the equilibrium position as it is shown in Fig. 2(a). The equation of motion of the string can be formulated applying the Lagrangian formalism. To achieve this, the kinetic and potential energy terms are needed to calculate. The string has a kinetic energy from its movement

$$T = \frac{1}{2}\varrho A \int \left(\frac{\partial \Psi}{\partial t}\right)^2 dx, \tag{12}$$

where Ψ is the displacement from the equilibrium position, ϱ is the density, A is the cross section of the string. The mass element is $dm = \varrho A dx$. The either of the potential energy terms comes from the small deformation (elongation) of the stretching which is

$$V = F\int \left[\sqrt{1 + \left(\frac{\partial \Psi}{\partial x}\right)^2} - 1\right] dx \sim \frac{1}{2}F\int \left(\frac{\partial \Psi}{\partial x}\right)^2 dx, \tag{13}$$

F is the stretching force. The other attractive potential energy term pertains to the little springs which is

$$V_s = \frac{1}{2}k_a \int \Psi^2 dx. \tag{14}$$

Here, k_a is the spring direction coefficient density along the string as is shown in Fig 2(a). The Lagrangian of the system can be formulated with the usual construction $L = T - V - V_s$, by which the Euler-Lagrange equation as equation of motion — a Klein-Gordon equation with positive "mass term" —

$$\frac{\partial^2 \Psi}{\partial t^2} - \frac{F}{\varrho A}\frac{\partial^2 \Psi}{\partial x^2} + \frac{k_a}{\varrho A}\Psi = 0. \tag{15}$$

can be deduced. Now, if a "repulsive" potential is imagined at the places of the springs shown in Fig. 2(b) then a Klein-Gordon type equation with negative "mass term" (Gambár & Márkus, 2008) is obtained

$$\frac{\partial^2 \Psi}{\partial t^2} - \frac{F}{\varrho A}\frac{\partial^2 \Psi}{\partial x^2} - \frac{k_a}{\varrho A}\Psi = 0. \tag{16}$$

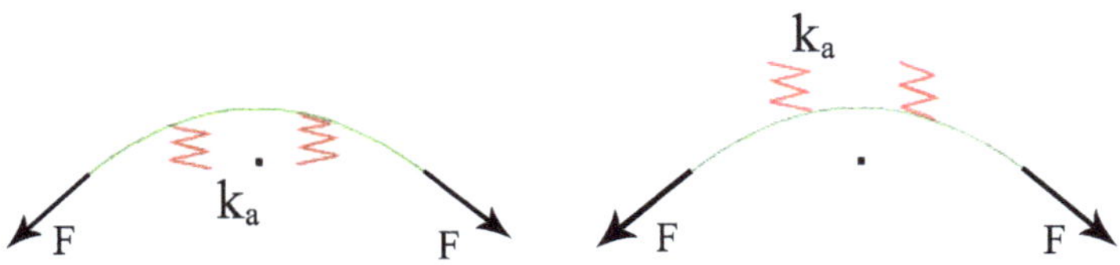

(a) A stretched string (green line) with an additional attractive interaction by the springs k_a

(b) A stretched string (green line) with an additional "repulsive" interaction by the springs k_a

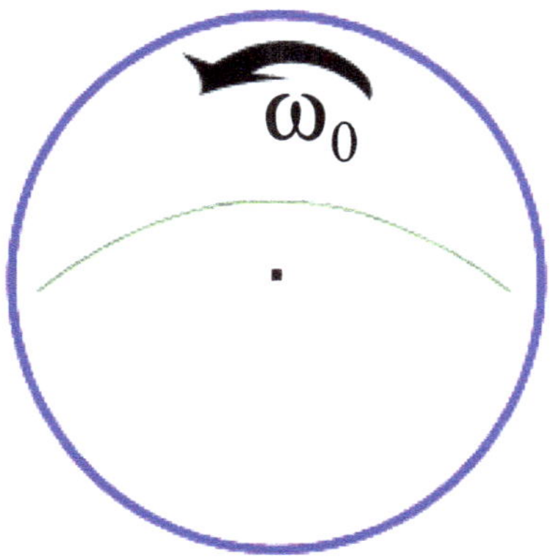

(c) A stretched string (green line) on a rotating disc; ω_0 is the angular velocity

Fig. 2. The three physical situations of the stretched string; the acting force is F for each cases. The equations of motion due to the attractive or "repulsive" interactions pertain to the different figures: Equation (15) for Fig. (a); Equation (16) for Fig. (b); Equation (18) for Fig. (c).

The structure of this equation is exactly the same as in the case of Lorentz invariant thermal energy propagation in Equation (5a). Since, it is clear from this mechanical example that the negative sign of the third term in Equation (16) pertains to a repulsive interaction, thus, this is the reson why the negative "mass term" may relate to a repulsive interaction in the relativistic

case in Equation (5a), in general. Maybe, it is complicated to prepare a device to ensure the repulsive interaction from little springs. However, if the stretched string is placed on the diameter of a rotating disk — shown in Fig. 2(c) that moves with the angular velocity ω_0, then the centrifugal force can produce the similar repulsive interaction.
The centrifugal potential of a point-like mass m moving on a circle with a radius r

$$-\frac{1}{2}mr^2\omega_0^2$$

can be generalized to the present case. This gives the potential V_{rot} pertaining to the rotational motion of the string

$$V_{rot} = -\frac{1}{2}\varrho A\omega_0^2 \int \Psi^2 dx. \tag{17}$$

The relevant Lagrangian is $L = T - V - V_{rot}$, by which the calculated equation of motion can be obtained

$$\frac{\partial^2\Psi}{\partial t^2} - \frac{F}{\varrho A}\frac{\partial^2\Psi}{\partial x^2} - \omega_0^2\Psi = 0. \tag{18}$$

The same mathematical structure can be immediately recognized comparing this equation with the Equations (5a) and (16). This means that these three equations must involve the similar physical behavior: the spinodal instability and the dynamic phase transition (Gambár, 2010). All together these examples clearly prove the physical reality of the Klein-Gordon equation with negative "mass term" in nature.
Finally, for the completeness the dispersion relation for Equation (18) can be also calculated

$$\Omega(k,\omega_0) = \sqrt{\frac{F}{\varrho A}k^2 - \omega_0^2}. \tag{19}$$

This formula shows again the same physical behavior clearly as it has been found in Equation (6a). The phase velocity is

$$w_{ph} = \frac{\Omega}{k} = \sqrt{\frac{F}{\varrho A} - \left(\frac{\omega_0}{k}\right)^2}. \tag{20}$$

It is easy to recognize that for small angular velocity ω_0 while

$$\sqrt{\frac{F}{\varrho A}} > \frac{\omega_0}{k} \tag{21}$$

is completed, then wave modes exist. The opposite case is when

$$\sqrt{\frac{F}{\varrho A}} < \frac{\omega_0}{k}, \tag{22}$$

there are no wave modes. The physical meaning is that, above a certain value of ω_0, the centrifugal force elongates the string to infinity, the string cannot have vibrating modes. The change in the propagation modes is an angular velocity controlled dynamic phase transition that divides the dissipative – non-dissipative transition like in Equations (7), (9a) and (9b) for the thermal case.

4. Inflationary cosmology with the dynamic temperature

It is a great challenge to experience and understand how the Lorentz invariant propagating thermal energy field φ can interact with other physical fields. In this way new physical relations, considerations and explanations may be expected for the relevant phenomena. As an advanced example, to point out the strength of the formulation, the thermal and cosmological inflaton fields are coupled within the Lagrangian framework (Márkus et al., 2009).

4.1 Linde's model of the inflaton field

In the present model the cosmological model is based on the Einstein's equation in the Friedman-Robertson-Walker metric. Now, the action S can be expressed as

$$S = \int \sqrt{-\tilde{g}} L_{FRW}\, d^4x, \tag{23}$$

where the expression $\sqrt{-\tilde{g}} = a^3$ is the Friedman-Robertson-Walker metric. Here, the $a(t) = R(t)/R_0$ is taken as the 'radius' of the universe. The Lagrange density function L_{FRW} of the inflaton field ϕ

$$L_{FRW} = \left(\frac{1}{2} \left(\frac{\partial \phi}{\partial t} \right)^2 - \frac{1}{2a^2} (\nabla \phi)^2 - V(\phi) \right) \tag{24}$$

is the starting point in the description; ∇ is the gradient operator. Then, the equation of motion for the inflaton can be calculated

$$\frac{\partial^2 \phi}{\partial t^2} - \frac{1}{a^2} \Delta \phi + 3H \frac{\partial \phi}{\partial t} = -\frac{\delta V(\phi)}{\delta \phi}, \tag{25}$$

where $\delta V(\phi)/\delta\phi$ means a functional derivative. The Hubble parameter $H(t)$ is defined by

$$H = \frac{\dot{a}}{a}. \tag{26}$$

The fate of the universe depends on the potential $V(\phi)$. The hybrid inflation model suggested by Linde (Felder et al., 1999; 2001; Linde, 1982; 1994) introduces an additional scalar field σ (in fact the Higgs field) into the effective potential

$$V(\sigma, \phi) = -\frac{1}{2a^2} (\nabla \phi)^2 + \frac{1}{2} m^2 \phi^2 + \frac{1}{2} g^2 \phi^2 \sigma^2 + \frac{1}{4\lambda} (M^2 - \lambda \sigma^2)^2. \tag{27}$$

Here, the first term on the right hand side pertains to the second term — the space derivate term — on the left hand side in Equation (25). The second term generates the inflation process, the third one couples the inflaton field to the introduced additional field σ and the last one produces mass generation through the spontaneous symmetry breaking. The canonical momentum of the inflaton field can be calculated

$$\Pi_\phi = \frac{\partial L_{FRW}}{\partial \dot{\phi}} = \dot{\phi}. \tag{28}$$

Then the Hamiltonian $\tilde{H}$ of the field which is the energy density can be obtained

$$\tilde{H} = \Pi_\phi \dot{\phi} - L_{FRW} = \left(\frac{1}{2} \left(\frac{\partial \phi}{\partial t} \right)^2 + \frac{1}{2a^2} (\nabla \phi)^2 + V(\phi) \right). \tag{29}$$

It is often used different notations for $\tilde{H}$

$$\tilde{H} = \varrho_\phi = T_{00}, \tag{30}$$

where T_{00} is called as the time-time component of the energy-momentum tensor. Furthermore, the Einstein's equation can be expressed in the FRW metric as

$$\left(\frac{\dot{a}}{a} \right)^2 = \frac{8\pi G}{3} \varrho, \tag{31}$$

where G is the gravitational constant and ϱ is the mass density. Substituting the energy density ϱ_ϕ and the Planck mass

$$M_{pl} = \sqrt{\frac{\hbar c}{8\pi G}} \tag{32}$$

into Equation (31) and applying Planck units, the Friedman's equation can be written in the following form

$$H^2 = \frac{1}{3M_{pl}^2} \varrho_\phi, \tag{33}$$

which corresponds to a flat universe. If it is assumed that the universe is growing homogeneously in the space we can neglect those terms where the spatial derivates (∇ and Δ) appear in Equation (25), then an ordinary differential equation can be obtained

$$\frac{d^2 \phi_0}{dt^2} + 3H \frac{d\phi_0}{dt} = -\frac{\delta V(\phi_0)}{\delta \phi_0}, \tag{34}$$

the 'field variable' ϕ_0 depends on the time parameter only. In this case the energy density ϱ_ϕ has a simplified form

$$\varrho_\phi = \left(\frac{1}{2} \left(\frac{d\phi_0}{dt} \right)^2 + V(\phi) \right), \tag{35}$$

by which the equation $H^2 = (1/3M_{pl}^2)\varrho_\phi$ naturally also remains valid, i.e.,

$$H^2 = \frac{1}{3M_{pl}^2} \left(\frac{1}{2} \left(\frac{d\phi_0}{dt} \right)^2 + V(\phi) \right). \tag{36}$$

Soon it will be seen that the above equations, (35) and (36), with the modifying effect of the thermal field φ_0 will become those equations which are going to be considered as the time-evolution equations of the inflaton field.

4.2 The coupling of the fields

The introduction of the dynamic temperature and the laws of thermodynamics into the theory of cosmology requires the same mathematical frame of the description. Now, the tool is ready to make this willing. The interaction of the thermal potential field φ [see Equation (11)] and the inflaton field ϕ [see Equation (24)] can be constructed by adding the Lagrangians of the different fields

$$L_{int} = \left(\frac{1}{2a^4}(\Delta\varphi)^2 + \frac{1}{2}\left(\frac{\partial^2\varphi}{\partial t^2}\right)^2 - \frac{1}{a^2}\frac{\partial^2\varphi}{\partial t^2}\Delta\varphi - \frac{1}{2}M_0^4\varphi^2\right) + \left(\frac{1}{2}\left(\frac{\partial\phi}{\partial t}\right)^2 - \frac{1}{2a^2}(\nabla\phi)^2 - V(\phi,\varphi)\right). \tag{37}$$

This Lagrangian L_{int} of the coupled inflaton-thermal field by the following interaction potential can also realize the spontaneous symmetry breaking

$$V(\phi,\varphi) = \frac{1}{2}m^2\phi^2 + \frac{1}{2}g_0^2\phi^2\varphi^2, \tag{38}$$

where m denotes the mass of the inflaton, and g_0 is the coupling constant, moreover, this description can involve the temperature of the inflaton field (Márkus et al., 2009). This fact is very interesting, since at this stage, there is no need for the Higgs field and the mass generation.

After all, applying the calculus of variation, two Euler-Lagrange equations as equations of motion are arisen from the variation with respect to the variables ϕ and φ

$$\frac{\partial^2\phi}{\partial t^2} - \frac{1}{a^2}\Delta\phi + 3\frac{\dot{a}}{a}\frac{\partial\phi}{\partial t} = -\frac{\delta V(\phi,\varphi)}{\delta\phi}, \tag{39}$$

and

$$\frac{1}{a^4}\Delta\Delta\varphi + \frac{\partial^4\varphi}{\partial t^4} + 6\frac{\dot{a}}{a}\frac{\partial^3\varphi}{\partial t^3} + \frac{1}{a^3}\frac{\partial^2(a^3)}{\partial t^2}\frac{\partial^2\varphi}{\partial t^2} - \frac{2}{a^2}\Delta\frac{\partial^2\varphi}{\partial t^2} - \frac{\ddot{a}}{a^3}\Delta\varphi - 2\frac{\dot{a}}{a^3}\Delta\frac{\partial\varphi}{\partial t} - M_0^4\varphi = \frac{\delta V(\phi,\varphi)}{\delta\varphi}. \tag{40}$$

An important remark is needed here. Since, for the cases when the Lagrangian contains second order time derivatives the Hamiltonian $\tilde{H}$ must be expressed as follows (Gambár & Márkus, 1994; Márkus & Gambár, 1991),

$$\tilde{H} = \frac{\partial\varphi}{\partial t}\frac{\partial L}{\partial\dot{\varphi}} - \frac{\partial\varphi}{\partial t}\frac{\partial}{\partial t}\frac{\partial L}{\partial\ddot{\varphi}} + \frac{\partial^2\varphi}{\partial t^2}\frac{\partial L}{\partial\ddot{\varphi}} - L. \tag{41}$$

By substituting the Lagrangian L_{int} from Equation (37), the Hamiltonian — energy density regarding the whole space with all interactions — can be calculated

$$\varrho_{\phi,\varphi} = \tilde{H} = -\frac{\partial\varphi}{\partial t}\frac{\partial^3\varphi}{\partial t^3} + \frac{\partial\varphi}{\partial t}\frac{\partial}{\partial t}\left(\frac{1}{a^2}\right)\Delta\varphi + \frac{1}{a^2}\frac{\partial\varphi}{\partial t}\frac{\partial}{\partial t}\Delta\varphi + \frac{1}{2}\left(\frac{\partial^2\varphi}{\partial t^2}\right)^2 - \frac{1}{2a^4}(\Delta\varphi)^2 + \frac{1}{2}M_0^4\varphi^2 + \frac{1}{2}\left(\frac{\partial\phi}{\partial t}\right)^2 + \frac{1}{2a^2}(\nabla\phi)^2 + V(\phi,\varphi). \tag{42}$$

In the case of a rapidly growing universe in a homogeneous space, the terms containing the operators ∇ and Δ can be omitted, thus the obtained field equations are simplified to the following coupled nonlinear ordinary differential equations:

$$\frac{d^2\phi_0}{dt^2}+3H\frac{d\phi_0}{dt}=-\left(m^2+g_0^2\varphi_0^2\right)\phi_0, \tag{43}$$

$$\frac{d^4\varphi_0}{dt^4}+6H\frac{d^3\varphi_0}{dt^3}=M_0^4\varphi_0+g_0^2\phi_0^2\varphi_0 \tag{44}$$

and

$$H^2=\frac{1}{3M_{pl}^2}\left[\frac{1}{2}\left(\frac{d^2\varphi_0}{dt^2}\right)^2-\frac{d\varphi_0}{dt}\frac{d^3\varphi_0}{dt^3}+\frac{1}{2}\left(\frac{d\phi_0}{dt}\right)^2+\frac{1}{2}M_0^4\varphi_0^2+\frac{1}{2}m^2\phi_0^2+\frac{1}{2}g_0^2\phi_0^2\varphi_0^2\right]. \tag{45}$$

Here, the field ϕ_0 and φ_0 depend on time only. The three coupled nonlinear ordinary differential equations, Equations (43), (44) and (45), can be considered as the equations of motion of the inflationary model. It is easy to recognize that Equation (45) can be considered as the modified version of Friedman's equation given in Equation (33). The temperature generated by the thermal field φ_0 can then be expressed as [see Equation (4a) and taking into account Equation (10) with Planck units]

$$T=\frac{d^2\varphi_0}{dt^2}+M_0^2\varphi_0. \tag{46}$$

4.3 On the time evolution of the fields

The mathematical and numerical examinations show that the solution of these coupled differential equations describes fairly well the time evolution of the inflationary universe including its thermodynamical behavior. Due to the complicated nonlinear Equations (43-45) the solutions can be achieved by numerical calculations for the time-dependence of the scalar fields and the dynamic temperature T. These equations are needed to solve simultaneously for the scalar field ϕ_0 and the thermal potential φ_0 first. After then the time evolution equation for the (thermo)dynamic temperature can be obtained.

In the present model there are two adjustable parameters, namely, the mass M_0 of the thermal field and the coupling constant g_0. The time scales of the temperature and the scalar inflaton field can be synchronized by the change of values for these two parameters. The mass of the scalar field m is chosen in the same order of magnitude as it is proposed by Linde Linde (1994), namely, $m=80GeV$. The two fitted parameters are $M_0=52.2GeV$ and $g_0=0.12GeV$. It is important to set relevant initial conditions to find reasonable numerical solutions for Equations (43) – (45). Thus, a big acceleration is assumed at the beginning of the expansion and the thermal field has a given initial value. This results an initial value for the temperature $T_0\sim 2.5\times10^6 GeV\sim 10^{19}K$. (Presently, the exact magnitude of the temperature has not too much importance, since another value can be obtained by rescaling, i.e., it does not touch the shape of the temperature function. However, it is sure, that this value is rather far from the theoretically possible $\sim 1.4\times10^{32}K$ value (Lima & Trodden, 1996; Márkus & Gambár, 2004).) In order to ensure the thermal and the inflaton field decay the first time derivatives of them are needed to be negative.

After finding a set of the numerical solutions, two main stages can be distinguished for the time evolution of the inflaton field ϕ_0. The first short period is when it decreases rapidly.

This follows the second rather long time interval in which the inflaton field oscillates with decreasing amplitude. Both of these processes can be recognized well in Fig. 3.

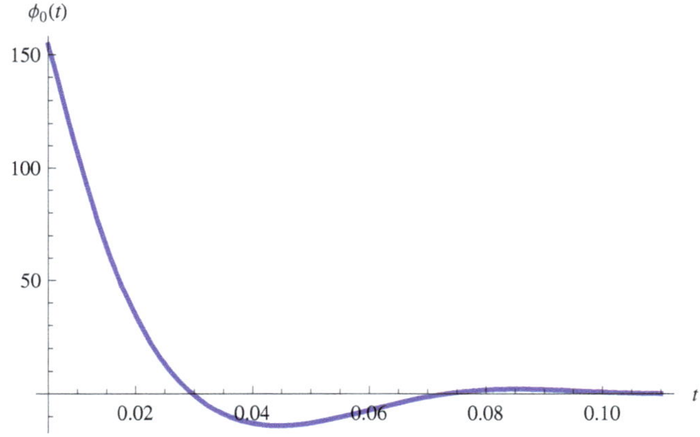

Fig. 3. The time evolution of the inflaton field $\phi_0(t)$ is shown. The short decreasing (deacying) period is followed by a rather long damped oscillating process. Time is in arbitrary units.

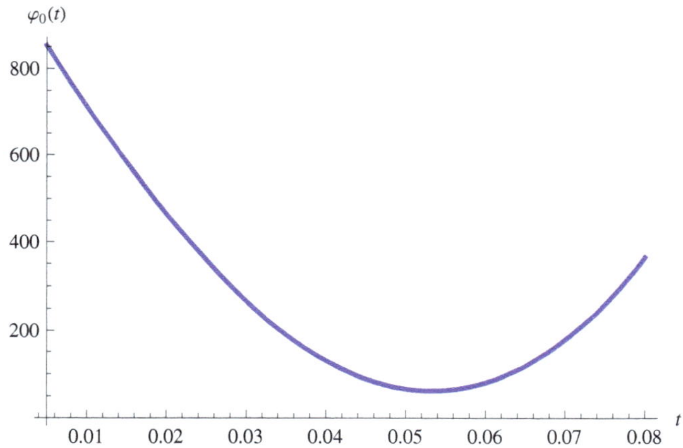

Fig. 4. The time evolution of the thermal field $\varphi_0(t)$. The field decays in the first period and reaches its minimal value. It begins to increase monotonically when the inflaton field $\phi_0(t)$ starts to oscillate. Time is in arbitrary units.

It is noticable that the above described behavior of the inflaton field is in line with Linde's cosmology model (Felder et al., 2002; Linde, 1982; 1990; 1994) based on a potential energy expression given by $V(\phi_0) = (m^2/2)\phi_0^2 + V_0$ with $V_0 > 0$ which is similar to Equation (38), here. The physically coupled thermal field φ_0 produces a completely different behavior. During inflation era, the field φ_0 decreases. Probably, the reason of this effect is strongly the radius and the volume increase of the universe. Once it reaches a minimum which happens about the same time when field ϕ_0 starts to oscillate. After then, the thermal field increases monotonically since the decaying inflaton field ϕ_0 with a time delay pumps up it as plotted in Fig. 4.

The temperature field T is coupled to the thermal field φ_0 by Equation (46), thus mathematically this can be obtained directly. The time evolution of the temperature can be followed in Fig. 5. In the first era of the inflation process the temperature decreases. After reaching its minimal value, which is at the same instantaneous of the minimum of the thermal field, it increases quite rapidly. This period of the cosmology is known as the reheating process of the universe. The present elaboration of the model can describe and reproduce to this stage of the life of the early universe.

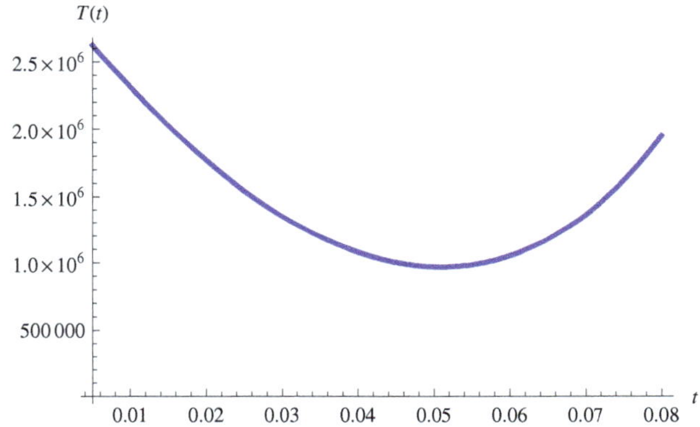

Fig. 5. The time evolution of the temperature field $T(t)$. The temperature follows the change of the thermal field φ_0. It decreases in the first period of the expansion while its reaches a minimal value. The, due to the pumping of the inflaton field ϕ_0 into the thermal field φ_0, the temperature starts increasing. This growing temperature period can be identified as the reheating process in Linde's cosmology model. Time is in arbitrary units.

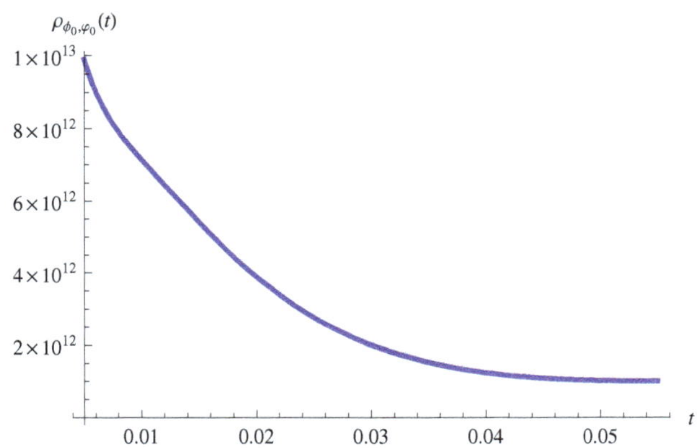

Fig. 6. The time evolution of the energy density $\rho_{\phi_0\varphi_0}(t)$. As it is expected the energy density decreases monotonically during the expansion. Time is in arbitrary units.

Since the whole energy of the universe is conserved during the expansion, the energy density is needed to decrease. This tendency can be seen in Fig. 6. Finally, the radius $a(t)$ of the universe is plotted in Fig. 7.

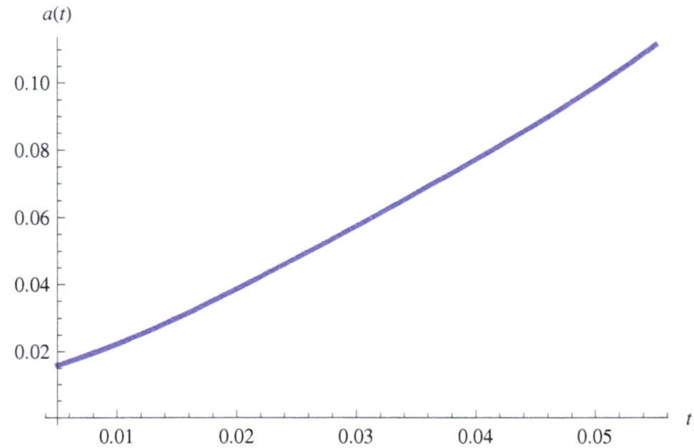

Fig. 7. The time evolution of the radius $a(t)$ of the universe. As it is expected the radius increases monotonically during the expansion. Time is in arbitrary units.

The presented model of the inflationary period is not complete in that sense that e.g., the Higgs mechanism is dropped by the elimination of the fourth term of the effective potential in Equation (27) comparing with the applied potential in Equation (38). However, hopefully, the strength of the theory can be read out from the most spectacular results: the thermal field can generate not only the spontaneous symmetry breaking involving the correct time evolution of the inflaton field, but it ensures a really dynamic Lorentz invariant thermodynamic temperature. The further development of this cosmological model would be to add the particle generator Higgs mechanism again.

5. Wheeler propagator of the Lorentz invariant thermal energy propagation

As it has been shown previously that the Lorentz invariant description involves different physically realistic propagation modes. However, the development of the theory is needed to learn more about propagation, the transition amplitude and the completeness of causality, i.e., the field equation in Equation (5a) does not violate the causality principle.

5.1 The Green function

A common way to examine these questions is based on the Green function method. Mathematically, the solution of the equation

$$\frac{1}{c^2}\frac{\partial^2 G}{\partial t^2} - \frac{\partial^2 G}{\partial x^2} - \frac{c^2 c_v^2}{4\lambda^2} G = -\delta^n(x - x') \tag{47}$$

for the Green function G is needed to find. The n-dimensional source function is $\delta^n(x - x') = \delta^{n-1}(\mathbf{r} - \mathbf{r}')\delta(t - t')$ which can be expressed by the delta function

$$\delta^n(x - x') = \frac{1}{(2\pi)^n} \int d^n k e^{ik(x - x')}. \tag{48}$$

Here, the vector $k = (\mathbf{k}, \omega_0)$ is n-dimensional; the $n-1$ dimensional $\mathbf{k}$ pertains to the space and the 1-dimensional ω_0 is to time. Moreover, the d'Alembert operator is

$$\Box = \frac{1}{c^2}\frac{\partial^2}{\partial t^2} - \Delta. \tag{49}$$

To shorten the formulations the following abbreviation is also introduced

$$m^2 = \frac{c^2 c_v^2}{4\lambda^2}. \tag{50}$$

Now, Equation (47) has a simpler form

$$(\Box - m^2)G = \delta^n(x - x'). \tag{51}$$

Since, the equality holds

$$(\Box - m^2)^{-1} e^{ik(x-x')} = -\frac{e^{ik(x-x')}}{k^2 - m^2}, \tag{52}$$

then we obtain

$$(\Box - m^2)^{-1}\delta^n(x - x') = -\frac{1}{(2\pi)^n}\int d^n k \frac{e^{ik(x-x')}}{k^2 - m^2}. \tag{53}$$

After all, the Green function can be formally expressed as

$$G(x, x') = \frac{1}{(2\pi)^n}\int d^n k \frac{e^{ik(x-x')}}{k^2 - m^2}. \tag{54}$$

To calculate this integral the zerus points of the denominator $k^2 - m^2 = \mathbf{p}^2 - p_0^2 - m^2 = 0$ are needed, from which

$$p_0 = \pm\sqrt{\mathbf{p}^2 - m^2}. \tag{55}$$

can be obtained. After then, the propagator should be expressed in proper way taking Equation (54)

$$G(p) = \frac{1}{\mathbf{p}^2 - p_0^2 - m^2}. \tag{56}$$

In the sense of the theory the retarded $G_{ret}(p) = 1/(\mathbf{p}^2 - p_0^2 - m^2)_{ret}$ and the advanced $G_{adv}(p) = 1/(\mathbf{p}^2 - p_0^2 - m^2)_{adv}$ propagators are needed to be expressed for the tachyons due to the presence of the imaginary poles. Now, the construction of the Wheeler propagator (Wheeler, 1945; 1949) can be expounded as a half sum of the above propagators

$$G(p) = \frac{1}{2}G_{adv}(p) + \frac{1}{2}G_{ret}(p). \tag{57}$$

5.2 The Bochner's theorem

The calculation of propagators is based on the Bochner's theorem (Bochner, 1959; Bollini & Giambiagi, 1996; Bollini & Rocca, 1998; 2004; Jerri, 1998). It states that if the function $f(x_1, x_2, ..., x_n)$ depends on the variable set $(x_1, x_2, ..., x_n)$ then its Fourier transformed is — without the factor $1/(2\pi)^{n/2}$ —

$$g(y_1, y_2, ..., y_n) = \int d^n x f(x_1, x_2, ..., x_n) e^{ix_i y_i} \quad (i = 1, ..., n). \tag{58}$$

However, it is useful to introduce the variables $x = (x_1^2 + x_2^2 + ... + x_n^2)^{1/2}$ and $y = (y_1^2 + y_2^2 + ... + y_n^2)^{1/2}$ instead of the original sets. Now, the examinations are restricted to the spherically symmetric functions $f(x)$ and $g(y)$. In these cases the above Fourier transform given by Equation (58) can be calculated by applying the Hankel (Bessel) transformation by which we obtain

$$g(y, n) = \frac{(2\pi)^{n/2}}{y^{n/2-1}} \int_0^\infty f(x) x^{n/2} J_{n/2-1}(xy) dx. \tag{59}$$

Here, J_α is a first kind α order Bessel function. Later it will be very useful to calculate the function f with causal functions depending on the momentum space p thus we write

$$f(x, n) = \frac{(2\pi)^{n/2}}{x^{n/2-1}} \int_0^\infty g(p) p^{n/2} J_{n/2-1}(xp) dp. \tag{60}$$

It can be seen that the singularity at the origin depends on n analytically.

5.3 Calculation of the Wheeler propagator

To obtain the Wheeler propagator, first, e.g., the integral in Equation (54) for the advanced propagator can be calculated

$$G_{adv}(x) = \frac{1}{(2\pi)^n} \int d^{n-1} p e^{i\mathbf{pr}} \int_{adv} dp_0 \frac{e^{-ip_0 x_0}}{\mathbf{p}^2 - p_0^2 - m^2}. \tag{61}$$

The path of integration runs parallel to the real axis and below both the poles for the advanced propagator. (For the retarded propagator the path runs above the poles.) Thus, considering the propagator $G_{adv}(p)$ for $x_0 > 0$ the path is closed on the lower half plane giving null result. In the opposite case, when $x_0 < 0$, there is a non-zero finite contribution of the residues at the poles

$$p_0 = \pm\omega = \sqrt{\mathbf{p}^2 - m^2} \quad if \quad \mathbf{p}^2 \geq m^2 \tag{62}$$

and

$$p_0 = \pm i\omega' = \sqrt{\mathbf{p}^2 - m^2} \quad if \quad \mathbf{p}^2 \leq m^2. \tag{63}$$

After applying the Cauchy's residue theorem for the integration with respect to p_0 we obtain an $n-1$ order integral

$$G_{adv}(x) = -\frac{H(-x_0)}{(2\pi)^{n-1}} \int d^{n-1} p e^{i\mathbf{pr}} \frac{sin[(\mathbf{p}^2 - m^2 + i0)^{\frac{1}{2}} x_0]}{(\mathbf{p}^2 - m^2 + i0)^{\frac{1}{2}}}, \tag{64}$$

where H is the Heaviside's function. The retarded propagator can be similarly obtained

$$G_{ret}(x) = \frac{H(x_0)}{(2\pi)^{n-1}} \int d^{n-1} p e^{i\mathbf{pr}} \frac{sin[(\mathbf{p}^2 - m^2 + i0)^{\frac{1}{2}} x_0]}{(\mathbf{p}^2 - m^2 + i0)^{\frac{1}{2}}}. \tag{65}$$

Considering the form of the propagator in Equation (57) and taking the propagators in Equations (64) and (65) we obtain the Wheeler-propagator

$$G(x) = \frac{Sgn(x_0)}{2(2\pi)^{n-1}} \int d^{n-1}pe^{i\mathbf{pr}} \frac{sin[(\mathbf{p}^2 - m^2 + i0)^{\frac{1}{2}} x_0]}{(\mathbf{p}^2 - m^2 + i0)^{\frac{1}{2}}}. \tag{66}$$

To evaluate the above propagators the integrals can be rewritten by the Hankel transformation based on Bochner's theorem [Equation (59)]

$$\frac{1}{(2\pi)^{n-1}} \int d^{n-1}pe^{i\mathbf{pr}} \frac{sin[(\mathbf{p}^2 - m^2 + i0)^{\frac{1}{2}} x_0]}{(\mathbf{p}^2 - m^2 + i0)^{\frac{1}{2}}} =$$

$$\frac{1}{(2\pi)^{\frac{n-1}{2}}} \frac{1}{x^{\frac{n-1}{2}-1}} \int_0^\infty p^{\frac{n-1}{2}} \frac{\sin(\mathbf{p}^2 - m^2)^{\frac{1}{2}} x_0}{(\mathbf{p}^2 - m^2)^{\frac{1}{2}}} J_{\frac{n-1}{2}-1}(xp)\, dp, \tag{67}$$

where $p = \sqrt{p_1^2 + p_2^2 + ... + p_{n-1}^2}$ and $r = \sqrt{x_1^2 + x_2^2 + ... + x_{n-1}^2}$. The following integrals (Gradshteyn & Ryzhik, 1994) are applied for the above calculations such as

$$\int_0^\infty dy\, y^{\gamma+1} \frac{\sin\left(a\sqrt{b^2 + y^2}\right)}{\sqrt{b^2 + y^2}} J_\gamma(cy) = \sqrt{\frac{\pi}{2}} b^{\frac{1}{2}+\gamma} c^\gamma (a^2 - c^2)^{-\frac{1}{4}-\frac{1}{2}\gamma} J_{-\gamma-\frac{1}{2}}(b\sqrt{a^2 - c^2}), \tag{68}$$

if $0 < c < a, \quad Re\, b > 0, \quad -1 < Re\,\gamma < 1/2$, and

$$\int_0^\infty dy\, y^{\gamma+1} \frac{\sin\left(a\sqrt{b^2 + y^2}\right)}{\sqrt{b^2 + y^2}} J_\gamma(cy) = 0, \tag{69}$$

if $0 < a < c, \quad Re\, b > 0, \quad -1 < Re\,\gamma < \frac{1}{2}$. The parameters of the model can be fitted by

$$a = x_0, \quad b = im = i\frac{cc_v}{2\lambda}, \quad c = r, \quad \gamma = \frac{n}{2} - \frac{3}{2}. \tag{70}$$

and we consider the relation between the Bessel functions

$$J_\alpha(ix) = i^\alpha I_\alpha(x), \tag{71}$$

where $I_\alpha(x)$ is the modified Bessel function. Now, we can express the advanced Wheeler propagator Equation (64) of the tachyonic thermal energy propagation

$$W_{adv}(x) = H(-x_0)\frac{\pi}{(2\pi)^{n/2}} \left(\frac{cc_v}{2\lambda}\right)^{\frac{n}{2}-1} (x_0^2 - r^2)_+^{\frac{1}{2}(1-\frac{n}{2})} I_{1-\frac{n}{2}} \left(\frac{cc_v}{2\lambda}(x_0^2 - r^2)_+^{\frac{1}{2}}\right). \tag{72}$$

The calculation for the retarded propagator can be similarly elaborated by Equations (65) and (67)

$$W_{ret}(x) = H(x_0)\frac{\pi}{(2\pi)^{n/2}} \left(\frac{cc_v}{2\lambda}\right)^{\frac{n}{2}-1} (x_0^2 - r^2)_+^{\frac{1}{2}(1-\frac{n}{2})} I_{1-\frac{n}{2}} \left(\frac{cc_v}{2\lambda}(x_0^2 - r^2)_+^{\frac{1}{2}}\right). \tag{73}$$

Comparing the results of Equations (72) and (73) it can be seen that we can write one common formula easily to express the complete propagator. Thus the Wheeler-propagator in the n dimensional space-time — remembering the construction in Equation (57) — is

$$W^{(n)}(x) = \frac{\pi}{2(2\pi)^{n/2}} \left(\frac{cc_v}{2\lambda}\right)^{\frac{n}{2}-1} (x_0^2 - r^2)_+^{\frac{1}{2}(1-\frac{n}{2})} I_{1-\frac{n}{2}} \left(\frac{cc_v}{2\lambda}(x_0^2 - r^2)_+^{\frac{1}{2}}\right). \tag{74}$$

The calculated Wheeler propagator in the $3+1$ dimensional space-time can be expressed for the thermal energy propagation

$$W^{(4)}(r, x_0) = \frac{1}{8\pi} \left(\frac{cc_v}{2\lambda}\right) (x_0^2 - r^2)^{-\frac{1}{2}} I_{-1} \left(\frac{cc_v}{2\lambda}(x_0^2 - r^2)^{\frac{1}{2}}\right). \tag{75}$$

The expected causality can be immediately recognized from the plot of the propagator in Fig. 8, since it differs to zero just within the light cone.

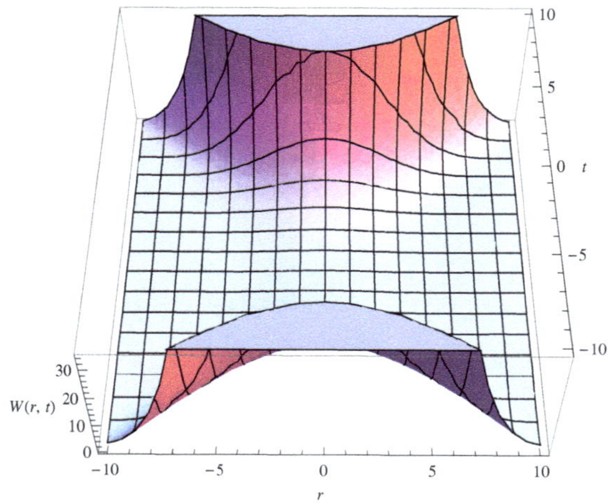

Fig. 8. The causal Wheeler propagator in the space-time — in arbitrary units — which is zero out of the light cone.

Finally, it is important to mention and emphasize that the participating particles of the above treated thermal energy propagation cannot be observable directly as Bollini's and Rocca's detailed studies (Bollini & Rocca, 1997a;b; Bollini et al., 1999) show. This is a consequence of the fact that the tachyons do not move as free particles, thus they can be considered as the mediators of the dynamic phase transition (Gambár & Márkus, 2007; Márkus & Gambár, 2010).

6. Summary and concluding remarks

This chapter of the book is dealing with the hundred years old open question of how it could be formulated and exploited the Lorentz invariant description of the thermal energy propagation. The relevant field equation as the leading equation of the theory providing the finite speed of action is a Klein-Gordon type equation with negative "mass term". It has been shown via the dispersion relations that the classical Fourier heat conduction equation is also involved, naturally. The tachyon solution of this kind of Klein-Gordon equation ensures that both wave-like (non-dissipative, oscillating) and the non-wave-like (dissipative, diffusive) signal propagations are present. The two propagation modes are divided by a spinodal instability pertaining to a dynamic phase transition. It is important to emphasize that in this

way, finally, the concept of the dynamic temperature has been introduced.
Then, a mechanical system is discussed to point out clearly that Klein-Gordon equations with the same mathematical structure and similar physical meaning can be found in the other disciplines of physics, too. The model involves a stretched string put on the diameter of a rotating disc. Collecting the kinetic and potential energy terms and formulating the Lagrange function of the problem, it has been shown that the equation of motion as Euler-Lagrange equation is exactly the above mentioned Klein-Gordon equation. The calculated dispersion relation points out unambiguously that the dynamics is similar to the case of Lorentz invariant heat conduction. The motion is vibrating (oscillating) below a system parameter dependent angular velocity, or diffusive (decaying) above this value.
The great challenge is to embed the concept of dynamic temperature into the general framework of physics. One of the aims via this step is to introduce the second law of thermodynamics by which the most basic law of nature may appear in the physical theories. Thus, such categories like dissipation, irreversibility, direction of processes can be handled directly within a description. This was the motivation to elaborate the coupling of the inflaton and the thermal field. As it can be concluded from the results, the introduced thermal field can generate the spontaneous symmetry breaking in the theory — without the Higgs mechanism — due to its property including the spinodal instability and the dynamic phase transition. The inflation decays into the thermal field by which the reheating process can start during the expansion of the universe. The time evolution of the inflation field is reproduced so well as it is known from the relevant cosmological models. It is important to emphasize that the thermal field generates a really dynamic temperature. A further progress could be achieved by the adding again the Higgs mechanism to generate massive particles in the space. This elaboration of the model remains for a future work.
Finally, it is an important step to justify that the above theory of thermal propagation completes the requirement of the causality. This question comes up due to the tachyon solutions. The arisen doubts can be eliminated in the knowledge of the propagator of the process. The relevant causal Wheeler propagator can be deduced by a longer, direct, analytic mathematical calculation applying the Bochner's theorem. The results clearly shows that the causality is completed since the propagator is within the light cone, i.e., the theory is consistent.
The presented theory of this chapter is put into the general framework of the physics coherently. These results mean a good base how to couple the thermodynamic field with the other fields of physics. Hopefully, it opens new perspectives towards in the understanding of irreversibility and dissipation in the field theoretical processes.

7. Acknowledgment

This work is connected to the scientific program of the " Development of quality-oriented and harmonized R+D+I strategy and functional model at BME" project. This project is supported by the New Hungary Development Plan (Project ID: TÁMOP-4.2.1/B-09/1/KMR-2010-0002).

8. References

Anderson C. D. R. & Tamma, K. K. (2006). Novel heat conduction model for bridging different space and time scales. *Physical Review Letters*, Vol. 96, No. 18, (May 2006) p. 184301, ISSN 0031-9007

Bochner, S. (1959). *Lectures on Fourier Integrals*, Princeton Univ., ISBN 0691079943, New Jersey, USA.

Bollini, C. G. & Giambiagi, J. J. (1996). Dimensional regularization in configuration space. *Physical Review D*, Vol. 53, No. 10, (May 1996) pp. 5761-5764, ISSN 0556-2821

Bollini, C. G. & Rocca, M. C. (1997). Vacuum state of the quantum string without anomalies in any number of dimensions. *Nuovo Cimento A*, Vol. 110, No. 4, (April 1997) pp. 353-361, ISSN 0369-3546

Bollini, C. G. & Rocca, M. C. (1997). Is the Higgs a visible particle? *Nuovo Cimento A*, Vol. 110, No. 4, (April 1997) pp. 363-367, ISSN 0369-3546

Bollini, C. G. & Rocca, M. C. (1998). Wheeler propagator. *International Journal of Theoretical Physics*, Vol. 37, No. 11, (November 1998) pp. 2877-2893, ISSN 0020-7748

Bollini, C. G., Oxman L. E. & M. C. Rocca, M. C. (1999). Coupling of tachyons to electromagnetism *International Journal of Theoretical Physics*, Vol. 38, No. 2, (February 1999) pp. 777-791, ISSN 0020-7748

Bollini, C. G. & Rocca, M. C. (2004). Convolution of Lorentz invariant ultradistributions and field theory. *International Journal of Theoretical Physics*, Vol. 43, No. 4, (April 2004) pp. 1019-1051, ISSN 0020-7748

Borsányi, Sz., Patkós, A., Polónyi, J. & Szép, Zs. (2000). Fate of the classical false vacuum. *Physical Review D*, Vol. 62, No. 8, (October 2000) p. 085013, ISSN 0556-2821

Borsányi, Sz., Patkós, A & Sexty, D. (2002). Goldstone excitations from spinodal instability. *Physical Review D*, Vol. 66, No. 2, (July 2002) p. 025014, ISSN 0556-2821

Borsányi, Sz., Patkós, A & Sexty, D. (2003). Nonequilibrium Goldstone phenomenon in tachyonic preheating. *Physical Review D*, Vol. 68, No. 6, (September 2003) p. 063512, ISSN 0556-2821

Cahill, D. G., Ford, W. K., Goodson, K. E., Mahan, G. D., Mahumdar, A., Maris, H. J., Merlin, R. & Phillpot, S. R. (2003). Nanoscale thermal transport. *Journal of Applied Physics*, Vol. 93, No. 2, (January 2003) pp. 793-818, ISSN 0021-8979

Chen, G. (2001). Ballistic-diffusive heat-conduction equations. *Physical Review Letters*, Vol. 86, No. 11, (March 2001) pp. 2297-2300, ISSN 0031-9007

Eckart, C. (1940). The thermodynamics of irreversible processes. III. Relativistic theory of the simple fluid. *Physical Review*, Vol. 58, No. 10, (November 1940) pp. 919-924

Felder, G., Kofman, L. & Linde, A. D. (1999). Inflation and preheating in nonoscillatory models. *Physical Review D*, Vol. 60, No. 10, (November 1999) p. 103505, ISSN 0556-2821

Felder, G., Kofman, L. & Linde, A. D. (2001). Tachyonic instability and dynamics of spontaneous symmetry breaking *Physical Review D*, Vol. 64, No. 12, (December 2001) p. 103505, ISSN 0556-2821

Felder, G., Frolov, A., L. Kofman, L. & Linde, A. (2002). Cosmology with negative potentials. *Physical Review D*, Vol. 66, No. 2, (July 2002) p. 023507, ISSN 0556-2821

Gambár, K. & Márkus, F. (1994). Hamilton-Lagrange formalism of nonequilibrium thermodynamics. *Physical Review E*, Vol. 50, No. 2, (August 1994) pp. 1227-1231, ISSN 1063-651X

Gambár, K. (2005). *Least action principle for dissipative processes*, pp. 245-266, in *Variational and Extremum Principles in Macroscopic Systems* (Eds. Sieniutycz, S. & Farkas, H.), Elsevier, ISBN 0080444881, Oxford.

Gambár, K. & Márkus, F. (2007). A possible dynamical phase transition between the dissipative and the non-dissipative solutions of a thermal process. *Physics Letters A*, Vol. 361, No. 4-5, (February 2007) pp. 283-286, ISSN 0375-9601

Gambár, K. & Márkus, F. (2008). A simple mechanical model to demonstrate a dynamical phase transition. *Reports on Mathematical Physics*, Vol. 62, No. 2, (October 2008) pp. 219-227, ISSN 0034-4877

Gambár, K. (2010). Change of the dynamics of the systems: dissipative – non-dissipative transition. *Informatika*, Vol. 12, No. 2, (2010) pp. 23-26, ISSN 1419-2527

Gradshteyn, S. & Ryzhik, I. M. (1994). *Tables of Integrals, Series, and Products*, Academic Press, ISBN 012294755X, San Diego.

Jerri, A. J. *The Gibbs Phenomenon in Fourier Analysis, splines, and wavelet approximations*, Kluwer, ISBN 0792351096, Dordrecht.

Joseph, D. D. & Preziosi, L. (1989). Heat waves. *Reviews of Modern Physics*, Vol. 61, No. 1, (January 1989) pp. 41-73, ISSN 0034-6861

Jou, D., Casas-Vázquez & Lebon, G. (2010). *Extended irreversible thermodynamics*, Springer, ISBN 9789048130733, New York, USA.

Lima, J. A. S. & Trodden , M. (1996). Decaying vacuum energy and deflationary cosmology in open and closed universes. *Physical Review D*, Vol. 53, No. 8, (April 1996) pp. 4280-4286, ISSN 0556-2821

Linde, A. D. (1982). A new inflationary universe scenario — A possible solution of the horizon, flatness, homogeneity, isotropy and primordial monopole problems. *Physics Letters B*, Vol. 108, No. 6, (February 1982) pp. 389-393, ISSN 0370-2693

Linde, A. D. (1990). *Particle Physics and Inflationary Cosmology* , Harwood Academic Publishers, ISBN 3718604892, Chur, Switzerland.

Linde, A. D. (1994). Hybrid inflation. *Physical Review D*, Vol. 49, No. 2, (January 1994) pp. 748-754, ISSN 0556-2821

Liu, W. & Asheghi, M. (2004). Phonon-boundary scattering in ultrathin single-crystal silicon layers. *Applied Physics Letters*, Vol. 84, No. 19, (May 2004) pp. 3819-3821, ISSN 0003-6951

Ma, S.-k. (1982). *Modern Theory of Critical Phenomena*, Addison-Wesley, ISBN 0805366717, California, USA.

Márkus, F. & Gambár, K. (1991). A variational principle in the thermodynamics. *Journal of Non-Equilibrium Thermodynamics*, Vol. 16, No. 1, (March 1991) pp. 27-31, ISSN 0340-0204

Márkus, F. & Gambár, K. (2004). Derivation of the upper limit of temperature from the field theory of thermodynamics. *Physical Review E*, Vol. 70, No. 5, (November 2004) p. 055102(R), ISSN 1063-651X

Márkus, F. (2005). *Hamiltonian formulation as a basis of quantized thermal processes*, pp. 267-291, in *Variational and Extremum Principles in Macroscopic Systems* (Eds. Sieniutycz, S. & Farkas, H.), Elsevier, ISBN 0080444881, Oxford, England.

Márkus, F. & Gambár, K. (2005). Quasiparticles in a thermal process. *Physical Review E*, Vol. 71, No. 2, (June 2005) p. 066117, ISSN 1539-3755

Márkus, F., Vázquez, F. & Gambár, K. (2009). Time evolution of thermodynamic temperature in the early stage of universe. *Physica A*, Vol. 388, No. 11, (June 2009) pp. 2122-2130, ISSN 0378-4371

Márkus, F. & Gambár, K. (2010). Wheeler propagator of the Lorentz invariant thermal energy propagation. *International Journal of Theoretical Physics*, Vol. 49, No. 9, (September 2010) pp. 2065-2073, ISSN 0020-7748

Morse, Ph. M. & Feshbach, H. (1953). *Methods of Theoretical Physics*, McGraw-Hill, New York, USA.

Sandoval-Villalbazo, A. & García-Colín L. S. (2000). The relativistic kinetic formalism revisited. *Physica A* Vol. 278, No. 3-4, (April 2000) pp. 428-439, ISSN 0378-4371

Schwab, K., Henriksen, E. A., Worlock, J. M. & Roukes, M. L. (2000). Measurement of the quantum of thermal conductance. *Nature*, Vol. 404, No. 6781, (April 2000) pp. 974-977, ISSN 0028-0836

Sieniutycz, S. (1994). *Conservation laws in variational thermo-hydrodynamics*, Kluwer Academic Publishers, ISBN 0792328027, Dordrecht, The Netherlands.

Sieniutycz, S. & Berry, R. S. (2002). Variational theory for thermodynamics of thermal waves. *Physical Review E*, Vol. 65, No. 4, (April 2002) p. 046132, ISSN 1063-651X

Vázquez, F., Márkus, F. & Gambár, K. (2009). Quantized heat transport in small systems: A phenomenological approach. *Physical Review E*, Vol. 79, No. 3, (March 2009) p. 031113, ISSN 1539-3755

Wheeler, J. A. & Feynman, R. P. (1945). Interaction with the absorber as the mechanism of radiation *Reviews of Modern Physics*, Vol. 17, No. 2-3, (April-July 1945) pp. 157-181

Wheeler, J. A. & Feynman, R. P. (1949). Classical electrodynamics in terms of direct interparticle action. *Reviews of Modern Physics*, Vol. 21, No. 3, (July 1949) pp. 425-433

Part 3

Coupling Between Heat Transfer and Electromagnetic or Mechanical Excitations

9

Energy Transfer in Pyroelectric Material

Xiaoguang Yuan[1] and Fengpeng Yang[2]
[1]*York University*
[2]*Shanghai Jiao Tong University*
[1]*Canada*
[2]*China*

1. Introduction

Smart materials are different from the usual materials and can sense their environment and respond, in the flexibility of its properties that can be significantly altered in a controlled fashion by external stimuli, such as stress, temperature, electric and magnetic fields. Fig. 1 shows the general relationship in smart materials among mechanical, electrical, and thermal fields. Such characteristics enable technology applications across a wide range of sectors including electronics, construction, transportation, agriculture, food and packaging, health care, sport and leisure, white goods, energy and environment, space, and defense.

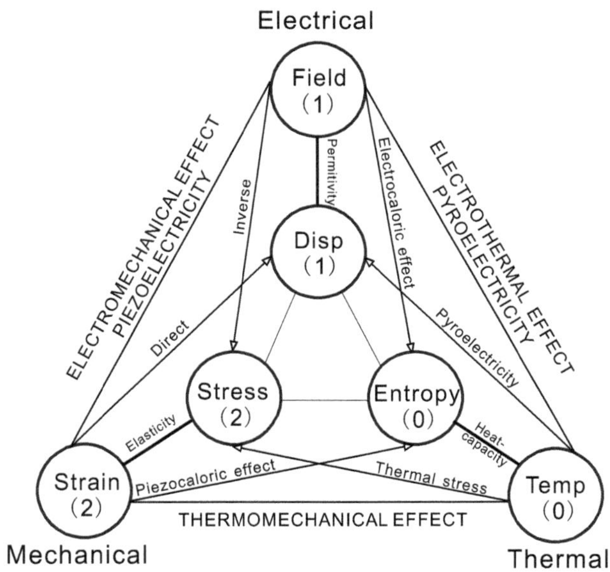

Fig. 1. The relationship among mechanical, electrical, and thermal fields.

The most widely used smart materials are piezoelectric ceramics, which expand or contract when voltage is applied. Pyroelectric material is a kind of smart materials and can be electrically polarized due to the temperature variation. Fig. 2 indicates the relationship between pyroelectrics and other smart materials. It follows that a pyroelectric effect cannot exist in a crystal possessing a center of symmetry. Among the 21 noncentrosymmetrical crystalline classes only 10 may theoretically show pyroelectric character, (Cady, 1946; Eringen & Maugin, 1990; Nelson, 1979). It has many applications which occur both in technology (i.e. infrared detection, imaging, thermometry, refrigeration, power conversion, memories, biology, geology, etc...) and science (atomic structure of crystals, anharmonicity of lattice vibrations etc...(Hadni, 1981)). Recently, advanced technical developments have increased the efficiency of devices by scavenging energy from the environment and transforming it into electrical energy. When thermal energy is considered and spatial thermal gradients are present, thermoelectric devices can be used. When thermal fluctuations are present, the pyroelectric effect can be considered, see (Cuadras et al., 2006; Dalola et al., 2010; Fang et al., 2010; Gael & et al., 2009; Guyomar et al., 2008; Khodayari et al., 2009; Olsen et al., 1984; Olsen & Evans, 1983; Shen et al., 2007; Sodano et al., 2005; Xie et al., 2009). The thermal wave, also called temperature wave, is also found to be a good method to probe in a remote way near surface boundaries, to measure layer thicknesses and to locate faults (Busse, 1991).

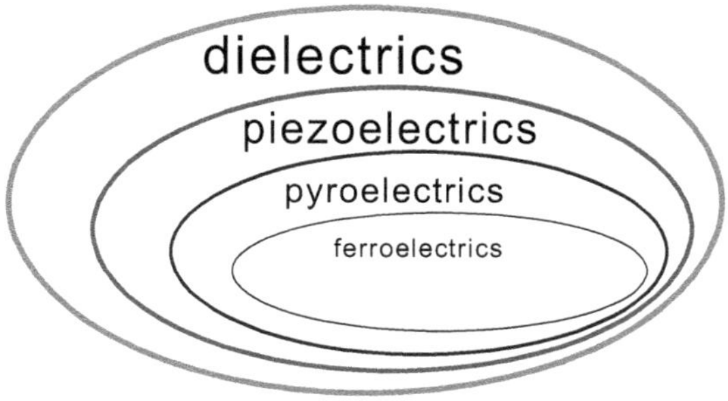

Fig. 2. The relationship of dielectrics, piezoelectrics, pyroelectrics and ferroelectrics.

Therefore, pyroelectric medium can be transformer among mechanical, electrical and thermal energies. It is with this feature in mind that we have to do research to cover the coupling even if only one type energy is needed. In this chapter the following works are performed to exploit pyroelectric material.

Firstly, the general theory of inhomogeneous waves in pyroelectric medium is addressed. Majhi (Majhi, 1995) studied the transient thermal response of a semi-infinite piezoelectric rod subjected to a local heat source along the length direction, by introducing a potential function and applying the Lord and Shulman theory. Sharma and Kumar (Sharma & Kumar, 2000) studied plane harmonic waves in piezo-thermoelastic materials. He, Tian and Shen (He et al., 2002) discussed various thermal shock problems of a piezoelectric plate. Baljeet (Baljeet, 2005) formulated the governing differential equations for generalized thermo-piezoelectric solid by using both L-S and G-L theories and found that the velocities of these plane waves depend upon properties of material and the angle of propagation. Sharma and Pal (Sharma & Pal, 2004) discussed the propagation of plane harmonic waves in transversely isotropic generalized piezothermoelastic materials and found four dispersive modes. The propagation of Rayleigh waves in generalized piezothermoelastic half-space is investigated by Sharma and Walis (Sharma & Walia, 2007). Topics of homogeneous and inhomogeneous waves, reflection/transmission and energy problems in pyroelectrics are firstly researched by authors (Kuang, 2009; 2010; Kuang & Yuan, 2010; Yuan, 2009; Yuan & Kuang, 2008; 2010).

The speciality of pyroelectric material lies in its relaxation in corresponding thermal field. Introduction of relaxation time into the heat conduction theory is about 50 years ago. Cattaneo (Cattaneo, 1958) and Vernotte (Vernotte, 1958) originally proposed the relaxation time for heat flux in the heat conduction theory, on basis of which the governing equations of thermoelasticity with relaxation time were deduced by Kaliski (Kaliski, 1965), and independently by Lord and Shulman (Lord & Shulman, 1967). Notwithstanding, this theory is usually called L-S theory. Several years later, Green and Lindsy (Green & Lindsay, 1972) gave another form of governing equations for thermoelasticity called G-L theory. Further, Joseph and Preziosi (Joseph & Preziosi, 1989) used two relaxation times: one for heat flux and the other for temperature gradient, and also obtained a system of equations of thermoelasicity. Kuang (Kuang, 2009; 2010) proposed an inertial entropy theory and got the governing equations for thermoelasticity which is different from L-S and G-L theories. For pyroelectrics the effects of relaxation times on wave velocities and attenuation are estimated by (Kuang, 2009; 2010; Yuan, 2009; Yuan & Kuang, 2008; 2010).

Taking account of the relaxation, we introduce the inhomogeneous wave into pyroelectric medium here. The difference from the homogeneous wave is that the wave propagation vector is not coincident with the attenuation vector. The attenuation angle, defined by the angle between wave propagation vector and attenuation vector, is found to be limited in the range of (-90°,90°). It is found that increasing the attenuation angle will introduce more dissipation and anisotropy. In our work, four wave modes are found in pyroelectric medium, which are temperature, quasitransverse I, II and quasilongitudinal due to the coupling state relationship. Though there is no independent wave mode for the electric field, it can still propagate with other wave modes. The variations of phase velocities and attenuations with propagation angle and attenuation angle are discussed. Phase velocity surfaces on anisotropic and isotropic planes are presented for different attenuation angle. It is found that attenuation angle almost doesn't influence the phase velocities of elastic waves in both anisotropic and isotropic planes. In contrast, the roles it plays on temperature wave are obvious. The effects of the positive and negative attenuation angles are not the same in anisotropic plane.

The propagation of a wave in any medium is associated with the movement of energy. Therefore, the energy process in pyroelectrics is researched for the first time.

The energy process especially the dissipation energy is one of the most important dynamic characteristics of continuous media. Many researches were conducted on this problem. Umov (Umov, 1874) introduced the concept of the energy flux vector and found the first integral of energy conservation equations of elasticity theory. Fedorov (Fedorov, 1968) used this theory and discussed the energy flux, energy density and the energy transport velocity of plane waves in the elastic theory. In paper of (Kiselev, 1982), the energy fluxes of complex fields in inhomogeneous media were considered. Based on Umov's theory of energy flux, he represented analogous results for complex fields which are characterized by the pair of complex vector fields. On the basis of the results, the Lagrangian density and Umov vector were derived. At the same time, the question of additivity of the Umov flux vectors of longitudinal and transverse waves was also discussed.

For the class of plane inhomogeneous waves propagating in linear viscoelastic media, Buchen (Buchen, 1971) gave a detailed description of the physical properties and energy associated with these inhomogeneous waves. The paralleled paper by Borcherdt (Borcherdt, 1973) adopted a different derivation from Buchen's and discussed the mathematical framework for describing plane waves in elastic and linear inelastic media. The expressions for the energy flux, energy densities, dissipated energy, stored energy were derived from an explicit energy conservation relation. Based on the motion equation and its integral form, Červený (Cerveny & Psencik, 2006) discussed three different types of energy fluxes in anisotropic dissipative media. The relationships among them, especially their applications in the interface between dissipative media, were researched in detail. In the field of piezoelectrics, Auld (Auld, 1973) derived the energy flux in the electromagnetic field and also its form in the piezoelectric media. Baesu (Baesu et al., 2003) considered non-magnetizable hyperelastic dielectrics which conduct neither heat nor electricity and also obtained the energy flux with the linearized theory.

In this chapter, the energy process in pyroelectric medium with generalized heat conduction theory is studied firstly. According to the derived energy conservation law, the energy densities, energy dissipated and energy flux are defined. Generally there are several type velocities in wave theory, such as phase velocity, group velocity and energy velocity. The phase velocity is related to the phase of the wave. Owing to damping, the usual definition of group velocity of waves become meaningless and this issue can be solved by considering the energy of the physical phenomenon of wave propagation (Mainardi, 1973). Regarding the propagation of the energy, the energy flux may be used in order to quantify the energy velocity vector and they have the same direction. The energy flux vector has a dynamical definition and consequently, polarization of the wave (the amplitudes of displacements, temperature and electric potential) is taken into account. In particular the phase velocity and energy velocity are compared in the results and discussion section.

We shall use the operation rules: the dot above a letter denotes the time derivative, the index following the comma in the subscript denotes the partial derivative with respect to relevant Cartesian coordinate, and the asterisk in the superscript denotes the complex conjugate.

2. The inhomogeneous waves in pyroelectric medium

2.1 The governing equations and state equations

The pyroelectric medium can be influenced by the mechanical, electric and thermal fields. These fields have their own governing equations. The physical quantities of pyroelectric medium in these fields are not independent, because they are related by the state equations. The known fundamental equations for the pyroelectric medium are listed as follows.

1. Mechanical field equations in $\Re^3$

   Equation of motion:

$$\sigma_{ij,j} + b_i = \rho \ddot{u}_i \tag{1}$$

   Geometric property:

$$\varepsilon_{ij} = \frac{1}{2}(u_{i,j} + u_{j,i}) \tag{2}$$

   where u_i is the displacement vector, σ_{ij} the stress tensor, b_i the body force per unit volume, ρ the density and ε_{ij} the strain tensor.

2. Electrical field equations under the quasi-static assumption $\Re^3$

   Gauss equation:

$$D_{i,i} = \varrho_e \tag{3}$$

   where D_i is the electric displacement. The absence of free charge requires $\varrho_e = 0$. In quasi-static approximation, the electric field $\mathbf{E}$ is derivable from a potential, that is

$$(\nabla \times \mathbf{E})_i = 0, E_i = -\varphi_{,i} \tag{4}$$

where φ is the scalar quasi-static electric potential.

3. Thermal field equations in $\Re^3$
   If the temperature disturbance $\theta \ll T_0$, the entropy equation is

$$\rho T_0 \dot{\eta} = -q_{i,i} \tag{5}$$

in which T_0 is the initial temperature, η is the entropy per unit volume. The thermal flux vector q_i is related to the temperature disturbance $\theta = T - T_0$ by

$$Lq_i = -\kappa_{ij}\theta_{,j} \tag{6}$$

in which L is an operator defined by

$$L = 1 + \tau\frac{\partial}{\partial t}$$

Equation (6) is called the generalized Fourier heat conduction equation. In these two equations, κ_{ij} indicates the heat conduction constant and τ is the relaxation time.
In the above individual field introduces physical quantities, and they are not independent and should satisfy the state equations, which play roles in two aspects: 1. physically they reflect the real world interactions among the three fields; 2. they are useful to formulate a solvable equation system mathematically. The constitutive equations (Yuan & Kuang, 2008) can be expressed by

$$\begin{aligned}
\sigma_{ij} &= c_{ijkl}\varepsilon_{kl} - e_{kij}E_k - \gamma_{ij}\theta \\
D_k &= e_{kij}\varepsilon_{ij} + \lambda_{ik}E_i + \xi_k\theta \\
\rho\eta &= \gamma_{ij}\varepsilon_{ij} + \xi_i E_i + \frac{\rho C\theta}{T_0}
\end{aligned} \tag{7}$$

In this system of equations, c_{ijkl} denotes the elastic stiffness; e_{kij} the piezoelectric tensor; γ_{ij} the thermo-mechanical tensor; ρ the density; λ_{ik} the dielectric permittivity tensor; ξ_k the pyroelectric constants'; T_0 the initial temperature; C is the specific heat capacity.
Inserting these state equations into Equations (1), (4) and (5) and using Equations (2) and (6), we obtain

$$\begin{aligned}
&c_{ijkl}u_{k,lj} + e_{kij}\varphi_{,kj} + \gamma_{ij}\theta_{,j} = \rho\ddot{u}_i \\
&e_{kij}u_{i,jk} - \lambda_{ik}\varphi_{,ik} + \xi_k\theta_{,k} = 0 \\
&T_0\gamma_{ij}\left(\dot{\varepsilon}_{ij} + \tau\ddot{\varepsilon}_{ij}\right) + T_0\xi_i\left(\dot{E}_i + \tau\ddot{E}_i\right) + \rho C\left(\dot{\theta} + \tau\ddot{\theta}\right) = \kappa_{ij}\theta_{,ij}
\end{aligned} \tag{8}$$

which is a system of equations in the unknown fundamental functions: the displacements u_k, the electric potential φ, the temperature disturbance θ. There are 7 equations in this system and also the same number of unknowns, therefore it can be solved.

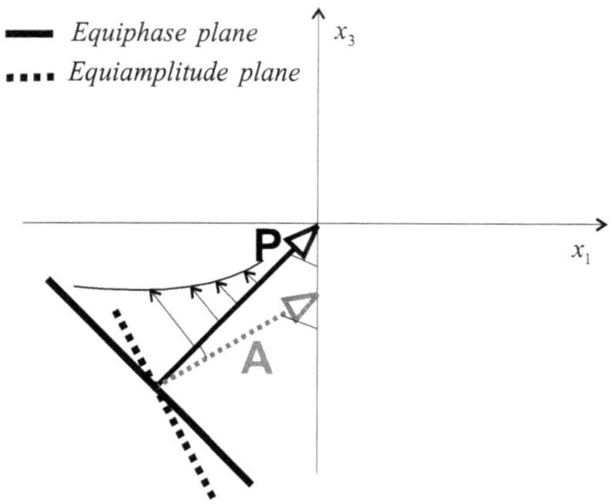

Fig. 3. Equiphase plane, equiamplitude plane and exponential variation of the amplitude along the phase propagation direction.

2.2 The fundamental concepts of inhomogeneous wave theory

When the wave vector is complex, generally speaking, the propagation direction (normal to the equiphase plane) is different from the attenuation direction (normal to the equiamplitude plane), see Fig. 3. Any plane wave can be expressed as

$$\begin{aligned} &f = f_0 e^{i(\mathbf{k}\cdot\mathbf{x}-\omega t)} = f_0 e^{i(k_m x_m - \omega t)}, \quad \mathbf{k} = [k_1, k_2]^T = \mathbf{P} + i\mathbf{A} \\ &\mathbf{P} = P\mathbf{n}, \quad \mathbf{A} = A\mathbf{m}, \quad k_j = P_j + iA_j, \quad k^2 = \mathbf{k}\cdot\mathbf{k} = P^2 - A^2 + 2i\mathbf{P}\cdot\mathbf{A} \end{aligned} \tag{9}$$

where $\mathbf{P}$ is the propagation vector, P is its module, and $\mathbf{n}$ is the unit vector along the propagation direction; $\mathbf{A}$ is the attenuation vector, A is its module, and $\mathbf{m}$ is the unit vector perpendicular to the plane of constant amplitude. When $\mathbf{n} = \mathbf{m}$, we call it homogeneous wave, otherwise inhomogeneous wave. Hereafter, we assume that θ, transportation angle, is the angle between $\mathbf{n}$ and x_2; γ, attenuation angle, is the angle between $\mathbf{n}$ and $\mathbf{m}$; and $\vartheta(= \theta + \gamma)$ is the angle between $\mathbf{m}$ and x_2. Using Equation (9), we obtain

$$\begin{aligned} &\mathbf{n} = [\sin\theta, \cos\theta]^T, \ \mathbf{m} = [\sin(\theta+\gamma), \cos(\theta+\gamma)]^T, \ \mathbf{n}\cdot\mathbf{m} = \cos\gamma \\ &k_1 = P_1 + iA_1 = Pn_1 + iAm_1, \ k_2 = P_2 + iA_2 = Pn_2 + iAm_2 \\ &P = \sqrt{P_1^2 + P_2^2}, \ A = \sqrt{A_1^2 + A_2^2} \end{aligned}$$

Due to $\mathbf{n} = \mathbf{m}$ and $\gamma = 0$ in homogeneous wave, we have $k_1 = (P + iA)\sin\theta$, $k_2 = (P + iA)\cos\theta$. Therefore, $\mathbf{k}$ is determined by one complex number and a real propagation angle θ, but in inhomogeneous wave $\mathbf{n} \neq \mathbf{m}$, we have to use four parameters (P, A, θ, γ) to determine wave vector.

Unlike propagation angle θ, γ has its boundary to guarantee the waves are of attenuation. On the basis of non-negative dissipation rate of linear viscoelastic media, Buchen (Buchen, 1971) verified that γ is in the range of 0° to 90° and the same conclusion can also be seen in reference (Borcherdt, 1973). In the present paper, the boundary of attenuation angle γ is determined by the condition that waves should be attenuate physically.

2.3 The propagation of inhomogeneous plane waves in infinite medium

For the solution to Equation 8, the general monochromatic plane waves are assumed as

$$\begin{aligned} u_k &= U_k \exp\left[i(x_i k_i - \omega t)\right] \\ \theta &= \Theta \exp\left[i(x_i k_i - \omega t)\right] \\ \varphi &= \Psi \exp\left[i(x_i k_i - \omega t)\right] \end{aligned} \tag{10}$$

where k_i is the complex-valued wave vector, ω is the circular frequency, t is the time variable and U_j, Θ and Ψ are generally the complex amplitudes (or polarizations) of displacements, temperature and electric potential respectively. The subscript i, k equal to 1, 2, 3. It is noted that in Equation (10), $\exp\left[i(x_i k_i - \omega t)\right]$ is used, which is different from homogeneous wave with $\exp\left[i(kn_i x_i - \omega t)\right]$. In other words, in the inhomogeneous wave, $k_i x_i$ can't be expressed as $kn_i x_i$.

Inserting Equation (10) into Equation (8) yields a system of Christoffel algebraic equations in amplitude vector $\mathbf{U}$

$$\mathbf{\Lambda}\left(\mathbf{k}, \omega\right) \mathbf{U} = \mathbf{0}, \qquad \mathbf{U} = \left[U_1, U_2, U_3, \Psi, \Theta\right]^T \tag{11}$$

$$\mathbf{\Lambda}\left(k, \omega, \mathbf{n}\right) = \begin{bmatrix} \Gamma_{11}\left(\mathbf{k}\right) - \rho\omega^2 & \Gamma_{12}\left(\mathbf{k}\right) & \Gamma_{13}\left(\mathbf{k}\right) & i\alpha_1^*\left(\mathbf{k}\right) & e_1^*\left(\mathbf{k}\right) \\ \Gamma_{21}\left(\mathbf{k}\right) & \Gamma_{22}\left(\mathbf{k}\right) - \rho\omega^2 & \Gamma_{23}\left(\mathbf{k}\right) & i\alpha_2^*\left(\mathbf{k}\right) & e_2^*\left(\mathbf{k}\right) \\ \Gamma_{31}\left(\mathbf{k}\right) & \Gamma_{32}\left(\mathbf{k}\right) & \Gamma_{33}\left(\mathbf{k}\right) - \rho\omega^2 & i\alpha_3^*\left(\mathbf{k}\right) & e_3^*\left(\mathbf{k}\right) \\ e_1^*\left(\mathbf{k}\right) & e_2^*\left(\mathbf{k}\right) & e_3^*\left(\mathbf{k}\right) & -i\xi_k k_k & \lambda^*\left(\mathbf{k}\right) \\ \gamma_1^*\left(\mathbf{k}\right)\omega & \gamma_2^*\left(\mathbf{k}\right)\omega & \gamma_3^*\left(\mathbf{k}\right)\omega & \kappa^*\left(\mathbf{k}\right) & \xi^*\left(\mathbf{k}\right) \end{bmatrix} \tag{12}$$

where

$$\begin{aligned} &\Gamma_{ik}\left(\mathbf{k}\right) = C_{ijkl} k_j k_l, e_i^*\left(\mathbf{k}\right) = e_{kij} k_k k_j, \gamma_i^*\left(\mathbf{k}\right) = T_0 \gamma_{ij} k_j\left(\omega - i\tau\omega^2\right) \\ &\xi^*\left(\mathbf{k}\right) = T_0 \xi_i k_i\left(-\omega + i\tau\omega^2\right), \lambda^*\left(\mathbf{k}\right) = \lambda_{ik} k_i k_k, \ \kappa^*\left(\mathbf{k}\right) = \kappa_{ij} k_i k_j - \rho C\left(i\omega + \tau\omega^2\right) \end{aligned} \tag{13}$$

Nontrivial solutions for U_i, Θ and Ψ require

$$\det \mathbf{\Lambda}\left(\mathbf{k}, \omega\right) = 0. \tag{14}$$

which is complex equation in wave vector $\mathbf{k}$ for given ω. Decomposing the equation into the real and imaginary parts, we can obtain a solvable equations in P and A:

$$\begin{cases} D_{\Re}(P, A) = 0 \\ D_{\Im}(P, A) = 0 \end{cases} \text{and } P, A \in 0 \cup \mathbb{R}^+ \tag{15}$$

Due to that the equations are very tedious, we would not present them in explicit forms. Equation (15) are nonlinear and coupling equations in (P, A). According to the definitions of P and A in Equation (9), the right solution of P and A should be real valued. Therefore, the domain of θ and γ are determined by the condition that P and A are nonnegative real numbers(only one direction of wave propagation is considered). The wave propagates with the velocity c_p $(=\omega/P)$, with non-negative value in attenuation A. This condition agrees with the Sommerfeld radiation condition; i.e., vanishing at infinity. When A and P are obtained for given θ and γ, we can use Equations (9) to determine the inhomogeneous wave vector $\mathbf{k}$. For each k_i, we can get a corresponding amplitude vector $\mathbf{U}$ with one undetermined component. Generally, there are four roots of **(P,A)** to Equation (15) corresponding to four wave vector $\mathbf{k}$. For every $\mathbf{k}$, $\mathbf{P}$ and $\mathbf{A}$, we have two components $\left(k_{i\alpha}, P_{k\alpha}, A_{k\alpha}\right)$, in which $i = 1, 2, 3, 4$ and $\alpha = 1, 2$. They are related to three elastic waves and one temperature wave; The electric field doesn't have its own wave mode, but, through the constitutive relations, it can propagate with other four wave modes. After P, A are solved, the phase velocity can be given by

$$c_p = \frac{\omega}{P} \tag{16}$$

and also the attenuation A.
Therefore, general solutions in pyroelectric medium equal to the sum of four wave modes, which are

$$\begin{gathered} u_k = \sum_{j=1}^{4} U_k^{(j)} e^{i\left(k_m^{(j)} x_m - \omega t\right)} = \sum_{j=1}^{4} U_k^{(j)} e^{i\left[\left(P^{(j)}\mathbf{n}^{(j)} + iA^{(j)}\mathbf{m}^{(j)}\right)\cdot \mathbf{x} - \omega t\right]} \\ \theta = \sum_{j=1}^{4} \Theta^{(j)} e^{i\left(k_m^{(j)} x_m - \omega t\right)} \quad \varphi = \sum_{j=1}^{4} \Psi^{(j)} e^{i\left(k_m^{(j)} x_m - \omega t\right)} \end{gathered} \tag{17}$$

in which j indicates the wave mode.

2.4 Quantitative analysis of pyroelectric media

The material under study is transversely isotropic $BaTiO_3$, in which the isotropic plane is x_1-x_2 and the anisotropic plane is x_1-x_3 plane. All the physical constants are rewritten with the help of Voigt notation, whose rule is that the subscript of a tensor is transformed by $\{11 \rightarrow 1,\ 22 \rightarrow 2,\ 33 \rightarrow 3,\ 23 \rightarrow 4,\ 31 \rightarrow 5,\ 12 \rightarrow 6\}$.

Coordinate index	(11)	(12)	(13)	(33)	(44)	(66)	(15)
Elastic moduli $E(10^{10}\text{Pa})$	15.0	6.6	6.6	14.6	4.4	4.3	
Piezoelectric Charge constant $e(\text{C/m}^2)$			-4.35	17.5			11.4
Electric permittivity $\lambda(10^{-9}\text{f/m})$	9.867			11.15			
Thermal expansion tensor $\alpha(10^{-6}\text{1/K})$	8.53		1.99				
Pyroelectric constant $\xi(10^{-4}\text{C/m}^2\text{K})$				5.53			
Thermal conductivity tensor $\kappa(\text{J/m}\cdot\text{K}\cdot\text{s})$	1.1	1.1		3.5			

Table 1. Material properties of $BaTiO_3$

The material constants of $BiTiO_3$ studied in this paper are shown in Table 1. The specific heat capacity C is 500 (J/K·Kg); the relaxation times $\tau = 10^{-10}$s for L-S theory; density $\rho = 5700$ kg/m^3; the prescribed circular frequency $\omega = 2\pi \times 10^6\ \text{s}^{-1}$; the thermo-mechanical coupling coefficients γ_{ij} are given by

$$\gamma_{11} = \gamma_{22} = (c_{11} + c_{12})\alpha_{11} + (c_{13} + e_{31})\alpha_{33},\ \gamma_{33} = 2c_{13}\alpha_{11} + (c_{33} + e_{33})\alpha_{33}$$

2.4.1 Determination of the boundary of attenuation angle

The condition in Equation (15) requires that the attenuation angle γ should be limited in the range of $(-90^0, 90^0)$ to get attenuate wave, by which we can obtain four wave modes: quasilongitudinal, quasitransverse I, II and temperature waves. This conclusion is consistent with previous researchers (Borcherdt, 1973; Buchen, 1971; Kuang, 2002). Their studies demonstrated that attenuation angle γ is confined in the range of $(0^0, 90^0)$ for isotropic viscoelastic medium. This result can be arrived at by ours, that the positive and negative attenuation angles come to the same results for isotropic medium. But the influences of positive and negative attenuation angles on waves in the anisotropic plane for the transverse material are different. Attenuation angle introduces more dissipation and anisotropy (Carcione & Cavallini, 1997).

2.4.2 The velocity surfaces

With the material constants shown in Table 1, the phase velocity surface sections are calculated. Fig. 4(a),(b) show the sections of phase velocity surfaces in the anisotropic x_1-x_3 plane and isotropic x_1-x_2 plane. It is demonstrated that the attenuation angle γ almost doesn't change the phase velocities of elastic waves, therefore only the case at $\gamma = 0$ is presented. The elastic wave velocity surfaces, including quasilongitudinal, quasitransverse I,II waves, show the anisotropic behaviors in the anisotropic x_1-x_3 plane. It is seen that, in Fig. 4(a), the quasi-longitudinal waves are the fastest, while the thermal wave are the slowest and the quasi-transversal waves stand in between them and all of them are related to propagation angle θ. Instead the role played by attenuation angle γ on temperature wave is obvious as shown in Fig. 4(b). The influences of the positive and negative attenuation angles are different in anisotropic x_1-x_3 plane, but both can reduce the velocity of temperature wave.
On the isotropic x_1-x_2 plane, Fig. 5(b) implies that the negative and positive attenuation angle have the same role. Velocities of all waves in isotropic plane don't depend on the propagation angle.

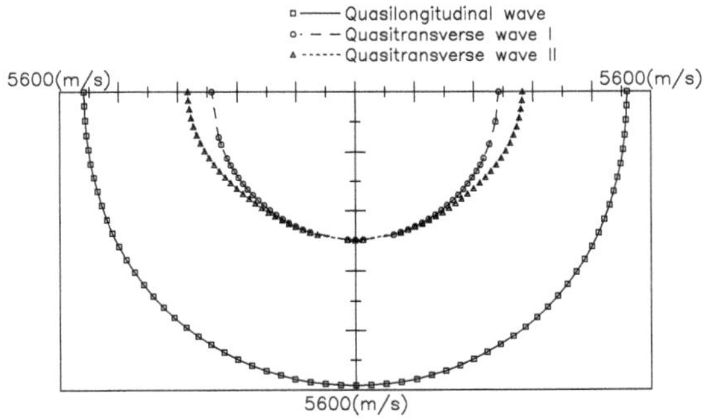

(a) Velocity surfaces of elastic waves.

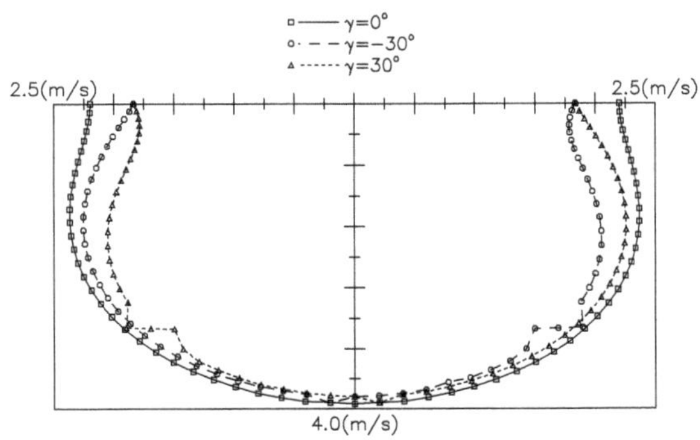

(b) Velocity surfaces of temperature wave.

Fig. 4. Sections of the velocity surfaces in (x_1,x_3) plane at different attenuation angle γ.

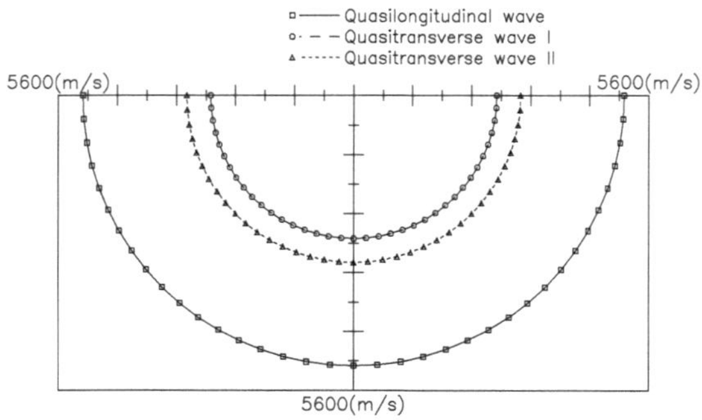

(a) Velocity surfaces of elastic waves.

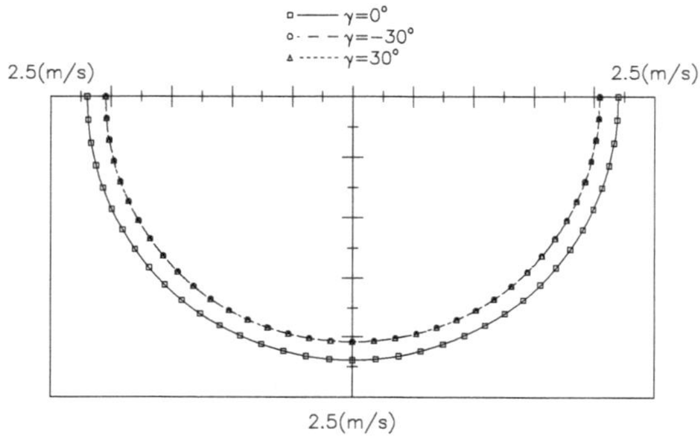

(b) Velocity surfaces of temperature waves.

Fig. 5. Sections of the velocity surfaces in (x_1,x_2) plane at different attenuation angle γ.

3. Dynamic energy balance law in pyroelectric medium

We shall formulate the energy balance laws as consequences of the governing equations presented in the previous section, see (Yuan, 2010).

I. We consider the scalar product of the velocity $\dot{u}_i$ with the motion equation. Multiplying Equation (1) by $\dot{u}_i$ results in

$$\sigma_{ij,j}\dot{u}_i + \rho b_i\dot{u}_i = \rho\ddot{u}_i\dot{u}_i$$

Taking account of the identity

$$(\sigma_{ij}\dot{u}_i)_{,j} = \sigma_{ij,j}\dot{u}_i + \sigma_{ij}\dot{u}_{i,j}$$

and, by considering a region Ω with surface element $\partial\Omega$ in the configuration of the body, applying the volume integral and Gaussian Theorem to the previous equation, we obtain

$$\int_{\partial\Omega} \sigma_{ij}\dot{u}_i n_j dS + \int_{\Omega} \rho b_i\dot{u}_i dV = \int_{\Omega} \sigma_{ij}\dot{u}_{i,j} dV + \int_{\Omega} \rho\ddot{u}_i\dot{u}_i dV$$

where n_j is the unit outward normal of dS. Let $t_i = \sigma_{ij}n_j$ in the surface integral, and substituting σ_{ij} with the constitutive equation Equation $(7)_1$ within the integral operator, this equation can be rewritten as

$$\int_{\partial\Omega} t_i\dot{u}_i dS + \int_{\Omega} \rho b_i \dot{u}_i dV = \int_{\Omega} \left(c_{ijkl} u_{k,l} \dot{u}_{i,j} - e_{kij} E_k \dot{u}_{i,j} - \gamma_{ij} \theta \dot{u}_{i,j} \right) dV + \int_{\Omega} \rho \ddot{u}_i \dot{u}_i dV \tag{18}$$

which can be of the form

$$\int_{\partial\Omega} t_i\dot{u}_i dS + \int_{\Omega} \rho b_i \dot{u}_i dV = \int_{\Omega} \left(\dot{W}_e - e_{kij} E_k \dot{u}_{i,j} - \gamma_{ij} \theta \dot{u}_{i,j} \right) dV + \int_{\Omega} \dot{K} dV \tag{19}$$

where

$$\dot{W}_e = c_{ijkl} u_{k,l} \dot{u}_{i,j} = \frac{1}{2}\frac{\partial}{\partial t}(c_{ijkl} u_{k,l} u_{i,j})$$

which represents the rate of mechanical potential energy density.

$$\dot{K} = \rho \ddot{u}_i \dot{u}_i = \frac{1}{2}\frac{\partial}{\partial t}(\rho \dot{u}_i \dot{u}_i)$$

which is the rate of kinetic energy density.

II. Multiplying φ by the time derivative of Equation (3), integrating the resulting expression over volume Ω and using the identity equation $\left(\dot{D}_k \varphi\right)_{,k} = \dot{D}_{k,k}\varphi + \dot{D}_k \varphi_{,k}$ and Gaussian Theorem, we have

$$-\int_{\partial\Omega} \varphi \dot{D}_k n_k dS - \int_{\Omega} E_k \dot{D}_k dV = 0$$

where n_k is the unit outward normal of dS.

Substitution the constitutive equation Equation $(7)_2$ into the above equation yields

$$-\int_{\partial\Omega} \varphi \dot{D}_k n_k dS - \int_{\Omega} E_k \left(e_{kij}\dot{\varepsilon}_{ij} + \lambda_{ik}\dot{E}_i + \xi_k \dot{\theta} \right) dV = 0$$

which is of the form

$$-\int_{\partial\Omega} \varphi \dot{D}_k n_k dS - \int_{\Omega} \left(e_{kij}\dot{\varepsilon}_{ij} E_k + \dot{W}_E + \xi_k \dot{\theta} E_k \right) dV = 0 \tag{20}$$

where the rate of electric energy density is defined as

$$\dot{W}_E = \frac{1}{2}\frac{\partial}{\partial t}(\lambda_{ik} E_i E_k)$$

The addition of Equation (19) and Equation (20) yields

$$\int_{\partial\Omega} t_i\dot{u}_i dS + \int_{\Omega} \rho b_i \dot{u}_i dV + \int_{\Omega} \gamma_{ij}\theta \dot{u}_{i,j} dV - \int_{\partial\Omega} \varphi \dot{D}_k n_k dS - \int_{\Omega} E_k \xi_k \dot{\theta} dV = \int_{\Omega} \left(\dot{W}_e + \dot{K} + \dot{W}_E \right) dV \tag{21}$$

III. Taking the time differential on Equation $(7)_3$ and using Equation (5), we get

$$T_0 \gamma_{ij} \dot{\varepsilon}_{ij} + T_0 \xi_i \dot{E}_i + \rho C \dot{\theta} = -q_{i,i}$$

Applying the operator L on both sides of this equation and using Equation (6) yields

$$\kappa_{ij}\theta_{,ij} - L(T_0\gamma_{ij}\dot{\varepsilon}_{ij} + T_0\xi_i\dot{E}_i + \rho C\dot{\theta}) = 0 \tag{22}$$

Multiplying Equation (22) by θ and apply volume integral on this expression, we obtain

$$\int_{\Omega} \kappa_{ij}\theta_{,ij}\theta dV - \int_{\Omega} T_0\gamma_{ij}L\left(\dot{\varepsilon}_{ij}\right)\theta dV - \int_{\Omega} T_0\xi_i L\left(\dot{E}_i\right)\theta dV - \int_{\Omega} \rho C L\left(\dot{\theta}\right)\theta dV = 0 \tag{23}$$

Using the identity $(\theta\theta_{,i})_{,j} = \theta_{,j}\theta_{,i} + \theta\theta_{,ij}$ and Gaussian Theorem, then we have

$$\int_{\Omega} \kappa_{ij}\theta_{,ij}\theta dV = \int_{\Omega} \kappa_{ij}\left[(\theta\theta_{,i})_{,j} - \theta_{,j}\theta_{,i}\right] dV = \int_{\partial\Omega} \kappa_{ij}n_j\theta\theta_{,i}dS - \int_{\Omega} \kappa_{ij}\theta_{,j}\theta_{,i}dV$$

Inserting this relation into Equation (23) and expanding the result by using the entropy equation Equation $(7)_3$, we get

$$\frac{1}{T_0}\int_{\partial\Omega} \kappa_{ij}\theta_{,i}\theta n_j dS - \frac{1}{T_0}\int_{\Omega} \kappa_{ij}\theta_{,j}\theta_{,i}dV = \int_{\Omega} \gamma_{ij}\dot{\varepsilon}_{ij}\theta dV + \int_{\Omega} \xi_i\dot{E}_i\theta dV + \int_{\Omega} \frac{\rho C}{T_0}\dot{\theta}\theta dV + \tau\int_{\Omega} \frac{\rho\ddot{\eta}\theta}{T_0}dV \tag{24}$$

Thus the rate of thermal energy density $\dot{W}_\theta$ can be expressed as

$$\dot{W}_\theta = \frac{\rho C}{T_0}\dot{\theta}\theta = \frac{1}{2}\frac{\partial}{\partial t}\left(\frac{\rho C}{T_0}\theta^2\right)$$

Combining Equation (21) and Equation (24) by eliminating $\int_{\Omega} \gamma_{ij}\dot{\varepsilon}_{ij}\theta dV$, finally we obtain

$$\begin{aligned} &\int_{\partial\Omega} t_i\dot{u}_i dS - \int_{\partial\Omega} \varphi\dot{D}_k n_k dS + \frac{1}{T_0}\int_{\partial\Omega} \kappa_{ij}\theta_{,i}\theta n_j dS \\ &= \int_{\Omega} \tau\frac{\rho\ddot{\eta}\theta}{T_0}dV + \frac{1}{T_0}\int_{\Omega} \kappa_{ij}\theta_{,j}\theta_{,i}dV + \int_{\Omega} \frac{\partial}{\partial t}\left(E_i\xi_i\theta\right)dV + \frac{\partial}{\partial t}\int_{\Omega} \left(W_e + K + W_E + W_\theta\right)dV \end{aligned} \tag{25}$$

which is the energy balance law for pyroelectric medium with generalized Fourier conduction law for arbitrary time dependent wave field.

As the general energy balance states:

$$\int_{\Omega} Q dV = -\oint_{\partial\Omega} P_i n_i dS - \frac{\partial}{\partial t}\int_{\Omega} W dV \tag{26}$$

which is the law governing the energy transformation. The physical significance of Equation (26) is that the rate of heat or dissipation energy Q equals to the reduction of the rate of entire energy $\dot{W}$ within the volume plus the reduction of this energy flux outward the surface bounding the volume. P_i is called the energy flux vector(also called the Poyting vector, Poyting-Umov vector) and its direction indicates the direction of energy flow at that point, the length being numerically equal to the amount of energy passing in unit time through unit area perpendicular to $\mathbf{P}$.

In this chapter, important conclusions can be made from Equation (25): the energy density W in the the pyroelectric medium:

$$
\begin{aligned}
W &= W_e + K + W_E + W_\theta \\
W_e &= \frac{1}{2}c_{ijkl}u_{k,l}u_{i,j},\ K = \frac{1}{2}\rho\dot{u}_i\dot{u}_i, \\
W_E &= \frac{1}{2}\lambda_{ik}E_iE_k,\ \ W_\theta = \frac{1}{2}\frac{\rho C}{T_0}\theta^2
\end{aligned} \tag{27}
$$

which is sum of the mechanical potential energy density W_e, the kinetic energy density K, the electric energy density W_E, the heat energy density W_θ .

The physical meaning of $E_i\xi_i\theta$ can be seen from constitutive equation in Equation $(7)_3$, from which $E_i\xi_i$ is found to contribute entropy. Therefore the result $E_i\xi_i\theta$, by its multiplication with temperature disturbance θ, is the dissipation due to the pyroelectric effect. Therefore Q the rate of energy dissipation per unit volume is represented by

$$
Q = \tau\frac{\rho\ddot{\eta}\theta}{T_0} + \frac{1}{T_0}\kappa_{ij}\theta_{,j}\theta_{,i} + \frac{\partial}{\partial t}\left(E_i\xi_i\theta\right) \tag{28}
$$

in which the energy dissipated by the heat conduction is $\frac{1}{T_0}\kappa_{ij}\theta_{,j}\theta_{,i}$, the dissipation energy generated by the relaxation is $\tau\frac{\rho\ddot{\eta}\theta}{T_0}$ and the last term is due to pyroelectric effect.

The energy flux vector(also called the Poyting vector, Poyting-Umov vector) P_i is defined as

$$
P_i = -\sigma_{ji}\dot{u}_j + \varphi\dot{D}_i - \kappa_{ij}\theta_{,j}\frac{\theta}{T_0} \tag{29}
$$

If the temperature effect is not taken account of, Equations (27), (29) can be degenerated into the forms in reference (Baesu et al., 2003).

3.1 Energy balance law for the real-valued inhomogeneous harmonic wave

In previous section, we derived the energy balance equation for the pyroelectric medium and defined the total energy, dissipation energy and energy flux vector explicitly. Keeping in mind that the real part is indeed the physical part of any quantity, and considering Equation (10), we can define the corresponding fundamental field functions as

$$
\begin{aligned}
u_i &= \frac{1}{2}\left[U_i\exp(ix_sk_s)\exp(i\omega t) + U_i^*\exp(-ix_sk_s^*)\exp(-i\omega t)\right] \\
\theta &= \frac{1}{2}\left[\Theta\exp(ix_ik_i)\exp(i\omega t) + \Theta^*\exp(-ix_ik_i^*)\exp(-i\omega t)\right] \\
\varphi &= \frac{1}{2}\left[\Psi\exp(ix_ik_i)\exp(i\omega t) + \Psi^*\exp(-ix_ik_i^*)\exp(-i\omega t)\right]
\end{aligned} \tag{30}
$$

which are the real-valued inhomogeneous harmonic waves assumed on the basis of the pair of complex vector fields for Equation (8).

The velocity of plane of constant phase is defined by

$$
\mathbf{v}_p = \omega\mathbf{P}/\left\|\mathbf{P}\right\|^2 \tag{31}
$$

and the maximum attenuation is $\|\mathbf{A}\|$, where $\|\ \|$ indicates the norm(or length) of a vector.

The quantities of the rate of energy density, the dissipation energy and the energy flux vector can be expressed by inserting Equation (30) into Equations (27), (28) and (29).

The mechanical potential energy density W_e

$$W_e = \frac{1}{2}c_{ijkl}\text{Re}\left[U_k k_l U_i^* k_j^*\right]\exp\left(-2x_s A_s\right) - \frac{1}{2}c_{ijkl}\text{Re}\left[U_k k_l U_i k_j \exp\left(2ix_s k_s\right)\exp\left(2i\omega t\right)\right] \quad (32)$$

The first term on the right-hand side of this equation is time-independent and the second term is time harmonic with frequency 2ω. The first term, expressed as $\langle W_e \rangle$ afterwards, represents the result of W_e averaged over one period. From now on, we shall use $\langle\ \rangle$ indicates the mean quantity over one period. The notation Re stands for the real part and Im the imaginary part. Similarly, the kinetic energy density K takes the form

$$K = \frac{1}{2}\rho\omega^2 U_i U_i^* \exp\left(-2x_s A_s\right) - \frac{1}{2}\rho\omega^2 \text{Re}\left[U_i U_i \exp\left(i2x_s k_s\right)\exp\left(i2\omega t\right)\right] \quad (33)$$

The electric energy density W_E

$$W_E = \frac{1}{2}\lambda_{ik}\text{Re}\left[k_i k_k^* \Psi\Psi^* \exp\left(-2x_s A_s\right)\right] - \frac{1}{2}\lambda_{ik}\text{Re}\left[k_i k_k \Psi\Psi \exp\left(i2x_s k_s\right)\exp\left(i2\omega t\right)\right] \quad (34)$$

The heat energy density W_θ

$$W_\theta = \frac{1}{2}\frac{\rho C}{T_0}\left\{\Theta^*\Theta \exp\left(-2x_s A_s\right) + \text{Re}\left[\Theta\Theta \exp\left(i2x_i k_i\right)\exp\left(i2\omega t\right)\right]\right\} \quad (35)$$

The rate of energy dissipation density

$$Q = Q^{(\kappa)} + Q^{(\tau)} + Q^{(\xi)}$$

where $Q^{(\kappa)}$ due to the heat conduction

$$Q^{(\kappa)} = \frac{1}{T_0}\kappa_{ij}\text{Re}\left[k_i k_j^* \Theta^*\Theta \exp\left(-2x_s A_s\right)\right] - \frac{1}{T_0}\kappa_{ij}\text{Re}\left[k_i k_j \Theta\Theta \exp\left(i2x_s k_s\right)\exp\left(i2\omega t\right)\right] \quad (36)$$

$Q^{(\tau)}$ because of the relaxation

$$\begin{aligned} Q^{(\tau)} &= \tau\frac{\rho}{T_0}\left(\gamma_{ij}\ddot{\varepsilon}_{ij}\theta + \xi_i \ddot{E}_i \theta + \frac{\rho C}{T_0}\ddot{\theta}\theta\right) \\ &= \tau\frac{\rho}{T_0}\omega^2\left\{\gamma_{ij}\text{Im}\left(U_i k_j \Theta^*\right)\exp\left(-2x_s A_s\right) + \gamma_{ij}\text{Im}\left[U_i k_j \Theta \exp\left(i2x_s k_s\right)\exp\left(i2\omega t\right)\right] + \right. \quad (37) \\ &\quad + \xi_i \text{Im}\left(k_i^* \Psi^* \Theta\right)\exp\left(-2x_s A_s\right) + \xi_i Im\left[k_i^* \Psi^* \Theta^* \exp\left(i2x_s k_s\right)\exp\left(i2\omega t\right)\right] + \\ &\quad \left. -\Theta\Theta^* \exp\left(-2x_s A_s\right) - \text{Re}\left[\Theta\Theta \exp i(2x_i k_i)\exp\left(i2\omega t\right)\right]\right\} \end{aligned}$$

At last, $Q^{(\xi)}$ attributed by the pyroelectric effect

$$Q^{(\xi)} = 2\xi_i Re(k_i \omega \Psi\Theta^*)\exp\left(-2x_s A_s\right) + 2\xi_i Re\left[(k_i \omega \Psi\Theta)\exp(2ix_s k_s)\exp\left(i2\omega t\right)\right] \quad (38)$$

The energy flux vector P_i consists of three different parts: $P_i^{(u)}$ is generated in the elastic field; $P_i^{(\varphi)}$ in the electric field; $P_i^{(\theta)}$ in the thermal field, which are expressed as

$$\begin{aligned} P_j^{(u)} &= -\sigma_{ji}\dot{u}_j \\ &= -\omega c_{jikl}\left\{\text{Re}\left(U_i^* U_k k_l\right)\exp\left(-2x_s A_s\right) + \text{Re}\left[U_i U_k k_l \exp\left(i2x_s k_s\right)\exp\left(i2\omega t\right)\right]\right\} + \quad (39) \\ &\quad \omega e_{kji}\left\{-\text{Re}\left(k_k U_i^* \Psi\right)\exp\left(-2x_s A_s\right) + \text{Re}\left[k_k \Psi U_i \exp\left(i2x_s k_s\right)\exp\left(i2\omega t\right)\right]\right\} + \\ &\quad \omega\gamma_{ji}\left[\left[\text{Im}\left(U_i^* \Theta\right)\exp\left(-2x_s A_s\right) - \omega \text{Im} U_i \Theta \exp\left(i2x_s k_s\right)\exp\left(i2\omega t\right)\right]\right] \end{aligned}$$

In the electric field, $P_j^{(\varphi)}$

$$
\begin{aligned}
P_j^{(\varphi)} &= \varphi \dot{D}_j \\
&= -\omega e_{jmn} \left\{ \mathrm{Re}\left(U_m k_n \Psi^*\right) \exp\left[-2x_s A_s\right] + \mathrm{Re}\left[\Psi U_m k_n \exp\left[i(2x_s k_s)\right] \exp i2\omega t\right] \right\} + \quad (40) \\
&\quad \omega \lambda_{mj} \left\{ \mathrm{Re}\left(k_m \Psi \Psi^*\right) \exp\left(-2x_s A_s\right) + \mathrm{Re}\left[k_m \Psi \Psi \exp\left(i2x_s k_s\right) \exp\left(i2\omega t\right)\right] \right\} + \\
&\quad \omega \xi_j \left\{ \mathrm{Im}\left(\Theta^* \Psi\right) \exp\left(-2x_s A_s\right) - \mathrm{Im}\left[\Theta \Psi \exp\left[i(2x_i k_i)\right] \exp\left(i2\omega t\right)\right] \right\}
\end{aligned}
$$

In the thermal field, $P_j^{(\theta)}$

$$
P_j^{(\theta)} = -\kappa_{ij}\theta_{,i}\frac{\theta}{T_0} = \frac{1}{T_0}\left\{ -\Theta\Theta^* \kappa_{ij} \mathrm{Im}\left(k_i\right) \exp\left(-2x_s A_s\right) + \mathrm{Im}\left[\kappa_{ij} k_i \Theta\Theta \exp\left(i2x_s k_s\right) \exp\left(i2\omega t\right)\right] \right\} \quad (41)
$$

It is to be noted that the mean quantities still satisfy Equation (25) of energy balance equation for pyroelectric medium.

Since the energy flux and the energy density have the dimensions of watt per square meter and joule per cubic meter respectively, their ratio gives a quantity with dimension of velocity. This energy velocity $\mathbf{v}_E$ is defined as the radio of the mean energy flux to the mean energy density over one period, that is

$$
\mathbf{v}_E = \langle \mathbf{P} \rangle / \langle W \rangle \quad (42)
$$

which corresponds to the average local velocity of energy transport. From an experimental point of view, it is more interesting to define velocity from averaged quantities (Deschamps et al., 1997).

We can substitute the expressions in Equations (32)-(35) and (39)-(41) into (42), which yields a lengthy formulation. Comparing the expression of phase velocity in Equation (31) with the energy velocity in Equation (42), it is obvious that they are different from each other in moduli as well as directions.

3.2 Results and discussion

According to previous studies, it is already known that there are waves of four modes, which are quasilongitudinal, quasitransverse I, II and temperature. In this section, we'd like to discuss phase velocity $\mathbf{v}_p$, energy velocity $\mathbf{v}_E$ related to the four mode waves. They are studied as functions defined in propagation angle θ and attenuation angle γ. After wave vector $\mathbf{k}$ is determined, Equations (31) and (42) yield the phase velocity and energy velocity respectively. The material constants under study is transversely isotropic material, see Section 2.4.

The variation of phase and energy velocity of quasilongitudinal wave is presented in Fig. 6 (a) which shows that the phase velocity does not vary with attenuation angle γ , while the corresponding energy velocity can be influenced by γ. With γ increasing, the energy velocity turns small. It is also noted that the phase velocity is a little bigger than the energy velocity for quasilongitudinal wave mode.

The case of temperature wave is shown in Fig. 6 (b). Different from quasilongitudinal wave, the phase velocity and energy velocity of temperature wave are influenced by propagation angle θ and attenuation angle γ. Both phase velocity and energy velocity decay with γ. For given γ, the phase velocity is also bigger than energy velocity.

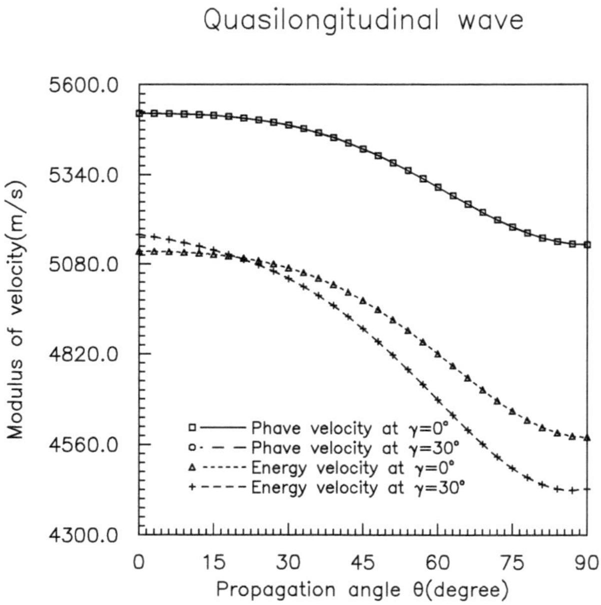

(a)

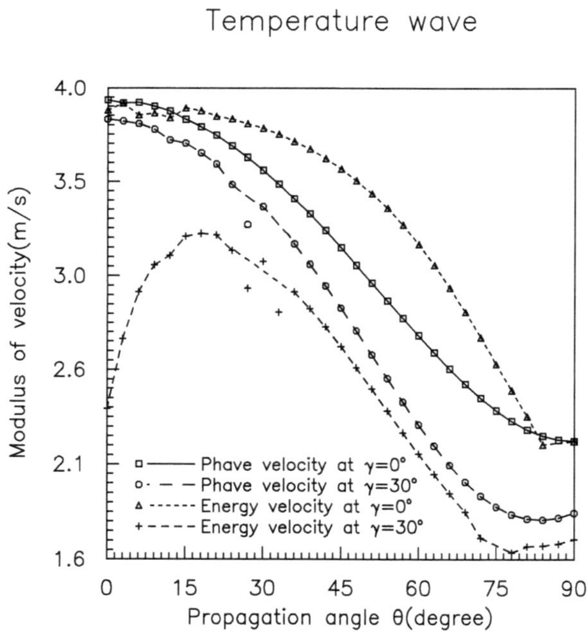

(b)

Fig. 6. Variations of velocity with propagation angle θ at γ=0°, 30°.

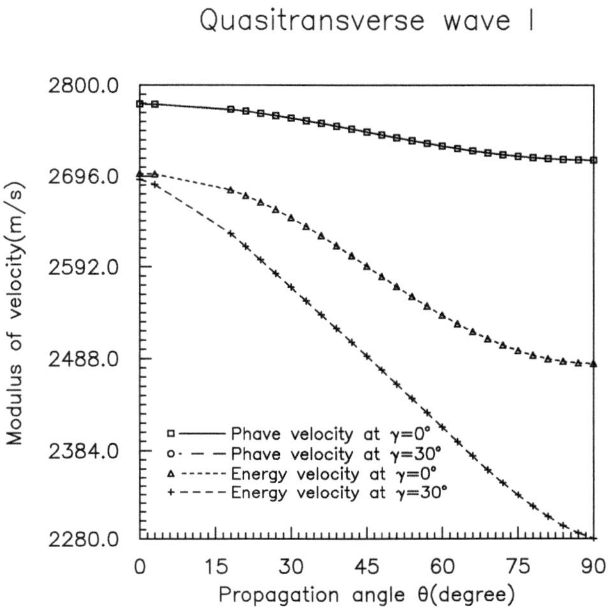

(a)

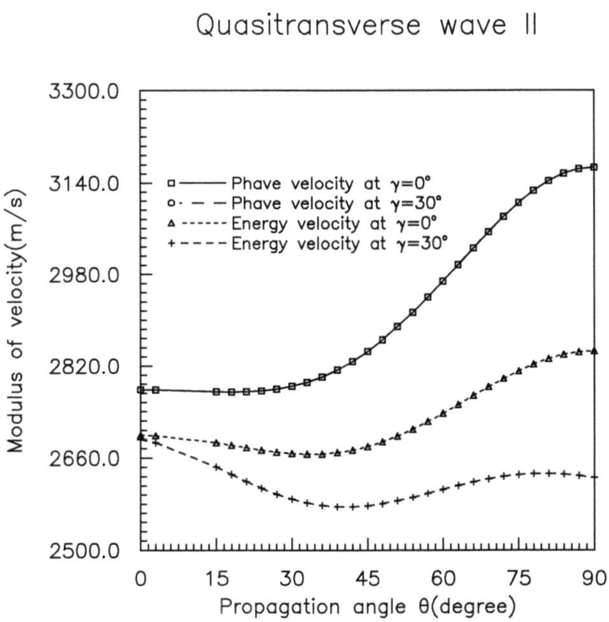

(b)

Fig. 7. Variations of velocity with propagation angle θ at γ=0°,30°.

Plots of the computed velocities of quasitransverse wave I and II are given in Fig. 7. The phase velocities of both wave modes are almost independent of γ and the energy velocity become small with γ increasing.

4. Conclusion

In this chapter, the energy process of the pyroelectric medium with generalized heat conduction theory is addressed in the framework of the inhomogeneous wave results originally. The characters of inhomogeneous waves lie in that its propagation direction is different from the biggest attenuation direction. The complex-valued wave vector is determined by four parameters. The range of attenuation angle should be confined in (-90°,90°) to make waves attenuate. Further analysis demonstrates that, in anisotropic plane, the positive and negative attenuation angle have different influences on waves, while, in the isotropic plane, they are the same. Based on the governing equations and state equations, the dynamic energy conservation law is derived. The energy transfer, in an arbitrary instant, is described explicitly by the energy conservation relation. From this relation, it is found that energy density in pyroelectric medium consists of the electric energy density, the heat energy density, the mechanical potential energy density, the kinetic energy density. The heat loss or dissipation energy is equal to the reduction of the entire energy within a fixed volume plus the reduction of this energy flux outward the surface bounding this volume. The dissipation energy in pyroelectric medium are attributed by the heat conduction, relaxation time and pyroelectric effect. The energy flux is obtained and it can not be influenced by the relaxation time. The phase velocity and energy velocity of four wave modes in pyroelectric medium are studied. Results demonstrate that the attenuation angle almost doesn't influence phase velocity of quasilongitudinal, quasitransverse I, II wave modes, while plays large role on the temperature wave. The energy velocities of the four wave modes all decay with the attenuation angle.

5. References

Auld, B. A. (1973). *Acoustic fields and waves in solids*, John Wiley & Sons, New York.

Baesu, E., Fortune, D. & Soós, E. (2003). Incremental behaviour of hyperelastic dielectrics and piezoelectric crystals, *Z. angew. Math. Phys.* 54: 160–178.

Baljeet, S. (2005). On the theory of generalized thermoelasticity for piezoelectric materials, *Applied Mathematics and Computation* 171: 398–405.

Borcherdt, R. D. (1973). Energy and plane waves in linear viscoelastic media, *Journal of Geophysical Research* 78(14): 2442–2453.

Buchen, P. W. (1971). Plane waves in linear viscoelastic media, *Geophys J R astr Soc* 23: 531–542.

Busse, G. (1991). Thermal waves for material inspection, *in* O. Leroy & M. A. Breazeale (eds), *Physical Acoustics: Fundermentals and Applications*, Plenum Press, Kortrijk, Belgium.

Cady, W. G. (1946). *Piezoelectricity*, An Introduction to the Theory and Applications of Electromechanical Phenomena in Crystals, McGraw-Hill, New York.

Carcione, J. M. & Cavallini, F. (1997). Forbidden directions for tm waves in anisotropic conducting media, *Antennas and Propagation, IEEE Transactions on* 45(1): 133–139.

Cattaneo, C. (1958). Sur une forme de l'équationéliminant le paradoxe d'une propagation instantanée, *Comptes Rend Acad Sc* 247: 431–432.

Cerveny, V. & Psencik, I. (2006). Energy flux in viscoelastic anisotropic media, *Geophys. J. Int.* 166(3): 1299–1317.

Cuadras, A., Gasulla, M., Ghisla, A. & Ferrari, V. (2006). Energy harvesting from pzt pyroelectric cells, *Instrumentation and Measurement Technology Conference, 2006. IMTC 2006. Proceedings of the IEEE*, pp. 1668–1672.

Dalola, S., Ferrari, V. & Marioli, D. (2010). Pyroelectric effect in pzt thick films for thermal energy harvesting in low-power sensors, *Procedia Engineering* 5: 685–688.

Deschamps, M., Poiree, B. & Poncelet, O. (1997). Energy velocity of complex harmonic plane waves in viscous fluids, *Wave Motion* 25: 51–60.

Eringen, A. C. & Maugin, G. A. (1990). *Electrodynamics of continua*, Vol. I, Springer-Verlag, New York.

Fang, J., Frederich, H. & Pilon, L. (2010). Harvesting nanoscale thermal radiation using pyroelectric materials, *Journal of Heat Transfer* 132(9): 092701–10.

Fedorov, F. I. (1968). *Theory of Elastic Waves in Crystals*, Plenum Press, New York.

Gael, S. & et al. (2009). On thermoelectric and pyroelectric energy harvesting, *Smart Materials and Structures* 18(12): 125006.

Green, A. E. & Lindsay, K. A. (1972). Thermoelasticity, *Journal of Elasticity* 2(1): 1–7.

Guyomar, D., Sebald, G., Lefeuvre, E. & Khodayari, A. (2008). Toward heat energy harvesting using pyroelectric material, *Journal of Intelligent Material Systems and Structures* .

Hadni, A. (1981). Applications of the pyroelectric effect, *J. Phys. E: Sci. Instrum.* 14: 1233–1240.

He, T., Tian, X. & Shen, Y. (2002). Two-dimensional generalized thermal shock problem of a thick piezoelectric plate of infinte extent, *Int. J. Eng. Sci.* 40: 2249–2264.

Joseph, D. & Preziosi, L. (1989). Heat waves, *Reviews of Modern Physics* 61: 41–73.

Kaliski, S. (1965). Wave equation of thermoelasticity, *Bull. Acad. Pol. Sci., Ser. Sci . Tech.* 13: 211–219.

Khodayari, A., Pruvost, S., Sebald, G., Guyomar, D. & Mohammadi, S. (2009). Nonlinear pyroelectric energy harvesting from relaxor single crystals, *Ultrasonics, Ferroelectrics and Frequency Control, IEEE Transactions on* 56(4): 693–699. 0885-3010.

Kiselev, A. P. (1982). Energy flux of elastic waves, *Journal of Mathematical Sciences* 19(4): 1372–1375.

Kuang, Z. B. (2002). *Nonlinear continuum mechanics*, Shanghai Jiaotong University Publishing House, Shanghai.

Kuang, Z. B. (2009). Variational principles for generalized dynamical theory of thermopiezoelectricity, *Acta Mechanica* 203(1): 1–11.

Kuang, Z. B. (2010). Variational principles for generalized thermodiffusion theory in pyroelectricity, *Acta Mechanica* 214(3): 275–289.

Kuang, Z. & Yuan, X. (2010). Reflection and transmission of waves in pyroelectric and piezoelectric materials, *Journal of Sound and Vibration* 330(6): 1111–1120.

Lord, H. W. & Shulman, Y. (1967). A generalized dynamical theory of thermoelasiticity, *J. Mech. Phys. Solids* 15: 299–309.

Mainardi, F. (1973). On energy velocity of viscoelastic waves, *Lettere Al Nuovo Cimento (1971-1985)* 6: 443–449.

Majhi, M. C. (1995). Discontinuities in generalized thermoelastic wave propagation in a semi-infinite piezoelectric rod, *J. Tech. Phys* 36: 269–278.

Nelson, D. F. (1979). *Electric, Optic and Acoustic Interactions in Dielectrics*, John Wiley, New York.

Olsen, R. B., Bruno, D. A., Briscoe, J. M. & Dullea, J. (1984). Cascaded pyroelectric energy converter, *Ferroelectrics* 59(1): 205–219.

Olsen, R. B. & Evans, D. (1983). Pyroelectric energy conversion: Hysteresis loss and temperature sensitivity of a ferroelectric material, *Journal of Applied Physics* 54(10): 5941–5944.

Sharma, J. N. & Kumar, M. (2000). Plane harmonic waves in piezo-thermoelastic materials, *Indian J. Eng. Mater. Sci.* 7: 434–442.

Sharma, J. N. & Pal, M. (2004). Propagation of lamb waves in a transversely isotropic piezothermoelastic plate, *Journal of Sound and Vibration* 270(4-5): 587–610.

Sharma, J. N. & Walia, V. (2007). Further investigations on rayleigh waves in piezothermoelastic materials, *Journal of Sound and Vibration* 301(1-2): 189–206.

Shen, D., Choe, S.-Y. & Kim, D.-J. (2007). Analysis of piezoelectric materials for energy harvesting devices under high-igi vibrations, *Japanese Journal of Applied Physics* 46(10A): 6755.

Sodano, H. A., Inman, D. J. & Park, G. (2005). Comparison of piezoelectric energy harvesting devices for recharging batteries, *Journal of Intelligent Material Systems and Structures* 16(10): 799–807.

Umov, N. A. (1874). *The equations of motion of the energy in bodies[in Russian]*, Odessa.

Vernotte, P. (1958). Les paradoxes de la théorie continue de l'équation de la chaleur, *Comptes Rend Acad Sc* 246: 3154.

Xie, J., Mane, X. P., Green, C. W., Mossi, K. M. & Leang, K. K. (2009). Performance of thin piezoelectric materials for pyroelectric energy harvesting, *Journal of Intelligent Material Systems and Structures* .

Yuan, X. (2009). *Wave theory in pyroeletric medium*, PhD thesis, Shanghai Jiaotong university.

Yuan, X. (2010). The energy process of pyroelectric medium, *Journal of Thermal Stresses* 33(5): 413–426.

Yuan, X. & Kuang, Z. (2008). Waves in pyroelectrics, *Journal of Thermal Stresses* 31(12): 1190–1211.

Yuan, X. & Kuang, Z. (2010). The inhomogeneous waves in pyroelectrics, *Journal of Thermal Stresses* 33: 172–186.

10

Heat Transfer and Reconnection Diffusion in Turbulent Magnetized Plasmas

A. Lazarian
Department of Astronomy, University of Wisconsin-Madison
USA

1. Introduction

It is well known that magnetic fields constrain motions of charged particles, impeding the diffusion of charged particles perpendicular to magnetic field direction. This modification of transport processes is of vital importance for a wide variety of astrophysical processes including cosmic ray transport, transfer of heavy elements in the interstellar medium, star formation etc. Dealing with these processes one should keep in mind that, in realistic astrophysical conditions, magnetized fluids are turbulent. In this review we single out a particular transport process, namely, heat transfer and consider how it occurs in the presence of the magnetized turbulence. We show that the ability of magnetic field lines to constantly change topology and connectivity is at the heart of the correct description of the 3D magnetic field stochasticity in turbulent fluids. This ability is ensured by fast magnetic reconnection in turbulent fluids and puts forward the concept of reconnection diffusion at the core of the physical picture of heat transfer in astrophysical plasmas. Appealing to reconnection diffusion we describe the ability of plasma to diffuse between different magnetized eddies explaining the advection of the heat by turbulence. Adopting the structure of magnetic field that follows from the modern understanding of MHD turbulence, we also discuss thermal conductivity that arises as electrons stream along stochastic magnetic field lines. We compare the effective heat transport that arise from the two processes and conclude that, in many astrophysically-motivated cases, eddy advection of heat dominates. Finally, we discuss the concepts of sub and superdiffusion and show that the subdiffusion requires rather restrictive settings. At the same time, accelerated diffusion or superdiffusion of heat perpendicular to the mean magnetic field direction is possible on the scales less than the injection scale of the turbulence.

2. Main idea and structure of the review

Heat transfer in turbulent magnetized plasma is an important astrophysical problem which is relevant to the wide variety of circumstancies from mixing layers in the Local Bubble (see Smith & Cox 2001) and Milky way (Begelman & Fabian 1990) to cooling flows in intracluster medium (ICM) (Fabian 1994). The latter problem has been subjected to particular scrutiny as observations do not support the evidence for the cool gas (see Fabian et al. 2001). This is suggestive of the existence of heating that replenishes the energy lost via X-ray emission. Heat transfer from hot outer regions is an important process to consider in this context.
It is well known that magnetic fields can suppress thermal conduction perpendicular to their direction. However, this is true for laminar magnetic field, while astrophysical plasmas are

generically turbulent (see Armstrong et al 1994, Chepurnov & Lazarian 2010). The issue of heat transfer in realistic turbulent magnetic fields has been long debated (see Bakunin 2005 and references therein).

Below we argue that turbulence changes the very nature of the process of heat transfer. To understand the differences between laminar and turbulent cases one should consider both motion of charged particles along turbulent magnetic fields and turbulent motions of magnetized plasma that also transfer heat. The description of both processes require the knowledge of the dynamics of magnetic field lines and the structure of the magnetic field lines in turbulent flows. The answers to these questions are provided by the theories of magnetic reconnection and magnetic turbulence. To provide the quantitative estimates of the heat transfer the review addresses both theories, discussing the generic process of reconnection diffusion which describes the diffusion induced by the action of turbulent motions in the presence of reconnection. We stress the fundamental nature of the process which apart from heat transfer is also important e.g. for removing magnetic field in star formation process (Lazarian 2005).

In §2 we discuss the omnipresence of turbulence in astrophysical fluids, introduce major ideas of MHD turbulence theory and turbulent magnetic reconnection in §3 and §4, respectively, relate the concept of reconnection diffusion to the processes of heat transfer in magnetized plasmas in §5. We provide detailed discussion of heat conductivity via streaming electrons in §6, consider heat advection by turbulent eddies in §7, and compare the efficiencies of the latter two processes in §8. Finally, we discuss heat transfer on scales smaller than the turbulence injection scale in §9 and provide final remarks in §10.

3. Magnetized turbulent astrophysical media

Astrophysical plasmas are known to be magnetized and turbulent. Magnetization of these fluids most frequently arises from the dynamo action to which turbulence is an essential component (see Schekochihin et al. 2007). In fact, it has been shown that turbulence in weakly magnetized conducting fluid converts about ten percent of the energy of the cascade into the magnetic field (see Cho et al. 2009). This fraction does not depend on the original magnetization and therefore magnetic fields will come to equipartition with the turbulent motions in about 10 eddy turnover times.

We deal with magnetohydrodynamic (MHD) turbulence which provides a correct fluid-type description of plasma turbulence at large scales[1]. Astrophysical turbulence is a direct consequence of large scale fluid motions experiencing low friction. This quantity is described by Reynolds number $Re \equiv LV/\nu$, where L is the scale of fluid motions, V is the velocity at this scale and ν is fluid viscosity. The Reynolds numbers are typically very large in astrophysical flows as the scales are large. As magnetic fields decrease the viscosity for the plasma motion perpendicular to their direction, Re numbers get really astronomically large. For instance, Re numbers of 10^{10} are very common for astrophysical flow. For so large Re the inner degrees of fluid motion get excited and a complex pattern of motion develops.

The drivers of turbulence, e.g. supernovae explosions in the interstellar medium, inject energy at large scales and then the energy cascades down to small scales through a hierarchy of eddies spanning up over the entire inertial range. The famous Kolmogorov picture (Kolmogorov 1941) corresponds to hydrodynamic turbulence, but, as we discuss further, a qualitatively similar turbulence also develops in magnetized fluids/plasmas.

---

[1] It is possible to show that in terms magnetic field wandering that is important, as we see below, for heat transfer the MHD description is valid in collisionless regime of magnetized plasmas (Eyink, Lazarian & Vishniac (2011).

Simulations of interstellar medium, accretion disks and other astrophysical environments also produce turbulent picture, provided that the simulations are not dominated by numerical viscosity. The latter requirement is, as we see below, is very important for the correct reproduction of the astrophysical reality with computers.
The definitive confirmation of turbulence presence comes from observations, e.g. observations of electron density fluctuations in the interstellar medium, which produce a so-called Big Power Law in the Sky (Armstrong et al. 1994, Chepurnov & Lazarian 2010), with the spectral index coinciding with the Kolmogorov one. A more direct piece of evidence comes from the observations of spectral lines. Apart from showing non-thermal Doppler broadening, they also reveal spectra of supersonic turbulent velocity fluctuations when analyzed with techniques like Velocity Channel Analysis (VCA) of Velocity Coordinate Spectrum (VCS) developed (see Lazarian & Pogosyan 2000, 2004, 2006, 2008) and applied to the observational data (see Padoan et al. 2004, 2009, Chepurnov et al. 2010) rather recently.
All in all, the discussion above was aimed at conveying the message that the turbulent state of magnetized astrophysical fluids is a rule and therefore the discussion of any properties of astrophysical systems should take this state into account. We shall show below that both magnetic reconnection and heat transfer in magnetized fluids are radically changed by turbulence.

4. Strong and weak Alfvenic turbulence

For the purposes of heat transfer, Alfvenic perturbations are most important. Numerical studies in Cho & Lazarian (2002, 2003) showed that the Alfvenic turbulence develops an independent cascade which is marginally affected by the fluid compressibility. This observation corresponds to theoretical expectations of the Goldreich & Sridhar (1995) theory that we briefly describe below (see also Lithwick & Goldreich 2001). In this respect we note that the MHD approximation is widely used to describe the actual magnetized plasma turbulence over scales that are much larger than both the mean free path of the particles and their Larmor radius (see Kulsrud 2004 and ref. therein). More generally, the most important incompressible Alfenic part of the plasma motions can described by MHD even below the mean free path (see Eyink et al. 2011 and ref. therein).
While having a long history of ideas, the theory of MHD turbulence has become testable recently due to the advent numerical simulations (see Biskamp 2003) which confirm (see Cho & Lazarian 2005 and ref. therein) the prediction of magnetized Alfvénic eddies being elongated in the direction of magnetic field (see Shebalin, Matthaeus & Montgomery 1983, Higdon 1984) and provided results consistent with the quantitative relations for the degree of eddy elongation obtained in Goldreich & Sridhar (1995, henceforth GS95).
The hydrodynamic counterpart of the MHD turbulence theory is the famous Kolmogorov theory of turbulence. In that theory, energy is injected at large scales, creating large eddies which correspond to large *Re* numbers and therefore do not dissipate energy through viscosity[2] but transfer energy to smaller eddies. The process continues till the cascade reaches the eddies that are small enough to dissipate energy over an eddy turnover time. In the absence of compressibility the hydrodynamic cascade of energy is $\sim v_l^2/\tau_{casc,l} = const$, where v_l is the velocity at the scale l and the cascading time for the eddies of size l is $\tau_{cask,l} \approx l/v_l$. From this the well known relation $v_l \sim l^{1/3}$ follows.

[2] Reynolds number $Re \equiv LV/\nu = (V/L)/(\nu/L^2)$ which is the ratio of the eddy turnover rate $\tau_{eddy}^{-1} = V/L$ and the viscous dissipation rate $\tau_{dis}^{-1} = \eta/L^2$. Therefore large *Re* correspond to negligible viscous dissipation of large eddies over the cascading time τ_{casc} which is equal to τ_{eddy} in Kolmogorov turbulence.

Modern MHD turbulence theory can also be understood in terms of eddies. However, in the presence of dynamically important magnetic field, eddies cannot be isotropic. Any motions bending magnetic field should induce a back-reaction and Alfven waves propagating along the magnetic field. At the same time, one can imagine eddies mixing magnetic field lines perpendicular to the direction of magnetic field. For the latter eddies the original Kolmogorov treatment is applicable resulting perpendicular motions scaling as $v_l l_\perp^{1/3}$, where $l_\perp$ denotes scales measured perpendicular to magnetic field and correspond to the perpendicular size of the eddy. These mixing motions induce Alfven waves which determine the parallel size of the magnetized eddy. The key stone of the GS95 theory is *critical balance*, i.e. the equality of the eddy turnover time $l_\perp / v_l$ and the period of the corresponding Alfven wave $\sim l_\| / V_A$, where $l_\|$ is the parallel eddy scale and V_A is the Alfven velocity. Making use of the earlier expression for v_l one can easily obtain $l_\| \sim l_\perp^{2/3}$, which reflects the tendency of eddies to become more and more elongated as energy cascades to smaller scales.

While the arguments above are far from being rigorous they correctly reproduce the basic scalings of magnetized turbulence when the velocity equal to V_A at the injection scale L. The most serious argument against the picture is the ability of eddies to perform mixing motions perpendicular to magnetic field. We shall address this issue in §3 but for now we just mention in passing that strongly non-linear turbulence does not usually allow the exact derivations. It is numerical experiments that proved the above scalings for incompressible MHD turbulence (Cho & Vishniac 2000, Maron & Goldreich 2001, Cho, Lazarian & Vishniac 2002) and for the Alfvenic component of the compressible MHD turbulence (Cho & Lazarian 2002, 2003, Kowal & Lazarian 2010).

It is important to stress that the scales $l_\perp$ and $l_\|$ are measured in respect to the system of reference related to the direction of the local magnetic field "seen" by the eddy. This notion was not present in the original formulation of the GS95 theory and was added in Lazarian & Vishniac (1999) (see also Cho & Vishniac 2000, Maron & Goldreich 2001, Cho et al. 2002). In terms of mixing motions that we mentioned above it is rather obvious that the free Kolmogorov-type mixing is possible only in respect to the local magnetic field of the eddy rather than the mean magnetic field of the flow.

GS95 theory assumes the isotropic injection of energy at scale L and the injection velocity equal to the Alfvén velocity in the fluid V_A, i.e. the Alfvén Mach number $M_A \equiv (\delta V / V_A) = 1$. This model can be easily generalized for both $M_A < 1$ and $M_A > 1$ at the injection (see Lazarian & Vishniac 1999 and Lazarian 2006, respectively). Indeed, if $M_A > 1$, instead of the driving scale L for one can use another scale, namely l_A, which is the scale at which the turbulent velocity gets equal to V_A. For $M_A \gg 1$ magnetic fields are not dynamically important at the largest scales and the turbulence at those scales follows the isotropic Kolmogorov cascade $v_l \sim l^{1/3}$ over the range of scales $[L, l_A]$. This provides $l_A \sim L M_A^{-3}$. If $M_A < 1$, the turbulence obeys GS95 scaling (also called "strong" MHD turbulence) not from the scale L, but from a smaller scale $l_{trans} \sim L M_A^2$ (Lazarian & Vishniac 1999), while in the range $[L, l_{trans}]$ the turbulence is "weak".

The properties of weak and strong turbulence are rather different. The weak turbulence is wave-like turbulence with wave packets undergoing many collisions before transferring energy to small scales[3]. On the contrary, the strong turbulence is eddy-like with cascading happening similar to Kolmogorov turbulence within roughly an eddy turnover time. One also should remember that the notion "strong" should not be associated with the amplitude of turbulent motions, but only with the strength of the non-linear interaction. As the weak

[3] Weak turbulence, unlike the strong one, allows an exact analytical treatment (Gaultier et al. 2002).

turbulence evolves, the interactions of wave packets increases as the ratio of the parallel to perpendicular scales of the packets increases making the turbulence strong. In this case, the amplitude of the perturbations may be very small.

While there ongoing debates whether the original GS95 theory should be modified to better describe MHD turbulence, we believe that, first of all, we do not have compelling evidence that GS95 is not adequate[4]. Moreover, the proposed additions to the GS95 model do not change the nature of the physical processes that we present below.

The quantitative picture of astrophysical turbulence sketched in this section gives us a way to proceed with the quantitative description of key processes necessary to describe heat transfer. The interaction of fundamental MHD modes within the cascade of compressible magnetized turbulence is described in Cho & Lazarian (2005), but this interaction is not so important for the processes of heat transfer that we discuss below.

5. Magnetic reconnection of turbulent magnetic flux

Magnetic reconnection is a fundamental process that violates magnetic flux being frozen in within highly conductive fluids. Intuitively one may expect that magnetic fields in turbulent fluids cannot be perfectly frozen in. Theory that we describe below provide quantitative estimates of the violation of frozen in condition within turbulent fluids.

We would like to stress that the we are discussing the case of dynamically important magnetic field, including the case of weakly turbulent magnetic field. The case of weak magnetic field which can be easily stretched and bended by turbulence at any scale up to the dissipation one is rather trivial and of little astrophysical significance[5]. At the same time, at sufficiently small scales magnetic fields get dynamically important even for superAlfvenic turbulence.

Within the picture of eddies mixing perpendicular to the local magnetic field that we provided in the previous section, it is suggestive that magnetized eddies can provide turbulent advection of heat similar to the ordinary hydrodynamic eddies. This is rather counter-intuitive notion in view of the well-entrenched idea of flux being frozen in astrophysical fluids. As it is explained in Eyink et al. (2011) the frozen-in condition is not a good approximation for the turbulent fluids[6]. The violation of the perfect frozenness of the magnetic field in plasmas also follows from LV99 model of reconnection (see discussion in Vishniac & Lazarian 1999).

A picture of two flux tubes of different directions which get into contact in 3D space is the generic framework to describe magnetic reconnection. The upper panel of Figure 1 illustrates why reconnection is so slow in the textbook Sweet-Parker model. Indeed, the model considers magnetic fields that are laminar and therefore the frozen-in condition for magnetic field is violated only over a thin layer dominated by plasma resistivity. The scales over which the resistive diffusion is important are microscopic and therefore the layer is very thin, i.e. $\Delta \ll L_x$, where L_x is the scale at which magnetic flux tubes come into contact. The latter

[4] Recent work by Beresnyak & Lazarian (2010) shows that present day numerical simulations are unable to reveal the actual inertial range of MHD turbulence making the discussions of the discrepancies of the numerically measured spectrum and the GS95 predictions rather premature. In addition, new higher resolution simulations by Beresnyak (2011) reveal the predicted $-5/3$ spectral slope.

[5] In the case of dynamically unimportant field, the magnetic dissipation and reconnection happens on the scales of the Ohmic diffusion scale and the effects of magnetic field on the turbulent cascade are negligible. However, turbulent motions transfer an appreciable portion of the cascading energy into magnetic energy (see Cho et al. 2010). As a result, the state of intensive turbulence with negligible magnetic field is short-lived.

[6] Formal mathematical arguments on how and why the frozen-in condition fails may be found in Eyink (2011).

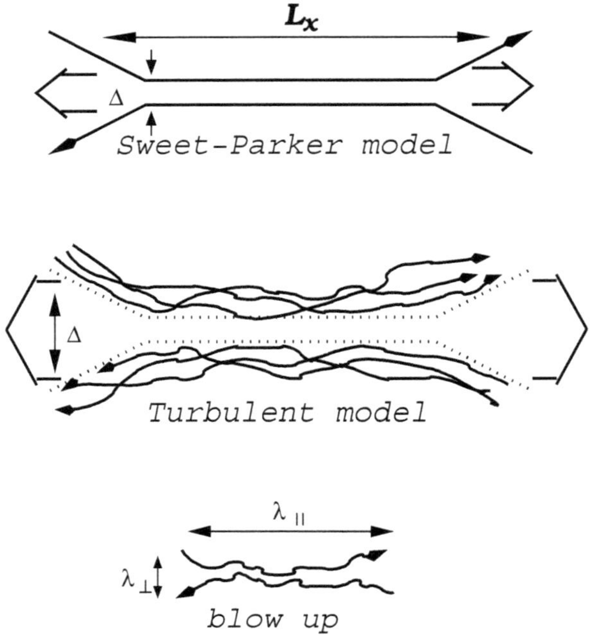

Fig. 1. *Upper panel*: Sweet-Parker reconnection. Δ is limited by resistivity and small. *Middle panel*: reconnection according to LV99 model. Δ is determined by turbulent field wandering and can be large. *Lower panel*: magnetic field reconnect over small scales. From Lazarian, Vishniac & Cho (2004).

is of the order of the diameter of the flux tubes and typically very large for astrophysical conditions. During the process of magnetic reconnection all the plasma and the shared magnetic flux[7] arriving over an astrophysical scale L_x should be ejected through a microscopic slot of thickness Δ. As the ejection velocity of magnetized plasmas is limited by Alfven velocity V_A, this automatically means that the velocity in the vertical direction, which is reconnection velocity, is much less than V_A.

The LV99 model generalizes the Sweet-Parker one by accounting for the existence of magnetic field line stochasticity (Figure 1 (lower panels)). The depicted turbulence is sub-Alfvenic with relatively small fluctuations of the magnetic field. At the same time turbulence induces magnetic field wandering. This wandering was quantified in LV99 and it depends on the intensity of turbulence. The vertical extend of wandering of magnetic field lines that at any point get into contact with the field of the other flux tube was identified in LV99 with the width of the outflow region. Note, that magnetic field wandering is a characteristic feature of magnetized turbulence in 3D. Therefore, generically in turbulent reconnection the outflow is no more constrained by the narrow resistive layer, but takes place through a much wider area Δ defined by wandering magnetic field lines. The extend of field wandering determines the reconnection velocity in LV99 model.

An important consequence of the LV99 reconnection is that as turbulence amplitude increases, the outflow region and therefore reconnection rate also increases, which entails the ability of

[7] Figure 1 presents only a cross section of the 3D reconnection layer. A shared component of magnetic field is going to be present in the generic 3D configurations of reconnecting magnetic flux tubes.

reconnection to change its rate depending on the level of turbulence. The latter is important both for understanding the dynamics of magnetic field in turbulent flow and for explaining flaring reconnection events, e.g. solar flares.

We should note that the magnetic field wandering is mostly due to Alfvenic turbulence. To describe the field wondering for weakly turbulent case LV99 extended the GS95 model for a subAlfvenic case. The same field wandering[8], as we discuss later, is important for heat transfer by electrons streaming along magnetic field lines.

The predictions of the turbulent reconnection rates in LV99 were successfully tested 3D numerical simualtions in Kowal et al. (2009) (see also Lazarian et al. 2010 for an example of higher resolution runs). This testing provided stimulated work on the theory applications, e.g. its implication for heat transfer. One should keep in mind that the LV model assumes that the magnetic field flux tubes can come at arbitrary angle, which corresponds to the existence of shared or guide field within the reconnection layer[9].

Alternative models of magnetic reconnection appeal to different physics to overcome the constraint of the Sweet-Parker model. In the Petcheck (1964) model of reconnection the reconnection layer opens up to enable the outflow which thickness does not depend on resistivity. To realize this idea inhomogeneous resistivity, e.g. anomalous resisitivity associated with plasma effects, is required (see Shay & Drake 1998). However, for turbulent plasmas, the effects arising from modifying the local reconnection events by introducing anomalous resistivity are negligible as confirmed e.g. in Kowal et al. (2009). Other effects, e.g. formation and ejection of plasmoids (see Shibata & Tanuma 2001, Lorreiro et al. 2008) which may be important for initially laminar environments are not likely to play the dominant role in turbulent plasmas either. Therefore in what follows dealing with turbulent transfer of hear we shall appeal to the LV99 model of reconnection.

6. Reconnection diffusion and heat transfer

In the absence of the frozen-in condition in turbulent fluids one can talk about reconnection diffusion in magnetized turbulent astrophysical plasmas. The concept of reconnection diffusion is based on LV99 model and was first discussed in Lazarian (2005) in terms of star formation[10]. However, reconnection diffusion is a much broader concept applicable to different astrophysical processes, including heat transfer in magnetized plasmas. In what follows we shall discuss several processes that enable heat transfer perpendicular to the mean magnetic field in the flow.

The picture frequently presented in textbooks may be rather misleading. Indeed, it is widely assumed that magnetic field lines always preserve their identify in highly conductive plasmas even in turbulent flows. In this situation the diffusion of charged particles perpendicular to magnetic field lines is very restricted. For instance, the mass loading of magnetic field lines

[8] As discussed in LV99 and in more details in Eyink et al. (2011) the magnetic field wandering, turbulence and magnetic reconnection are very tightly related concepts. Without magnetic reconnection, properties of magnetic turbulence and magnetic field wandering would be very different. For instance, in the absence of fast reconnection, the formation of magnetic knots arising if magnetic fields were not able to reconnect would destroy the self-similar cascade of Alfvenic turbulence. The rates predicted by LV99 are exactly the rates required to make Goldreich-Sridhar model of turbulence self-consistent.

[9] The model in LV99 is three dimensional and it is not clear to what extend it can be applied to 2D turbulence (see discussion in ELV11 and references therein). However, the cases of pure 2D reconnection and 2D turbulence are of little practical importance.

[10] Indeed, the issue of flux being conserved within the cloud presents a problem for collapse of clouds with strong magnetic field. These clouds also called subcritical were believed to evolve with the rates determined by the relative drift of neutrals and ions, i.e. the ambipolar diffusion rate.

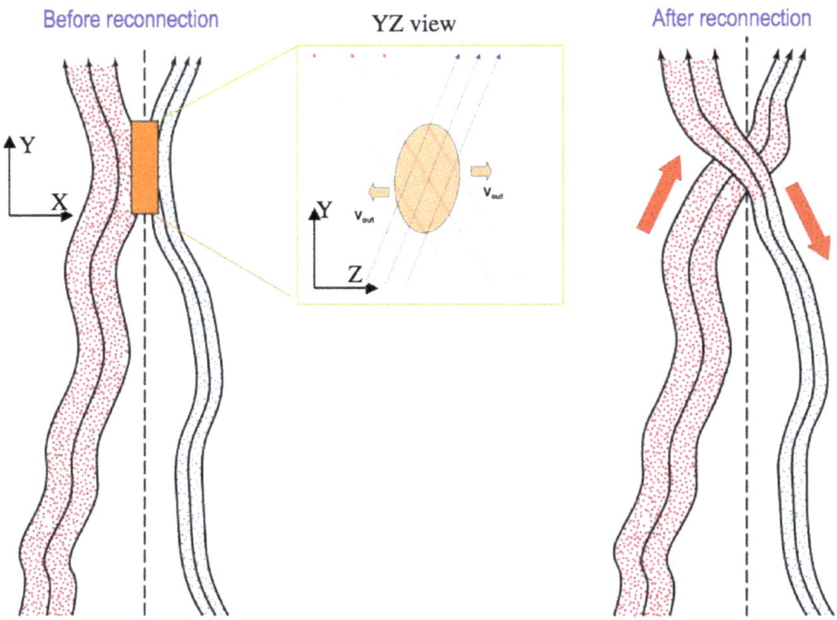

Fig. 2. Diffusion of plasma in inhomogeneous magnetic field. 3D magnetic flux tubes get into contact and after reconnection plasma streams along magnetic field lines. *Right panel*: XY projection before reconnection, upper panel shows that the flux tubes are at angle in X-Z plane. *Left Panel*: after reconnection.

does not change to a high degree and density and magnetic field compressions follow each other. All these assumptions are violated in the presence of reconnection diffusion.

We shall first illustrate the reconnection diffusion process showing how it allows plasma to move perpendicular to the mean inhomogeneous magnetic field (see Figure 2). Magnetic flux tubes with entrained plasmas intersect each other at an angle and due to reconnection the identity of magnetic field lines change. Before the reconnection plasma pressure P_{plasma} in the tubes is different, but the total pressure $P_{plasma} + P_{magn}$ is the same for two tubes. After reconnection takes place, plasma streams along newly formed magnetic field lines to equalize the pressure along two new flux tubes. The diffusion of plasmas and magnetic field takes place. The effect of this process is to make magnetic field and plasmas more homogeneously distributed in the absence of the external fields[11]. In terms of heat transfer, the process mixes up plasma at different temperatures if the temperatures of plasma volumes along different magnetic flux tubes were different.

If turbulence had only one scale of motions its action illustrated by Figure 2 would create every flux tube columns of hot and cold gas exchanging heat with each other through the diffusion of charged particles along magnetic field lines. This is not the case, however, for a turbulence

[11] If this process acts in the presence of gravity, as this is the case of star formation, the heavy fluid (plasma) will tend to get to the gravitating center changing the mass to flux ratio, which is important to star formation processes. In other words, reconnection diffusion can do the job that is usually associated with the action of ambipolar diffusion (see numerical simulations in Santos de Lima et al. (2010).

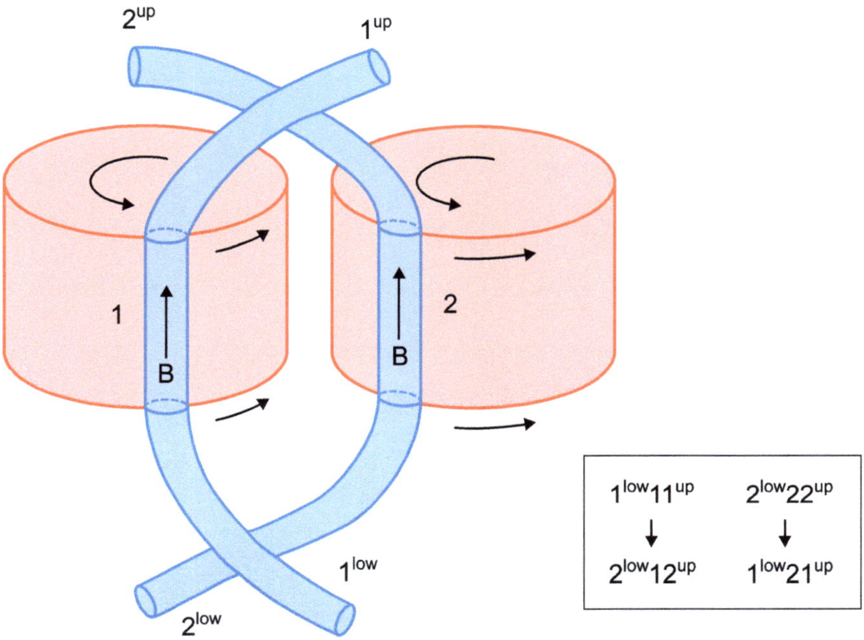

Fig. 3. Exchange of plasma between magnetic eddies. Eddies carrying magnetic flux tubes interact through reconnection of the magnetic field lines belonging to two different eddies. This enables the exchange of matter between eddies and induces a sort of turbulent diffusivity of matter and magnetic field.

with an extended inertial cascade. Such a turbulence would induce mixing depicted in Figure 2 on every scale, mixing plasma at smaller and smaller scales.

When plasma pressure along magnetic field flux tubes is the same, the connection of flux tubes which takes place in turbulent media as shown in Figure 3 is still important for heat transfer. The reconnected flux tubes illustrate the formation of the wandering magnetic field lines along which electron and ions can diffuse transporting heat. For the sake of simplicity, we shall assume that electrons and ions have the same temperature. In this situation, the transfer of heat by ions is negligible and for the rest of the presentation we shall talk about the transport of heat by electrons moving along wandering field lines[12].

Consider the above process of reconnection diffusion in more detail. The eddies 1 and 2 interact through the reconnection of the magnetic flux tubes associated with eddies. LV99 model shows that in turbulent flows reconnection happens within one eddy turnover time, thus ensuring that magnetic field does not prevent free mixing motions of fluid perpendicular to the local direction of magnetic field. As a result of reconnection, the tube $1^{low}11^{up}$ transforms into $2^{low}12^{up}$ and a tube $2^{low}22^{up}$ transforms into $1^{low}21^{up}$. If eddy 1 was

[12] This is true provided that the current of diffusing hot electrons is compensated by the current of oppositely moving cold electrons, the diffusivity of electrons along wandering magnetic field lines is dominant compared with the diffusivity and heat transfer by protons and heavier ions. If there is no compensating current, electrons and ions are coupled by electric field and have to diffuse along wandering magnetic fields together and at the same rate. This could be the case of diffusion of plasmas into neutral gas. However, we do not discuss these complications here

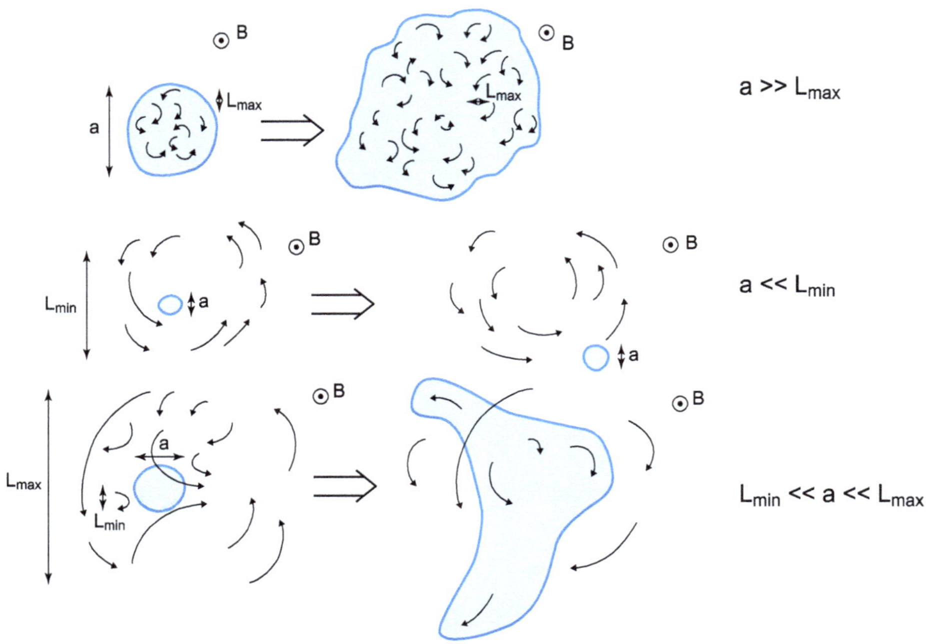

Fig. 4. Heat diffusion depends on the scale of the hot spot. Different regimes emerge depending on the relation of the hot spot to the sizes of maximal and minimal eddies present in the turbulence cascade. Mean magnetic field *B* is directed perpendicular to the plane of the drawing. Eddies perpendicular to magnetic field lines correspond to Alfvenic turbulence. The plots illustrate heat diffusion for different regimes. *Upper* plot corresponds to the heat spot being less than the minimal size of turbulent eddies; *Middle plot* corresponds to the heat spot being less than the damping scale of turbulence; *Lower plot* corresponds to the heat spot size within the inertial range of turbulent motions.

associated with hotter plasmas and eddy 2 with colder plasmas, then the newly formed magnetic flux tubes will have both patches of hot and cold plasmas. For the hierarchy of eddies the shedding of entrained plasmas into hot and cold patches along the same magnetic field lines allows electron conductivity to remove the gradients, conducting heat. This is the process of turbulent advection of heat in magnetized plasmas.

The difference between the processes depicted in Figures 2 and 3 is due to the fact that the process in Figure 2 is limited by the thermal velocity of particles, while the process in Figure 3 depends upon the velocity of turbulent eddies only. In actual plasmas in the presence of temperature gradients plasmas along different elementary flux tubes will have different temperature and therefore two processes will take place simultaneously.

Whether the motion of electrons along wandering magnetic field lines or the dynamical mixing induced by turbulence is more important depends on the ratio of eddy velocity to the sonic one, the ratio of the turbulent motion scale to the mean free path of electrons and the degree of plasma magnetization. Strong magnetization both limits the efficiency of turbulent mixing perpendicular to magnetic field lines and the extent to which plasma streaming along magnetic field lines moves perpendicular to the direction of the mean field. However, but reduction of heat transfer efficiency is different for the two processes. We provide quantitative treatment of these processes in the next section.

An interesting example of practical interest is related to the diffusion of heat from a hot spot. This case of reconnection diffusion is illustrated by Figure 4. In this situation heat transfer depends on whether the scale of turbulent motions is larger or smaller than the hot spot. Consider this situation in more detail. Turbulence is characterized by its injection scale L_{max}, its dissipation scale L_{min} and its inertial range $[L_{min}, L_{max}]$. The heat transfer depends on what scales we consider the process. Figure 4 illustrates our point. Consider a hot spot of the size a in turbulent flow and consider Alfvenic eddies perpendicular to magnetic field lines. If turbulent eddies are much smaller than a, which is the case when $a \gg L_{min}$ they extend the hot spot acting in a random walk fashion. For eddies much larger than the hot spot, i.e. $a \ll L_{min}$ they mostly advect hot spot. If a is the within the inertial range of turbulent motions, i.e. $L_{min} < a < L_{max}$ then a more complex dynamics of turbulent motions is involved. This is also the case where the field wandering arising from these motions is the most complex. Turbulent motions with the scale comparable with the hot spot induce a process of the accelerated Richardson diffusion (see more in §10).

In terms of practical simulation of reconnection diffusion effects, it is important to keep in mind that the LV99 model predicts that the largest eddies are the most important for providing outflow in the reconnection zone and therefore the reconnection will not be substantially changed if turbulence does not have an extended inertial range. In addition, LV99 predicts that the effects of anomalous resistivity arising from finite numerical grids do not change the rate of turbulent reconnection. We note that both effects were successfully tested in Kowal et al. (2009).

7. Heat conduction through streaming of electrons

7.1 General considerations

As magnetic reconnection was considered by many authors even more mysterious than the heat transfer in plasmas, it is not surprising that the advection of heat by turbulent eddies was not widely discussed. Instead for many year the researchers preferred to consider heat transfer by plasma conductivity along turbulent magnetic field lines (see Chandran & Cowley 1998, Malyshkin & Kulsrud 2001). This conductivity is mostly due to electrons streaming along magnetic field lines. Turbulent magnetic field lines allow streaming electrons to diffuse perpendicular to the mean magnetic field and spread due to the magnetic field wandering that we discussed earlier. Therefore the description of magnetic field wandering obtained in LV99 is also applicable for describing the processes of heat transfer.

We start with the case of trans-Alfvenic turbulence considered by Narayan & Medvedev (2001, henceforth NM01). They appeal to magnetic field wandering and obtained estimates of thermal conductivity by electrons for the special case of turbulence velocity V_L at the energy injection scale L that is equal to the Alfven velocity V_A. As we discussed earlier this special case is described by the original GS95 model and the Alfven Mach number $M_A \equiv (V_L/V_A) = 1$. We note that this case is rather restrictive, as the intracuster medium (ICM) is superAlfvenic, i.e. $M_A > 1$, while other astrophysical situations, e.g. solar atmosphere, are subAlfvenic, i.e. $M_A < 1$. Different phases of interstellar medium (ISM) (see Draine & Lazarian 1998 and Yan, Lazarian & Draine 2004 for lists of idealized ISM phases) present the cases of both superAlfvenic and subAlfvenic turbulence.

As we discussed above, the generalization of GS95 model of turbulence for subAlfvenic case is provided in LV99. This was employed in Lazarian (2006) to describe heat conduction for magnetized turbulent plasmas with $M_A < 1$. In addition, Lazarian (2006) considered heat

conduction by tubulence with $M_A > 1$ as well as heat advection by turbulence and compares the efficiencies of electron heat conduction and the heat transfer by turbulent motions.
Let us initially disregard the dynamics of fluid motions on diffusion, i.e. consider diffusion induced by particles moving along wandering turbulent magnetic field lines, which motions we disregard for the sake of simplicity. Magnetized turbulence with a dynamically important magnetic field is anisotropic with eddies elongated along (henceforth denoted by $\|$) the direction of local magnetic field, i.e. $l_\perp < l_\|$, where $\perp$ denotes the direction of perpendicular to the local magnetic field. Consider isotropic injection of energy at the outer scale L and dissipation at the scale $l_{\perp,min}$. This scale corresponds to the minimal dimension of the turbulent eddies.
Turbulence motions induce magnetic field divergence. It is easy to notice (LV99, NM01) that the separations of magnetic field lines at small scales less than the damping scale of turbulence, i.e. for $r_0 < l_{\perp,min}$, are mostly influenced by the motions at the smallest scale. This scale $l_{\perp,min}$ results in Lyapunov-type growth $\sim r_0 \exp(l/l_{\|,min})$. This growth is similar to that obtained in earlier models with a single scale of turbulent motions (Rechester & Rosenbluth 1978, henceforth RR78, Chandran & Cowley 1998). Indeed, as the largest shear that causes field line divergence is due to the marginally damped motions at the scale around $l_{\perp,min}$ the effect of larger eddies can be neglected and we are dealing with the case of single-scale "turbulence" described by RR78.
The electron Larmor radius presents the minimal perpendicular scale of localization. Thus it is natural to associate r_0 with the size of the cloud of electrons of the electron Larmor radius $r_{Lar,particle}$. Applying the original RR78 theory (see also Chandran & Cowley 1998) they found that the electrons should travel over the distance

$$L_{RR} \sim l_{\|,min} \ln(l_{\perp,min}/r_{Lar,e}) \tag{1}$$

to get separated by $l_{\perp,min}$.
Within the single-scale "turbulent model" which formally corresponds to $Lss = l_{\|,min} = l_{\perp,min}$ the distance L_{RR} is called Rechester-Rosenbluth distance. For the ICM parameters the logarithmic factor in Eq. (1) is of the order of 30, and this causes 30 times decrease of thermal conductivity for the single-scale models[13].
The single-scale "turbulent model" is just a toy model to study effects of turbulent motions. One can use this model, however, to describe what is happening below the scale of the smallest eddies. Indeed, the shear and, correspondingly, magnetic field line divergence is maximal for the marginally damped eddies at the dissipation scale. Thus for scales less than the damping scale the action of the critically damped eddies is dominant.
In view of above, the realistic multi-scale turbulence with a limited (e.g. a few decades) inertial range the single scale description is applicable for small scales up to the damping scale. The logarithmic factor stays of the same order but instead of the injection scale L_{ss} for the single-scale RR model, one should use $l_{\|,min}$ for the actual turbulence. Naturally, this addition does not affect the thermal conductivity, provided that the actual turbulence injection scale L is much larger than $l_{\|,min}$. Indeed, for the electrons to diffuse isotropically they should spread from $r_{Lar,e}$ to L. Alfvenic turbulence operates with field lines that are sufficiently stiff, i.e. the deviation of the field lines from their original direction is of the order unity at scale L and less for smaller scales. Therefore to get separated from the initial distance of $l_{\perp,min}$ to a distance L (see Eq. (5) with $M_A = 1$), at which the motions get uncorrelated, the electrons

[13] For the single-scale model $L_{RR} \sim 30L$ and the diffusion over distance Δ takes L_{RR}/Lss steps, i.e. $\Delta^2 \sim L_{RR}L$, which decreases the corresponding diffusion coefficient $\kappa_{e,single} \sim \Delta^2/\delta t$ by the factor of 30.

should diffuse the distance slightly larger (as field lines are not straight) than $\sqrt{2}L$. This is much larger than the extra travel distance $\sim 30 l_{\|,min}$ originating from sub-diffusive behavior at scales less than the turbulence damping scale. Explicit calculations in NM01 support this intuitive picture.

7.2 Diffusion for $M_A > 1$

Turbulence with $M_A > 1$ evolves along hydrodynamic isotropic Kolmogorov cascade, i.e. $V_l \sim V_L(l/L)^{1/3}$ over the range of scales $[L, l_A]$, where

$$l_A \approx L(V_A/V_L)^3 \equiv L M_A^{-3}, \tag{2}$$

is the scale at which the magnetic field gets dynamically important, i.e. $V_l = V_A$. This scale plays the role of the injection scale for the GS95 turbulence, i.e. $V_l \sim V_A(l_\perp/l_A)^{1/3}$, with eddies at scales less than l_A geting elongated in the direction of the local magnetic field. The corresponding anisotropy can be characterized by the relation between the semi-major axes of the eddies

$$l_\| \sim L(l_\perp/L)^{2/3} M_A^{-1}, \quad M_A > 1, \tag{3}$$

where $\|$ and $\perp$ are related to the direction of the local magnetic field. In other words, for $M_A > 1$, the turbulence is still isotropic at the scales larger to l_A, but develops $(l_\perp/l_A)^{1/3}$ anisotropy for $l < l_A$.

If particles (e.g. electrons) mean free path $\lambda \gg l_A$, they stream freely over the distance of l_A. For particles initially at distance $l_{\perp,min}$ to get separated by L, the required travel is the random walk with the step l_A, i.e. the mean-squared displacement of a particle till it enters an independent large-scale eddy $\Delta^2 \sim l_A^2(L/l_A)$, where L/l_A is the number of steps. These steps require time $\delta t \sim (L/l_A) l_A / C_1 v_e$, where $v_{particle}$ is electron thermal velocity and the coefficient $C_1 = 1/3$ accounts for 1D character of motion along magnetic field lines. Thus the electron diffusion coefficient is

$$\kappa_e \equiv \Delta^2/\delta t \approx (1/3) l_A v_e, \quad l_A < \lambda, \tag{4}$$

which for $l_A \ll \lambda$ constitutes a substantial reduction of diffusivity compared to its unmagnetized value $\kappa_{unmagn} = \lambda v_e$. We assumed in Eq. (4) that $L \gg 30 l_{\|,min}$ (see §2.1).

For $\lambda \ll l_A \ll L$, $\kappa_e \approx 1/3 \kappa_{unmagn}$ as both the L_{RR} and the additional distance for electron to diffuse because of magnetic field being stiff at scales less than l_A are negligible compared to L. For $l_A \to L$, when magnetic field has rigidity up to the scale L, it gets around $1/5$ of the value in unmagnetized medium, according to NM01.

7.3 Diffusion for $M_A < 1$

It is intuitively clear that for $M_A < 1$ turbulence should be anisotropic from the injection scale L. In fact, at large scales the turbulence is expected to be *weak*[14] (see Lazarian & Vishniac 1999, henceforth LV99). Weak turbulence is characterized by wavepackets that do not change their $l_\|$, but develop structures perpendicular to magnetic field, i.e. decrease $l_\perp$. This cannot proceed indefinitely, however. At some small scale the GS95 condition of *critical balance*, i.e. $l_\|/V_A \approx l_\perp/V_l$, becomes satisfied. This perpendicular scale l_{trans} can be obtained substituting the scaling of weak turbulence (see LV99) $V_l \sim V_L(l_\perp/L)^{1/2}$ into the critical balance condition.

[14] The terms "weak" and "strong" turbulence are accepted in the literature, but can be confusing. As we discuss later at smaller scales at which the turbulent velocities decrease the turbulence becomes *strong*. The formal theory of weak turbulence is given in Galtier et al. (2000).

This provides $l_{trans} \sim LM_A^2$ and the corresponding velocity $V_{trans} \sim V_L M_A$. For scales less than l_{trans} the turbulence is *strong* and it follows the scalings of the GS95-type, i.e. $V_l \sim V_L(L/l_\perp)^{-1/3} M_A^{1/3}$ and

$$l_\parallel \sim L(l_\perp/L)^{2/3} M_A^{-4/3}, \quad M_A < 1. \tag{5}$$

For $M_A < 1$, magnetic field wandering in the direction perpendicular to the mean magnetic field (along y-axis) can be described by $d\langle y^2\rangle/dx \sim \langle y^2\rangle/l_\parallel$ (LV99), where[15] $l_\parallel$ is expressed by Eq. (5) and one can associate $l_\perp$ with $2\langle y^2\rangle$

$$\langle y^2\rangle^{1/2} \sim \frac{x^{3/2}}{3^{3/2}L^{1/2}} M_A^2, \quad l_\perp < l_{trans} \tag{6}$$

For weak turbulence $d\langle y^2\rangle/dx \sim LM_A^4$ (LV99) and thus

$$\langle y^2\rangle^{1/2} \sim L^{1/2}x^{1/2}M_A^2, \quad l_\perp > l_{trans}. \tag{7}$$

Fig. 5 confirms the correctness of the above scaling numerically.
Eq. (6) differs by the factor M_A^2 from that in NM01, which reflects the gradual suppression of thermal conductivity perpendicular to the mean magnetic field as the magnetic field gets stronger. Physically this means that for $M_A < 1$ the magnetic field fluctuates around the well-defined mean direction. Therefore the diffusivity gets anisotropic with the diffusion coefficient parallel to the mean field $\kappa_{\parallel,particle} \approx 1/3\kappa_{unmagn}$ being larger than coefficient for diffusion perpendicular to magnetic field $\kappa_{\perp,e}$.
Consider the coefficient $\kappa_{\perp,e}$ for $M_A \ll 1$. As NM01 showed, particles become uncorrelated if they are displaced over the distance L in the direction perpendicular to magnetic field. To do this, a particle has first to travel L_{RR} (see Eq. (1)), where Eq. (5) relates $l_{\parallel,min}$ and $l_{\perp,min}$. Similar to the case in §2.1, for $L \gg 30 l_{\parallel,min}$, the additional travel arising from the logarithmic factor is negligible compared to the overall diffusion distance L. At larger scales electron has to diffuse $\sim L$ in the direction parallel to magnetic field to cover the distance of LM_A^2 in the direction perpendicular to magnetic field direction. To diffuse over a distance R with random walk of LM_A^2 one requires $R^2/L^2M_A^4$ steps. The time of the individual step is $L^2/\kappa_{\parallel,e}$. Therefore the perpendicular diffusion coefficient is

$$\kappa_{\perp,e} = R^2/(R^2/[\kappa_{\parallel,e}M_A^4]) = \kappa_{\parallel,e}M_A^4, \quad M_A < 1, \tag{8}$$

An essential assumption there is that the particles do not trace their way back over the individual steps along magnetic field lines, i.e. $L_{RR} << L$. Note, that for M_A of the order of unity this is not accurate and one should account for the actual 3D displacement. This introduces the change by a factor of order unity (see above).

8. Transfer of heat through turbulent motions

As we discussed above, turbulent motions themselves can induce advective transport of heat. Appealing to LV99 model of reconnection one can conclude that turbulence with $M_A \sim 1$ should be similar to hydrodynamic turbulence, i.e.

$$\kappa_{dynamic} \approx C_{dyn} L V_L, \quad M_A > 1, \tag{9}$$

[15] The fact that one gets $l_{\parallel,min}$ in Eq. (1) is related to the presence of this scale in this diffusion equation.

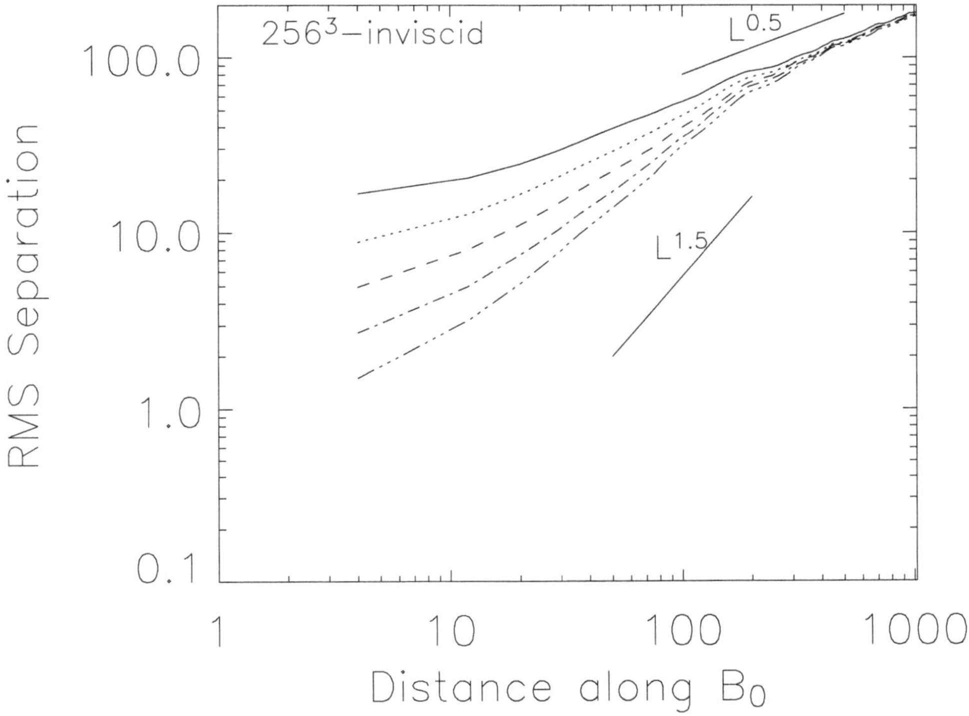

Fig. 5. Root mean square separation of field lines in a simulation of inviscid MHD turbulence, as a function of distance parallel to the mean magnetic field, for a range of initial separations. Each curve represents 1600 line pairs. The simulation has been filtered to remove pseudo-Alfvén modes, which introduce noise into the diffusion calculation. From Lazarian, Vishniac & Cho 2004.

where $C_{dyn} \sim 0(1)$ is a constant, which for hydro turbulence is around $1/3$ (Lesieur 1990). This was confirmed in Cho et al. (2003) (see Figure 6 and also Cho & Lazarian 2004) where MHD calculations were performed for transAlfvenic turbulence with $M_A \sim 1$. As large scale eddies of superAlfvenic turbulence are essentially hydrodynamic, the correspondence between the ordinary hydrodynamic heat advection and superAlfvenic one should only increase as M_A increases.

If we deal with heat transport, for fully ionized non-degenerate plasmas we assume $C_{dyn} \approx 2/3$ to account for the advective heat transport by both protons and electrons[16]. Thus eq. (9) covers the cases of both $M_A > 1$ up to $M_A \sim 1$. For $M_A < 1$ one can estimate $\kappa_{dynamic} \sim d^2\omega$, where d is the random walk of the field line over the wave period $\sim \omega^{-1}$. As the weak turbulence at scale L evolves over time $\tau \sim M_A^{-2}\omega^{-1}$, $\langle y^2 \rangle$ is the result of the random walk

[16] This becomes clear if one uses the heat flux equation $q = -\kappa_c \nabla T$, where $\kappa_c = nk_B\kappa_{dynamic/electr}$, n is *electron* number density, and k_B is the Boltzmann constant, for both electron and advective heat transport.

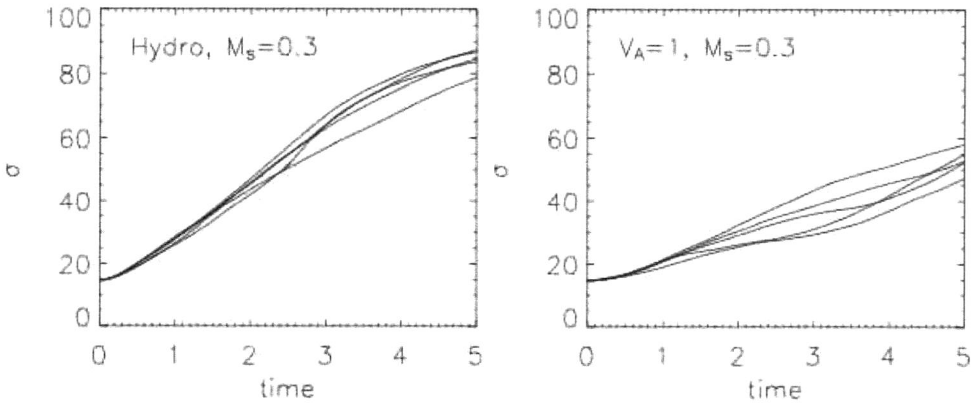

Fig. 6. Comparison of the heat diffusion with time for hydro turbulence (left panel) and MHD transAlfvenic turbulence (right panel). Different curves correspond to different runs. From Cho et al. (2003).

with a step d, i.e. $\langle y^2 \rangle \sim (\tau\omega)d^2$. According to eq.(6) and (7), the field line is displaced over time τ by $\langle y^2 \rangle \sim LM_A^4 V_A \tau$. Combining the two one gets $d^2 \sim LM_A^3 V_L \omega^{-1}$, which provides $\kappa_{dynamic}^{weak} \approx C_{dyn} LV_L M_A^3$, which is similar to the diffusivity arising from strong turbulence at scales less than l_{trans}, i.e. $\kappa_{dynamic}^{strong} \approx C_{dyn} l_{trans} V_{trans}$. The total diffusivity is the sum of the two, i.e. for plasma

$$\kappa_{dynamic} \approx (\beta/3) LV_L M_A^3, \quad M_A < 1, \tag{10}$$

where $\beta \approx 4$.

9. Relative importance of two processes

9.1 General treatment

Figure 7 illustrates the existing ideas on processes of heat conduction in astrophysical plasmas. They range from the heat insulation by unrealistically laminar magnetic field (see panel (a)), to heat diffusion in turbulent magnetic field (see panel (b)) and to heat advection by turbulent flows (see panel (c)). The relative efficiencies of the two latter processes depend on parameters of turbulent plasma.

In thermal plasma, electrons are mostly responsible for thermal conductivity. The schematic of the parameter space for $\kappa_{particle} < \kappa_{dynamic}$ is shown in Fig 8, where the the Mach number M_s and the Alfven Mach number M_A are the variables. For $M_A < 1$, the ratio of diffusivities arising from fluid and particle motions is $\kappa_{dynamic}/\kappa_{particle} \sim \beta\alpha M_S M_A (L/\lambda)$ (see Eqs. (8) and (10)), the square root of the ratio of the electron to proton mass $\alpha = (m_e/m_p)^{1/2}$, which provides the separation line between the two regions in Fig. 2, $\beta\alpha M_s \sim (\lambda/L) M_A$. For $1 < M_A < (L/\lambda)^{1/3}$ the mean free path is less than l_A which results in $\kappa_{particle}$ being some fraction of κ_{unmagn}, while $\kappa_{dynamic}$ is given by Eq. (9). Thus $\kappa_{dynamic}/\kappa_{particle} \sim \beta\alpha M_s (L/\lambda)$, i.e. the ratio does not depend on M_A (horisontal line in Fig. 2). When $M_A > (L/\lambda)^{1/3}$ the mean free path of electrons is constrained by l_A. In this case $\kappa_{dynamic}/\kappa_{particle} \sim \beta\alpha M_s M_A^3$ (see Eqs. (9) and (4)) . This results in the separation line $\beta\alpha M_s \sim M_A^{-3}$ in Fig. 8.

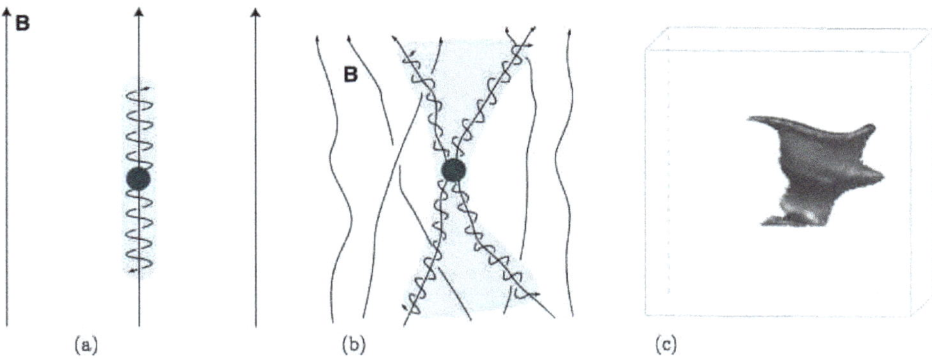

Fig. 7. (a) The textbook description of confinement of charged particles in magnetic fields; (b) diffusion of particles in turbulent fields; (c) advection of heat from a localized souce by eddies in MHD numerical simulations. From Cho & Lazarian 2004.

9.2 Application to ICM plasmas

Consider plasmas in clusters of galaxies to illustrate the relative importance of two processes of heat transfer. Below we shall provide evidence that *magnetized* Intracluster Medium (ICM) is turbulent and therefore our considerations above should be applicable.

It is generally believed that ICM plasma is turbulent. However, naive estimates of diffusivity for collisionless plasma provide numbers which may cast doubt on this conclusion. Indeed, in unmagnatized plasma with the ICM temperatures $T \sim 10^8$ K and and density 10^{-3} cm^{-3} the kinematic viscosity $\eta_{unmagn} \sim v_{ion}\lambda_{ion}$, where v_{ion} and λ_{ion} are the velocity of an ion and its mean free path, respectively, would make the Reynolds number $Re \equiv LV_L/\eta_{unmagn}$ of the order of 30. This is barely enough for the onset of turbulence. For the sake of simplicity we assume that ion mean free path coincides with the proton mean free path and both scale as $\lambda \approx 3T_3^2 n_{-3}^{-1}$ kpc, where the temperature $T_3 \equiv kT/3$ keV and $n_{-3} \equiv n/10^{-3}$ cm^{-3}. This provides λ of the order of 0.8–1 kpc for the ICM (see NM01). We shall argue that the above low estimate of *Re* is an artifact of our neglecting magnetic field.

In general, a single value of *Re* uniquely characterizes hydrodynamic flows. The case of magnetized plasma is very different as the diffusivities of protons parallel and perpendicular to magnetic fields are different. The diffusion of protons perpendicular to the local magnetic field is usually very slow. Such a diffusion arises from proton scattering. Assuming the maximal scattering rate of an proton, i.e. scattering every orbit (the so-called Bohm diffusion limit) one gets the viscosity perpendicular to magnetic field $\eta_\perp \sim v_{ion} r_{Lar,ion}$, which is much smaller than η_{unmagn}, provided that the ion Larmor radius $r_{Lar,ion} \ll \lambda_{ion}$. For the parameters of the ICM this allows essentially inviscid fluid motions[17] of magnetic lines parallel to each other, e.g. Alfven motions.

[17] A *regular* magnetic field $B_\lambda \approx (2mkT)^{1/2}c/(e\lambda)$ that makes $r_{Lar,ion}$ less than λ and therefore $\eta_\perp < v_{unmagn}$ is just 10^{-20} G. *Turbulent* magnetic field with many reversals over $r_{Lar,ion}$ does not interact efficiently with a proton, however. As the result, the protons are not constrained until l_A gets of the order of $r_{Lar,ion}$. This happens when the turbulent magnetic field is of the order of $2 \times 10^{-9}(V_L/10^3\text{km/s})$ G. At this point, the step for the random walk is $\sim 2 \times 10^{-6}$ pc and the Reynolds number is 5×10^{10}.

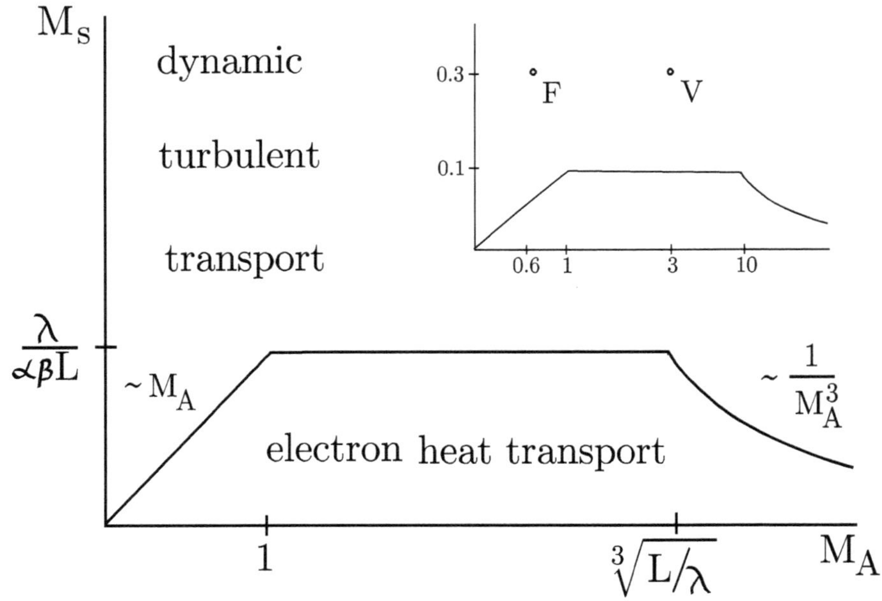

Fig. 8. Parameter space for particle diffusion or turbulent diffusion to dominate: application to heat transfer. Sonic Mach number M_s is ploted against the Alfven Mach number M_A. The heat transport is dominated by the dynamics of turbulent eddies is above the curve (area denoted "dynamic turbulent transport") and by thermal conductivity of electrons is below the curve (area denoted "electron heat transport"). Here λ is the mean free path of the electron, L is the driving scale, and $\alpha = (m_e/m_p)^{1/2}$, $\beta \approx 4$. *Example of theory application*: The panel in the right upper corner of the figure illustrates heat transport for the parameters for a cool core Hydra cluster (point "F"), "V" corresponds to the illustrative model of a cluster core in Ensslin et al. (2005). Relevant parameters were used for L and λ. From Lazarian (2006).

In spite of the substantial progress in understading of the ICM (see Enßlin, Vogt & Pfrommer 2005, henceforth EVP05, Enßlin & Vogt 2006, henceforth EV06 and references therein), the basic parameters of ICM turbulence are known within the factor of 3 at best. For instance, the estimates of injection velocity V_L varies in the literature from 300 km/s to 10^3 km/s, while the injection scale L varies from 20 kpc to 200 kpc, depending whether the injection of energy by galaxy mergers or galaxy wakes is considered. EVP05 considers an *illustrative* model in which the magnetic field with the 10 μG fills 10% of the volume, while 90% of the volume is filled with the field of $B \sim 1$ μG. Using the latter number and assuming $V_L = 10^3$ km/s, $L = 100$ kpc, and the density of the hot ICM is 10^{-3} cm^{-3}, one gets $V_A \approx 70$ km/s, i.e. $M_A > 1$. Using the numbers above, one gets $l_A \approx 30$ pc for the 90% of the volume of the hot ICM, which is much less than λ_{ion}. The diffusivity of ICM plasma gets $\eta = v_{ion} l_A$ which for the parameters above provides $Re \sim 2 \times 10^3$, which is enough for driving superAlfvenic turbulence at the outer scale L. However, as l_A increases as $\propto B^3$, Re gets around 50 for the field of 4 μG, which is at the border line of exciting turbulence[18]. However, the regions with higher magnetic fields

[18] One can imagine dynamo action in which superAlfvenic turbulence generates magnetic field till l_A gets large enough to shut down the turbulence.

(e.g. 10 μG) can support Alfvenic-type turbulence with the injection scale l_A and the injection velocities resulting from large-scale shear $V_L(l_A/L) \sim V_L M_A^{-3}$.
For the regions of $B \sim 1$ μG the value of l_A is smaller than the mean free path of electrons λ. According to Eq. (4) the value of κ_{electr} is 100 times smaller than $\kappa_{Spitzer}$. On the contrary, $\kappa_{dynamic}$ for the ICM parameters adopted will be $\sim 30\kappa_{Spitzer}$, which makes the heat transfer by turbulent motions the dominant process. This agrees well with the observations in Voigt & Fabian (2004). Fig. 2 shows the dominance of advective heat transfer for the parameters of the cool core of Hydra A ($B = 6$ μG, $n = 0.056$ cm^{-3}, $L = 40$ kpc, $T = 2.7$ keV according to EV06), point "F", and for the illustrative model in EVP05, point "V", for which $B = 1$ μG (see also Lazarian 2006).
Note that our stationary model of MHD turbulence is not directly applicable to transient wakes behind galaxies. The ratio of the damping times of the hydro turbulence and the time of straightening of the magnetic field lines is $\sim M_A^{-1}$. Thus, for $M_A > 1$, the magnetic field at scales larger than l_A will be straightening gradually after the hydro turbulence has faded away over time L/V_L. The process can be characterized as injection of turbulence at velocity V_A but at scales that increase linearly with time, i.e. as $l_A + V_A t$. The study of heat transfer in transient turbulence and magnetic field "regularly" stretched by passing galaxies is an interesting process that requires further investigation.

10. Richardson diffusion and superdiffusion on small scales

All the discussion above assumed that we deal with diffusion within magnetized plasmas over the scales much larger than the turbulence injection scale L. Below we show that on the scales less than L we deal with non-stationary processes.

10.1 Richardson-type advection of heat

The advection of heat on scales less than the turbulent injection scale L happens through smaller scale eddies. Thus the earlier estimate of turbulent diffusion of heat in terms of the injection velocity and the injection scale does not apply. In the lab system of reference the transfer of heat is difficult to describe and one should use the Lagrangian description.
One can consider two-particle turbulent diffusion or Richardson diffusion by dealing with the separation $\boldsymbol{\ell}(t) = \mathbf{x}(t) - \mathbf{x}'(t)$ between a pair of Lagrangian fluid particles (see Eyink et al. 2011). It was proposed by Richardson (1926) that this separation grows in turbulent flow according to the formula

$$\frac{d}{dt}\langle \ell_i(t)\ell_j(t)\rangle = \langle \kappa_{dynanic,ij}(\boldsymbol{\ell})\rangle \tag{11}$$

with a scale-dependent eddy-diffusivity $\kappa_{dynamic}(\ell)$. In hydrodynamic turbulence Richardson deduced that $\kappa_{dynamic}(\ell) \sim \varepsilon^{1/3}\ell^{4/3}$ (see Obukhov 1941) and thus $\ell^2(t) \sim \varepsilon t^3$. An analytical formula for the 2-particle eddy-diffusivity was derived by Batchelor (1950) and Kraichnan (1966):

$$\kappa_{dynamic,ij}(\boldsymbol{\ell}) = \int_{-\infty}^{0} dt \langle \delta U_i(\boldsymbol{\ell},0)\delta U_j(\boldsymbol{\ell},t)\rangle \tag{12}$$

with $\delta U_i(\boldsymbol{\ell},t) \equiv U_i(\mathbf{x}+\boldsymbol{\ell},t) - U_i(\mathbf{x},t)$ the relative velocity at time t of a pair of fluid particles which were at positions $\mathbf{x}$ and $\mathbf{x}+\boldsymbol{\ell}$ at time 0.
How can one understand these results? Consider a hot spot of the size l in a turbulent flow. The spot is going to be mostly expanded by turbulent eddies of size l. The turbulent velocity $u(l) = \frac{d}{dt}l(t)$ for Kolmogorov turbulence is proportional to $l^{1/3}$. Performing formal integration one gets an asymptotic solution for large time scales $l^2(t) \sim t^3$, which corresponds

to the Richardson diffusion law. Physically, as the hot spot extends, it is getting sheared by larger and eddies, which induce the accelerated expansion of the hot spot.

For magnetic turbulence the Kolmogorov-like description is valid for motions induced by strong Alfvenic turbulence in the direction perpendicular to the direction of the local magnetic field[19]. Thus we expect that Richardson diffusion to be applicable to the magnetized turbulence case.

10.2 Superdiffusion of heat perpendicular to mean magnetic field

The effects related to the diffusion of heat via electron streaming along magnetic field lines are different when the problem is considered at scales $\gg L$ and $\ll L$. This difference is easy to understand as on small scales magnetized eddies are very elongated, which means that the magnetic field lines are nearly parallel. However, as electrons diffuse into larger eddies, the dispersion of the magnetic field lines in these eddies gets bigger and the diffusion perpendicular to the mean magnetic field increases[20]

SuperAlfvenic turbulence:

On scales $k_{\|}^{-1} < l_A$, i.e., on scales at which magnetic fields are strong enough to influence turbulent motions, the mean deviation of a field in a distance $k_{\|}^{-1} = \delta z$ is given by LV99 as

$$< (\delta x)^2 >^{1/2} = \frac{([\delta z] M_A)^{3/2}}{3^{3/2} L^{1/2}}, \quad M_A > 1 \tag{13}$$

Thus, for scales much less than L (see also Yan & Lazarian 2008)

$$\kappa_{e,\perp} \approx \left(\frac{\delta x}{\delta z}\right)^2 \kappa_{e,\|} \sim \frac{[\delta z] M_A^3}{3^3 L} \kappa_{e,\|} \sim \kappa_{\|} (k_{\|} l_A)^{-1}, \quad M_A > 1, \tag{14}$$

which illustrates the non-stationary regime of *superdiffusion*, where the diffusion coefficient changes with the scale $k_{e,\|}^{-1}$.

SubAlfvenic turbulence:

On scales larger than l_{tr}, the turbulence is weak. The mean deviation of a field in a distance δz is given by Lazarian (2006):

$$< (\delta x)^2 >^{1/2} = \frac{[\delta z]^{3/2}}{3^{3/2} L^{1/2}} M_A^2, \quad M_A < 1. \tag{15}$$

For the scales $L > k_{\|}^{-1} = \delta z$ we combine Eq. (15) with

$$\delta z = \sqrt{kappa_{e,\|} \delta t} \tag{16}$$

and get for scales much less than L

$$\kappa_{e,\perp} \approx \frac{\delta x^2}{\delta t} = \frac{\kappa_{e,\|} \delta z}{3^3 L} M_A^4 \sim \kappa_{e,\|} (k_{\|} L)^{-1} M_A^4, \tag{17}$$

[19] The local magnetic field direction fluctuates in the lab system of reference. Thus the results of the diffusion in the lab system are less anisotropic.

[20] Below we consider turbulent scales that are larger than the electron mean free path λ_e. Heat transfer at smaller scale is not a diffusive process, but happens at the maximal rate determined by the particle flux nv_{th} provided that we deal with scales smaller than l_A. The perpendicular to magnetic field flux is determined by the field line deviations on the given scale as we discussed above (see also LV99).

which for a limiting case of $k_{e,\|} \sim L^{-1}$ coincides up to a factor with the Eq. (8).
Eqs. (14) and (17) certify that the perpendicular diffusion at scales much less than the injection scale accelerates as z grows.

10.3 Comparison of processes

Both processes of heat transport at the scales less than the turbulence injection scale are different from the diffusion at large scales as the rate of transport depends on the scale. However, the description of heat transport by electrons is more related to the measurements in the lab system. This follows from the fact that the dynamics of magnetic field lines is not important for the process and it is electrons which stream along wandering magnetic field lines. Each of these wandering magnetic field lines are snapshot of the magnetic field line dynamics as it changes through magnetic reconnection its connectivity in the ambient plasma. Therefore the description of heat transfer is well connected to the lab system of reference. On the contrary, the advection of heat through the Richardson diffusion is a process that is related to the Langrangian description of the fluid. Due to this difference the direct comparison of the efficiency of processes is not so straightforward.
For example, if one introduces a localized hot spot, electron transport would produce heating of the adjacent material along the expanding cone of magnetic field lines, while the turbulent advection would not only spread the hot spot, but also advect it by the action of the largest eddies.

11. Outlook on the consequences

Magnetic thermal insulation is a very popular concept in astrophysical literature dealing with magnetized plasmas. Our discussion above shows that in many cases this insulation is very leaky. This happens due to ubiquitous astrophysical turbulence which induces magnetic field wandering and interchange of pieces of magnetized plasma enabled by turbulent motions. Both processes are very closely related to the process of fast magnetic reconnection of turbulent magnetic field (LV99).
As a result, instead of an impenetrable wall of laminar ordered magnetic field lines, the actual turbulent field lines present a complex network of tunnels along which electrons can carry heat. As a result, the decrease of heat conduction amounts to a factor in the range of $1/3$ for mildly superAlfvenic turbulence to a factor $\sim 1/5$ for transAlfvenic turbulence. The cases when heat conductivity by electrons may be suppressed to much greater degree include highly superAlfvenic turbulence and highly subAlfvenic turbulence. In addition, turbulent motions induce heat advection which is similar to turbulent diffusivity of unmagnetized fluids.
The importance of magnetic reconnection cannot be stressed enough in relation to the process of heat transfer in magnetized plasmas. As a consequence of fast magnetic reconnection plasma does not stay entrained on the same magnetic field lines, as it is usually presented in textbooks. On the contrary, magnetic field lines constantly change their connectivity and plasma constantly samples newly formed magnetic field lines enabling efficient diffusion. Therefore we claim that the advection of heat by turbulence is an example of a more general process of reconnection diffusion. It can be noticed parenthetically that the turbulent advection of heat is a well knows process. However, for decades the discussion of the process avoided in astrophysical literature due the worries of the effect of reconnection that inevitably should accompany it. The situation has changed with better understanding of magnetic reconnection in turbulent environments (LV99). It worth pointing out that our

estimates indicate that in many astrophysicaly important cases, e.g. for ICM, the advective heat transport by dynamic turbulent eddies dominates thermal conductivity.

Having the above processes in hand, one can describe heat transport within magnetized astrophysical plasmas. For instance, we discussed the heat transfer by particle and turbulent motions for $M_A < 1$ and $M_A > 1$. It is important that we find that turbulence can both enhance diffusion and suppress it. We showed that when λ gets larger than l_A the conductivity of the medium $\sim M_A^{-3}$ and therefore the turbulence *inhibits* heat transfer, provided that $\kappa_e > \kappa_{dynamic}$. Along with the plasma effects that we mention below, this effect can, indeed, support sharp temperature gradients in hot plasmas with weak magnetic field.

As discussed above, rarefied plasma, e.g. ICM plasma, has large viscosity for motions parallel to magnetic field and marginal viscosity for motions that induce perpendicular mixing. Thus fast dissipation of sound waves in the ICM does not contradict the medium being turbulent. The later may be important for the heating of central regions of clusters caused by the AGN feedback (see Churasov et al. 2001, Nusser, Silk & Babul 2006 and more references in EV06). Note, that models that include both heat transfer from the outer hot regions and an additional heating from the AGN feedback look rather promissing (see Ruszkowkski & Begelman 2002, Piffaretti & Kaastra 2006). We predict that the viscosity for 1 μG regions is less than for 10 μG regions and therefore heating by sound waves (see Fabian et al. 2005) could be more efficient for the latter. Note, that the plasma instabilities in collisionless magnetized ICM arising from compressive motions (see Schekochihin & Cowley 2006, Lazarian & Beresnyak 2006) can resonantly scatter particles and decrease λ. This decreases further κ_e compared to κ_{unmagn} but increases Re. In addition, we disregarded mirror effects that can reflect electrons back[21] (see Malyshkin & Kulsrud 2001 and references therein), which can further decrease κ_e. While there are many instabilities that are described in plasmas with temperature gradient, many of those are of academic interest, as they do not take into account the existence of ambient turbulence.

For years the attempts to describe heat transfer in magnetized plasma were focused on finding the magic number which would be the reduction factor characterizing the effect of magnetic field on plasmas' diffusivity. Our study reveals a different and more complex picture. The heat transfer depends on sonic and Alfven Mach numbers of turbulence and the corresponding diffusion coefficient vary substantially for plasmas with different level of magnetization and turbulent excitation. In different astrophysical environments turbulence can both inhibit or enhance diffusivity depending on the plasma magnetization and turbulence driving.

The issues of "subdiffusivity" or magnetic field retracing their paths was a worrisome issue that for years impeded the progress in understanding heat transport in plasmas. We claim that the retracing does happen, but on the scales which are of the order of the eddies at the dissipation scale. As an electron has a finite Larmor radius in the retracing the same magnetic field line it experiences the deviations from its original trajectory. On the scale less than the dissipation scale these deviations grow from the electron Larmor radius in accordance with Lyapunov exponents, but on larger scale the separation is determined by field wandering only and does not depend on the Larmor radius. Thus the effect of retracing for heat transfer in real-world astrophysical turbulence with a substantial separation of the turbulence injection scale and dissipation scales is marginal.

On the contrary, the issue of "superdiffusivity" may be important for heat transfer on the scales less than the turbulence injection scale. Richardson diffusion or more correctly its anisotropic analog present in magnetized plasma (see Eyink et al. 2011) is an example of

[21] Many of these papers do not use realistic models of turbulence and therefore overestimate the effect of electron reflection.

superdiffusion induced by eddies of increasing size. A similar effect is also true for magnetic field line wandering. The effect of "superdiffusive" heat transfer requires additional studies.
It is worth mentioning that another parameter that determines the heat flux into the magnetized volume is the area of the contact of plasmas with different temperatures. For instance, if the magnetic flux is "shredded", i.e. consists of numerous separated individual flux tubes, then the heating of plasma within magnetized tubes may be more efficient. For instance, Fabian et al. (2011) appealed to reconnection diffusion of ambient plasma into "shredded" magnetic flux of NGC1275 in Perseus cluster in order to explain heating and ionization of the magnetic filaments.
In view of the discussion above one can conclude that realistically turbulent magnetic fields do not completely suppress heat conductivity of astrophysical plasmas. The decrease of thermal conductivity depends on the Alfven Mach number of turbulence. At the same time, turbulent motions enhance heat transport via heat advection. In special situations, e.g. in very weakly turbulent magnetic field, the transport of heat in plasmas may still be slow.
Acknowledgments The research is supported by the NSF grant AST 0808118 and the Center for Magnetic Self Organization in Laboratory and Astrophysical Plasmas (CMSO).

12. References

[1] Bakunin, O.G. 2005, Plasma Phys. and Contr. Fus. 47, 1857
[2] Begelman, M. & Fabian, A. 1990, MNRAS, 244, 26
[3] Beresnyak, A. 2011, Phys. Rev. Lett., 106, 075001
[4] Beresnyak, A., & Lazarian, A. 2010, ApJL, 722, L110
[5] Biskamp, D. 2003, *Magnetohydrodynamic Turbulence.* (Cambridge: CUP)
[6] Chandran, B. & Cowley, S. 1998, Phys. Rev. Lett., 80, 307
[7] Chepurnov, A., Lazarian, A., Stanimirović, S., Heiles, C., & Peek, J. E. G. 2010, ApJ, 714, 1398
[8] Cho J., Lazarian A., 2005, Theoret. Comput. Fluid Dynamics, 19, 127
[9] Cho, J., & Lazarian, A. 2004, Journal of Korean Astronomical Society, 37, 557
[10] Cho, J. & Lazarian, A. 2002, Phys. Rev. Lett., 88, 5001
[11] Cho, J., Lazarian, A., Honein, A., Kassions, S., & Moin, P. 2003, ApJ, 589, L77
[12] Cho, J., Vishniac, E. T., Beresnyak, A., Lazarian, A., & Ryu, D. 2009, ApJ, 693, 1449
[13] Churazov, E., Bruggen, M., Kaiser, C., Bohringer, H., & Forman W. 2001, ApJ, 554, 261
[14] Draine, B. T., & Lazarian, A. 1998, ApJ, 508, 157
[15] Galtier, S., Nazarenko, S., Newel, A. & Pouquet, A. 2000, J. Plasma Phys., 63, 447
[16] Enßlin, T., Vogt, C. & Pfrommer, C. 2005, in The Magnetized Plasma in Galaxy Evolution, Eds. K.T. Chyzy, K. Otminowska-Mazur, M. Soida and R.-J. Dettmar, Jagielonian University, Kracow, p. 231
[17] Enßlin, T., & Vogt, C. 2006, ApJ,
[18] Eyink, G. L. 2011,Phys. Rev. E 83, 056405
[19] Eyink, G., Lazarian, A., & Vishniac E. 2011, ApJ, submitted
[20] Fabian, A.C. 1994, ARA& A, 32, 277
[21] Fabian, A.C., Mushotzky, R.F., Nulsen, P.E.J., & Peterson, J.R. 2001, MNRAS, 321, L20
[22] Fabian, A.C., Reynolds, C.S., Taylor, G.B. & Dunn, R.J. 2005, MNRAS, 363, 891
[23] Fabian, A. C., Sanders, J. S., Williams, R. J. R., Lazarian, A., Ferland, G. J., & Johnstone, R. M. 2011, ApJ, in press, arXiv:1105.1735
[24] Goldreich, P. & Sridhar, S. 1995, ApJ, 438, 763
[25] Higdon J. C., 1984, ApJ, 285, 109
[26] Kota, J. & Jokipii, J. 2000, ApJ, 531, 1067

[27] Kowal, G., & Lazarian, A. 2010, ApJ, 720, 742
[28] Kowal, G., Lazarian, A., Vishniac, E. T., & Otmianowska-Mazur, K. 2009, ApJ, 700, 63
[29] Kulsrud R., 2004, Plasma Physics for Astrophysics, Princeton, NJ, Princeton University Press
[30] Lazarian, A. 2007, ApJ, 660, 173
[31] Lazarian, A. 2006, ApJ, 645, 25
[32] Lazarian, A. & Beresnyak, A. 2006, MNRAS, 373, 1195
[33] Lazarian, A., Kowal, G., Vishniac, E., & de Gouveia Dal Pino, E. 2011, Planetary and Space Science, 59, 537
[34] Lazarian, A., & Pogosyan, D. 2008, ApJ, 686, 350
[35] Lazarian, A., & Pogosyan, D. 2006, ApJ, 652, 1348
[36] Lazarian, A., & Pogosyan, D. 2000, ApJ, 537, 720
[37] Lazarian, A., Vishniac, E. & Cho, J. 2004, ApJ, 603, 180
[38] Lazarian, A. & Vishniac, E. 1999, ApJ, 517, 700
[39] Lithwick, Y., & Goldreich, P. 2001, ApJ, 562, 279
[40] Malyshkin, L. & Kulsrud, R. 2001, ApJ, 549, 402
[41] Lesieur, M. 1990, *Turbulence in fluids : stochastic and numerical modelling*, 2nd. rev. ed. (Dordrecht; Kluwer)
[42] Narayan, R., & Medvedev M. 2001, ApJ, 562, L129
[43] Nusser, A., Silk, J., & Babul, A. 2006, MNRAS, 373, 739
[44] Padoan, P., Juvela, M., Kritsuk, A., & Norman, M. L. 2009, ApJL, 707, L153
[45] Petschek, H.E. Magnetic field annihilation. The Physics of Solar Flares, AAS-NASA Symposium (NASA SP-50), ed. WH. Hess (Greenbelt, MD: NASA) 425
[46] Piffaretti, R., & Kaastra, J. S. 2006, A&A, 453, 423
[47] Rechester, A., & Rosenbluth, M. 1978, Phys. Rev. Lett., 40, 38
[48] Richardson, L. F. 1926, Proc. R. Soc. London, Ser. A, 110, 709
[49] Ruzkowski, M. & Begelman, M.C. 2002, ApJ, 581, 223
[50] Santos-Lima, R., Lazarian, A., de Gouveia Dal Pino, E. M., & Cho, J. 2010, ApJ, 714, 442
[51] Schekochihin, A. A., Iskakov, A. B., Cowley, S. C., McWilliams, J. C., Proctor, M. R. E., & Yousef, T. A. 2007, New Journal of Physics, 9, 300
[52] Shay, M. A. & Drake, J. F., 1998, Geophys. Res. Letters Geophysical Research Letters, 25, 3759-3762
[53] Shebalin J.V., Matthaeus W.H., Montgomery D.C., 1983, J. Plasma Phys., 29, 525
[54] Shibata, K., & Tanuma, S. 2001, Earth, Planets, and Space, 53, 473
[55] Smith, R. & Cox, D. 2001, ApJS, 134, 283
bibitem[Vishniac & Lazarian (1999)]VishniacLazarian99 Vishniac, E. & Lazarian, A. 1999, in: Plasma Turbulence and Energetic Particles in Astrophysics; Proceedings of the International Conference, Cracow, Poland, 5-10 September, 1999 eds. M. Ostrowski & R. Schlickeiser.
[56] Voigt, L.M. & Fabian, A.C. 2004, MNRAS, 347, 1130
[57] Yan, H., & Lazarian, A. 2008, ApJ, 673, 942
[58] Yan, H., Lazarian, A., & Draine, B. T. 2004, ApJ, 616, 895
[59] Webb, G., Zank, G., Kaghashvili, E., & Roux J. 2006, ApJ, 651, 211

11

Steady-State Heat Transfer and Thermo-Elastic Analysis of Inhomogeneous Semi-Infinite Solids

Yuriy Tokovyy[1] and Chien-Ching Ma[2]
[1]*Pidstryhach Institute for Applied Problems of Mechanics and Mathematics National Academy of Sciences of Ukraine, Lviv,*
[2]*Department of Mechanical Engineering, National Taiwan University Taipei*
[1]*Ukraine*
[2]*Taiwan ROC*

1. Introduction

The advancement in efficient modeling and methodology for thermoelastic analysis of structure members requires a variety of the material characteristics to be taken into consideration. Due to the critical importance of such analysis for adequate determination of operational performance of structures, it presents a great deal of interest for scientists in both academia and industry. However, the assumption that the material properties depend on spatial coordinates (material inhomogeneity) presents a major challenge for analytical treatment of relevant heat conduction and thermoelasticity problems. The main difficulty lies in the need to solve the governing equations in the differential form with variable coefficients which are not pre-given for arbitrary dependence of thermo-physical and thermo-elastic material properties on the coordinate. Particularly, for functionally graded materials, whose properties vary continuously from one surface to another, it is impossible, except for few particular cases, to solve the mentioned problems analytically (Tanigawa, 1995). The analytical, semi-analytical, and numerical methods for solving the heat conduction and thermoelasticity problems in inhomogeneous solids attract considerable attention in recent years. The overview of the relevant literature is given in our publications (Tokovyy & Ma, 2008, 2008a, 2009, 2009a). On the other hand, determination of temperature gradients, stresses and deformations is usually an intermediate step of a complex engineering investigation. Therefore analytical methods are of particular importance representing the solutions in a most convenient form. The great many of existing analytical methods are developed for particular cases of inhomogeneity (e.g., in the form of power or exponential functions of a coordinate, etc.). The methods applicable for wider ranges of material properties are oriented mostly on representation the inhomogeneous solid as a composite consisting of tailored homogeneous layers. However, such approaches are inconvenient for applying to inhomogeneous materials with large gradients of inhomogeneity due to the weak convergence of the solution with increasing the number of layers.
A general method for solution of the elasticity and thermoelasticity problems in terms of stresses has been developed by Prof. Vihak (Vigak) and his followers in (Vihak, 1996; Vihak

et al., 1998, 2001, 2002; Vigak, 1999; Vigak & Rychagivskii, 2000, 2002). The method consists in construction of analytical solutions to the problems of thermoelasticity based on direct integration of the original equilibrium and compatibility equations. Originally the equilibrium equations are in terms of stresses, and they do not depend on the physical stress-strain relations, as well as on the material properties. At the same time the general equilibrium relates all the stress-tensor components. This enables one to express all the stresses in terms of the governing stresses. The compatibility equations in terms of strain are then reduced to the governing equations for the governing stresses. When these equations are solved, all the stress-tensor components can be found by means of the aforementioned expressions. In addition, the method enables the derivation of: *a*) fundamental integral equilibrium and compatibility conditions for the imposed thermal and mechanical loadings and the stresses and strains; *b*) one-to-one relations between the stress-tensor and displacement-vector components. Therefore, when the stress-tensor components are found, then the displacement-vector components are also found automatically. Such relations have been derived for the case of one-dimensional problem for a thermoelastic hollow cylinder (Vigak, 1999a) and plane problems for elastic and thermoelastic semi-plane (Vihak & Rychahivskyy, 2001; Vigak, 2004; Rychahivskyy & Tokovyy, 2008).

Since application of this method rests upon the direct integration of the equilibrium equations, the proposed solution scheme offers ample opportunities for efficient analysis of inhomogeneous solids. In contrast to homogeneous materials, the compatibility equations in terms of stresses are with variable coefficients. This causes that the governing equations, obtained on the basis of the compatibility ones, appear as integral equations of Volterra type. By following this solution strategy, the one-dimensional thermoelasticity problem for a radially-inhomogeneous cylinder has been analyzed (Vihak & Kalyniak, 1999; Kalynyak, 2000). In the same manner, the two-dimensional elasticity and thermoelasticity problems for inhomogeneous cylinders, strips, planes and semi-planes were solved in (Tokovyy & Rychahivskyy, 2005; Tokovyy & Ma, 2008, 2008a, 2009, 2009a). The same method has also been extended for three-dimensional problems (Tokovyy & Ma, 2010, 2010a). Application of this method for analysis of inhomogeneous solids exhibits several advantages. First of all, this method is unified for various kinds of inhomogeneity and different shape of domain and it does not impose any restriction on the material properties. Moreover, when applying the resolvent-kernel algorithm for solution of the governing Volterra-type integral equation, the solutions of corresponding elasticity and thermoelasticity problems for inhomogeneous solids appear in the form of explicit functional dependences on the thermal and mechanical loadings, which makes them to be rather usable for complex engineering analysis.

Herein, we consider an application of the direct integration method for analysis of thermoelastic response of an inhomogeneous semi-plane within the framework of linear uncoupled thermoelasticity (Nowacki, 1962). The solution of this problem consists of two stages: *i*) solution of the in-plane steady-state heat conduction problem under certain boundary conditions, and *ii*) solution of the plane thermal stress problem due to the above determined temperature field and appropriate boundary conditions. Solution of both problems is reduced to the governing Volterra integral equation. By making use of the resolvent-kernel solution technique, the governing equation is solved and the solution of the original problem is presented in an explicit form. Due to the later result, the one-to-one relationships are set up between the tractions and displacements on the boundary of the inhomogeneous semi-plane. Using these relations and the solution in terms of stresses, we find solutions for the boundary value problems with displacements or mixed conditions imposed on the boundary. It is shown that these solutions are correct if the tractions satisfy the necessary equilibrium conditions, the displacements meet the integral compatibility conditions, and the heat sources and heat flows satisfy the integral condition of thermal balance.

2. Analysis of the steady-state heat conduction problem in an inhomogeneous semi-plane

In this section, we consider an application of the direct integration method for solution of the in-plane steady-state (stationary) heat conduction problem for a semi-plane whose thermal conductivity is an arbitrary function of the depth-coordinate. Having applied the Fourier integral transformation to the differential heat conduction equation with variable coefficients, this equation is reduced to the Volterra-type integral equation, which then is solved by making use of the resolvent-kernel technique. As a result, the temperature distribution is found in an explicit functional form that can be efficiently used for analysis of thermal stresses and displacements in the semi-plane.

2.1 Formulation of the heat conduction problem

Let us consider the two-dimensional heat conduction problem for semi-plane $D=\{(x,y)\in(-\infty,\infty)\times[0,\infty)\}$ in the dimensionless Cartesian coordinate system (x, y). In assumption of the isotropic material properties, the problem is governed by the following heat conduction equation (Hetnarski & Reza Eslami, 2008)

$$\frac{\partial}{\partial x}\left(k(y)\frac{\partial T(x,y)}{\partial x}\right)+\frac{\partial}{\partial y}\left(k(y)\frac{\partial T(x,y)}{\partial y}\right)=-q(x,y), \tag{1}$$

where $T(x,y)$ is the steady-state temperature distribution, $k(y)$ stands for the coefficient of thermal conduction, and $q(x, y)$ denotes the quantity of heat generated by internal heat sources in semi-plane D. When the coefficient of thermal conduction is constant, then equation (1) presents the classical equation of quasi-static heat conduction (Nowacki, 1962; Carslaw & Jaeger, 1959)

$$\Delta T(x,y)=-W(x,y), \tag{2}$$

where $\Delta=\partial^2/\partial x^2+\partial^2/\partial y^2$ and $W(x,y)=q(x,y)/k$ denotes the density of internal sources of heat. In the steady-state case, the temperature $T(x,y)$ can be determined from equation (1) for $k(y)$ or (2) for $k=\text{const}$ under certain boundary condition imposed at $y=0$ (Nowacki, 1962). We consider the boundary condition in one of the following forms:

a. the temperature distribution is prescribed on the boundary

$$T(x,y)=T_0(x), \quad y=0; \tag{3}$$

b. the heat flux over the limiting line $y=0$ is prescribed on the boundary

$$\frac{\partial T(x,y)}{\partial y}=\Phi_0(x), \quad y=0; \tag{4}$$

c. the heat exchange condition is imposed on the boundary

$$\frac{\partial T(x,y)}{\partial y} + \alpha_0 T(x,y) = \beta_0, \quad y = 0. \tag{5}$$

Here α_0 and β_0 are constants, $T_0(x)$ and $\Phi_0(x)$ are given functions. Assuming that the temperature field, heat fluxes, and the density of heat sources vanish with $|x|, y \to \infty$, we consider finding the solution to equation (1) or (2) under either of the boundary conditions (3) - (5) and the supplementary conditions of integrability of the functions in question in their domain of definition.

2.2 Solution of the stated heat conduction problem by reducing to the Volterra-type integral equation

In the case when $k = \text{const}$, it has been shown (Rychahivskyy & Tokovyy, 2008) that for construction of a correct solution to equation (2) with boundary condition (4), the following necessary condition

$$\iint_D W(x,y)\mathrm{d}x\mathrm{d}y = \int_{-\infty}^{\infty} \Phi_0(x)\mathrm{d}x \tag{6}$$

is to be fulfilled. This condition of thermal balance postulates that the resultant heat flux trough the boundary $y = 0$ is equal to the resultant action of internal heat sources within domain D. In the case of boundary conditions (3) or (5), the right-hand side of condition (6) should be replaced by the integral of the heat flux at $y = 0$, which is determined by the temperature. Due to this reason, condition (6) can be regarded as an efficient tool for verification of the solution correctness.

Note that condition (6) is natural for steady-state thermal processes in bounded solids. However, it is not intuitive for non-stationary thermal regimes since then only certain distribution of the temperature field is possible inside the solid implying that the heating of the solid until an average temperature is not achievable. Thus, condition (6) for a semi-plane is a consequence of application of solid mechanics to the oversimplified geometrical model.

By denoting

$$\Phi_x(x,y) = k(y)\frac{\partial T(x,y)}{\partial x}, \quad \Phi_y(x,y) = k(y)\frac{\partial T(x,y)}{\partial y}$$

in equation (1), when $k = k(y)$, and following the strategy presented in (Rychahivskyy & Tokovyy, 2008), it can be shown that condition (6) holds for the case of inhomogeneous material. In addition, the resultant of the temperature is necessarily equal to zero

$$\iint_D T(x,y)\mathrm{d}x\mathrm{d}y = 0 \tag{7}$$

for both homogeneous and inhomogeneous cases.

Let us construct the solution to equation (1) under boundary conditions (3), (4), or (5) by taking conditions (6) and (7) into consideration. Having applied the Fourier integral transformation (Brychkov & Prudnikov, 1989)

$$\overline{f}(y;s)=\int_{-\infty}^{\infty}f(x,y)\exp(-isx)\mathrm{d}x \tag{8}$$

to the aforementioned equation and boundary conditions, we arrive at the following second order ODE

$$\frac{\mathrm{d}^2\overline{T}(y;s)}{\mathrm{d}y^2}-s^2\overline{T}(y;s)=-\frac{1}{k(y)}\left(\overline{q}(y;s)+\frac{\mathrm{d}k(y)}{\mathrm{d}y}\frac{\mathrm{d}\overline{T}(y;s)}{\mathrm{d}y}\right) \tag{9}$$

that is accompanied with one of the following boundary conditions:

$$\overline{T}(y;s)=\overline{T}_0(s),\quad y=0; \tag{10}$$

$$\frac{\mathrm{d}\overline{T}(y;s)}{\mathrm{d}y}=\overline{\Phi}_0(s),\quad y=0; \tag{11}$$

$$\frac{\mathrm{d}\overline{T}(y;s)}{\mathrm{d}y}+\alpha_0\overline{T}(y;s)=\beta_0,\quad y=0. \tag{12}$$

Here s is a parameter of the integral transformation, $i^2=-1$, $f(x,y)\in L(D)$. For the sake of brevity, the parameter s will be omitted from the arguments of functions in the following text.

A general solution to equation (9) in semi-plane D can be given in the form

$$\overline{T}(y)=C\exp(-|s|y)+\frac{1}{2|s|}\int_0^{\infty}\frac{\overline{q}(\eta)}{k(\eta)}\exp(-|s|\,|y-\eta|)\mathrm{d}\eta$$

$$+\frac{1}{2|s|}\int_0^{\infty}\frac{1}{k(\eta)}\frac{\mathrm{d}k(\eta)}{\mathrm{d}\eta}\frac{\mathrm{d}\overline{T}(\eta)}{\mathrm{d}\eta}\exp(-|s|\,|y-\eta|)\mathrm{d}\eta\,, \tag{13}$$

where C is a constant of integration and $|\cdot|$ denotes the absolute-value function. By applying the integration by parts to the last integral in equation (13), we can obtain the following Volterra-type integral equation of second kind:

$$\overline{T}(y)=\left(C-\frac{\overline{T}(0)}{2|s|k(0)}\frac{\mathrm{d}k(0)}{\mathrm{d}y}\right)\exp(-|s|y)+q^{*}(y)+\int_0^{\infty}\overline{T}(\eta)K(y,\eta)\mathrm{d}\eta. \tag{14}$$

Here

$$q^{*}(y)=\frac{1}{2|s|}\int_0^{\infty}\frac{\overline{q}(\eta)}{k(\eta)}\exp(-|s|\,|y-\eta|)\mathrm{d}\eta,$$

$$K(y,\eta)=-\frac{1}{2|s|}\frac{\mathrm{d}}{\mathrm{d}\eta}\left(\frac{1}{k(\eta)}\frac{\mathrm{d}k(\eta)}{\mathrm{d}\eta}\exp(-|s|\,|y-\eta|)\right)$$

$$=\frac{1}{2|s|}\left(\left(\frac{1}{k(\eta)}\frac{\mathrm{d}k(\eta)}{\mathrm{d}\eta}-|s|\operatorname{sgn}(y-\eta)\right)\frac{\mathrm{d}k(\eta)}{\mathrm{d}\eta}-\frac{\mathrm{d}^2k(\eta)}{\mathrm{d}\eta^2}\right)\frac{\exp(-|s|\,|y-\eta|)}{k(\eta)}. \quad (15)$$

To solve integral equation (14), different algorithms can be employed, for instance, the Picard's process of successive approximations (Tricomi, 1957; Kalynyak, 2000; Tokovyy & Ma, 2008a), the operator series method (Bartoshevich, 1975), the Bubnov-Galerkin method (Fedotkin et al., 1983), a numerical procedure based on trapezoidal integration and a Newton-Raphson method (Frankel, 1991), iterative-collocation method (Hącia, 2007), discretization method, special kernels method and projection-iterative method (Domke & Hącia, 2007), spline approximations (Kushnir et al., 2002), the quadratic-form method (Belik et al. 2008), the greed methods (Peng & Li, 2010). Herein we employ the resolvent-kernel algorithm (Pogorzelski, 1966; Porter & Stirling, 1990) in the same manner as it has been done in (Tokovyy & Ma, 2008, 2009a). This method allows us to obtain the explicit-form analytical solution that is convenient for analysis. As a result, the transformation of temperature appears as

$$\bar{T}(y)=\left(C-\frac{\bar{T}(0)}{2|s|k(0)}\frac{\mathrm{d}k(0)}{\mathrm{d}y}\right)\varphi(y)+\theta(y), \quad (16)$$

where

$$\varphi(y)=\exp(-|s|y)+\int_0^\infty \exp(-|s|\eta)\Re(y,\eta)\mathrm{d}\eta, \quad (17)$$

$$\theta(y)=q^*(y)+\int_0^\infty q^*(\eta)\Re(y,\eta)\mathrm{d}\eta, \quad (18)$$

and the resolvent-kernel is determined by the recurring kernels as

$$\Re(y,\eta)=\sum_{n=0}^{\infty}K_{n+1}(y,\eta), \quad (19)$$

$$K_1(y,\eta)=K(y,\eta),\quad K_{n+1}(y,\eta)=\int_0^\infty K(y,\zeta)K_n(\zeta,\eta)\mathrm{d}\zeta,\quad n=1,2,\ldots$$

Note that expression (16) is advantageous in comparison with the analogous solutions constructed by means of the aforementioned techniques for solution of the Volterra integral equations. First of all, solution (16) is obtained in explicit functional form. This fact can be efficiently used for complex analysis involving solution of thermoelasticity problem. Next, the resolvent (19) is expressed only through the kernel (15) of integral equation (14) ("intrinsic" properties of an integral equation) and is non-dependent of the free term ("external" properties of an integral equation). Consequently, being computed once for certain kernel (which means for certain material properties, obviously), resolvent (19) can be employed for various kinds of thermal loading.

To determine the unknown constant C in equation (16), one of conditions (10) - (12) should be employed. Insertion of (16) into condition (10) yields

$$C = \frac{\overline{T}_0}{\varphi(0)}\left(1 + \frac{\varphi(0)}{2|s|k(0)}\frac{\mathrm{d}k(0)}{\mathrm{d}y}\right) - \frac{\theta(0)}{\varphi(0)},$$

and then the temperature can be given as

$$\overline{T}(y) = \frac{\overline{T}_0 - \theta(0)}{\varphi(0)}\varphi(y) + \theta(y). \tag{20}$$

In the case of boundary condition (11), the constant C appears as

$$C = \overline{q}_0\left(\frac{\mathrm{d}\varphi(0)}{\mathrm{d}y}\right)^{-1} + \frac{\overline{T}(0)}{2|s|k(0)}\frac{\mathrm{d}k(0)}{\mathrm{d}y} - \frac{\mathrm{d}\theta(0)}{\mathrm{d}y}\left(\frac{\mathrm{d}\varphi(0)}{\mathrm{d}y}\right)^{-1}.$$

Then the temperature can be given as

$$\overline{T}(y) = \left(\overline{q}_0 - \frac{\mathrm{d}\theta(0)}{\mathrm{d}y}\right)\left(\frac{\mathrm{d}\varphi(0)}{\mathrm{d}y}\right)^{-1}\varphi(y) + \theta(y). \tag{21}$$

In the case of boundary condition (12), the constant C takes the form

$$C = \left(\beta_0 - \alpha_0\theta(0) - \frac{\mathrm{d}\theta(0)}{\mathrm{d}y}\right)\left(\alpha_0\varphi(0) + \frac{\mathrm{d}\varphi(0)}{\mathrm{d}y}\right)^{-1} + \frac{\overline{T}(0)}{2|s|k(0)}\frac{\mathrm{d}k(0)}{\mathrm{d}y},$$

and, consequently,

$$\overline{T}(y) = \left(\beta_0 - \alpha_0\theta(0) - \frac{\mathrm{d}\theta(0)}{\mathrm{d}y}\right)\left(\alpha_0\varphi(0) + \frac{\mathrm{d}\varphi(0)}{\mathrm{d}y}\right)^{-1}\varphi(y) + \theta(y). \tag{22}$$

Having determined the expressions for the temperature field in the form (20), (21), or (22) and applying the formula

$$f(x,y) = \frac{1}{2\pi}\int_{-\infty}^{\infty}\overline{f}(y;s)\exp(isx)\mathrm{d}s \tag{23}$$

of inverse Fourier transformation (Brychkov & Prudnikov, 1989), we can obtain the expressions for temperature field in semi-plane D.

Note that according to the resolvent-kernel theory (Verlan & Sizikov, 1986), the recurring kernels K_{n+1} tend to zero as $n \to \infty$. Thus, for practical computations, the series in expression (19) can be truncated. Consequently,

$$\Re(y,\eta) \approx \Re_N(y,\eta) = \sum_{n=0}^{N} K_{n+1}(y,\eta), \tag{24}$$

where N is a natural number which depends on required accuracy of calculation.

2.3 Numerical analysis

To verify the obtained solution to the heat conduction problem, let us examine the case, when the semi-plane is heated by a single concentrated internal heat source

$$q(x,y) = q_0 \delta(x)\delta(y - y_0). \tag{25}$$

Meanwhile the boundary $y = 0$ remains of the constant temperature, $T_0 = 0$. Here q_0 is a constant dimensional parameter; $\delta(\cdot)$ denotes the Dirac delta-function. In this case, the temperature should be computed on the basis of expression (20). The coefficient of thermal conductivity is assumed to be in the following form

$$k(y) = k_0 \exp(\gamma y), \tag{26}$$

where k_0 and γ are constants. Note that for $\gamma = 0$, the thermal conductivity in the form (26) is constant, that corresponds to the case of homogeneous material. Then, on the basis of expression (19), the resolvent $\Re(y,\eta) \equiv 0$ and thus expression (20) presents an exact analytical solution

$$\frac{\overline{T}(y)k_0}{q_0} = \frac{1}{2|s|}\left(\exp(-|s|\,|y - y_0|) - \exp(-|s|(y + y_0))\right). \tag{27}$$

Application of the Fourier inversion (23) to formula (27) yields the expression for the temperature in the homogeneous semi-plane, as follows:

$$\frac{k_0}{q_0}T(x,y) = \frac{1}{4\pi}\ln\frac{x^2 + (y + y_0)^2}{x^2 + (y - y_0)^2}. \tag{28}$$

The full-field distributions of the temperature (28) and the components of corresponding heat flux are depicted in Fig. 1 for $y_0 = 1$. Distribution of the temperature (28) versus the

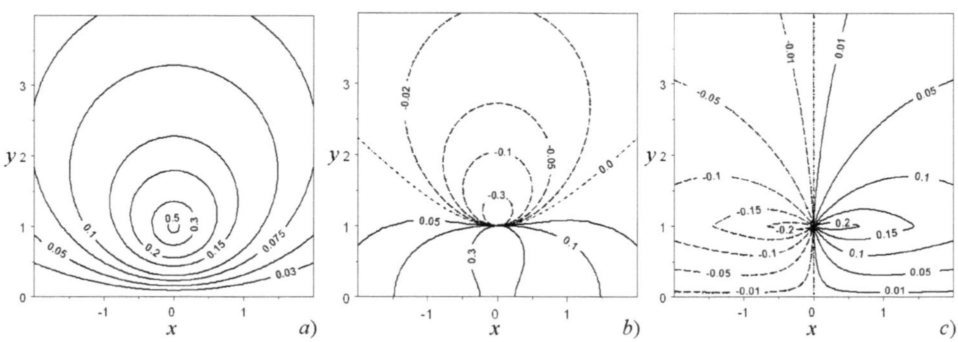

Fig. 1. Full-field distributions of (*a*) the dimensionless temperature $\dfrac{k_0 T(x,y)}{q_0}$, and (*b*) transversal $\dfrac{k_0}{q_0}\dfrac{\partial T(x,y)}{\partial y}$ and (*c*) longitudinal $\dfrac{k_0}{q_0}\dfrac{\partial T(x,y)}{\partial x}$ components of the heat flux for $y_0 = 1$

variable y is shown in Fig. 2 for $x = 0.0; 0.5$ and different values $y_0 = 1.0;2.0;3.0;4.0$. As we can observe in both figures, the thermal state is symmetrical with respect to the line $x = 0$. The temperature vanish when mowing away from the location of the heat source $(0, y_0)$. When approaching the boundary $y = 0$, the temperature vanish faster than in the opposite direction (due to satisfaction of the boundary condition). When the location of the heat source is moving away from the boundary, then the thermal state tends to one symmetrical with respect to the line $y = y_0$ (Fig. 2) due to the lowering influence of the boundary (in analogy to the case of an infinite plane).

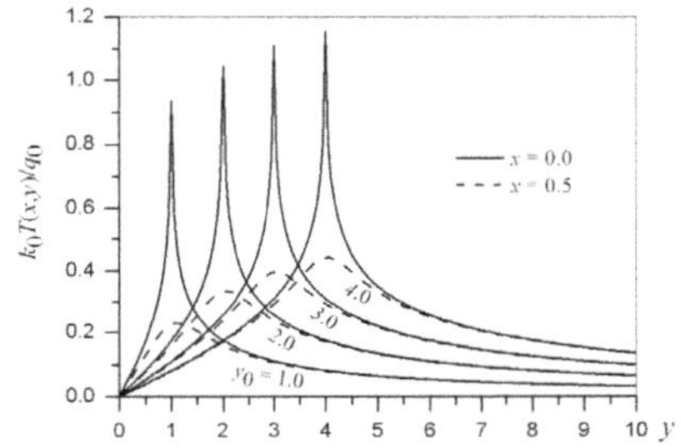

Fig. 2. Distribution of the temperature (28) versus coordinate y for $x = 0.0;0.5$

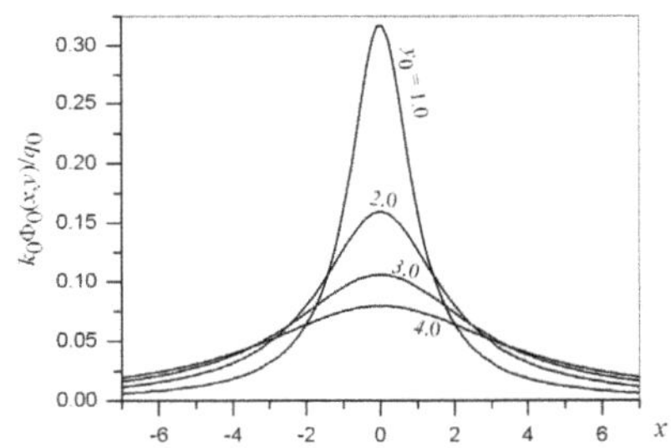

Fig. 3. The heat flux (29) for different values of $y_0 = 1.0;2.0;3.0;4.0$

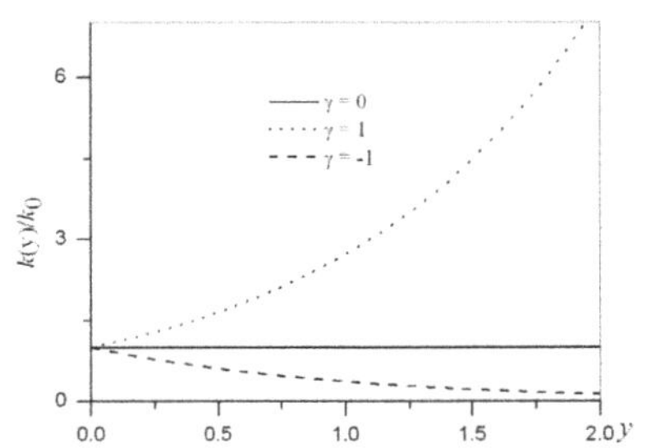

Fig. 4. Dependence of the thermal conductivity on the coordinate y for different values of γ

For the obtained temperature, the heat flux trough the boundary $y = 0$ can be found as

$$\frac{k_0 \Phi_0}{q_0} = \frac{y_0}{(x^2 + y_0^2)\pi}. \tag{29}$$

By taking formulae (25) and (29) into consideration, it is easy to see that condition (6) is satisfied for the considered case. Distribution of the heat flux (29) for different values of y_0 is shown in Fig. 3. As we can see, the heat flux over the boundary is locally distributed with the maximum value at $x = 0$ which decreases as the heat source is further from the boundary.

Now let $\gamma \neq 0$ in (26). For this case, the exact solution can be constructed by following the technique presented in (Ma & Lee, 2009; Ma & Chen, 2011). According to this technique, the exact solution to the problem (1), (3), (25) can be found in the form

$$\frac{k_0}{q_0}\overline{T}(y) = -\frac{q_0}{\sqrt{\gamma^2 - 4s^2}} \times \begin{cases} \left(\exp\left(y_0\gamma^+/2\right) - \exp\left(-y_0\gamma^-/2\right)\right)\exp\left(-y\gamma^+/2\right), & y \geq y_0, \\ \left(\exp\left(y\gamma^-/2\right) - \exp\left(-y\gamma^+/2\right)\right)\exp\left(-y_0\gamma^-/2\right), & y < y_0, \end{cases} \tag{30}$$

where $\gamma^{\pm} = \sqrt{\gamma^2 - 4s^2} \pm \gamma$. To obtain the distribution of temperature due to Fourier transform (30), the inversion formula (23) can be applied. The distributions of obtained temperature and corresponding heat flux are examined for different values of the parameter of inhomogeneity: $\gamma = 1$, $\gamma = -1$, and, for comparison with above-discussed homogeneous case, $\gamma = 0$ (Fig. 4). For $\gamma = 1$, the thermal conductivity grows exponentially from 1 to infinity; for $\gamma = -1$ it decreases from 1 to 0.

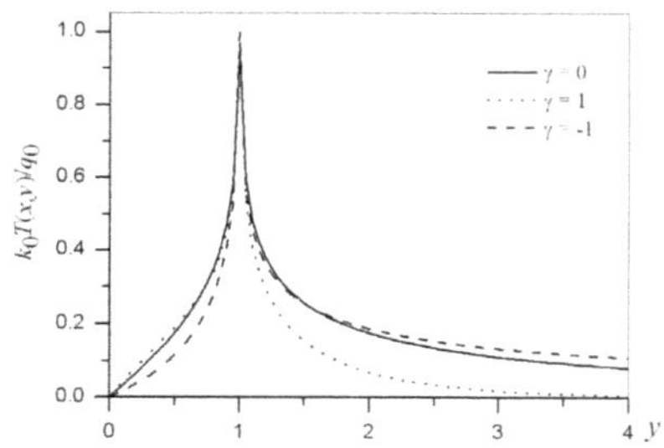

Fig. 5. Distribution of the temperature due to transformant (30) versus coordinate y at $x = 0.0$ for $\gamma = 0, \pm 1$

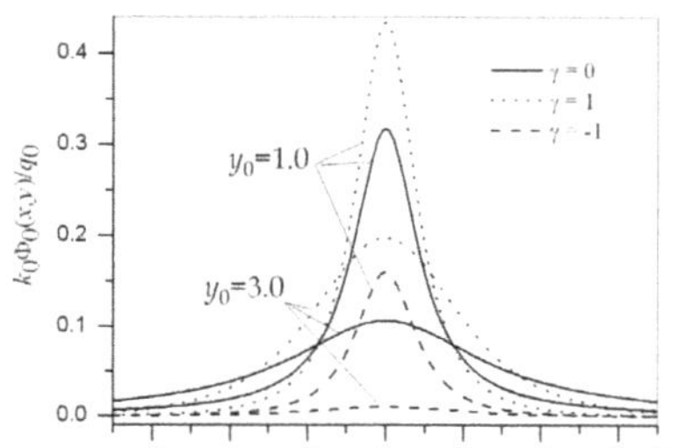

Fig. 6. Distribution of the heat flux across the boundary $y = 0$ for $y_0 = 1.0; 3.0$, $\gamma = 0; \pm 1$

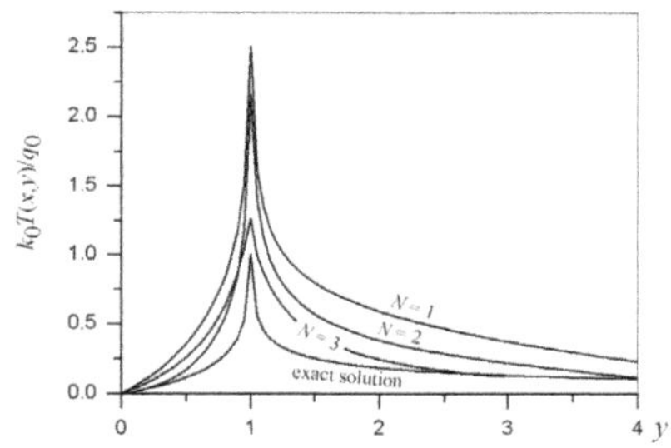

Fig. 7. Distribution of the heat flux for $y_0 = 1.0;3.0$, $\gamma = 0,\pm 1$

The effect of inhomogeneity on the temperature and heat flux over the boundary $y=0$ is shown in Figs. 5 and 6, respectively. As we observe in Fig. 5, the temperature vanishes with $y \to \infty$ as faster as the parameter of inhomogeneity γ is greater, whereas for $0 \le y < y_0$ vice-verse. Consequently, the heat flux over the boundary $y=0$ is greater for greater values of γ (Fig. 6).

Now we consider application of formula (20) for computation of the temperature in the inhomogeneous semi-plane. We employ formula (24) instead of (19) in (17) and (18). Distribution of the temperature computed by formula (20) for different values of parameter N in (24) is shown in Fig. 7. With growing N, the result naturally tends to the exact solution (30) and for $N=5$ they coincide. This result shows that expression (24) provides sufficiently good approximation for the resolvent $\Re(y,\eta)$ by holding few terms only.

3. Analysis of thermal stresses in an inhomogeneous semi-plane

In this section, the technique for solving the plane thermoelasticity problem for an isotropic inhomogeneous semi-plane with boundary conditions for stresses or displacements, as well as mixed boundary conditions, is developed by establishing one-to-one relations between boundary tractions and displacements. This technique is based on integration of the Cauchy relations to express displacements in terms of strains. Then, by taking the physical strain-stress relations into consideration, the displacements are expressed through the stress-tensor components. Finally, by making use of the explicit-form analytical solution to the corresponding problem with boundary tractions, the displacements on the boundary can be expressed through the tractions. The technique for establishment of the one-to-one relations between the tractions and displacements on the boundary, as well as for deriving the necessary equilibrium and compatibility conditions in the case of homogeneous semi-plane has been developed in (Rychahivskyy & Tokovyy, 2008).

3.1 Formulation of the problem

Let us consider the plane quasi-static thermoelasticity problem in inhomogeneous semi-plane D. In absence of body forces, this problem is governed (Nowacki, 1962) by the equilibrium equations

$$\frac{\partial \sigma_x(x,y)}{\partial x} + \frac{\partial \tau_{xy}(x,y)}{\partial y} = 0, \qquad \frac{\partial \tau_{xy}(x,y)}{\partial x} + \frac{\partial \sigma_y(x,y)}{\partial y} = 0, \tag{31}$$

the compatibility equation in terms of strains

$$\frac{\partial^2 \varepsilon_x(x,y)}{\partial y^2}+\frac{\partial^2 \varepsilon_y(x,y)}{\partial x^2}=\frac{\partial^2 \gamma_{xy}(x,y)}{\partial x \partial y}, \tag{32}$$

the physical thermoelasticity relations

$$\varepsilon_x(x,y)=\frac{1}{E^*(y)}\sigma_x(x,y)-\frac{\nu^*(y)}{E^*(y)}\sigma_y(x,y)+\alpha^*(y)T(x,y),$$

$$\varepsilon_y(x,y)=\frac{1}{E^*(y)}\sigma_y(x,y)-\frac{\nu^*(y)}{E^*(y)}\sigma_x(x,y)+\alpha^*(y)T(x,y), \tag{33}$$

$$\gamma_{xy}(x,y)=\frac{1}{G(y)}\tau_{xy}(x,y),$$

and the geometrical Cauchy relations

$$\varepsilon_x(x,y)=\frac{\partial u(x,y)}{\partial x},\quad \varepsilon_y=\frac{\partial v(x,y)}{\partial y},\quad \gamma_{xy}=\frac{\partial u(x,y)}{\partial y}+\frac{\partial v(x,y)}{\partial x}. \tag{34}$$

Here $\sigma_x,\sigma_y,\tau_{xy}$ and $\varepsilon_x,\varepsilon_y,\gamma_{xy}$ denote the stress- and strain-tensor components, respectively;

$$E^*(y)=\begin{cases}\dfrac{E(y)}{1-\nu^2(y)}, & \text{plane strain,}\\ E(y), & \text{plane stress,}\end{cases}\qquad \nu^*(y)=\begin{cases}\dfrac{\nu(y)}{1-\nu(y)}, & \text{plane strain,}\\ \nu(y), & \text{plane stress,}\end{cases}$$

$$\alpha^*(y)=\begin{cases}\alpha(y)(1+\nu(y)), & \text{plane strain,}\\ \alpha(y), & \text{plane stress,}\end{cases}$$

$E(y)$ denotes the Young modulus, $\nu(y)$ stands for the Poisson ratio; $G(y)=\dfrac{E(y)}{2(1+\nu(y))}$ is the shear modulus, $\alpha(y)$ denotes the coefficient of linear thermal expansion; $u(x,y)$ and $v(x,y)$ are the dimensionless displacements; $T(x,y)$ is the temperature field that is given or determined in the form (20), (21), or (22) by means of the technique proposed in the previous section.

We shall construct the solutions of the set of equations (31)-(34) for each of the three versions of boundary conditions prescribed on the line $y=0$:

a. in terms of stresses

$$\sigma_y(x,y)=-p(x),\qquad \tau_{xy}(x,y)=q(x),\qquad y=0; \tag{35}$$

b. in terms of displacements

$$u(x,y)=u_0(x),\qquad v(x,y)=v_0(x),\qquad y=0; \tag{36}$$

c. mixed conditions, when one of the following couples of relations

$$\begin{aligned}
&\sigma_y(x,y) = -p(x), v(x,y) = v_0(x), y = 0;\\
&\sigma_y(x,0) = -p(x), v(x,0) = v_0(x), y = 0;\\
&\tau_{xy}(x,0) = q(x), u(x,0) = u_0(x), y = 0;\\
&\tau_{xy}(x,0) = q(x), v(x,0) = v_0(x), y = 0
\end{aligned} \tag{37}$$

is imposed on the boundary. The boundary tractions and displacements, those are mentioned in conditions (35)–(37), as well the temperature field, vanish with $|x| \to \infty$, $y \to \infty$. We consider finding the solutions (stresses and displacements) of the stated boundary value problems.

3.2 Construction of the solutions

3.2.1 Case *A*: Boundary condition in terms of external tractions

Let us consider the construction of solution to the problem (31) - (34) under boundary conditions (35) with given tractions $p(x)$ and $q(x)$. The boundary displacements $u_0(x)$ and $v(x)$ are unknown and, thus, they should be determined in the process of solution. By following the solution strategy (Tokovyy & Ma, 2009), the stress-tensor components can be expressed trough the in-plane total stress $\bar{\sigma} = \bar{\sigma}_x + \bar{\sigma}_y$ as

$$\begin{aligned}
\bar{\sigma}_x &= \bar{\sigma} - \bar{\sigma}_y,\\
\bar{\sigma}_y &= -\bar{p}\exp(-|s|y) + \frac{|s|}{2}\int_0^\infty \bar{\sigma}(\xi)\left(\exp(-|s||y-\xi|) - \exp(-|s|(y+\xi))\right)d\xi,\\
\bar{\tau}_{xy} &= \frac{i|s|}{s}\bar{p}\exp(-|s|y) - \frac{is}{2}\int_0^\infty \bar{\sigma}(\xi)\left(\exp(-|s||y-\xi|)\operatorname{sgn}(y-\xi) - \exp(-|s|(y+\xi))\right)d\xi.
\end{aligned} \tag{38}$$

In turn, the total stress can be found as a solution of the Volterra-type integral equation of second kind:

$$\bar{\sigma}(y) = E^*(y)\left(\bar{p}P(y) + A\exp(-|s|y) - \alpha^*(y)\bar{T}(y)\right) + \int_0^\infty \bar{\sigma}(\eta)M(y,\eta)d\eta, \tag{39}$$

where

$$M(y,\eta) = -\frac{E^*(y)}{8}\int_0^\infty \frac{d^2}{d\xi^2}\left(\frac{1}{G(\xi)}\right)\left(\exp\left(-|s|(|\xi-\eta|+|y-\xi|)\right) - \exp\left(-|s|(\xi+\eta+|y-\xi|)\right)\right)d\xi,$$

$$P(y) = \frac{1}{4|s|}\int_0^\infty \frac{d^2}{d\xi^2}\left(\frac{1}{G(\xi)}\right)\exp\left(-|s|(\xi+|y-\xi|)\right)d\xi,$$

and the constant of integration A is to be found from the following integral condition

$$\int_0^\infty \bar{\sigma}(y)\exp(-|s|y)dy = -\frac{\bar{p}}{|s|} - \frac{i\bar{q}}{s}.$$

To solve equation (39), we employ the resolvent-kernel solution technique (Tokovyy & Ma, 2009a) with the following resolvent

$$\aleph(y,\eta)=\sum_{n=0}^{\infty}N_{n+1}(y,\eta),$$

$$N_1(y,\eta)=M(y,\eta),\quad N_{n+1}(y,\eta)=\int_0^{\infty}M(y,\zeta)N_n(\zeta,\eta)\mathrm{d}\zeta,\quad n=1,2,\ldots$$

As a result, the in-plane total stress appears in the form

$$\bar{\sigma}(y)=\bar{p}\Pi(y)+\Theta(y)+Af_A(y), \tag{40}$$

where

$$A=-\frac{1}{f_A(0)}\left(\bar{p}\left(\frac{1}{|s|}+\Pi(0)\right)+\frac{i\bar{q}}{s}+\Theta(0)\right),$$

$$\Pi(y)=E^*(y)P(y)+\int_0^{\infty}E^*(\eta)P(\eta)\aleph(y,\eta)\mathrm{d}\eta,$$

$$\Theta(y)=-\alpha^*(y)E^*(y)\bar{T}(y)-\int_0^{\infty}\alpha^*(\eta)E^*(\eta)\bar{T}(\eta)\aleph(y,\eta)\mathrm{d}\eta,$$

$$f_A(y)=E^*(y)\exp(-|s|y)+\int_0^{\infty}E^*(\eta)\exp(-|s|\eta)\aleph(y,\eta)\mathrm{d}\eta.$$

Having determined the total stress $\bar{\sigma}$ by formula (40), the stress-tensor components can be computed by means of formulae (38). The displacement-vector components $u(x,y)$ and $v(x,y)$, as well as the boundary displacement $u_0(x)$ and $v_0(x)$, can be also determined by the stresses by means of correct integration of the Cauchy relations (34).

3.2.2 Integration of the Cauchy relations and determination of the displacement-vector components in the inhomogeneous semi-plane due to the given boundary tractions

By taking the boundary conditions (36) with unknown boundary displacements $u_0(x)$ and $v_0(x)$ into account, the first and second relations of (34) yield

$$\begin{aligned} u(x,y)&=\frac{1}{2}\int_{-\infty}^{\infty}\varepsilon_x(\xi,y)\operatorname{sgn}(x-\xi)\mathrm{d}\xi,\\ v(x,y)&=\frac{v_0}{2}+\frac{1}{2}\int_0^{\infty}\varepsilon_y(x,\eta)\operatorname{sgn}(y-\eta)\mathrm{d}\eta. \end{aligned} \tag{41}$$

By letting $x\to\pm\infty$ in the first equation of (41), we derive the integral condition

$$\int_{-\infty}^{\infty}\varepsilon_x(x,y)\mathrm{d}x=0, \tag{42}$$

which is necessary for compatibility of strains. Analogously, by letting $y=0$ in the second equation of (41), the condition

$$\int_0^\infty \varepsilon_y(x,y)\mathrm{d}y = -v_0(x) \tag{43}$$

can be obtained. By substitution of expressions (41) into the third formula of (34), we derive following equation

$$2\gamma_{xy}(x,y) - \frac{\mathrm{d}v_0(x)}{\mathrm{d}x} = \int_{-\infty}^{\infty} \frac{\partial \varepsilon_x(\xi,y)}{\partial y}\mathrm{sgn}(x-\xi)\mathrm{d}\xi$$

$$+\int_0^\infty \frac{\partial \varepsilon_y(x,\eta)}{\partial x}\mathrm{sgn}(y-\eta)\mathrm{d}\eta, \tag{44}$$

which presents the condition of compatibility for strains. It is easy to see that by differentiation by variables x and y, equation (44) can be reduced to the classical compatibility equation (32). However, for the equivalence of these two equations, the following fitting condition

$$\frac{\mathrm{d}v_0(x)}{\mathrm{d}x} = \gamma_{xy}(x,0) - \frac{1}{2}\int_{-\infty}^{\infty} \frac{\partial \varepsilon_x(\xi,0)}{\partial y}\mathrm{sgn}(x-\xi)\mathrm{d}\xi \tag{45}$$

is to be fulfilled. This condition is obtained by integration of equation (32) over x and y with conditions (36) and (43) in view and comparison of the result to equation (44).
To determine the displacement-vector components, we can employ formulae (41) with conditions (42), (43), and (45) in view. Having applied the Fourier transformation (8) to the mentioned equations, we arrive at the formulae

$$\begin{gathered}\overline{u}(y) = -\frac{i}{s}\overline{\varepsilon}_x(y),\\ \overline{v}(y) = \frac{1}{2}\int_0^\infty \overline{\varepsilon}_y(\eta)\big(\mathrm{sgn}(y-\eta)-1\big)\mathrm{d}\eta.\end{gathered} \tag{46}$$

Putting the first and second physical relations of (33) along with (38) and (40) into the obtained formulae yields the following expressions:

$$\begin{gathered}\overline{u}(y) = \overline{p}\Pi_u(y) + \Theta_u(y) + Af_u(y),\\ \overline{v}(y) = \overline{p}\Pi_v(y) + \Theta_v(y) + Af_v(y),\end{gathered} \tag{47}$$

where

$$\Pi_u(y) = -\frac{i}{s}\left(\frac{\Pi(y)}{E^*(y)} - \frac{|s|}{4G(y)}\int_0^\infty \Pi(\xi)\big(\exp(-|s|\,|y-\xi|) - \exp(-|s|(y+\xi))\big)\mathrm{d}\xi + \frac{\exp(-|s|y)}{2G(y)}\right),$$

$$\Theta_u(y) = -\frac{i}{s}\left(\frac{\Theta(y)}{E^*(y)} - \frac{|s|}{4G(y)}\int_0^\infty \Theta(\xi)\big(\exp(-|s|\,|y-\xi|) - \exp(-|s|(y+\xi))\big)\mathrm{d}\xi + \alpha^*(y)\overline{T}(y)\right),$$

$$f_u(y) = -\frac{i}{s}\left(\frac{f_A(y)}{E^*(y)} - \frac{|s|}{4G(y)}\int_0^\infty f_A(\xi)\big(\exp(-|s|\,|y-\xi|) - \exp(-|s|(y+\xi))\big)\mathrm{d}\xi\right),$$

$$\Pi_v(y) = \frac{1}{2}\int_0^\infty (\mathrm{sgn}(y-\eta)-1)\left(-\frac{\exp(-|s|\eta)}{2G(\eta)} - \frac{\nu^*(\eta)}{E^*(\eta)}\Pi(\eta)\right.$$

$$\left.+\frac{|s|}{4G(y)}\int_0^\infty \Pi(\xi)\big(\exp(-|s|\,|\eta-\xi|) - \exp(-|s|(\eta+\xi))\big)\mathrm{d}\xi\right)\mathrm{d}\eta,$$

$$\Theta_v(y) = \frac{1}{2}\int_0^\infty (\mathrm{sgn}(y-\eta)-1)\left(\alpha^*(\eta)\overline{T}(\eta) - \frac{\nu^*(\eta)}{E^*(\eta)}\Theta(\eta)\right.$$

$$\left.+\frac{|s|}{4G(y)}\int_0^\infty \Theta(\xi)\big(\exp(-|s|\,|\eta-\xi|) - \exp(-|s|(\eta+\xi))\big)\mathrm{d}\xi\right)\mathrm{d}\eta,$$

$$f_v(y) = -\frac{1}{2}\int_0^\infty (\mathrm{sgn}(y-\eta)-1)\left(\frac{\nu^*(\eta)}{E^*(\eta)}f_A(\eta)\right.$$

$$\left.-\frac{|s|}{4G(y)}\int_0^\infty f_A(\xi)\big(\exp(-|s|\,|\eta-\xi|) - \exp(-|s|(\eta+\xi))\big)\mathrm{d}\xi\right)\mathrm{d}\eta.$$

Formulae (47) present the expression for determination of the displacement-vector components in the inhomogeneous semi-plane due to given external tractions $\overline{p}$ and $\overline{q}$, and the temperature field $\overline{T}(y)$.

3.2.3 One-to-one relations between the tractions and displacements on the boundary

Putting $y=0$ into (45) and (46), we obtain the relations

$$\overline{u}_0 = -\frac{i}{s}\overline{\varepsilon}_x(0),$$
$$is\overline{v}_0 = \overline{\varepsilon}_{xy}(0) + \frac{i}{s}\frac{\mathrm{d}\overline{\varepsilon}_x(0)}{\mathrm{d}y}.$$

Having substituted the corresponding physical relations (33) into the latter relations, we arrive at the following one-to-one relations

$$\overline{u}_0 = a_{11}\overline{p} + a_{12}\overline{q} + b_1,$$
$$\overline{v}_0 = a_{21}\overline{p} + a_{22}\overline{q} + b_2 \tag{48}$$

between the tractions and displacements on the boundary of semi-plane D. Here

$$a_{11} = \frac{i}{s}\left(\frac{1}{|s|E^*(0)} - \frac{1}{2G(0)}\right),\quad a_{12} = -\frac{1}{s^2E^*(0)},\quad b_1 = -\frac{i}{s}\alpha^*(0)\overline{T}(0),$$

$$a_{21}=\frac{1}{s^2}\frac{\mathrm{d}}{\mathrm{d}y}\left(-\frac{\Pi(y)}{E^*(y)}+\left(\frac{1}{|s|}+\Pi(0)\right)\frac{f_A(y)}{f_A(0)E^*(y)}-\frac{\exp(-|s|y)}{2G(y)}\right.$$

$$\left.+\frac{|s|}{4G(y)}\int_0^\infty\left(\Pi(\xi)-\left(\frac{1}{|s|}+\Pi(0)\right)\frac{f_A(\xi)}{f_A(0)}\right)(\exp(-|s|\,|y-\xi|)-\exp(-|s|(y+\xi)))\mathrm{d}\xi\right)\Bigg|_{y=0},$$

$$a_{21}=\frac{i}{sG(0)}+\frac{i}{s^2 f_A(0)}\frac{\mathrm{d}}{\mathrm{d}y}\left(\frac{1}{s}\frac{f_A(y)}{E^*(y)}\right.$$

$$\left.-\frac{|s|}{4sG(y)}\int_0^\infty f_A(\xi)(\exp(-|s|\,|y-\xi|)-\exp(-|s|(y+\xi)))\mathrm{d}\xi\right)\Bigg|_{y=0},$$

$$b_2=\frac{1}{s^2}\frac{\mathrm{d}}{\mathrm{d}y}\left(\frac{\Theta(0)}{E^*(y)}\frac{f_A(y)}{f_A(0)}-\frac{\Theta(y)}{E^*(y)}+\alpha^*(y)\overline{T}(y)\right.$$

$$\left.+\frac{|s|}{4G(y)}\int_0^\infty\left(\Theta(\xi)-\frac{\Theta(0)}{f_A(0)}f_A(\xi)\right)(\exp(-|s|\,|y-\xi|)-\exp(-|s|(y+\xi)))\mathrm{d}\xi\right)\Bigg|_{y=0}.$$

The obtained expressions of (48) allow us to determine the displacements on the boundary through the given tractions, and vice-versa.

3.2.4 Case *B*: Boundary condition in terms of displacement

Consider the problem of thermoelasticity (31) - (34), (36), where the boundary displacements $u_0(x)$ and $v_0(x)$ are given, meanwhile, the corresponding boundary tractions $p(x)$ and $q(x)$ are to be determined. By solving (48) with respect to $\overline{p}$ and $\overline{q}$, we find the transforms of tractions on the boundary through the displacements as

$$\overline{p}=\frac{a_{22}}{\omega}\overline{u}_0-\frac{a_{12}}{\omega}\overline{v}_0+\frac{a_{12}b_2-a_{22}b_1}{\omega},$$
$$\overline{q}=-\frac{a_{21}}{\omega}\overline{u}_0+\frac{a_{11}}{\omega}\overline{v}_0+\frac{a_{21}b_1-a_{11}b_2}{\omega}, \qquad (49)$$

where $\omega=a_{11}a_{22}-a_{12}a_{21}$. Having determined the boundary tractions (49), we can find the stress-tensor components by formulae (38), (40), and the displacement-vector components by formulae (47).

3.2.5 Case *C*: Solution of the problem with mixed boundary conditions

Finally, we consider the thermoelasticity problem (31) - (34) in the semi-plane D, when mixed boundary conditions of either the type (37) are imposed on the boundary. For four versions of the mixed boundary conditions (37), by making use of one of the relations (48), we express the Fourier transform of the unknown traction in terms of the given functions on the boundary and the temperature; inserting the expression into (38) and (40), we calculate the stresses and eventually the displacements by formula (47).

4. Conclusions

Using the method of direct integration, the explicit-form analytical solutions have been found for the equations of in-plane heat conduction and plane thermoelasticity problems in an inhomogeneous semi-plane provided the tractions, displacement, and mixed conditions are prescribed on the boundary. Due to the fact that the application of technique for reducing the aforementioned equations to the governing Volterra-type integral equations with further employment of the resolvent-kernel solution algorithm provides us with the explicit-form solutions of the thermoelasticity problems, the one-to-one relations between the tractions and the displacements on the boundary of the semi-plane are derived. Making use of these relations, we have reduced quasi-static boundary value problems of the plane theory of thermoelasticity with displacement or mixed boundary conditions to the solution of the problem when the tractions are prescribed on the boundary. Application of this technique does not impose any restrictions for the functions prescribing the material properties (besides existence of corresponding derivatives, at least, in generalized sense). But from mechanical point of view, it can be concluded, that the material properties should be in agreement with the model of continua mechanics assuring strain-energy within the necessary restrictions.

5. Acknowledgment

The authors gratefully acknowledge the financial support of this research by the National Science Council (Republic of China) under Grant NSC 99-2221-E002-036-MY3.

6. References

Bartoshevich, M.A. (1975). A Heat-Conduction Problem. *Journal of Engineering Physics and Thermophysics*, Vol.28, No.2, (February 1975), pp. 240-244, ISSN 1062-0125

Belik, V.D., Uryukov, B.A., Frolov, G.A., & Tkachenko, G.V. (2008). Numerical-Analytical Method of Solution of a Nonlinear Unsteady Heat-Conduction Equation. *Journal of Engineering Physics and Thermophysics*, Vol.81, No.6, (November 2008), pp. 1099-1103, ISSN 1062-0125

Brychkov, Yu.A. & Prudnikov, A.P. (1989). *Integral Transforms of Generalized Functions*, Gordon & Breach, ISBN: 2881247059, New York, USA

Carslaw, H.S. & Jaeger, J.C. (1959). *Conduction of Heat in Solids*, Oxford University Press, ISBN: 978-0-19-853368-9, London, England

Domke, K. & Hącia, L. (2007). Integral Equations in Some Thermal Problems, *International Journal of Mathematics and Computers IN Simulation*, Vol.1, No.2, pp. 184-188, ISSN: 1998-0159

Fedotkin, I.M., Verlan, E.V., Chebotaresku, I.D., & Evtukhovich, S.V. (1983). Heat Conductivity of the Adjoining Plates with a Plane Heat Source Between Them. *Journal of Engineering Physics and Thermophysics*, Vol.45, No.3, (September 1983), pp. 1071-1075, ISSN 1062-0125

Frankel, J.I. (1991). A Nonlinear Heat Transfer Problem: Solution of Nonlinear, Weakly Singular Volterra Integral Equations of the Second Kind. *Engineering Analysis with Boundary Elements*, Vol.8, No. 5, (October 1991) 231-238, ISSN 0955-7997

Hącia, L. (2007). Iterative-Collocation Method for Integral Equations of Heat Conduction Problems, *Numerical Methods and Applications, Lecture Notes in Computer Science*,Vol.4310, pp. 378-385, ISSN 0302-9743

Hetnarski, R.B. & Reza Eslami, M. (2008). *Thermal Stresses (Solid Mechanics and Its Applications)*, Springer, ISBN 1402092466, New York, USA

Kalynyak, B.M. (2000). Integration of Equations of One-Dimensional Problems of Elasticity and Thermoelasticity for Inhomogeneous Cylindrical Bodies, *Journal of Mathematical Sciences*, Vol.99, No.5, (May 2000) pp. 1662-1670, ISSN 1072-3374

Kushnir, R.M., Protsyuk, B.V., & Synyuta, V.M. (2002). Temperature Stresses and Displacements in a Multilayer Plate with Nonlinear Conditions of Heat Exchange. *Materials Science*, Vol.38, No.6, (November 2002) pp. 798-808, ISSN 1068-820X

Ma, C.-C. & Chen, Y.-T. (2011). Theoretical Analysis of Heat Conduction Problems of Nonhomogeneous Functionally Graded Materials for a Layer Sandwiched Between Two Half-Planes. *Acta Mechanica*, Online first, doi: 10.1007/s00707-011-0498-7, ISSN 0001-5970

Ma, C.-C. & Lee, J.-M. (2009). Theoretical Analysis of In-Plane Problem in Functionally Graded Nonhomogeneous Magnetoelectroelastic Bimaterials. *International Journal of Solids and Structures*, Vol.46, No.24, (December 2009), pp. 4208-4220, ISSN 0020-7683

Nowacki, W. (1962). *Thermoelasticity*, Pergamon Press, Oxford

Peng, X. L. & Li, X.F. (2010). Transient Response of Temperature and Thermal Stresses in a Functionally Graded Hollow Cylinder. *Journal of Thermal Stresses*, Vol.33, No.5, (August 2010), pp. 485 - 500, ISSN 0149-5739

Pogorzelski, W. (1966). *Integral Equations and their Applications*. Vol 1. Pergamon Press, ISBN 978-0080106625, New York, USA

Porter, D. & Stirling, D.S.G. (1990). *Integral Equations. A Practical Treatment, from Spectral Theory to Applications*. Cambridge University Press, ISBN 0521337429, Cambridge, England.

Rychahivskyy, A.V. & Tokovyy, Yu.V. (2008). Correct Analytical Solutions to the Thermoelasticity Problems in a Semi-Plane. *Journal of Thermal Stresses*, Vol.31, No.11, (November 2008), pp. 1125 - 1145, ISSN 0149-5739

Tanigawa, Y. (1995). Some Basic Thermoelastic Problems for Nonhomogeneous Structural Materials. *Applied Mechanics Reviews*, Vol.48, No.6, (June 1995), pp. 287-300, ISSN 0003-6900

Tokovyy, Y.V. & Ma, C.-C. (2008). Thermal Stresses in Anisotropic and Radially Inhomogeneous Annular Domains. *Journal of Thermal Stresses*, Vol.31, No.9 (September 2008), pp. 892 – 913, ISSN 0149-5739

Tokovyy,Yu.V. & Ma,C.-C. (2008a). Analysis of 2D Non-Axisymmetric Elasticity and Thermoelasticity Problems for Radially Inhomogeneous Hollow Cylinders, *Journal of Engineering Mathematics*, Vol.61,No.2-4, (August 2008), pp. 171-184, ISSN 0022-0833

Tokovyy, Yu. & Ma, C.-C. (2009). Analytical Solutions to the 2D Elasticity and Thermoelasticity Problems for Inhomogeneous Planes and Half-Planes. *Archive of Applied Mechanics*, Vol.79, No.5, (May 2009), pp. 441-456, ISSN 0939-1533

Tokovyy, Yu. & Ma, C.-C. (2009a). An Explicit-Form Solution to the Plane Elasticity and Thermoelasticity Problems for Anisotropic and Inhomogeneous Solids. *International Journal of Solids and Structures*, Vol.46, No.21, (October 2009), pp. 3850-3859, ISSN 0020-7683

Tokovyy, Y.V. & Ma, C.-C. (2010). General Solution to the Three-Dimensional Thermoelasticity Problem for Inhomogeneous solids. *Multiscaling of Synthetic and Natural Systems with Self-Adaptive Capability, Proceedings of the 12th International Congress in Mesomechanics "Mesomechanics 2010"*, ISBN 978-986-02-3909-6, Taipei, Taiwan ROC, June 2010

Tokovyy, Y.V. & Ma, C.-C. (2010a). Elastic and Thermoelastic Analysis of Inhomogeneous Structures. *Proceedings of 3rd Engineering Conference on Advancement in Mechanical and Manufacturing for Sustainable Environment "EnCon2010"*, ISBN 978-967-5418-10-5, Kuching, Sarawak, Malaysia, April 2010

Tokovyy, Y.V. & Rychahivskyy, A.V. (2005). Reduction of Plane Thermoelasticity Problem in Inhomogeneous Strip to Integral Volterra Type Equation. *Mathematical Modelling and Analysis*, Vol.10, No.1, pp. 91-100, ISSN 1392-6292

Trikomi, F.G. (1957). *Integral Equations*. Interscience Publishers. Inc., New York, USA

Verlan, A.F. & Sizikov, V.S. (1986). *Integral Equations: Methods, Algorithms and Codes* (in Russian), Naukova Dumka, Kiev, Ukraine

Vigak, V.M. (1999). Method for Direct Integration of the Equations of an Axisymmetric Problem of Thermoelasticity in Stresses for Unbounded Regions. *International Applied Mechanics*, Vol.35, No.3, (March 1999), pp. 262-268, ISSN 1063-7095

Vigak, V.M. (1999a). Solution of One-Dimensional Problems of Elasticity and Thermoelastic-city in Stresses for a Cylinder, *Journal of Mathematical Sciences*, Vol.96, No.1, (August 1999), pp. 2887-2891, ISSN 1072-3374.

Vigak, V.M. (2004). Correct Solutions of Plane Elastic Problems for a Semi-Plane. *International Applied Mechanics*, Vol.40, No.3, (March 2004), pp. 283-289, ISSN 1063-7095

Vigak, V.M. & Rychagivskii, A.V. (2000). The Method of Direct Integration of the Equations of Three-Dimensional Elastic and Thermoelastic Problems for Space and a Halfspace. *International Applied Mechanics*, Vol.36, No.11, (November 2000), pp. 1468-1475, ISSN 1063-7095

Vihak, V. & Rychahivskyy, A. (2001). Bounded Solutions of Plane Elasticity Problems in a Semi-Plane, *Journal of Computational and Applied Mechanics*, Vol.2, No.2, pp. 263-272, ISSN 1586-2070

Vigak, V.M. & Rychagivskii, A.V. (2002). Solution of a Three-Dimensional Elastic Problem for a Layer, *International Applied Mechanics*, Vol.38, No.9, (September 2002), pp. 1094-1102, ISSN 1063-7095

Vihak, V.M. (1996). Solution of the Thermoelastic Problem for a Cylinder in the Case of a Two-Dimensional Nonaxisymmetric Temperature Field. *Zeitschrift für Angewandte Mathematik und Mechanik*, Vol.76, No.1, pp. 35-43, ISSN: 0044-2267

Vihak, V.M. & Kalyniak, B.M. (1999). Reduction of One-Dimensional Elasticity and Thermoelasticity Problems in Inhomogeneous and Thermal Sensitive Solids to the Solution of Integral Equation of Volterra Type, *Proceedings of the 3rd International Congress "Thermal Stresses-99"*, ISBN 8386991577, Cracow, Poland, June 1999

Vihak, V.M., Povstenko, Y.Z., & Rychahivskyy A.V. (2001). Integration of Elasticity and Thermoelasticity Equations in Terms of Stresses. *Journal of Mechanical Engineering*, Vol.52, No.4, pp. 221-234,

Vihak, V., Tokovyi, Yu., & Rychahivskyy, A. (2002). Exact Solution of the Plane Problem of Elasticity in a Rectangular Region. *Journal of Computational and Applied Mechanics*, Vol.3, No.2, pp. 193-206, ISSN 1586-2070

Vihak, V.M., Yuzvyak, M.Y., & Yasinskij, A.V. (1998). The Solution of the Plane Thermo-elasticcity Problem for a Rectangular Domain, *Journal of Thermal Stresses*, Vol.21, No.5, (May 1998), pp. 545-561, ISSN 0149-5739

12

Self-Similar Hydrodynamics with Heat Conduction

Masakatsu Murakami
Institute of Laser Engineering, Osaka University
Japan

1. Introduction

1.1 Dimensional analysis and self-similarity

Dimensional and similarity theory provides one with the possibility of prior qualitative-theoretical analysis and the choice of a set for characteristic dimensionless parameters. The theory can be applied to the consideration of quite complicated phenomena and makes the processing of experiments much easier. What is more, at present, the competent setting and processing of experiments is inconceivable without taking into account dimensional and similarity reasoning. Sometimes at the initial stage of investigation of certain complicated phenomena, dimensional and similarity theory is the only possible theoretical method, though the possibilities of this method should not be overestimated. The combination of similarity theory with considerations resulting from experiments or mathematical operations can sometimes lead to significant results. Most often dimensional and similarity theory is very useful for theoretical as well for practical use. All the results obtained with the help of this theory can be obtained quite easily and without much trouble.

A phenomenon is called self-similar if the spatial distributions of its properties at various moments of time can be obtained from one another by a similarity transformation. Establishing self-similarity has always represented progress for a researcher: self-similarity has simplified computations and the representation of the characteristics of phenomena under investigation. In handling experimental data, self-similarity has reduced what would seem to be a random cloud of empirical points so as to lie on a single curve of surface, constructed using self-similar variables chosen in some special way. Self-similarity enables us to reduce its partial differential equations to ordinary differential equations, which substantially simplifies the research. Therefore with the help of self-similar solutions researchers have attempted to find the underlying physics. Self-similar solutions also serve as standards in evaluating approximate methods for solving more complicated problems.

Scaling laws, which are obtained as a result of the dimensional analysis and other methods, play an important role for understanding the underlying physics and applying them to practical systems. When constructing a full-scale system in engineering, numerical simulations will be first made in most cases. Its feasibility should be then demonstrated experimentally with a reduced-scale system. For astrophysical studies, for instance, such scaling considerations are indispensable and play a decisive role in designing laboratory experiments. Then one should know how to design such a miniature system and how to judge whether two experimental results in different scales are hydrodynamically equivalent or similar to each other. Lie group analysis (Lie, 1970), which is employed in the present chapter, is not only a powerful method to seek self-similar solutions of partial differential equations (PDE) but also a unique and most adequate technique to extract the group invariance properties of such a PDE system. Lig group analysis and dimensional analysis are useful methods to find self-similar solutions in a complementary manner.

An instructive example of self-similarity is given by an idealized problem in the mathematical theory of linear heat conduction: Suppose that an infinitely stretched planar space ($-\infty < x < \infty$) is filled with a heat-conducting medium. At the initial instant $t = 0$ and at the origin of the coordinate $x = 0$, a finite amount of heat E is supplied instantaneously. Then the propagation of the temperature Θ is described by

$$\frac{\partial \Theta}{\partial t} = \kappa \frac{\partial^2 \Theta}{\partial x^2}, \tag{1}$$

where κ is the constant heat diffusivity of the medium. Then the temperature Θ at an arbitrary time t and distance from the origin x is given by

$$\Theta = \frac{E}{c\sqrt{4\pi\kappa t}} \exp\left(-\frac{x^2}{4\kappa t}\right), \tag{2}$$

where c is the specific heat of the medium. As a matter of fact, it is confirmed with the solution (2) that the integrated energy over the space is kept constant regardless of time:

$$\int_{-\infty}^{\infty} c\,\Theta(x,t)dx = E \tag{3}$$

The structure of Eq. (2) is instructive: There exist a temperature scale $\Theta_0(t)$ and a linear scale $x_0(t)$, both depending on time,

$$\Theta_0(t) = \frac{E}{c\sqrt{4\pi\kappa t}}, \quad x_0(t) = \sqrt{\kappa t}, \tag{4}$$

such that the spatial distribution of temperature, when expressed in these scales, ceases to depend on time at least in appearance:

$$\frac{\Theta}{\Theta_0} = f(\xi), \quad f(\xi) = \exp\left(-\frac{\xi^2}{4}\right), \quad \xi = \frac{x}{x_0}. \tag{5}$$

Suppose that we are faced with a more complex problem of mathematical physics in two independent variables x and t, requiring the solution of a system of partial differential equations on a variable $u(x,t)$ of the phenomenon under consideration. In this problem, self-similarity means that we can choose variable scales $u_0(t)$ and $x_0(t)$ such that in the new scales, $u(x,t)$ can be expressed by functions of one variable:

$$u = u_0(t)U(\xi), \quad \xi = x/x_0(t) \tag{6}$$

The solution of the problem thus reduces to the solution of a system of ordinary differential equations for the function $U(\xi)$.

At a certain point of analysis, dimensional consideration called Π-theorem plays a crucial role in a complementary manner to the self-similar method. Suppose we have some relationship defining a quantity a as a function of n parameters $a_1, a_2, \dots, a_n$:

$$a = f(a_1, a_2, \dots a_n)\,. \tag{7}$$

If this relationship has some physical meaning, Eq. (7) must reflect the clear fact that although the numbers $a_1, a_2, \dots, a_n$ express the values of corresponding quantities in a definite system of units of measurement, the physical law represented by this relation does not depend on the arbitrariness in the choice of units. To explain this, we shall divide the quantities $a, a_1, a_2, \dots a_n$ into two groups. The first group, $a_1, \dots a_k$, includes the governing quantities with independent dimensions (for example, length, mass, and time). The second group, $a, a_{k+1}, \dots a_n$, contains quantities whose dimensions can be expressed in terms of dimensions of the quantities of the first group. Thus, for example, the quantity a has the dimensions of the product $a_1^p a_2^q \cdots a_k^r$, the quantity a_{k+1} has the dimensions of the product $a_1^{p_{k+1}} a_2^{q_{k+1}} \cdots a_k^{r_{k+1}}$, etc. The exponents $p, q, \dots$ are obtained by a simple arithmetic. Thus the quantities,

$$\Pi = \frac{a}{a_1^p a_2^q \cdots a_k^r}, \quad \Pi_1 = \frac{a_{k+1}}{a_1^{p_{k+1}} a_2^{q_{k+1}} \cdots a_k^{r_{k+1}}}, \quad \dots, \quad \Pi_{n-k} = \frac{a_n}{a_1^{p_n} a_2^{q_n} \cdots a_k^{r_n}} \tag{8}$$

turn out to be dimensionless, so that their values do not depend how one choose the units of measurement. This fact follows that the dimensionless quantities can be expressed in the form,

$$\Pi = \Phi(\Pi_1, \Pi_2, \dots, \Pi_{n-k})\,, \tag{9}$$

where no dimensional quantity is contained. What should be stressed is that in the original relation (7), $n+1$ dimensional quantities $a, a_1, a_2, \dots, a_n$ are connected, while in the reduced relation (9), $n-k+1$ dimensionless quantities Π, $\Pi_1, \Pi_2, \dots, \Pi_{n-k}$ are connected with k quantities being reduced from the original relation.

We now apply dimensional analysis to the heat conduction problem considered above. Below we shall use the symbol [a] to give its dimension, as Maxwell first introduced, in terms of the unit symbols for length, mass, and time by the letters L, M, and T, respectively. For example, velocity v has its dimension $[v] = L/T$. Then the physical quantities describing the present system have following dimensions,

$$[x] = L,\ \ [t] = T,\ \ [\kappa] = L^2T^{-1},\ \ [E] = ML^2T^{-2},\ \ [c\,\Theta] = ML^3T^{-2}. \tag{10}$$

From Eq. (10), in which five dimensional quantities ($n+1=5$) under the three principal dimensions ($k=3$ for L, M, and T), one can construct the following dimensionless system with two dimensionless parameters Π and ξ $(=\Pi_1)$:

$$\Pi = f(\xi), \quad \Pi = \frac{c\,\Theta\sqrt{\kappa t}}{E}\,, \quad \xi = \frac{x}{\sqrt{\kappa t}}\,, \tag{11}$$

where f is unknown function. Substituting Eq. (11) for Eq. (1), one obtains,

$$f'' + \tfrac{1}{2}(f + \xi f') = 0\,, \tag{12}$$

where the prime denotes the derivative with respect to ξ; also the transform relation from partial to ordinary derivatives

$$\frac{\partial f(\xi)}{\partial t} = -\frac{\xi}{2t} f'(\xi), \quad \frac{\partial f(\xi)}{\partial x} = \frac{1}{\sqrt{\kappa t}} f'(\xi), \tag{13}$$

are used. With the help of the boundary condition, $f'(0) = 0$, and Eq. (3), Eq. (12) is integrated to give

$$f(\xi) = \frac{1}{\sqrt{4\pi}} \exp\left(-\frac{\xi^2}{4}\right). \tag{14}$$

Thus Eqs. (11) and (14) reproduce the solution of the problem, Eq. (2).
What is described above is the simple and essential scenario of the approach in terms of self-similar solution and dimensional analysis, more details of which can be found, for example, in Refs. (Lie, 1970; Barenblatt, 1979; Sedov, 1959; Zel'dovich & Raizer, 1966). In the following subsections, we show three specific examples with new self-similar solutions, as reviews of previously published papers for readers' further understanding how to use the dimensional analysis and to find self-similar solutions: The first is on plasma expansion of a limited mass into vacuum, in which two fluids composed of cold ions and thermal electrons expands via electrostatic field (Murakami et al, 2005). The second is on laser-driven foil acceleration due to nonlinear heat conduction (Murakami et al, 2007). Finally, the third is an astrophysical problem, in which self-gravitation and non-linear radiation heat conduction determine the temporal evolution of star formation process in a self-organizing manner (Murakami et al, 2004).

2. Isothermally expansion of laser-plasma with limited mass

2.1 Introduction

Plasma expansion into a vacuum has been a subject of great interest for its role in basic physics and its many applications, in particular, its use in lasers. The applied laser parameter spans a wide range, $10^{10} \le \hat{I}_L \hat{\lambda}_L^2 \le 10^{19}$, where $\hat{I}_L$ is the laser intensity in the units of W/cm^2 and $\hat{\lambda}_L$ is the laser wavelength normalized by $1\,\mu m$. For $\hat{I}_L \hat{\lambda}_L^2 \ge 10^{14}$, generation of fast ions is governed by hot electrons with an increase in $\hat{I}_L \hat{\lambda}_L^2$. In this subsection, we focus on rather lower intensity range, $10^{10} \le \hat{I}_L \hat{\lambda}_L^2 \le 10^{14}$, where the effect of hot electrons is negligibly small and background cold electrons can be modeled by one temperature. Typical examples of applications for this range are laser driven inertial confinement fusion (Murakami et al., 1995; Murakami & Iida, 2002) and laser-produced plasma for an extreme ultra violet (EUV) light source (Murakami et al, 2006). As a matter of fact, the experimental data employed below for comparison with the analytical model were obtained for the EUV study. Theoretically, this topic had been studied only through hydrodynamic models until the early 1990s. In such theoretical studies, a simple planar (SP) self-similar solution has often been used (Gurevich et al, 1966). In the SP model, a semi-infinitely stretched planar plasma is considered, which is initially at rest with unperturbed density ρ_0. At $t = 0$, a rarefaction wave is launched at the edge to penetrate at a constant sound speed c_s into the unperturbed uniform plasma being accompanied with an isothermal expansion. The density and velocity profiles of the expansion are given by (Landau & Lifshitz, 1959) $\rho = \rho_0 \exp[-(1 + x/c_s t \text{ cst})]$ and $v = c_s + x/t$, respectively. The solution is indeed quite useful when using relatively short laser pulses or thick targets such that the density scale can be kept constant throughout the process.

However, in actual laser-driven plasmas, a shock wave first penetrates the unperturbed target instead of the rarefaction wave. Once this shock wave reaches the rear surface of a finite-sized target and the returning rarefaction wave collides with the penetrating rarefaction wave, the entire region of the target begins to expand, and thus the target disintegration sets in. If the target continues to be irradiated by the laser even after the onset of target disintegration, the plasma expansion and the resultant ion energy spectrum are expected to substantially deviate from the physical picture given by the SP solution. Figure 1 demonstrates a simplified version of the physical picture mentioned above with temporal evolution of the density profile obtained by hydrodynamic simulation for an isothermal expansion. A spherical target with density and temperature profiles being uniform is employed as an example. In Fig. 1, the density is always normalized to unity at the center, and the labels assigned to each curve denote the normalized time $t/(R_0 c_s)$, where R_0 is the initial radius. The horizontal Lagrange coordinate is normalized to unity at the plasma edge. It can be discerned from Fig. 1 that the profile rapidly develops in the early stage for $t/(R_0 c_s) \leq 1$. After the rarefaction wave reflects at the center, the density distribution asymptotically approaches its final self-similar profile (the thick curve with label " ∞ "), which is expressed in the Gaussian form, $\rho \propto \exp[-(r/R)^2]$ as will be derived below. The initial and boundary conditions employed in Fig. 1 are substantially simplified such that the laser-produced shock propagation and resultant interactions with the rarefaction wave are not described. However, the propagation speeds of the shock and rarefaction waves are always in the same order as the sound speed c_s of the isothermally expanding plasma. Therefore the physical picture shown in Fig. 1 is expected to be qualitatively valid also for

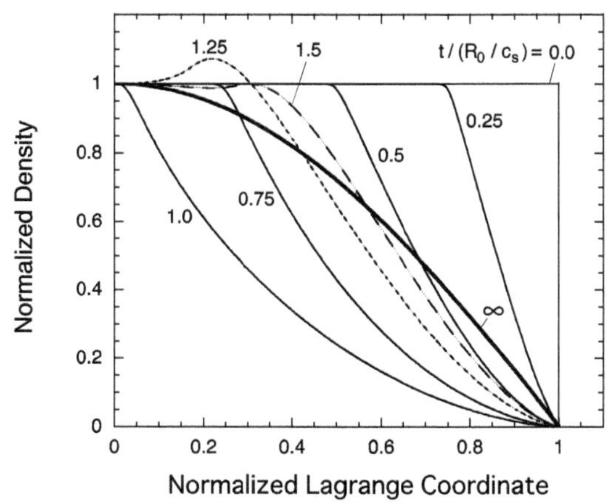

Fig. 1. Temporal evolution of the density profile of a spherical isothermal plasma, which is normalized by that at the center; R_0 and c_s are the initial radius and the sound speed, respectively. After the rarefaction wave reflects at the center, the density distribution asymptotically approaches its final self-similar profile (the thick curve with " ∞ ").

realistic cases. Below, we present a self-similar solution for the isothermal expansion of limited masses (Murakami et al., 2005). The solution explains plasma expansions under relatively long laser pulses or small-sized targets so that the solution responds to the above argument on target disintegration. Note that other self-similar solutions of isothermal plasma expansion have been found for laser-driven two-fluid expansions in light of ion acceleration physics (Murakami & Basko, 2006) and heavy-ion-driven cylindrical x-ray converter (Murakami et al., 1990), though they are not discussed here.

2.2 Isothermal expansion

The plasma is assumed to be composed of cold ions and electrons described by one temperature T_e, which is measured in units of energy as follows. Furthermore, the electrons are assumed to obey the Boltzmann statistics,

$$n_e = n_{ec} \exp(e\Phi/T_e), \tag{15}$$

where $n_{ec}(t)$ is the temporal electron density at the target center, e is the elementary charge, and $\Phi(r,t)$ is the electrostatic potential, the zero-point of which is set at the target center, i.e., $\Phi(0,t) = 0$. The potential Φ satisfies the Poisson equation,

$$\frac{1}{r^{\alpha-1}}\frac{\partial}{\partial r}\left(r^{\alpha-1}\frac{\partial\Phi}{\partial r}\right) = 4\pi e(n_e - Zn_i), \tag{16}$$

where Z is the ionization state; the superscript α stands for the applied geometry such that α = 1, 2, and 3 correspond to planar, cylinder, and spherical geometry, respectively. Throughout the present analysis, the electron temperature T_e and the ionization state Z are assumed to be constant in space and time.

An ion in the plasma is accelerated via the electrostatic potential in the form,

$$\frac{\partial v}{\partial t} + v\frac{\partial v}{\partial r} = -\frac{Ze}{m_i}\frac{\partial\Phi}{\partial r}, \tag{17}$$

where m_i is the ion mass and v is the ion velocity. Note that, in the following, we consider such a system that the plasma has quasi-neutrality, i.e., $n_e \approx Zn_i$, where n_i and n_e are the number densities of the ions and the electrons, respectively. Equations (15) and (17) are combined to derive a single-fluid description,

$$\frac{\partial v}{\partial t} + v\frac{\partial v}{\partial r} = -\frac{c_s^2}{\rho}\frac{\partial \rho}{\partial r}, \tag{18}$$

where $c_s = \sqrt{ZT_e/m_i}$ is the sound speed. Also, a fluid element with mass density $\rho(r,t) = m_i n_i$ satisfies the following mass conservation law,

$$\frac{\partial\rho}{\partial t} + \frac{1}{r^{\alpha-1}}\frac{\partial}{\partial r}(r^{\alpha-1}\rho v) = 0. \tag{19}$$

We now seek a self-similar solution to Eqs. (18) and (19) on $\rho(r,t)$ and $v(r,t)$ under the similarity ansatz,

$$v = \dot{R}\xi \,, \;\; \xi \equiv \frac{r}{R}, \tag{20}$$

$$\rho = \rho_{00} \left(\frac{R}{R_0}\right)^{-\alpha} G(\xi), \tag{21}$$

where $R(t)$ stands for a time-dependent characteristic system size, and ξ is the dimensionless similarity coordinate; the over-dot in Eq. (20) denotes the derivative with respect to time; $\rho_{00} \equiv \rho(0,0)$ and $R_0 \equiv R(0)$ are the initial central density and the size, respectively; $G(\xi)$ is a positive unknown function with the normalized boundary condition $G(0) = 1$. Then, Eqs. (15) and (21) give

$$n_e \approx n_{ec}(t)G(\xi) \approx Z\frac{\rho_{00}}{m_i}\left(\frac{R}{R_0}\right)^{-\alpha} G(\xi), \tag{22}$$

Under the similarity ansatz, Eqs. (20) and (21), the mass conservation, Eq. (19), is automatically satisfied. Substituting Eqs. (20) and (21) for Eq. (18), and making use of the derivative rules, $\partial/\partial r = R^{-1}\,(d/d\xi)$ and $\partial/\partial t = -\xi\dot{R}R^{-1}\,(d/d\xi)$, one obtains the following ordinary differential equations in the form of variable separation,

$$\frac{R\ddot{R}}{c_s^2} = -\frac{G'}{\xi G} = \psi_0 \,, \tag{23}$$

where $\psi_0(>0)$ is a separation constant, and the prime denotes the derivative with respect to ξ. Without losing generality, the constant ψ_0 can be set equal to an arbitrary numerical value, because this is always possible with a proper normalization of R and ξ. Here, just for simplicity, we set $\psi_0 = 2$ in Eq. (23). Then the spatial profile of the density, $G(\xi)$, is straightforwardly obtained under $G(0) = 1$ in the form (True et al., 1981; London & Rosen, 1986),

$$G(\xi) = \exp(-\xi^2)\,. \tag{24}$$

As was seen in Fig. 1, the density profile of isothermally expanding plasma with a limited mass is found to approach asymptotically the solution, Eq. (24), even if it has a different profile in the beginning. Meanwhile, $R(t)$ in Eq. (23) cannot be given explicitly as a function of time but has the following integrated forms,

$$\dot{R} = 2c_s\sqrt{\ln(R/R_0)}\,, \tag{25}$$

$$\frac{c_s t}{R_0} = \frac{1}{2}\int_1^{R/R_0} \frac{dx}{\sqrt{\ln x}}\,, \tag{26}$$

where in obtaining Eqs. (25) and (26), the system is assumed to be initially at rest, i.e., $\dot{R}(0) = 0$. Here it should be noted that Eqs. (23) - (26) do not explicitly include the geometrical index α, and therefore they apply to any geometry.

Based on the solution given above, some other important quantities are derived as follows. First, the total mass of the system M_0 is conserved and given with the help of Eqs. (21) and (24) in the form,

$$M_0 = (4\pi)_\alpha \rho_{00} R_0^\alpha \int_0^\infty \xi^{\alpha-1} \exp(-\xi^2)\, d\xi = \left(\sqrt{\pi} R_0\right)^\alpha \rho_{00}\,, \tag{27}$$

with

$$(4\pi)_\alpha \equiv \begin{cases} 2, (\alpha = 1) \\ 2\pi, (\alpha = 2) \\ 4\pi, (\alpha = 3) \end{cases} = \frac{2\pi^{\alpha/2}}{\Gamma(\alpha/2)}\,, \tag{28}$$

where Γ is the Gamma function. Although the quantitative meaning of $R(t)$ was somewhat unclear when first introduced in Eq. (20), it can be now clearly understood by relating it to the temporal central density, $\rho_c(t) \equiv \rho(0,t) \approx m_i n_{ec}(t)/Z$, with the help of Eqs. (21) and (27) in the form,

$$R(t) = \frac{1}{\sqrt{\pi}} \left(\frac{M_0}{\rho_c(t)}\right)^{1/\alpha}. \tag{29}$$

Additionally the potential Φ and corresponding electrostatic field $E = -\nabla\Phi$ are obtained from Eqs. (15), (21), (22), and (24) in the following forms,

$$\frac{e\Phi}{T_e} = -\xi^2\,, \tag{30}$$

$$\frac{eE}{T_e} = \frac{2\xi}{R}\,. \tag{31}$$

The above field quantities contrast well with the fields of the SP solution obtained for a semi-infinitely stretched planar plasma: $e\Phi/T_e = -1 - x/c_s t$ and $eE/T_e = 1/c_s t$ for $x/c_s t \geq -1$ and $t > 0$. It is here worth emphasizing that the electrostatic field increases linearly with ξ for the present model, while it is constant in space for the SP model. Furthermore, the kinetic energy of the system $E_k(t)$ is given with the help of Eqs. (20), (21) and (27) by

$$E_k = \frac{(4\pi)_\alpha}{2} \rho_{00} R_0^\alpha \dot{R}^2 \int_0^\infty \xi^{\alpha+1} \exp(-\xi^2)\, d\xi = \frac{\alpha}{4} M_0 \dot{R}^2, \tag{32}$$

while the internal (thermal) energy of the system $E_i(t)$ is kept constant,

$$E_i = \frac{3 M_0 Z T_e}{2 m_i} = \frac{3}{2} M_0 c_s^2\,. \tag{33}$$

Correspondingly, the power required to keep the isothermal expansion, $P(t) = dE_k/dt$, is given from Eqs. (23), (25), and (32) in the form,

$$P/P_0 = \sqrt{\ln(R/R_0)}\,/(R/R_0\,), \tag{34}$$

where $P_0 = 2\alpha M_0 c_s^3/R_0$.

The ion energy spectrum is a physical quantity of high interest. In the present model, the kinetic energy of an ion in flight directly relates its location, in other words, the further an ion is located, the faster it flies. Then, the number of ions contained in an infinitesimally narrow area of the similarity coordinate between ξ and $\xi + d\xi$ is given by

$$dN = (4\pi)_\alpha n_{i00} R_0^\alpha \xi^{\alpha-1} \exp(-\xi^2)\, d\xi\,, \tag{35}$$

where $n_{i00} = \rho_{00}/m_i$ is the initial number density of the ions at the center. Meanwhile, the kinetic energy of an ion at ξ is $\varepsilon = m_i \dot{R}^2 \xi^2/2$, and therefore

$$d\varepsilon = m_i \dot{R}^2 \xi d\xi\,. \tag{36}$$

From Eqs. (35) and (36), the ion energy spectrum is obtained,

$$\frac{d\hat{N}}{d\hat{\varepsilon}} = \frac{\hat{\varepsilon}^{(\alpha-2)/2}\exp(-\hat{\varepsilon})}{\Gamma(\alpha/2)}\,, \tag{37}$$

where $\hat{N} \equiv N/N_0$ and $\hat{\varepsilon} \equiv \varepsilon/\varepsilon_0$ are normalized quantities with

$$\varepsilon_0(t) = m_i \dot{R}^2/2\,, \tag{38}$$

$$N_0 = (\sqrt{\pi} R_0)^\alpha n_{i00}\,. \tag{39}$$

It should be noted that, for $\alpha = 3$, the energy spectrum, Eq. (37), coincides with the well-known Maxwellian energy distribution; this is not just a coincidence because an isotropically heated mass always has such a distribution.

Although the spectrum, Eq. (37), is for the ion number density, another spectrum for the energy density, $dE_k/d\varepsilon$, is an even more interesting quantity. It can be easily obtained quite in the same manner as for $dN/d\varepsilon$ taking the specific kinetic energy $v^2/2$ into account:

$$\frac{dE_k}{d\hat{\varepsilon}} = \frac{\varepsilon_0 N_0}{\Gamma(\alpha/2)} \hat{\varepsilon}^{\alpha/2} \exp(-\hat{\varepsilon})\,. \tag{40}$$

The peak value of Eq. (40) is attained at $\hat{\varepsilon} = \alpha/2$, which is three times higher than that of Eq. (37) for the spherical case ($\alpha = 3$).

2.3 Comparison with experiments

We apply the analytical model to two different laser experiments focusing on the ion energy spectrum. The two experimental results were separately obtained under different conditions by means of the time-of-flight method. In both cases, the laser conditions were almost the same, i.e., the wavelength $\lambda_L = 1.06\,\mu m$, the irradiation intensity $I_L = 0.5 - 1.0 \times 10^{11}\, W/cm^2$, and the pulse length $\tau_L \sim 10\, ns$ with a sufficiently large F-number of a focal lens. Moreover, the target thicknesses were $R_0 \sim 10\,\mu m$. Once the key laser parameters, I_L and τ_L, are given, the other basic parameters required for the model analysis are calculated. For example, the plasma temperature is roughly estimated from the power balance, $\eta_a I_L \approx 4\rho_{cr} c_s^3$ (Murakami & Meyer-ter-Vhen, 1991), where η_a is the absorption efficiency and ρ_{cr} is the critical mass density:

$$T_e\,[eV] = 27(A/Z)^{1/3} \hat{\lambda}_L^{4/3} \left(\eta_a \hat{I}_L\right)^{2/3}, \tag{41}$$

where A is the ion mass number. The corresponding sound speed turns out to be in the order of $10^6 cm/s$, and the disintegration time $\sim 2R_0/c_s$ (recall Fig. 1) is calculated to be about 1 ns ($\ll \tau_L \sim 10\ ns$). The normalized radius R/R_0 at the laser turn-off is obtained by Eq. (26) as a function of the normalized time $\tau_L/(R_0/c_s)$. In addition, the scale length of the plasma expansion is $c_s\tau_L \sim 100\ \mu m (\gg R_0 \sim 10\ \mu m)$. Therefore, the present self-similar analysis is considered to be applicable to the experiments under consideration. From the above key numerical values, the characteristic ion kinetic energy at the laser turn-off defined by Eq. (38) is roughly estimated to be $\varepsilon_0 = 2.5 - 3.5$ keV.

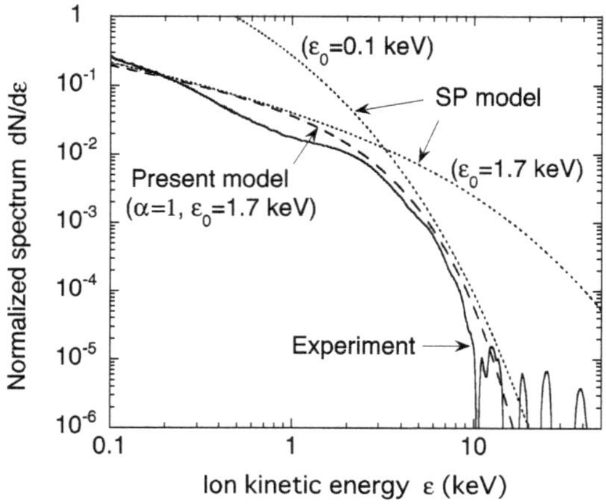

Fig. 2. Comparison of the experimental result (solid line) and the analytical curve (dashed line) obtained by Eq. (37) under planar geometry. Dotted curves for reference are obtained by the SP model, Eq. (42).

In the first case, a laser beam was irradiated on a spherical target with diameter of 500 μm, which was composed of 8μm-thick plastic shell coated by a 100 nm-thick tin (Sn) layer. In this case, the plasma expansion during the laser irradiation can be regarded as quasi-planar, because the plasma scale ~100 μm is appreciably smaller than the laser spot size ~500 μm. As mentioned in the introduction, the purpose of the Sn-coat was to observe the characteristics of the EUV light and energetic ion fluxes emitted from the Sn plasma. The detector was tuned to observe massive Sn ions in the direction of 30 degrees with respect to the beam axis. Figure 2 shows the ion energy spectrum comparing the experimental result (solid line) and the analytical curve (dashed line) obtained by Eq. (37) with a fitted numerical factor $\varepsilon_0 = 1.7$ keV and $\alpha = 1$ (planar geometry). With respect to the vertical axis, the physical quantities are properly normalized such that the peak values stay in the order of unity. The fluctuated structure of the experimental data for $\varepsilon > 10$ keV cannot be clearly judged as concerns whether the signals simply span the region with less precision of diagnosis, or whether they should be attributed to other causes such as carbon ions, protons, and photons. In Fig. 2, two other curves (dotted lines) are also plotted for comparison. They are obtained by the SP model (Mora, 2003),

$$\frac{dN}{d\hat{\varepsilon}} \propto \frac{\exp(-\sqrt{\hat{\varepsilon}})}{\sqrt{\hat{\varepsilon}}}, \tag{42}$$

where $\varepsilon_0 = 1.7$ keV and $\varepsilon_0 = 0.1$ keV are used to draw the fitted curves to relatively low and high energy regions, respectively. It can be seen that it is hard to reproduce the experimental result by Eq. (42). The essential difference of the two analytical models is attributed to their density profiles, i.e., $\rho \propto \exp(-\xi^2)$ for the present model and $\rho \propto \exp(-\xi)$ for the SP model. This can be elaborated on as follows: The pressure scale decreases with time all over the region in the present model, while it is kept constant in time in the SP model. Therefore, the ions in the former model are less accelerated due to the pdV work than those in the latter model.

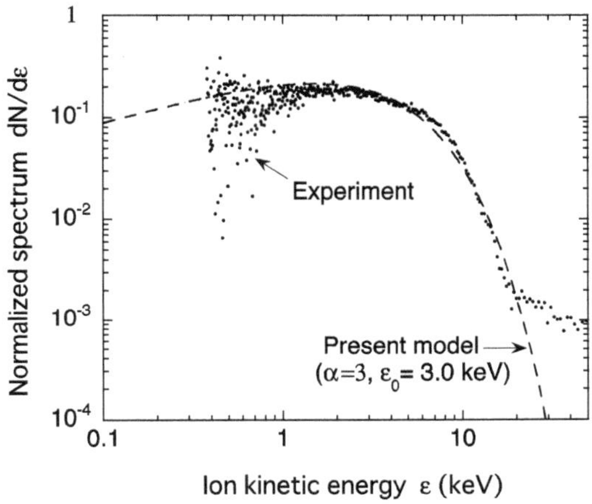

Fig. 3. Comparison of the experimental result (dots) and the analytical curve (dashed line) obtained by Eq. (37) under spherical geometry.

In the second case, a laser beam was irradiated from a single side with a liquid-Xe jet ejected through a nozzle with diameter of 30 μm. The focal spot size was also 30 μm in diameter. Therefore, the resultant plasma expansion was very likely unsymmetrical. In this case, however, the specific mass can expand into much larger space three-dimensionally than in the first case, and thus is regarded as a quasi-spherical expansion ($\alpha = 3$). Figure 3 shows the experimental result and an analytical curve obtained by Eq. (37) with a fitted numerical factor $\varepsilon_0 = 3.0$ keV. Again, with respect to the vertical axis, the physical quantities are properly normalized such that the peak values stay in the order of unity. The ion fluxes were observed at an angle of 45 degrees with respect to the laser beam axis. The experimental signals strongly fluctuate at energies close to the lowest detection limit at around $\varepsilon \sim 400$ eV, but are otherwise well reproduced by the analytical curve.

3. Laser-driven nonstationary accelerating foil due to nonlinear heat conduction

3.1 Introduction

When one side of a thin planar foil is heated by an external heat source, typically by laser or thermal x-ray radiation, the heated material quickly expands into vacuum with its density being reduced drastically - this phenomenon is called "ablation". In inertial confinement fusion (ICF) research, for example, it is indispensable to correctly understand the shell acceleration due to ablation. Thereby self-similar solutions play a crucial role in the analysis and prediction of the detailed behavior of the shell acceleration. Although some analytical models have been proposed to study the shell acceleration due to mass ablation (Gitomer et al., 1977; Takabe et al., 1983; Kull, 1989, 1991), most of them have assumed a stationary ablation layer. Pakula and Sigel (1985), for example, reported a self-similar solution for the ablative heat wave. In the solution, however, the ablation surface is ideally treated such that the density goes to infinity, and the surface does not accelerate. Below, we present a new self-similar solution (Murakami et al, 2007), which describes non-stationary acceleration dynamics of a planar foil target ablatively driven by non-linear heat transfer. The most striking differences from the other models are that the target has a decreasing mass with a peak density, and that it has a distinct shell/vacuum boundary, where the density and the temperature converge to null.

3.2 Basic equations and similarity ansatz

Suppose that a planar shell is being accelerated in the positive direction of the x-axis in an inertial laboratory frame via the recoil force due to the ablation. The characteristic scale length of the shell $D(t)$ decreases with time. Let us assume that the shell is burnt out at the origin of the coordinates, i.e., $D(0) = 0$ at $x = 0$. One can always find such an inertial frame by appropriately choosing relative position and velocity to another reference inertial frame. In this case the shell velocity is initially ($t < 0$) negative, its absolute value gradually decreases due to the positive acceleration, and finally the burned-out shell halts at $(x, t) = (0,0)$. The fluid system is then described by the following equations:

$$\frac{\partial \rho}{\partial t} + \frac{\partial (\rho v)}{\partial x} = 0\,, \tag{43}$$

$$\frac{\partial v}{\partial t} + v\frac{\partial v}{\partial x} = -\frac{\partial p}{\partial x}, \tag{44}$$

$$\rho\left(\frac{\partial \epsilon}{\partial t} + v\frac{\partial \epsilon}{\partial x}\right) + p\frac{\partial v}{\partial x} = \frac{\partial}{\partial x}\left(\kappa\frac{\partial T}{\partial x}\right), \tag{45}$$

where ρ is the mass density, v is the flow velocity, ϵ is the specific internal energy, T is the temperature in units of energy, and κ is the thermal conductivity. We assume an ideal gas equation of state in the form,

$$p = \rho T,\ \epsilon = T/(\gamma - 1), \tag{46}$$

where γ is the specific heats ratio. We assume that the thermal conductivity is expressed in the following power-law form with m, n, and κ_0 being constants,

$$\kappa = \kappa_0 T^n/\rho^m. \tag{47}$$

We introduce the following well-known similarity ansatz (Guderley, 1942; Lie, 1970) to eliminate the temporal dependence of the system and thus to find a self-similar solution:

$$\eta = x/D(t)\,,\ D(t) = A(-t)^{\alpha}\,,\ \alpha \geq 1\,, \tag{48}$$

$$v = \alpha A(-t)^{\alpha-1}u(\eta)\,, \tag{49}$$

$$T = (\alpha A)^2(-t)^{2(\alpha-1)}\theta(\eta)\,, \tag{50}$$

$$\rho = B(-t)^{\beta} g(\eta)\,, \qquad \beta = \frac{2(n-1)(\alpha-1)-1}{1+m}, \tag{51}$$

where η is the self-similar variable; $u(\eta)$, $\theta(\eta)$, and $g(\eta)$ stand for the self-similar profiles of the velocity, temperature, and density, respectively; α, A and B are arbitrary constants. In most of numerical calculations in this paper, we employ $\alpha = 2$ (constant acceleration), $(m,n) = (0, 5/2)$ (electron heat conductivity) and $\gamma = 5/3$ as a reference case. The constraint, $\alpha \geq 1$, in Eq. (48) stems from Eqs. (49) and (50) in order that v and T do not diverge to infinity as $t \to 0$. The limiting value, $\alpha = 1$, corresponds to a special case, where the characteristic scale of v and T are kept constant in time, while $\alpha = (2n-1)/2(n-1) = 4/3$ corresponds to another special case, where the density scale does not change in time, i.e., $\beta = 0$ [see Eq. (51)].

Using ansatz (48) - (51), Eqs. (43) - (45) are reduced to the following set of ordinary differential equations:

$$(u+\eta)g' + \left(u' - \frac{\beta}{\alpha}\right)g = 0\,, \tag{52}$$

$$(u+\eta)\,u' + (\alpha^{-1}-1)u + (g\theta)'/g = 0\,, \tag{53}$$

$$(\gamma-1)^{-1}[(u+\eta)\,\theta' + 2(\alpha^{-1}-1)\theta] + \theta u' = Kg^{-1}(g^{-m}\theta^n\theta')' \tag{54}$$

where the prime denotes the derivative with respect to η, and

$$K = \kappa_0 \alpha^{2n-1} A^{2n-2} B^{-1-m} \tag{55}$$

is a dimensionless parameter. Solving Eqs. (52) and (53) algebraically for g' and u', one finds that a singular point appears when $u + \eta = \pm\sqrt{\theta}$ (more details on the singular point will be given later). Let η_s, u_s, g_s, and θ_s be their values at the singular point. Here we introduce re-normalized variables, ξ, $U(\xi)$, $G(\xi)$, and $\Theta(\xi)$:

$$\xi = \frac{\eta - \eta_s}{\sqrt{\theta_s}}, \qquad \xi = \frac{u - \eta_s}{\sqrt{\theta_s}}, \qquad G = \frac{g}{g_s}, \qquad \Theta = \frac{\theta}{\theta_s}, \tag{56}$$

At the singular point, $\xi = 0$, the re-normalized variables are specified to be

$$U(0) = -1, \quad G(0) = 1, \quad \Theta(0) = 1, \tag{57}$$

where we employ the flow direction such that $u_s + \eta_s = -\sqrt{\theta_s}$. Equations (10) - (12) are then transformed to

$$(U+\xi)\,G' + (U' - \beta/\alpha)G = 0\,, \tag{58}$$

$$(U+\xi)\,U' + (\alpha^{-1}-1)U + (G\Theta)'/G + K_1 = 0\,, \tag{59}$$

$$(\gamma-1)^{-1}[(U+\xi)\,\Theta' + 2(\alpha^{-1}-1)\Theta] + \Theta U' = K_2 G^{-1}(G^{-m}\Theta^n\Theta')' \tag{60}$$

where the prime hereafter denotes the derivative with respect to ξ, and

$$K_1 = (1-\alpha^{-1})\eta_s/\sqrt{\theta_s}\,,\; K_2 = K\theta_s^{n-1}g_s^{-m-1}\,, \tag{61}$$

are dimensionless constants representing the gravity (acceleration) and the heat conductivity, respectively. Thus the system is clearly defined by Eqs. (57) - (60). Equations (58) and (59) yield

$$G' = \frac{\Delta_2}{\Delta_1}G\,, \qquad U' = \frac{\beta}{\alpha} - (U+\xi)\frac{\Delta_2}{\Delta_1}, \tag{62}$$

where

$$\Delta_1 = (U+\xi)^2 - \Theta\,, \tag{63}$$

$$\Delta_2 = \left(\frac{\beta}{\alpha}\right)(U+\xi) + (\alpha^{-1}-1)U + \Theta' + K_1\,. \tag{64}$$

It is clear that G′ and U′ in Eq. (62) are singular when $\Delta_1 = 0$. This singular point corresponds to the sonic point, where the flow velocity relative to the surface $\xi = const$ is equal to the local isothermal sound speed. An integrated curve which is physically acceptable is expected to pass this singular sonic point smoothly, the condition of which is given by

$$\Delta_1 = \Delta_2 = 0\,. \tag{65}$$

Since $\xi = 0$ is the singular point, one should start numerical integration at its infinitesimally adjacent point. One then needs the four derivatives $G'(0)$, $U'(0)$, $\Theta'(0)$, and $\Theta''(0)$, which are fully provided by relation (65). At $\xi = 0$, the derivatives of Eq. (62) are reduced from L'Hopital's theorem to

$$G' = \frac{\Delta_2'}{\Delta_1'}\,, \quad U' = \frac{\beta}{\alpha} + \frac{\Delta_2'}{\Delta_1'}\,. \tag{66}$$

Thus all the four derivatives at the sonic point are explicitly obtained from Eqs. (57) - (60), and (66).

The present system has another singular point at the vacuum interface, the coordinate at which, $\xi = \xi_v$, is an eigenvalue of the system. On the vacuum interface the relative flow velocity to the free surface vanishes, i.e., $U(\xi_v) + \xi_v = 0$, which can also be interpreted as the definition of the free surface. Moreover at $\xi = \xi_v$ the pressure and thus the density are expected to vanish coherently, because practically no heat conduction prevails in this front region (typically characterized such that $G \gg 1$, $\Theta \ll 1$, and $(U+\xi)^2 \ll \Theta$) and thus the specific entropy is kept constant in time. It is then shown that Eqs. (16) and (18) (neglecting the heat conduction) have the adiabatic integral with an arbitrary constant c_0 (Zel'dovich & Raizer, 1966):

$$\Theta(U+\xi)^{\mu}G^{\mu+1-\gamma} = c_0, \qquad \mu \equiv \frac{2(1-\alpha)+\beta(\gamma-1)}{\alpha+\beta}. \tag{67}$$

The vacuum interface is a singular point of the adiabatic flow of the saddle type (Sanz et al., 1988), where the spatial profiles in the vicinity of $\xi = \xi_v$ is worked out from Eqs. (58) - (60) to a first-order approximation in $(\xi_v - \xi)$:

$$\Theta \approx \frac{\big((\gamma+1)\alpha-2\big)(\alpha K_1+(\alpha-1)\xi_v)}{(\alpha+\beta)\gamma}(\xi_v-\xi), \tag{68}$$

$$U+\xi \approx -\frac{\gamma+1-2\alpha^{-1}}{\gamma}(\xi_v-\xi) \tag{69}$$

$$G \approx c_1(\xi_v-\xi)^{\nu}, \quad \nu \equiv \frac{-\alpha+\beta\gamma+2}{\alpha(\gamma+1)-2}, \tag{70}$$

where c_1 is an arbitrary constant; $c_1 \approx G_a(\xi_v-\xi_a)^{-\nu}$ for a relatively high aspect shell, i.e., $G_a/(\xi_v-\xi_a) \gg 1$, where G_a and ξ_a are their corresponding values at the density peak; G_a and ξ_a are also eigenvalues of the system as will be given below together with ξ_v. In particular, under constant acceleration ($\alpha=2$), the velocity becomes constant, $U=-\xi_v$, and $G \propto (\xi_v-\xi_a)$ apart from a linear temperature profile in space, as one can predict from Eqs. (69) and (70).

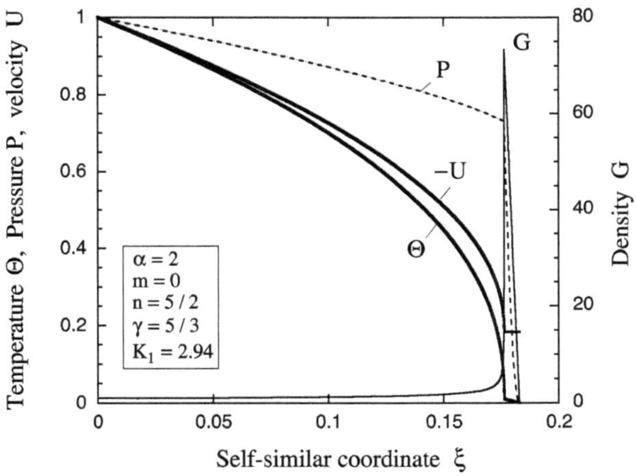

Fig. 4. Eigenstructure of the accelerated shell under a constant gravity ($\alpha=2$).

3.3 Two dimensional eigenvalue problem and numerical results

Although one can start the numerical integration at $\xi=0$ toward the positive direction of ξ-axis, it soon turns out that such numerical integrations produce physically unacceptable pictures under an arbitrary set of the values of K_1 and K_2 such that $G \to \infty$ on its way in the integration without showing the converging behavior, Eqs. (68) - (70), at the vacuum boundary. Therefore the present system is supposed to be an eigenvalue problem, in which only some special combinations of K_1 and K_2 can produce the converging behavior expected as a physically meaningful solution (Murakami et al., 2004).

Figure 4 shows such an eigenstructure numerically obtained for the density G, the temperature Θ, the velocity U, and the pressure $P=G\Theta$ under the fixed parameters given in Fig. 4. As mentioned earlier, the spatial profiles thus obtained strikingly contrast with ones for the stationary ablation models (Gitomer et al., 1977; Takabe et al., 1983; Kull, 1989, 1991).

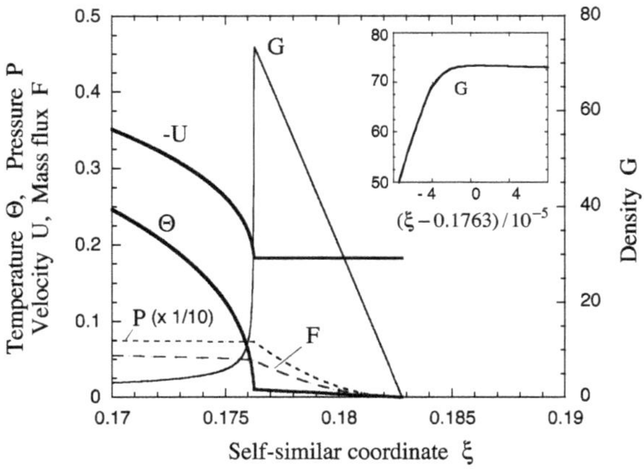

Fig. 5. Magnified view of Fig.4 around the ablation surface.

Figure 5 shows the magnified view around the ablation surface of Fig. 4, in which the mass flux relative to the surface with $\xi = const$, $F \equiv -(U+\xi)G$, is additionally depicted. Surprisingly the predicted profiles, (68) - (70), apply not only to the vicinity of the vacuum boundary but also to almost all the region beyond the ablation surface ($\xi > 0.1763$). This in turn supports the earlier argument that the heat conduction in the shell is practically negligible. It should also be noted that at $\xi = \xi_a$ the physical quantities seemingly have a sharp jump in their derivatives. However, all those quantities change smoothly but on a very narrow range, which can be observed in the further magnified view for G in the upper right corner in Fig. 5. The characteristic scale length of the drastic change in the physical quantities can be roughly estimated from Eq. (60) to be $\Delta\xi_a \sim \Theta_a^n/|U_a|\, G_a^{1+m} \sim \mathcal{O}(10^{-5})$ as can be observed in Fig. 5.

4. Gravitational collapse of radiatively cooling sphere in view of star-formation

4.1 Introduction

Self-similar solutions play a crucial role in astrophysics as well. Below we describe a spherically contracting system observed in the star formation processes, in which the effect of radiative heart conduction is expected to play an important role. In such a system, substantial dissociation and ionization of molecules and atoms proceed with time, and the isothermal assumption used in the so-called LP model (Larson, 1969; Penston, 1969) becomes inappropriate. A solution introduced here (Murakami et al., 2004) can be clearly placed in a thermodynamic perspective as follows: The LP model with the isothermal assumption means infinitely large heat conductivity, i.e., $Pe \to 0$, where Pe denotes the Péclet number. Meanwhile, there are a number of works based on the perfect adiabaticity, i.e., $Pe \to \infty$, which corresponds to zero heat conductivity (Sedov, 1959; Barenblatt, 1979; Antonova, 2000). These are two opposed limiting cases, with which the analytical and numerical treatment are substantially simplified, and the energy conservation law is often expressed in an integrated form or neatly installed in the equation of motion. In contrast, we explicitly leave the radiative conduction term in the hydrodynamic system to handle its nonlinear effect.

An important feature of the present subsection, which is essentially different from the conventional ones obtained under the isothermal or adiabatic assumptions, is that all the scales of the physical quantities are uniquely determined as a function of time only. This is clear from the following argument: When discussing self-similarity within the one-dimensional framework, one needs four physical quantities to produce a dimensionless parameter as a basic self-similar variable, where the system is contained in the class of systems of the so-called MLT fundamental units of measurement. Radius r, time t, and the gravitational constant G, are apparently the first three quantities in a spherically contracting system under self-gravity. The fourth quantity is, for example, the temperature for an isothermal system, or the specific entropy for an adiabatic system. Such quantities cannot be specified in the absolute value, and therefore they can serve as an external control parameter of individual systems. In the present system, however, the fourth quantity is the heat diffusion conductivity, ν_0; the numerical value of which is quite unique, once the conductive mechanism is specified. Therefore ν_0 can never be a control parameter, and the resultant behavior of the system is unique.

4.2 Basic equation and similarity ansatz

The one-dimensional spherical gas-dynamical equations with both self-gravity and diffusivity are

$$\frac{\partial \rho}{\partial t}+\frac{1}{r^2}\frac{\partial}{\partial r}(r^2\rho u)=0, \tag{71}$$

$$\frac{\partial u}{\partial t}+u\frac{\partial u}{\partial r}=-\frac{1}{\rho}\frac{\partial p}{\partial r}-\frac{\partial \phi}{\partial r}, \tag{72}$$

$$\frac{1}{r^2}\frac{\partial}{\partial r}\left(r^2\frac{\partial \phi}{\partial r}\right)=4\pi G\rho, \tag{73}$$

$$\rho\left(\frac{\partial \epsilon}{\partial t}+u\frac{\partial \epsilon}{\partial r}\right)+\frac{p}{r^2}\frac{\partial}{\partial r}(r^2 u)=\frac{1}{r^2}\frac{\partial}{\partial r}\left(r^2\nu\frac{\partial T}{\partial r}\right) \tag{74}$$

where p is the pressure, ρ the density, ϵ the specific internal energy, u the flow velocity, and ϕ the gravitational potential. We assume the ideal gas equation of state (EOS) in the form,

$$\frac{(Z+1)k_B}{\mu}T=\frac{p}{\rho}=(\gamma-1)\epsilon, \tag{75}$$

where k_B is the Boltzmann constant, μ the mean atomic mass, and γ the specific heats ratio; Z is the ionization state, and $Z=1$ is assumed for hydrogen plasma. Equation (74), described by the one-temperature model, includes the non-linear heat diffusion term on the right hand side, where we assume a power-law dependence for the diffusion coefficient, $\nu=\nu_0 T^n/\rho^m$, with ν_0, m, and n being constants. For normal physical values, $n>0$ and $m>0$ are assumed. With an intention to apply our solution primarily to the case of radiative heat diffusion, we can express ν as $\nu=(16\sigma_{SB}T^3)/3\kappa_R$ where $\kappa_R=\kappa_0\rho^m/T^{n-3}$ is the Rosseland

mean opacity, σ_{SB} is the Stefan-Boltzmann constant, and $\kappa_0 = 16\sigma_{SB}/3\nu_0$ is a constant. In the formulae given below, we keep the generality in terms of the parameters, m, n, and γ, but also show specific forms using the values of the reference set at the same time: $m = 2$, $n = 13/2$, and ν_0 describing the opacity due to inverse bremsstrahlung in a fully ionized hydrogen plasma (Zel'dovich & Raizer, 1966) together with $\gamma = 5/3$.
To find a self-similar solution, we here introduce the following group transformation,

$$r = \lambda\hat{r},\quad t = \lambda^a\hat{t}, u = \lambda^b\hat{u}, T = \lambda^c\hat{T}, \rho = \lambda^d\hat{\rho}, \phi = \lambda^e\hat{\phi} \tag{76}$$

where the hats denote the physical quantities in the scaled system related by the scale factor λ with the parent system without hats. The constants, a, b, c, d, and e, can be appropriately determined by substituting Eq. (76) for Eqs. (71) - (74) such that the transformed system is kept symmetric and thus has the same structure as the original one based on the Lie's idea (Lie, 1970):

$$1 - a = b = c/2 = 1 + d/2 = e/2 = (1 * 2m)/(3 + 2m - 2n). \tag{77}$$

For the reference case, $m = 2$ and $n = 13/2$, Eq. (77) gives $a = 11/6$, $b = -5/6$, $c = e = -5/3$, and $d = -11/3$. Equation (77), together with the following similarity ansatz, enables the removal of the temporal dependence from Eqs. (71) - (74),

$$R(t) = A\ |t|^{1/a},\ \xi \equiv r/R(t), \tag{78}$$

$$u = \frac{A}{a}|t|^{b/a}v(\xi), \tag{79}$$

$$T = \left(\frac{A}{a}\right)^2 |t|^{c/a}\tau(\xi), \tag{80}$$

$$\rho = B|t|^{-2}g(\xi), \tag{81}$$

$$\frac{\partial\phi}{\partial r} = \frac{ABG}{\xi^2}|t|^{(c-1)/a}\Omega(\xi), \qquad \Omega(\xi) \equiv 4\pi\int_0^{\xi} g(\xi)d\xi, \tag{82}$$

where $R(t)$ is the temporal characteristic scale length of the system; A and B are positive constants defining the scales of the radius and the density, respectively. Note that the relation, $d/a = -2$, is used for the similarity ansatz of the density in Eq. (84), which holds regardless of the values of m and n. Furthermore, it should be noted that, at a glance, the ansatz for u and T given in Eqs. (82) and (83) seem to be bounded with each other with the similar front factors, A/a and $(A/a)^2$, respectively. However, these factors are chosen just for simplicity, and u and T are kept independent of each other, because the functions, $v(\xi)$ and $\tau(\xi)$, are left free until they are self-consistently determined as the solution of the eigenvalue problem as shown below. In this paper, we consider a contracting fluid system for $t < 0$ which collapses at $t = 0$, and therefore $|t| = -t$. Then, Eqs. (71), (72), and (74) are respectively reduced to the following ordinary differential equations,

$$-(\pm\xi - v)g' + \left(\pm d + v' + \frac{2v}{\xi}\right)g = 0, \tag{83}$$

$$\pm bv - (\pm\xi - v)v' + \frac{(g\tau)'}{g} + \frac{K_1\,\Omega}{\xi^2} = 0, \tag{84}$$

$$\frac{\pm c\tau - (\pm\xi - v)\tau'}{\gamma - 1} + \left(v' + \frac{2v}{\xi}\right)\tau = K_2\frac{(\xi^2 g^{-m}\tau^n\tau')'}{g\xi^2}, \tag{85}$$

where the prime denotes the derivative with respect to ξ, and concerning the double signs, $(\pm)$, the upper (plus) and lower (minus) sign correspond to $t > 0$ and $t < 0$, respectively. Since Eq. (73) is automatically satisfied, its reduced form does not appear in the above set of equations. Thus, the present system is characterized by the two positive dimensionless parameters, K_1 and K_2, defined by $K_1 = a^2GB$ and $K_2 = (\nu_0/aB^{m+1})(A/a)^{2n-2}$. It can be interpreted that K_1 and K_2 are introduced for simplicity instead of A and B. Equations (83) - (85) are second-order ordinary equation system for g, τ, and v, and the obvious boundary conditions are

$$v(0) = 0,\ g(0) = 1,\ \tau(0) = 1,\ (g\tau)'_{\xi=0} = 0. \tag{86}$$

The last relation means that there is no pressure gradient at the center.

4.3 The self-similar solution as two dimensional eigenvalue problem

At first glance, the ODE system, Eqs. (83) - (85), together with the boundary condition (86), seem to be closed mathematically. However, one can easily find that numerical integration of the system produces a physically unacceptable picture under an arbitrary set of the values for K_1 and K_2 such that the temperature suddenly diverges to infinity at a finite radius. Since the physical quantities are expected to change smoothly in space, it is conjectured that some special values of K_1 and K_2, which are still unknown, can give such a physically acceptable picture. Therefore the present system is supposed to be a two-dimensional eigenvalue problem, which is essentially different from the one-dimensional eigenvalue problems of the previous work.

To determine a unique set of parameters, K_1 and K_2, we need two more physical conditions. The first one is quite an orthodox prescription, in which the right integration curve smoothly passes through the singular point, which is located somewhere at a finite distance from the center. On this singular point, the fluid velocity is equal to the local sound speed. The second parameter is less obvious compared with the first one, but still seems natural enough, namely, that both the density and the temperature converge to zero simultaneously with an increase in radius. The numerical calculation is started from the center, and therefore it is necessary to make clear the asymptotic behavior of the solution in the vicinity of the center as follows.

For the central region, the asymptotic behaviors of the above physical quantities are obtained by inserting the following ansatz,

$$g = 1 - g_0\xi^2,\ \tau = 1 - \tau_0\xi^2,\ v = -v_0\xi,\ (\xi \ll 1), \tag{87}$$

into Eqs. (83) - (85), where g_0, τ_0, and v_0 are unknown positive constants, where we make use of the symmetry at the center and thus employed only the lowest quadratic terms for g and τ. After some manipulation, the constants are obtained,

$$v_0 = \frac{2(n-1)}{3(n-m-3/2)}, \tau_0 = \frac{(\gamma-1)n-m-\gamma+1/2}{3(\gamma-1)(n-m-3/2)K_2} g_0 = 2\pi K_1 - \frac{\tau_0 v_0^2}{4} \tag{88}$$

As can be seen in Eq. (88), in order to conduct the numerical calculation starting from the center, K_1 and K_2 must both be specified as trial values, which are expected to converge to their genuine eigenvalues after numerical iteration. Figure 6 shows the first step of the solving process, or how a right eigenvalue, K_2, is obtained on the g-τ plane, where $K_1 = 0.64$ is fixed just as a trial value. As can be seen in Figure 6, there exists an appropriate value of K_2, with which the integrated curve smoothly passes through the singular point, while the other integrated curves deviate from the right curve as the integration proceeds toward the singular point, resulting in an unacceptable physical picture. In this manner, an appropriate eigenvalue K_2 can be determined as a function of arbitrary K_1.

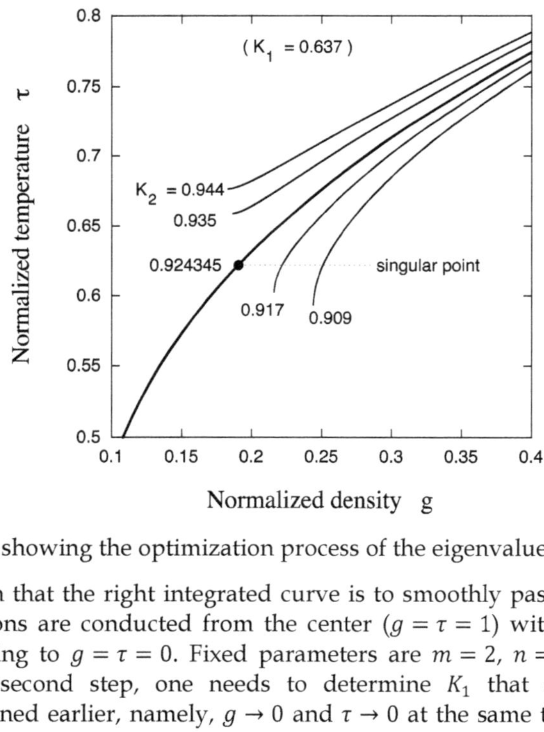

Fig. 6. g - τ diagram showing the optimization process of the eigenvalue, K_2.

Under the condition that the right integrated curve is to smoothly pass through a singular point, the integrations are conducted from the center ($g = \tau = 1$) with the radius toward infinity corresponding to $g = \tau = 0$. Fixed parameters are $m = 2$, $n = 13/2$, $\gamma = 5/3$, and $K_1 = 0.64$. As the second step, one needs to determine K_1 that satisfies the second requirement mentioned earlier, namely, $g \to 0$ and $\tau \to 0$ at the same time. Figure 7 shows how the right eigenvalue, K_1, is determined on the g-τ plane, where each curve is already optimized such that it passes through each singular point. As a result, it turns out that there exists a unique pair of the eigenvalues of K_1 and K_2, which satisfies the both requirements.
Figure 8 shows the eigenstructure for the temperature, $\tau \propto T$ the density, $g \propto \rho$, the velocity, $v \propto u$, and the heat flux, $\boldsymbol{q} \propto -\nu \nabla T$, under the eigenvalues of the reference system thus obtained, where the curves are assigned with labels corresponding to the original physical

quantities just for readers' comprehension. The behavior of the velocity for $v \to \infty$ may seem physically unacceptable at least in a rigorous sense. As a matter of fact, however, there are a number of examples for implosions and explosions in which the velocity profile is approximately linear with the radius (Sedov, 1959; Bernstein, 1978). In addition, the physical condition at enough large radii ($\xi \gg 1$) will not affect the core dynamics for an intermediate time period. Therefore, when we restrict our considerations to a finite closed volume containing the core, the present self-similar solution is expected to be an approximation of the core evolution at higher densities and temperatures.

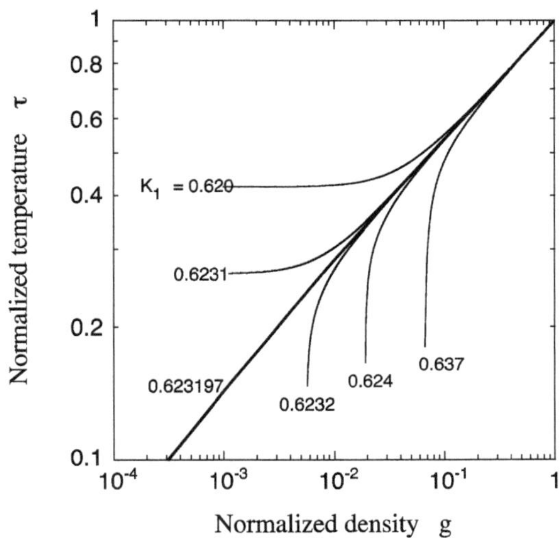

Fig. 7. g - τ diagram showing the optimization process of the eigenvalue, K_1.

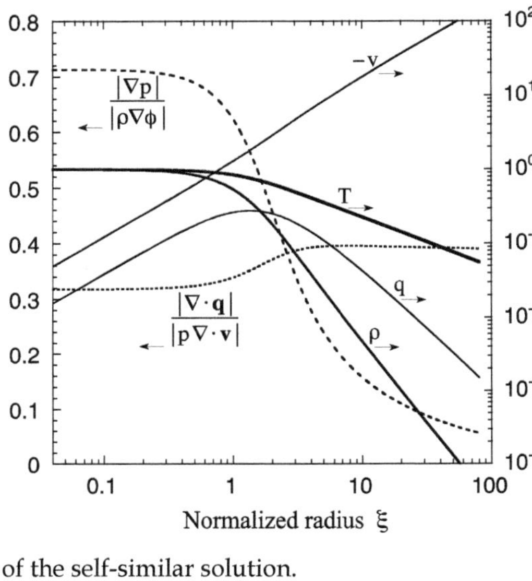

Fig. 8. Eigenstructure of the self-similar solution.

Under the condition that the right integrated curve is to converge to $g = \tau = 0$, each curve has already optimized with respect to K_2 as was shown in Fig. 6. Other fixed parameters are the same as in Fig. 6.
The normalized physical quantities are obtained as a result of the two-dimensional eigenvalue problem with fixed parameters, $m = 2, n = 13/2$, and $\gamma = 5/3$.

5. Conclusions

The crucial role of dimensional analysis and self-similarity are discussed in the introduction and the three subsequent examples. Self-similar solutions for individual cases have been demonstrated to be derivable by applying the Lie group analysis to the set of PDE for the hydrodynamic system, taking nonlinear heat conductivity into account as the decisive physical ingredient. The scaling laws for thermally conductive fluids are conspicuously different from those for adiabatic fluids (not discussed in the present chapter; see references by Murakami et al., 2002, 2005 for details). The former has one freedom less than the latter due to the additional constraint of thermal conductivity. If a thermo-hydrodynamic system comprises multiple heat conduction mechanisms, self-similarity cannot be expected in a vigorous sense except for special cases. However, self-similarity and scaling laws can always be found at least in an approximate manner, by shedding light on the dominant conduction mechanism, which should give the basis of system design and diagnostics for scaled experiments for individual cases. The necessity of dimensional analysis and finding self-similar solutions is encountered in many problems over wide ranges of research. The simple general scheme and the examples mentioned in this chapter will help the reader who encounters a similar situation in his or her investigation find the underlying physics and prepare further theoretical and experimental setup.

6. References

Antonova, R.N. & Kazhdan, Y.M. (2000). "A self-similar solution for spherically symmetric gravitational collapse " *Astronomy Letters*, Vol. 26, pp. 344 - 355.

Barenblatt, G.I. (1979). *Similarity, Self-Similarity, and Intermediate Asymptotics* (New York: Consultants Bureau).

Basko, M.M. & Murakami, M., (1998). "Self-similar implosions and explosions of radiatively cooling gaseous masses" Phys. Plasma, Vol. 5, pp. 518 – 528.

Bernstein, I.B. & Book, D.L. (1978). "Rayleigh-Taylor instability of a self-similar spherical expansion" *Astrophysical Journal,* Vol. 225, pp. 633 – 640.

Gitomer, S.J.; Morse, R.L. & Newberger, B.S. (1977). "Structure and scaling laws of laser-driven ablative implosions", Phys. Fluids Vol. 12, pp. 234 - 238.

Guderley, G. (1942) "Starke kugelige und zylindrische Verdichtungsstöße in der Nähe des Kugelmittelpunktes bzw. der Zylinderachse" Luftfahrtforschung Vol. 19, pp. 302–312.

Gurevich, A.V.; Parrska, L.V. & Pitaevsk, L.P. (1966). "Self-similar motion of rarefied plasma" *Sov. Phys. JETP,* Vol. 22, pp. 449 - &.

Kull, H.J. (1989). "Incompressible Description of Rayleigh-Taylor Instabilities in Laser-Ablated Plasmas" *Phys. Fluids,* Vol. B1, pp.170 - 182.

Kull, H.J. (1991). "Theory of Rayleigh-Taylor Instability" *Phys. Reports,* Vol.206, pp.197 - 325.

Landau, L.D. & Lifshitz, E.M. (1959). *Fluid Mechanics* (New York: Pergamon).
Larson, R.B. (1969). "Numerical calculations of the dynamics of collapsing proto-star" *Mon. Not. R. Astr. Soc.*, Vol. 145, pp. 271-&.
Lie, S. (1970). *Theorie der Transformationsgruppen* (New York: Chelsea).
London, R.A. & Rosen, M.D. (1986) "Hydrodynamics of Exploding Foil X-ray Lasers" Phys. Fluids, Vol. 29, pp. 3813 - 3822.
Mora, P. (2003). "Plasma Expansion into a Vacuum" Phys. Rev. Lett. Vol.90, 185002 (pp. 1 - 4).
Murakami, M.; Meyer-ter-Vehn, J. & Ramis, R. (1990). "Thermal X-ray Emission from Ion-Beam-Heated Matter" J. X-ray Sci. Technol., Vol. 2, pp. 127 - 148.
Murakami, M. & Meyer-ter-Vehn, J. (1991) "Indirectly Driven Targets for Inertial Confinement Fusion" *Nucl. Fusion*, Vol. 31, pp. 1315 – 1331.
Murakami, M., Shimoide, M., and Nishihara, K. (1995). "Dynamics and stability of a stagnating hot spot" *Phys. Plasmas*, Vol.2, pp. 3466 - 3472.
Murakami, M. & Iida, S., (2002). "Scaling laws for hydrodynamically similar implosions with heat conduction", *Phys. Plasmas*, Vol.9, pp.2745 - 2753.
Murakami, M.; Nishihara, K. & Hanawa, T. (2004). "Self-similar Gravitational Collapse of Radiatively Cooling Spheres", *Astrophysical Journal*, Vol. 607, pp.879 - 889.
Murakami, M.; Kang, Y.-G.; Nishihara, K.; Fujioka, S. & Nishimura, H. (2005). "Ion energy spectrum of expanding laser-plasma with limited mass", *Phys. Plasmas,* Vol.12, pp. 062706 (1-8).
Murakami, M. & M. M. Basko (2006). "Self-similar expansion of finite-size non-quasi-neutral plasmas into vacuum: Relation to the problem of ion acceleration", *Phys. Plasmas,* Vol. 13, pp. 012105 (1-7).
Murakami, M.; Fujioka, S.; Nishimura, H.; Ando, T.; Ueda, N.; Shimada, Y. & Yamaura, M. (2006). "Conversion efficiency of extreme ultraviolet radiation in laser-produced plasmas", *Phys. Plasmas*, Vol.13, pp. 033107 (1-8).
Murakami, M.; Sakaiya, T. & Sanz, J. (2007). "Self-similar ablative flow of nonstationary accelerating foil due to nonlinear heat conduction", *Phys. Plasmas,* Vol. 14, pp. 022707 (1-7).
Pakula, R. & Sigel, R., (1985). "Self-similar expansion of dense matter due to heat-transfer by nonlinear conduction " *Phys. Fluids,* Vol. 28, pp. 232 - 244.
Penston, M.V. (1969). "Dynamics of Self-Gravitating Gaseous Spheres – III Analytical Results in the Free-Fall of Isothermal Cases" *Mon. Not. R. astr. Soc.*, Vol. 144, pp. 425 - 448.
Sedov, L.I. (1959). *Similarity and Dimentional Methods in Mechanichs* (New York : Academic).
Sanz, J.; Nicolás, J.A.; Sanmartín, J.R. & Hilario, J. (1988). "Nonuniform target illumination in the deflagration regime: Thermal smoothing", *Phys. Fluids,* Vol. 31, pp. 2320 – 2326.
Takabe, H; Montierth, L. & Morse, R.L. (1983). "Self-consistent Eigenvalue Analysis of Raileigh-Taylor Instability in an Ablating Plasma", *Phys. Fluids*, Vol. 26, pp. 2299 - 2307.
True, M.A.; Albritton, J.R. & Williams, E.A. (1981). "Fast Ion Production by Suprathermal Electrons in Laser Fusion Plasmas", *Phys. Fluids* Vol. 24, pp. 1885 - 1893.

Zel'dovich, Ya.B. & Raizer, Yu.P. (1966). *Physics of Shock Waves and High Temperature Hydrodynamic Phenomena* (New York: Academic Press).

Part 4

Numerical Methods

13

Meshless Heat Conduction Analysis by Triple-Reciprocity Boundary Element Method

Yoshihiro Ochiai
Kinki University
Japan

1. Introduction

The main advantage of the boundary element method (BEM) formulation for the solution of boundary value problems results from the localization of unknowns on the boundary of the analyzed domain. The necessary condition for a pure boundary formulation is the knowledge of the fundamental solution of the governing differential operator. In addition to the reduction of the dimensionality, other advantages of the BEM formulation include good conditioning of the discretized equations, high accuracy and the stability of numerical computations because of the utilization of fundamental solutions. Sometimes, domain integrals are also involved in integral equation formulations; in such cases, the advantageousness of the BEM formulation is partially decreased. The most frequent reasons for the occurrence of domain integrals are body sources, nonlinear constitutive laws and nonvanishing initial conditions in time-dependent problems (Partridge et al., 1992, Sladek and Sladek, 2003, Tanaka et al., 2003).

Since the fundamental solution for a diffusion operator is available in closed form, one can attempt to achieve a pure boundary integral formulation for transient heat conduction problems considered within the linear theory. This can be easily achieved provided that the initial temperature and/or heat sources are distributed uniformly. Then, one can convert the domain integrals of the fundamental solution into boundary integrals using the higher-order polyharmonic fundamental solutions (Nowak, 1989, 1994). As regards the discretization of the time variable, two time-marching schemes are appropriate in formulations with time-dependent fundamental solutions. In one of them, the integration is performed from the initial time to the current time, while in the second scheme the integration is considered within a single time step, taking the temperature at the end of the previous time step as the initial value (pseudo-initial) at the current time step (Ochiai, 2006). Although the domain integral of the uniform initial temperature can be avoided in the first time-marching scheme, the number of boundary integrals increases with increasing number of time steps even in this special case. On the other hand, the spatial integrations are performed only once and are used at each time step in the second scheme provided that a constant length of the time steps is used. The time-marching scheme with integration within a single time step increases the efficiency of numerical integration over boundary elements. The integral formulation as well as the triple-reciprocity approximation are derived in this chapter. The higher-order polyharmonic fundamental solutions and their time integrals are shown in the Appendies. The numerical examples given concern the investigation of the accuracy of the proposed BEM formulation using the triple-reciprocity approximation of either pseudo-initial temperatures or body heat sources.

In this chapter, the steady and unsteady problems in the one-, two- and three-dimensional cases are discussed. In the triple-reciprocity BEM, the distributions of heat generation and initial temperature are interpolated using two Poisson equations. These two Poisson equations are solved using boundary integral equations. This interpolation method is very important in the triple-reciprocity BEM. This numerical process is particularly focused on this chapter.

2. Basic equations

2.1 Steady heat conduction

Point and line heat sources can easily be treated by the conventional BEM. In this study an arbitrarily distributed heat source W_1^S is treated. In steady heat conduction problems, the temperature T under an arbitrarily distributed heat source W_1^S is obtained by solving the following equation (Carslaw, 1938):

$$\nabla^2 T = \frac{-W_1^s}{\lambda}, \tag{1}$$

where λ is thermal conductivity. Denoting heat generation by $W_1^S(q)$, the boundary integral equation for the temperature in the case of steady heat conduction is given by (Brebbia, 1984)

$$\mathrm{cT(P)} = \int_\Gamma \{\mathrm{T_1(P,Q)}\frac{\partial T(Q)}{\partial n} - \frac{\partial T_1(P,Q)}{\partial n}T(Q)\}d\Gamma(Q) + \lambda^{-1}\int_\Omega T_1(P,q)W_1^S(q)d\Omega, \tag{2}$$

where $c = 0.5$ on the smooth boundary and $c = 1$ in the domain. The notations Γ and Ω represent the boundary and domain, respectively. The notations p and q become P and Q on the boundary.

In one-dimensional problems, the fundamental solution $T_1(p,q)$ in Eq. (2) used for steady temperature analyses and its normal derivative are given by

$$T_1(p,q) = -\frac{1}{2}r \tag{3}$$

$$\frac{\partial T_1(p,q)}{\partial n} = -\frac{1}{2}\frac{\partial r}{\partial n}. \tag{4}$$

In two-dimensional problems,

$$T_1(p,q) = \frac{1}{2\pi}\ln(\frac{1}{r}) \tag{5}$$

$$\frac{\partial T_1(p,q)}{\partial n} = \frac{-1}{2\pi r}\frac{\partial r}{\partial n}, \tag{6}$$

and in three-dimensional problems,

$$T_1(p,q) = \frac{1}{4\pi r} \tag{7}$$

$$\frac{\partial T_1(p,q)}{\partial n} = \frac{-1}{4\pi r^2}\frac{\partial r}{\partial n}, \tag{8}$$

where r is the distance between the observation point p and the loading point q. As shown in Eq. (2),when arbitrary heat generation $W_1^S(q)$ exists in the domain, a domain integral is necessary.

In the triple-reciprocity BEM, the distribution of heat generation is interpolated using integral equations. Using these interpolated values, a heat conduction problem with arbitrary heat generation can be solved without internal cells by the triple-reciprocity BEM. The conventional BEM requires internal cells for the domain integral. The internal cells decrease the advantageousness of the BEM, in which the arrangement of data is simple. In the triple-reciprocity BEM, the fundamental solution of lower order is used. The triple-reciprocity BEM requires internal points similarly to the dual reciprocity method (DRM) (Partridge, 1992) as shown in Fig. 1, although the boundary values W_f need not be given analytically.

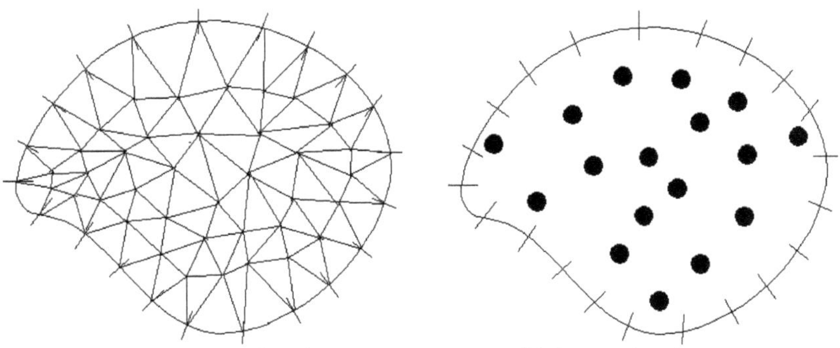

(a) Internal cells (b) Internal points

Fig. 1. Triple-reciprocity BEM.

2.2 Interpolation of heat generation

The distribution of heat generation W is interpolated using integral equations to transform the domain integral into the boundary integral. The deformation of a thin plate is utilized to interpolate the distribution of the heat source W_1^S, where superscript S indicates a surface distribution. The following equations can be used for interpolation (Ochiai, 1995a-c, 1996a, b):

$$\nabla^2 W_1^S = -W_2^S, \tag{9}$$

$$\nabla^2 W_2^S = -\sum_{m=1}^{M} W_3^P(q_m), \tag{10}$$

where W_3^P is a Dirac-type function, which has a value at only one point. The term W_2^S in Eq. (9) corresponds to the sum of curvatures $\partial^2 W_1^S / \partial x^2$ and $\partial^2 W_1^S / \partial y^2$. From Eqs. (9) and (10), the following equation can be obtained:

$$\nabla^4 W_1^S = \sum_{m=1}^{M} W_3^P(q_m). \tag{11}$$

This equation is the same type of equation as that for the deformation w_1^S of a thin plate with point load P, which is

$$\nabla^4 w_1^S = \sum_{m=1}^{M} \frac{P_m}{D}, \tag{12}$$

where the Poisson's ratio is ν =0 and the flexural rigidity is D=1. A natural spline originates from the deformation of a thin beam, which is used to interpolate the one-dimensional distribution, as shown in Fig. 2. In this chapter, the deformation of a thin plate is utilized to interpolate the two-dimensional distribution W_1^S. The deformation w_1^S is given, and the force of point load P is unknown and is obtained inversely from the deformation of the fictitious thin plate, as shown in Fig. 3. The term W_2^S corresponds to the moment of the beam.

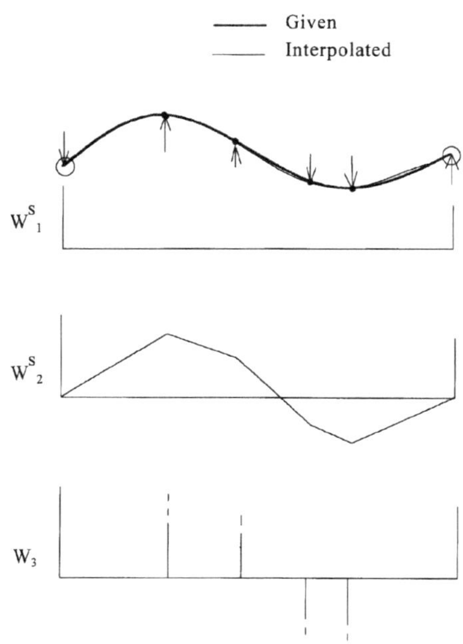

Fig. 2. Interpolation using thin beam with unknown point load.

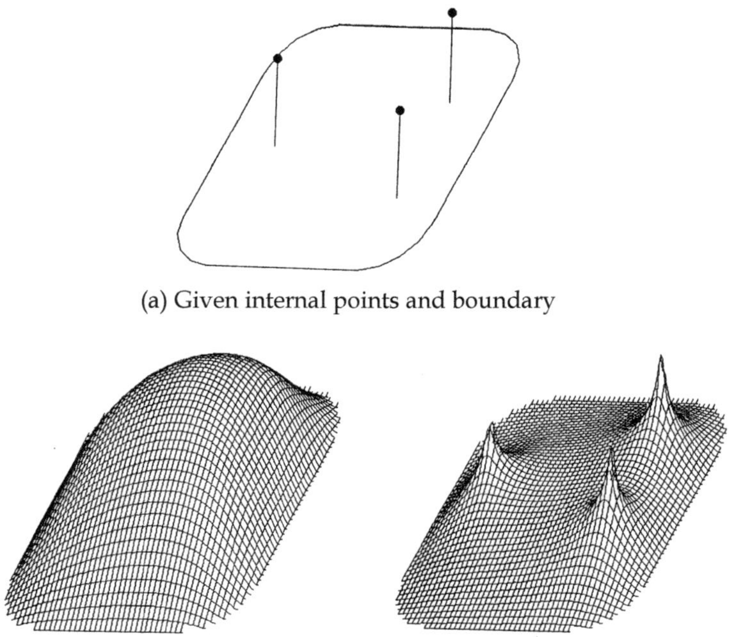

(a) Given internal points and boundary

(b) Obtained shape W_1^S (c) Sum of curvatures W_2^S

Fig. 3. Interpolation using fictitious thin plate with unknown point load.

The moment W_2^S on the boundary is assumed to be 0, the same as in the case of the natural spline. This means that the thin plate is simply supported. In this method, the distribution of heat generation is assumed to be that for a freeform surface (Ochiai, 1995c). Equations (9) and (10) are similar to the equation used to generate a freeform surface using integral equations.

2.3 Representation of heat generation by integral equations

The distribution of heat generation is represented by an integral equation. The following harmonic function $T_1(p,q)$ and biharmonic function $T_2(p,q)$ are used for interpolation (Ochiai, 1999-2003).

$$T_1(p,q) = \frac{1}{2\pi}[\ln(\frac{1}{r}) + B] \tag{13}$$

$$T_2(p,q) = \frac{r^2}{8\pi}[\ln(\frac{1}{r}) + B + 1] \tag{14}$$

B is an arbitrary constant. $T_1(p,q)$ and $T_2(p,q)$ have the relationship

$$\nabla^2 T_2(p,q) = T_1(p,q)\,. \tag{15}$$

Let the number of W_3^P be M. The heat generation W_1^S is given by Green's theorem and Eqs. (9), (10) and (15) as

$$\begin{aligned} cW_1^S(P) &= \int_\Gamma \{T_1(P,Q)\frac{\partial W_1^S(Q)}{\partial n} - \frac{\partial T_1(P,Q)}{\partial n} W_f^S(Q)\} d\Gamma(Q) + \sum_{m=1}^{M} T_2(P,q) W_3^P(q_m) \\ &= \sum_{f=1}^{2} (-1)^{f+1} \int_\Gamma \left\{ T_f(P,Q)\frac{\partial W_f^S(Q)}{\partial n} - \frac{\partial T_f(P,Q)}{\partial n} W_f^S(Q) \right\} d\Gamma - \sum_{m=1}^{M} T_2(P,q_m) W_3^P(q_m), \end{aligned} \tag{16}$$

where $c = 0.5$ on the smooth boundary and $c = 1$ in the domain. Moreover, W_2^S in Eq. (10) is similarly given by

$$cW_2^S(P) = \int_\Gamma \left\{ T_1(P,Q)\frac{\partial W_2^S(Q)}{\partial n} - \frac{\partial T_1(P,Q)}{\partial n} W_2^S(Q) \right\} d\Gamma + \sum_{m=1}^{M} T_1(P,q_m) W_3^P(q_m) . \tag{17}$$

The integral equations (16) and (17) are used to interpolate the distribution. The thin plate spline $F(p, q)$ used to make a freeform surface is defined as (Dyn, 1987, Micchelli, 1986)

$$F(p,q)=r^2 \ln(r). \tag{18}$$

Equations (14) and (18) include the same type of function. Assuming $W_1^S(Q) = 0$, the values of W_3^P and $\partial W_f^S / \partial n$ are obtained using Eqs. (16) and (17).

2.4 Polyharmonic functions

The polyharmonic function $T_f(p,q)$ is defined by

$$\nabla^2 T_{f+1} = T_f . \tag{19}$$

Therefore, $T_f(p,q)$ for the Kth dimensional case can be obtained using the next equation

$$T_f = \int \frac{1}{r^{K-1}} [\int r^{K-1} T_{f-1} dr] dr . \tag{20}$$

From Eq. (20), $T_f(p,q)$ for the two-dimensional case can be obtained using the next equation

$$T_f = \int \frac{1}{r} [\int r T_{f-1} dr] dr . \tag{21}$$

The function $T_f(p,q)$ and its normal derivative for the two-dimensional case are explicitly expressed as

$$T_f(p,q) = \frac{r^{2(f-1)}}{2\pi[(2f-1)!!]^2} [\ln(\frac{1}{r}) + B + \mathrm{sgn}(f-1) \sum_{e=1}^{f-1} \frac{1}{e}], \tag{22}$$

$$\frac{\partial T_f(p,q)}{\partial n} = \frac{r^{2f-3}}{2\pi[(2f-2)!!]^2}[2(f-1)\{\ln(\frac{1}{r})+B\} \ -1+2(f-1)\sum_{e=1}^{f-1}\frac{1}{e}]\frac{\partial r}{\partial n}\ , \tag{23}$$

where $(2f-1)!! = (2f-1)(2f-3)(2f-5)....1$, and sgn() is the sign function.
For the one-dimensional case,

$$T_f(p,q) = -\frac{1}{2}\frac{r^{2f-1}}{(2f-1)!}\ , \tag{24}$$

$$\frac{\partial T_f(p,q)}{\partial n} = \frac{1}{2}\frac{r^{2f-2}}{(2f-2)!}\frac{\partial r}{\partial n}\ . \tag{25}$$

For the three-dimensional case,

$$T_f(p,q) = \frac{r^{2f-3}}{4\pi(2f-2)!}\ , \tag{26}$$

$$\frac{\partial T_f(p,q)}{\partial n} = \frac{(2f-3)r^{2f-4}}{4\pi(2f-2)!}\frac{\partial r}{\partial n}\ . \tag{27}$$

Equations (16) and (17) are similar to the equation used to generate a freeform surface using integral equations (Ochiai, 1995c).
Using Green's theorem three times and Eqs. (9), (10) and (19), Eq. (2) becomes

$$CT(P) = \int_\Gamma \{T_1(P,Q)\frac{\partial T(Q)}{\partial n} - \frac{\partial T_1(P,Q)}{\partial n}T(Q)\}d\Gamma(Q)$$

$$-\lambda^{-1}\int_\Gamma \{T_2(P,Q)\frac{\partial W_1^S(Q)}{\partial n} \ - \frac{\partial T_2(P,Q)}{\partial n}W_1^S(Q)\}d\Gamma(Q) \ +\lambda^{-1}\int_\Omega \ T_2(P,q)\nabla^2 W_1^S(q)d\Omega$$

$$= \int_\Gamma \{T_1(P,Q)\frac{\partial T(Q)}{\partial n} - \frac{\partial T_1(P,Q)}{\partial n}T(Q)\}d\Gamma(Q)$$

$$-\lambda^{-1}\int_\Gamma \{T_2(P,Q)\frac{\partial W_1^S(Q)}{\partial n} \ - \frac{\partial T_2(P,Q)}{\partial n}W_1^S(Q)\}d\Gamma(Q) \ -\lambda^{-1}\int_\Omega \ \nabla^2 T_3(P,q)W_2^S(q)d\Omega$$

$$= \int_\Gamma \{T_1(P,Q)\frac{\partial T(Q)}{\partial n} - \frac{\partial T_1(P,Q)}{\partial n}T(Q)\}d\Gamma(Q)$$

$$+\lambda^{-1}\sum_{f=1}^{2} \ (-1)^f \int_\Gamma \{T_{f+1}(P,Q)\frac{\partial W_f^S(Q)}{\partial n} \ - \frac{\partial T_{f+1}(P,Q)}{\partial n}W_f^S(Q)\}d\Gamma(Q) \ +\lambda^{-1}\sum_{m=1}^{M} T_3(P,q_m)W_3^P(q_m) \tag{28}$$

2.5 Interpolation for 3D case

In the three-dimensional case, the following equations are used for smooth interpolation:

$$\nabla^2 W_1^S(q) = -W_2^S(q)\,, \tag{29}$$

$$\nabla^2 W_2^S(q) = -\sum_{m=1}^{M} W_{3A}^P(q_m)\,, \tag{30}$$

where the function W_{3A}^P expresses the state of a uniformly distributed polyharmonic function in a spherical region with radius A. Figure 4 shows the shape of the polyharmonic functions; the biharmonic function T_2 is not smooth at $r = 0$. In the three-dimensional case, smooth interpolation cannot be obtained solely by using the biharmonic function T_2. To obtain smooth interpolation, a polyharmonic function with volume distribution T_{2A} is introduced. The function T_{fA} shown in Fig. 5 is defined as (Ochiai, 2005)

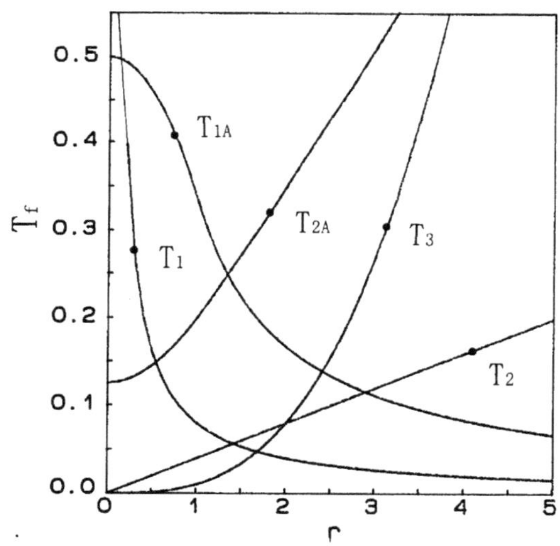

Fig. 4. Polyharmonic functions (T_f, T_{fA})

$$T_{fA}(p,q) = \int_0^A [\int_0^{2\pi} \{\int_0^{\pi} T_f(p,q) a^2 \sin\theta d\theta\} d\varphi] da\,, \tag{31}$$

where Ω_A is a spherical region with radius A, and S is the surface of a spherical shell with radius a. The function T_{fA} can be easily obtained using the relationships $r^2 = R^2 + a^2 - 2aR\cos\theta$ and $dr = aR\sin\theta d\theta$ as shown in Fig. 5. Therefore,

$$\sin\theta d\theta = \frac{r}{aR} dr\,. \tag{32}$$

This function is written using r instead of R, similarly to Eqs. (26) and (27), although the function obtained from Eq. (31) is a function of R. The newly defined function T_{fA} can be explicitly written as

$$T_{fA}(p,q)=\frac{1}{2r(2f+1)!}\{(2fA-r)(r+A)^{2f}+(2fA+r)(r-A)^{2f}\}\quad r>A \tag{33}$$

$$T_{fA}(p,q)=\frac{1}{2r(2f+1)!}\{(2fA-r)(A+r)^{2f}-(2fA+r)(A-r)^{2f}\}\quad r\leq A\,. \tag{34}$$

In Fig. 5, A=1. The newly defined functions T_{fA} used in the chapter can be explicitly written as

$$T_{1A}=\frac{A^3}{3r}\qquad r>A \tag{35}$$

$$T_{1A}=\frac{3A^2-r^2}{6}\quad r\leq A \tag{36}$$

$$T_{2A}=\frac{A^3}{6r}(r^2+\frac{A^2}{5})\quad r>A \tag{37}$$

$$T_{2A}=-\frac{r^4-10r^2A^2-15A^4}{120}\qquad r\leq A \tag{38}$$

$$T_{3A}=\frac{A^3(35r^4+42r^2A^2+3A^4)}{2520r}\quad r>A \tag{39}$$

$$T_{3A}=\frac{-r^6+21r^4A^2+105r^2A^4+35A^6}{5040}\quad r\leq A\,. \tag{40}$$

The heat generation W_1^S is given by Green's theorem and Eqs. (29)-(31) as

$$cW_1^S(P)=\sum_{f=1}^{2}(-1)^{f+1}\int_\Gamma\{T_f(P,Q)\frac{\partial W_f^S(Q)}{\partial n}-\frac{\partial T_f(P,Q)}{\partial n}W_f^S(Q)\}d\Gamma(Q)-\sum_{m=1}^{M}T_{2A}(P,q_m)W_{3A}^P(q_m) \tag{41}$$

Moreover, W_2^S in Eq. (30) is similarly given by

$$cW_2^S(P)=\int_\Gamma\{T_1(P,Q)\frac{\partial W_2^S(Q)}{\partial n}-\frac{\partial T_1(P,Q)}{\partial n}W_2^S(Q)\}d\Gamma(Q)+\sum_{m=1}^{M}T_{1A}(P,q_m)W_{3A}^P(q_m)\,. \tag{42}$$

Equations (41) and (42) are similar to the equation used to generate the freeform surface using integral equations. Using Green's theorem three times and Eqs. (29), (30) and (15), Eq. (2) becomes

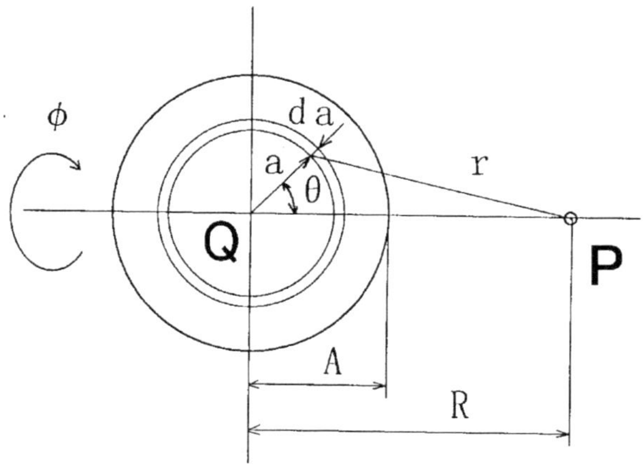

Fig. 5. Notations in three-dimensional problem

$$cT(P) = \int_{\Gamma}\{T_1(P,Q)\frac{\partial T(Q)}{\partial n} - \frac{\partial T_1(P,Q)}{\partial n}T(Q)\}d\Gamma(Q)$$

$$-\lambda^{-1}\sum_{f=1}^{2}(-1)^f\int_{\Gamma}\{T_{f+1}(P,Q)\frac{\partial W_f^S(Q)}{\partial n} - \frac{\partial T_{f+1}(P,Q)}{\partial n}W_f^S(Q)\}d\Gamma(Q) + \lambda^{-1}\sum_{m=1}^{M}T_{3A}(P,q_m)W_{3A}^P(q_m) \quad (43)$$

In the same manner, a polyharmonic function with surface distribution T_{fB} is defined as (Ochiai, 2009)

$$T_{Bf}(p,q) = \int_0^{2\pi}(\int_0^{\pi}T_f(p,q)A^2\sin\theta\, d\theta)d\varphi\,. \quad (44)$$

The newly defined function T_{fB} can be explicitly written as

$$T_{fB}(p,q) = \frac{A\{(r+A)^{2f-1}-(r-A)^{2f-1}\}}{2(2f-1)!r} \quad r > A\ , \quad (45)$$

$$T_{fB}(p,q) = \frac{A\{(A+r)^{2f-1}-(A-r)^{2f-1}\}}{2(2f-1)!r} \quad r \le A\,. \quad (46)$$

Additionally, the temperature gradient is given by differentiating Equation (28), and expressed as:

$$\frac{\partial T(P)}{\partial x_i} = \int_{\Gamma}\{\frac{\partial T_1(P,Q)}{\partial x_i}\frac{\partial T(Q)}{\partial n} - \frac{\partial^2 T_1(P,Q)}{\partial x_i\partial n}T(Q)\}d\Gamma(Q)$$

$$+\lambda^{-1}\sum_{f=1}^{2}(-1)^f\int_{\Gamma}\{\frac{\partial T_{f+1}(P,Q)}{\partial x_i}\frac{\partial W_f^S(Q)}{\partial n} - \frac{\partial^2 T_{f+1}(P,Q)}{\partial x_i\partial n}W_f^S(Q)\}d\Gamma(Q)$$

$$+\lambda^{-1}\sum_{m=1}^{M}\frac{\partial T_3(P,q_m)}{\partial x_i}W_3^P(q_m) \tag{47}$$

The function $\partial T_f(p,q)/\partial x_i$ and its normal derivative for the two-dimensional case are explicitly expressed as

$$\frac{\partial T_f(p,q)}{\partial x_i}=\frac{r^{2f-3}}{2\pi[(2f-2)!!]^2}[2(f-1)\{\ln(\frac{1}{r})+B\}\ -1+2(f-1)\sum_{e=1}^{f-1}\frac{1}{e}]\frac{\partial r}{\partial x_i}, \tag{48}$$

$$\frac{\partial^2 T_f(p,q)}{\partial x_i \partial n}=\frac{r^{2f-4}}{2\pi[(2f-2)!!]^2}\{[n_i+2(f-2)r_{,i}r_{,i}n_j][2(f-1)\{\ln(\frac{1}{r})+B\}$$

$$-1+2(f-1)\sum_{e=1}^{f-1}\frac{1}{e}]-2(f-1)r_{,i}r_{,i}n_j\} \tag{49}$$

2.6 Basic equations for unsteady heat conduction

In unsteady heat conduction problems with heat generation $W_1^S(q,t)$, the temperature T is obtained by solving

$$\nabla^2 T+\frac{W_1^S}{\lambda}=\kappa^{-1}\frac{\partial T}{\partial t}, \tag{50}$$

where κ and t are the thermal diffusivity and time, respectively. Denoting an arbitrary time and the pseudo-initial temperature by τ and $T^0(q,0)$, respectively, the boundary integral equation for the temperature in the case of unsteady heat conduction is expressed as (Wrobel, 2002)

$$cT(P,t)=-\kappa\int_0^t\int_\Gamma[T(Q,\tau)\frac{\partial T_1^*(P,Q,t,\tau)}{\partial n}\ -\frac{\partial T(P,Q)}{\partial n}T_1^*(P,Q,t,\tau)]d\Gamma d\tau$$

$$+\int_\Omega T_1^*(P,q,t,0)T^0(q,0)d\Omega+\frac{\kappa}{\lambda}\int_0^t\int_\Omega T_1^*(P,q,t,\tau)W_1^S(q,\tau)d\Omega d\tau, \tag{51}$$

where c=0.5 on the smooth boundary and c=1 in the domain. The notations Γ and Ω represent the boundary and domain, respectively. The notations p and q become P and Q on the boundary. In the case of K-dimensional problems, the time-dependent fundamental solution $T_1^*(p,q,t,\tau)$ in Eq. (51) used for the unsteady temperature analyses and its normal derivative are given by

$$T_1^*(p,q,t,\tau)=\frac{1}{4\pi\kappa(t-\tau)^{K/2}}\exp(-a), \tag{52}$$

$$\frac{\partial T_1^*(p,q,t,\tau)}{\partial n} = \frac{-r}{8\pi\kappa^2(t-\tau)^{K/2+1}} \frac{\partial r}{\partial n} \exp(-a)\,, \tag{53}$$

where

$$a = \frac{r^2}{4\kappa(t-\tau)}\,. \tag{54}$$

Here, r is the distance between the observation point p and the loading point q. As shown in Eq. (51), when an arbitrary pseudo-initial temperature distribution $T^0(q,0)$ exists in the domain, a domain integral is necessary. Therefore, the triple-reciprocity BEM (Ochiai, 2001) is used to avoid internal cells.

This study reveals that the problem of unsteady heat conduction with many time steps can be solved effectively by the triple-reciprocity BEM. Two different numerical procedures can be employed for the numerical solution of Eq. (51). One method requires internal cells. At the end of each time step, the temperature at a sufficient number of internal points must be computed for use as the initial temperature in the next time step. The other method uses the history of boundary values, making internal cells unnecessary, if the initial temperature can be assumed to be 0. However, the CPU time required for calculation increases rapidly with increasing number of time steps. In the presented method, the temperature distributions in some time steps are assumed to be pseudo-initial and are interpolated using integral equations and internal points.

2.7 Interpolation of time-dependent value

Heat generation $W_1^S(q,t)$ is assumed to vary within each time step in accordance with the time interpolation function such thatφηγϕδσαζξχϖβνμ

$$W_1^S(q,t) = \mathbf{W}_1^S\ \phi\ , \tag{55}$$

where ϕ is the time interpolation function. Let us now assume a linear variation of $W_1^S(q,t)$,

$$\psi_1 = \frac{t_f - t}{\Delta t_f}\,,\ \psi_2 = \frac{t - t_{f-1}}{\Delta t_f}\,, \tag{56}$$

where $\Delta t_f = t_f - t_{f-1}$.

The following equations can be used to obtain time-dependent values of heat generation $W_1(q,t_f)$:

$$\nabla^2 W_1^S(q,t_f) = -W_2^S(q,t_f)\,, \tag{57}$$

$$\nabla^2 W_2^S(q,t_f) = -\sum_{m=1}^{M} W_3^P(q_m,t_f)\,. \tag{58}$$

An interpolation method for the pseudo-initial temperature distribution using the boundary integral equations that avoids the use of internal cells is next shown. The pseudo-initial temperature $T^0(q,0)$ in Eq. (51) is represented as $T^{0S}(q,0)$.
The following equations can be used for the two-dimensional interpolation (Ochiai, 2001):

$$\nabla^2 T_1^{0S}(q,0) = -T_2^{0S}(q,0)\,, \tag{59}$$

$$\nabla^2 T_2^{0S}(q,0) = -\sum_{m=1}^{M} T_3^{0P}(q_m,0)\,. \tag{60}$$

The term T_2^{0S} in Eq. (59) corresponds to the sum of the curvatures $\partial^2 T_1^{0S}/\partial x^2$ and $\partial^2 T_1^{0S}/\partial y^2$. The term T_2^{0S} is the unknown strength of a Dirac-type function. From Eqs. (59) and (60), the following equation can be obtained.

$$\nabla^4 T_1^{0S}(q,0) = \sum_{m=1}^{M} T_3^{0P}(q,0) \tag{61}$$

In this study, the deformation of an imaginary thin plate is utilized to interpolate the two-dimensional distribution T_1^{0S}. The deformation T_1^{0S} is given, but the force of the point load T_3^{0P} is unknown. T_3^{0P} is obtained inversely from the deformation T_1^{0S} of the fictitious thin plate, as shown in Fig. 3. T_2^{0S} corresponds to the moment of the thin plate. The moment T_2^{0S} on the boundary is assumed to be 0, which is the same as that in the natural spline. This indicates that the fictitious thin plate is simply supported.
Using Green's second identity and Eqs. (59), (60) and (15), we obtain (Ochiai, 2001)

$$cT_1^{0S}(P,0) = \sum_{f=1}^{2}(-1)^{f+1}\int_\Gamma \{T_f(P,Q)\frac{\partial T_f^{0S}(Q,0)}{\partial n}$$

$$-\frac{\partial T_f(P,Q)}{\partial n}T_f^{0S}(Q,0)\}d\Gamma - \sum_{m=1}^{M} T_2(P,q_m)T_3^{0P}(q_m,0)\,, \tag{62}$$

where M is the number of T_3^{0P}. Moreover, T_2^{0S} in Eq. (60) is similarly given by

$$cT_2^{0S}(P,0) = \int_\Gamma \{T_1(P,Q)\frac{\partial T_2^{0S}(Q,0)}{\partial n} - \frac{\partial T_1(P,Q)}{\partial n}T_2^{0S}(Q,0)\}d\Gamma + \sum_{m=1}^{M} T_1(P,q_m)T_3^{0P}(q_m,0)\,. \tag{63}$$

The integral equations (62) and (63) are used to interpolate the pseudo-initial temperature distribution T_1^{0S}. On the other hand, the polyharmonic function $T_f^*(p,q,t,\tau)$ in the unsteady heat conduction problem is defined by

$$\nabla^2 T^*_{f+1}(p,q,t,\tau) = T^*_f(p,q,t,\tau) \,. \tag{64}$$

Using Green's theorem twice and Eqs. (54)- (57) and (61), Eq. (51) becomes

$$T_3^{0P} \; T^*_{f+1}(p,q,t,\tau) = \int \frac{1}{r}\Big[\int r T^*_f(p,q,t,\tau)dr\Big]dr$$

$$+\frac{\kappa}{\lambda}\sum_{f=1}^{2}(-1)^f \int_0^t \int_\Gamma [T^*_{f+1}(P,Q,t,\tau)\frac{\partial W_f^S(Q,\tau)}{\partial n} - \frac{\partial T^*_{f+1}(P,Q,t,\tau)}{\partial n} W_f^S(Q,\tau)]d\Gamma d\tau$$

$$+\frac{\kappa}{\lambda}\sum_{m=1}^{M}\int_0^t W^P_{3(m)}(q,\tau)T_3^{*A}(P,q,t,\tau)d\tau + \sum_{f=1}^{2}(-1)^f \int_\Gamma [T^*_{f+1}(P,Q,t,0)\frac{\partial T_f^{0S}(Q,0)}{\partial n}$$

$$-\frac{\partial T^*_{f+1}(P,Q,t,0)}{\partial n} T_f^{0S}(Q,0)]d\Gamma + \sum_{m=1}^{M} T_3^{0P}(q_m,0)T_3^*(P,q_m,t,0) \,. \tag{65}$$

Therefore, it is clear that temperature analysis without the use of a domain integral is possible, provided that the initial temperature T_1^0 is interpolated using Eqs. (62) and (63). In practice, T_1^{0S} and $\partial T_2^S / \partial n$ are obtained using results from the previous time step; however, T_2^{0S}, $\partial T_2^S / \partial n$ and T_3^{0P} in Eq. (65) are not obtained in this way.

2.8 Polyharmonic function for unsteady state

The two-dimensional polyharmonic function $T^*_f(p,q,t,\tau)$ in Eq. (65) is determined as

$$T^*_{f+1}(p,q,t,\tau) = \int \frac{1}{r}\Big[\int r T^*_f(p,q,t,\tau)dr\Big]dr \,. \tag{66}$$

$T^*_f(p,q,t,\tau)$ in the unsteady state and its normal derivative are concretely given by

$$T_2^*(p,q,t,\tau) = \frac{1}{4\pi}\big[E_1(a) + \ln(a) + C\big], \tag{67}$$

$$\frac{\partial T_2^*(p,q,t,\tau)}{\partial n} = \frac{1}{2\pi r}\frac{\partial r}{\partial n}[1-\exp(-a)] \,, \tag{68}$$

$$T_3^*(p,q,t,\tau) = \frac{r^2}{16\pi}[E_1(a) + \ln(a) + C + \frac{1-\exp(-a)}{a} + \frac{E_1(a)+\ln(a)+C}{a} - 2], \tag{69}$$

$$\frac{\partial T_3^*(p,q,t,\tau)}{\partial n} = \frac{r}{8\pi}\frac{\partial r}{\partial n}[E_1(a) + \ln(a) + C - 1 + \frac{1-\exp(-a)}{a}], \tag{70}$$

where E 1 () is the exponential integral function and C is Euler's constant.

Numerical solutions are obtained using the interpolation functions for time and space. If a constant time interpolation and time step $(t_k - t_{k-1})$ are used, the time integral can be treated analytically. The time integrals for $T_f^*(p,q,t,\tau)$ are given as follows:

$$\int_{t_f}^{t_F} T_1^*(p,q,t,\tau)d\tau = \frac{1}{4\kappa\pi} E_1(a_f), \tag{71}$$

$$\int_{t_f}^{t_F} \frac{\partial T_1^*(p,q,t,\tau)}{\partial n} d\tau = \frac{1}{2\kappa\pi r} \frac{\partial r}{\partial n} \exp(-a_f), \tag{72}$$

where

$$a_f = \frac{r^2}{4\kappa(t_F - t_f)}. \tag{73}$$

Assuming that functions $T(Q,\tau)$ and $\partial T(Q,\tau)/\partial n$ remain constant over time in each time step, Eq. (65) can be written in matrix form. Replacing $T(Q,\tau)$ and $\partial T(Q,\tau)/\partial n$ with vectors **T** f and **Q** f , respectively, and discretizing Eq. (65), we obtain (Brebbia ,1984)

$$\sum_{f=1}^{F} \mathbf{H}_{\mathbf{fF}} \mathbf{T}_{\mathbf{f}} = \sum_{f=1}^{F} \mathbf{G}_{\mathbf{fF}} \mathbf{Q}_{\mathbf{f}} + \mathbf{B}_{\mathbf{0}} , \tag{74}$$

where $\mathbf{B}_0$ represents the effect of the pseudo-initial temperature. Adopting a constant time step throughout the analysis, the coefficients of the matrix at several time steps need to be computed and stored only once.

If there is heat generation, the following time integrals are used (Ochiai, 2001).

$$\int_{t_f}^{t_F} T_2^*(p,q,t,\tau)d\tau = \frac{r^2}{16\kappa\pi}\{E_1(a_f) + \frac{1}{a_f}[E_1(a_f) + \ln(a_f) + C + 1 - \exp(-a_f)]\} \tag{75}$$

$$\int_{t_f}^{t_F} \frac{\partial T_2^*(p,q,t,\tau)}{\partial n} d\tau = \frac{r}{8\kappa\pi} \frac{\partial r}{\partial n}[\frac{1-\exp(a_f)}{a_f} + E_1(a_f)] \tag{76}$$

$$\int_{t_f}^{t_F} T_3^*(p,q,t,\tau)d\tau = \frac{r^4}{256\kappa\pi}\{E_1(a_f) + \frac{1}{a_f}[4E_1(a_f) + 4\ln(a_f) + 4C + 1 - \exp(-a_f)] + \frac{1}{a_f^2}[2E(a_f)$$

$$+2\ln(a_f) + 2C - 2a_f + 3 - 3\exp(-a_f) - 5a_f]\} \tag{77}$$

$$\int_{t_f}^{t_F} \frac{\partial T_3^*(p,q,t,\tau)}{\partial n} d\tau = \frac{r^3}{64\kappa\pi} \frac{\partial r}{\partial n}\{\frac{1-\exp(a_f) - a_f}{a_f^2} + E_1(a_f)$$

$$+\frac{1}{a_f}[2E_1(a_f) + 2\ln(a_f) + 2C + 1 - \exp(a_f)]\} \tag{78}$$

Additionally, the temperature gradient is given by differentiating Equation (65), and expressed as:

$$\frac{\partial T(p,t)}{\partial x_i} = -\kappa\int_0^t\int_\Gamma [T(Q,\tau)\frac{\partial^2 T_1^*(p,Q,t,\tau)}{\partial x_i \partial n} - \frac{\partial T(Q,\tau)}{\partial n}\frac{\partial T_1^*(p,Q,t,\tau)}{\partial x_i}]d\Gamma d\tau$$

$$+\kappa\sum_{f=1}^{2}(-1)^f\int_0^t\int_\Gamma [\frac{\partial T_{f+1}^*(p,Q,t,\tau)}{\partial x_i}\frac{\partial W_f(Q,\tau)}{\partial n} - \frac{\partial^2 T_{f+1}^*(p,Q,t,\tau)}{\partial x_i \partial n}W_f(Q,\tau)]d\Gamma d\tau$$

$$+\kappa\sum_{m=1}^{M}\int_0^t W_3^P(q_m,\tau)\frac{\partial T_3^*(p,q_m,t,\tau)}{\partial x_i}d\tau + \sum_{f=1}^{2}(-1)^f\int_\Gamma [\frac{\partial T_{f+1}^*(p,Q,t,0)}{\partial x_i}\frac{\partial T_f^0(Q,0)}{\partial n}$$

$$-\frac{\partial^2 T_{f+1}^*(p,Q,t,0)}{\partial x_i \partial n}T_f^0(Q,0)]d\Gamma d\tau + \sum_{m=1}^{M}T_3^{0P}(q_m,0)\frac{\partial T_3^*(p,q_m,t,0)}{\partial x_i}d\tau \quad (79)$$

The derivative of the polyharmonic function $T_f^*(P,q,t,\tau)$ and the normal derivative with respect to x_i in Eq.(79) are explicitly given by

$$\frac{\partial T_1^*(p,q,t,\tau)}{\partial x_i} = \frac{-rr,_i}{8\pi\kappa^2(t-\tau)^2}\exp(-a) , \quad (80)$$

$$\frac{\partial^2 T_1^*(p,q,t,\tau)}{\partial x_i \partial n} = \frac{-1}{8\pi\kappa^2(t-\tau)^2}[n_i\exp(-a) + 2ar,_i\frac{\partial r}{\partial n}\exp(-a)] , \quad (81)$$

$$\frac{\partial T_2^*(p,q,t,\tau)}{\partial x_i} = \frac{r,_i}{2\pi r}[1-\exp(-a)] , \quad (82)$$

$$\frac{\partial^2 T_2^*(p,q,t,\tau)}{\partial x_i \partial n} = \frac{1}{2\pi r^2}\{n_i[1-\exp(-a)] - 2r,_i\frac{\partial r}{\partial n}[1-\exp(-a) - a\cdot\exp(-a)]\} , \quad (83)$$

$$\frac{\partial T_3^*(p,q,t,\tau)}{\partial x_i} = \frac{rr,_i}{8\pi}\left\{E_1(a) + \ln(a) + C - 1 + \frac{1-\exp(-a)}{a}\right\} , \quad (84)$$

$$\frac{\partial^2 T_3^*(p,q,t,\tau)}{\partial x_i \partial n} = \frac{1}{8\pi}\{n_i[E_1(a) + \ln(a) + C + 1 - \frac{1-\exp(-a)}{a}] + 2r,_i\frac{\partial r}{\partial n}[1 - \frac{1-\exp(-a)}{a}]\} , \quad (85)$$

where $r,_i = \partial r / \partial x_i$. The time integrals for $\partial T_f^* / \partial x_i$ and $\partial^2 T_f^*(P,q,t,\tau) / \partial x_i \partial n$ in Eq. (79) are given as follows:

$$\int_{t_f}^{t_F}\frac{\partial T_1^*(p,q,t,\tau)}{\partial x_i}d\tau = \frac{-r,_i}{2\kappa\pi r}\exp(-a_f) , \quad (86)$$

$$\int_{t_f}^{t_F} \frac{\partial^2 T_1^*(p,q,t,\tau)}{\partial x_i \partial n} d\tau = \frac{-1}{2\pi\kappa r^2}[n_i - 2r_{,i}\frac{\partial r}{\partial n}(1+a_f)]\exp(-a_f) , \tag{87}$$

$$\int_{t_f}^{t_F} \frac{\partial T_2^*(p,q,t,\tau)}{\partial x_i} d\tau = \frac{r}{8\kappa\pi}\frac{\partial r}{\partial x_i}[\frac{1-\exp(-a_f)}{a_f} + E_1(a_f)] , \tag{88}$$

$$\int_{t_f}^{t_F} \frac{\partial^2 T_2^*(p,q,t,\tau)}{\partial x_i \partial n} d\tau = \frac{1}{8\kappa\pi}\left\langle n_i\{E_1(a_f) + \frac{1}{a_f}[1-\exp(-a_f)]\} - 2r_{,i}\frac{\partial r}{\partial n}\frac{1}{a_f}[1-\exp(-a_f)]\right\rangle , \tag{89}$$

$$\int_{t_f}^{t_F} \frac{\partial T_3^*(p,q,t,\tau)}{\partial x_i} d\tau = \frac{r^3 r_{,i}}{64\kappa\pi}\{\frac{1-\exp(-a_f)-a_f}{a_f^2} + E_1(a_f) + \frac{1}{a_f}[2E_1(a_f) + 2\ln(a_f) + 2C + 1 - \exp(-a_f)]\} , \tag{90}$$

$$\int_{t_f}^{t_F} \frac{\partial^2 T_3^*(p,q,t,\tau)}{\partial x_i \partial n} d\tau = \frac{r^2}{64\kappa\pi}\left\langle n_i\{E_1(a_f) + \frac{2}{a_f}[E_1(a_f) + \ln(a_f) + C] + \frac{1}{a_f{}^2}[1-\exp(-a_f) - a_f\exp(-a_f)]\} + 2r_{,i}\frac{\partial r}{\partial n}\{E_1(a_f) + \frac{2}{a_f}[1-\exp(-a_f)] - \frac{1}{a_f{}^2}[1 - a_f - a_f\exp(-a_f)]\}\right\rangle . \tag{91}$$

3. Numerical examples

To verify the accuracy of the present method, unsteady heat conduction in a circular region with radius *a*, as shown in Fig. 6, is treated with a boundary temperature given by

$$T = T_H[1-\cos(\omega t)] . \tag{92}$$

We assume an initial temperature $T0 = 0\,^\circ\mathrm{C}$, and R denotes the distance from the center of the circular region. A two-dimensional state, in which there is no heat flow in the direction perpendicular to the plane of the domain, is assumed. Figure 6 also shows the internal points used for interpolation. A thermal diffusivity of $\kappa = 16$ mm $2/\mathrm{s}$ and a radius of a=10 mm are assumed. $T_H = 10\,^\circ\mathrm{C}$ in Eq. (92) and a frequency of $\omega = \pi/2$ rad/s are also assumed. The BEM results at R=0 and R=8 mm and the exact values are compared in Fig. 7. The exact solution for the temperature distribution is given by (Carslaw, 1938)

$$T(R,t) = T_H[1 - \frac{ber\sqrt{\frac{\omega}{\kappa}}R\,ber\sqrt{\frac{\omega}{\kappa}}a + bei\sqrt{\frac{\omega}{\kappa}}R\,bei\sqrt{\frac{\omega}{\kappa}}a}{ber^2\sqrt{\frac{\omega}{\kappa}}a + bei^2\sqrt{\frac{\omega}{\kappa}}a}\cos\omega t$$

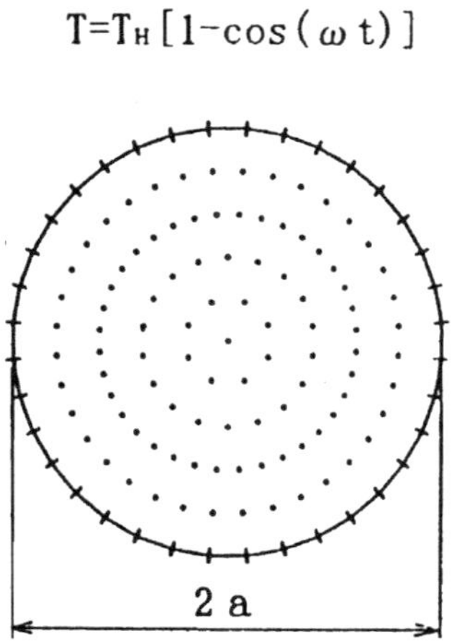

Fig. 6. Circular region with temperature change at the boundary.

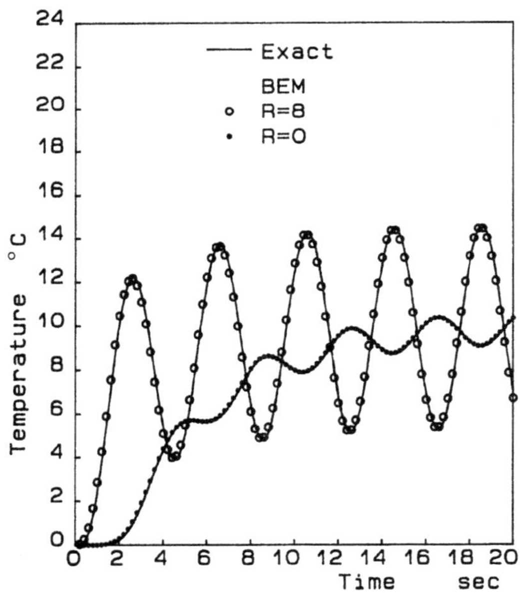

Fig. 7. Temperature history in circular region.

$$+\frac{ber\sqrt{\frac{\omega}{\kappa}}R\,bei\sqrt{\frac{\omega}{\kappa}}a - ber\sqrt{\frac{\omega}{\kappa}}a\,bei\sqrt{\frac{\omega}{\kappa}}R}{ber^2\sqrt{\frac{\omega}{\kappa}}a + bei^2\sqrt{\frac{\omega}{\kappa}}a}\sin\omega t - \frac{2\kappa^2}{a}\sum_{s=1}^{\infty}\exp(-\kappa\alpha_s^2 t)\frac{J_0(\alpha_s R)\alpha_s^3}{J_0'(\alpha_s a)(\kappa^2\alpha_s^4+\omega^2)}] \qquad (93)$$

where *ber*( ) and *bei*( ) are Kelvin functions, and α_s (s=1, 2, ...) are the roots of $J_0(a\alpha)=0$. Constant elements are used for boundary and time interpolation.

Appendix A (3D)

The higher-order functions for 3D unsteady heat conduction are

$$T_2^*(p,q,t,\tau) = \frac{1}{2\pi^{3/2} r}\{-\gamma(1.5,a) + a^{1/2}[1-\exp(-a)]\} = \frac{-1}{2\pi^{3/2} r}\gamma(0.5,a) \quad \text{(A-1)}$$

$$\frac{\partial T_2^*(p,q,t,\tau)}{\partial n} = \frac{1}{2\pi^{3/2} r^2}\gamma(1.5,a)\frac{\partial r}{\partial n} \quad \text{(A-2)}$$

$$T_3^* = \frac{r}{12\pi^{3/2}}\{-3\gamma(1.5,a) + 6a^{-1/2}\gamma(2,a) - 3a^{-1}\gamma(2.5,a) + a^{1/2} + 3\gamma(1.5,a)\frac{1}{a} - 3a^{-1/2}[1-\exp(-a)]\}$$

$$= \frac{r}{4\pi^{3/2}}\{-\gamma(0.5,a) + 2\gamma(1.5,a)\frac{1}{a} - 2a^{-1/2}[1-\exp(-a)]\} \quad \text{(A-3)}$$

$$\frac{\partial T_3^*}{\partial n} = \frac{1}{4\pi^{3/2}}[\gamma(0.5,a) - \frac{1}{a}\gamma(1.5,a)]\frac{\partial r}{\partial n}. \quad \text{(A-4)}$$

where $\gamma(,)$ is an incomplete gamma function of the first kind (Abramowitz, 1970) and $r_{,i} = \partial r / \partial x_i$. Using Eqs. (44) and (A-3), the polyharmonic function with a surface distribution is obtained as follows:

$$T_{3B}^* = \frac{2A(kt)^{3/2}}{3\pi^{1/2} r}\{-2u_2^{3/2}\gamma(1.5,u_2) + 2u_1^{3/2}\gamma(1.5,u_1) + 2\gamma(3,u_2) - 2\gamma(3,u_1) + 6u_2\gamma(2,u_2) - 6u_1\gamma(2,u_1)$$
$$-6u_2^{1/2}\gamma(2.5,u_2) + 6u_1^{1/2}\gamma(2.5,u_1) + \frac{1}{2}u_2^2 - \frac{1}{2}u_1^2$$
$$+6u_2^{1/2}\gamma(1.5,u_2) - 6u_1^{1/2}\gamma(1.5,u_1) - 6\gamma(2,u_2) + 6\gamma(2,u_1) - 3u_2 + 3u_1 - 3\exp(-u_2) + 3\exp(-u_1)\}, \quad \text{(A-5)}$$

where

$$u_1 = \frac{(r-A)^2}{4\kappa(t-\tau)} \quad \text{(A-6)}$$

$$u_2 = \frac{(r+A)^2}{4\kappa(t-\tau)}. \quad \text{(A-7)}$$

The time integral of Eq. (62) can be obtained as follows:

$$\int_{t_f}^{t_F} T_1^*(p,q,t,\tau)d\tau = \frac{1}{4\kappa\pi^{3/2} r}\Gamma(0.5,a_f) \quad \text{(A-8)}$$

$$\int_{t_f}^{t_F} \frac{\partial T_1^*(p,q,t,\tau)}{\partial n}d\tau = \frac{1}{2\kappa\pi^{3/2} r^2}\frac{\partial r}{\partial n}\Gamma(1.5,a_f) \quad \text{(A-9)}$$

$$\int_{t_f}^{t_F} T_2^*(p,q,t,\tau)d\tau = \frac{r}{8\kappa\pi^{3/2}}[\gamma(1.5,a_f)\frac{-1}{a_f} - \Gamma(0.5,a_f) + \frac{2}{a_f^{1/2}} - \Gamma(-0.5,a_f)]$$

$$= \frac{r}{8\kappa\pi^{3/2}}[\gamma(0.5,a_f)\ \frac{1}{a_f} - \Gamma(-0.5,a_f)] \tag{A-10}$$

$$\int_{t_f}^{t_F} \frac{\partial T_2^*(p,q,t,\tau)}{\partial n} d\tau = \frac{1}{8\kappa\pi^{3/2}} \frac{\partial r}{\partial n}[\gamma(1.5,a_f)\frac{1}{a_f} + \Gamma(0.5,a_f)] \tag{A-11}$$

$$\int_{t_f}^{t_F} T_3^*(p,q,t,\tau)d\tau = \frac{r^3}{96\kappa\pi^{3/2}}[-6\gamma\ (1.5,a_f)\frac{1}{a_f} - \Gamma(0.5,a_f)\ + 8\gamma\ (2,a_f)\frac{1}{a_f^{3/2}}\ - 3\gamma(2.5,a_f)\ \frac{1}{a_f^{2}}$$
$$+ \frac{4}{a_f^{1/2}}\ + 3\gamma(1.5,a_f)\frac{1}{a_f^{2}} + 3\Gamma(-0.5,a_f)\ - \frac{4}{a_f^{3/2}} + 6\Gamma(-1.5,a_f)] \tag{A-12}$$

$$\int_{t_f}^{t_F} \frac{\partial T_3^*(p,q,t,\tau)}{\partial n} d\tau\ = \frac{r^2}{96\kappa\pi^{3/2}} \frac{\partial r}{\partial n}[-6\gamma\ (1.5,a_f)\frac{1}{a_f} - 3\Gamma(0.5,a_f)\ + 3\gamma\ (2.5,a_f)\frac{1}{a_f^{2}}\ + \frac{8}{a_f^{1/2}}$$
$$-\gamma(1.5,a_f)\frac{3}{a_f^{2}} - 3\Gamma(-0.5,a_f)]\ , \tag{A-13}$$

where

$$a_f = \frac{r^2}{4\kappa(t_F - t_f)} \tag{A-14}$$

and $\Gamma(,)$ is an incomplete gamma function of the second kind (Abramowitz, 1970). The time integral of Eq. (A-5) can be obtained as follows:

$$\int_{t_f}^{t_F} T_{3B}^*(p,q,t,\tau)d\tau\ = \frac{A}{48\pi^{1/2}r} \frac{(r-A)^5}{\kappa}\{2\gamma(1.5,a_{1f})\frac{1}{a_f} + \frac{1}{5}\Gamma(0.5,a_{1f})\ - \frac{4}{5}\gamma(3,a_{1f})\ \frac{1}{a_{1f}^{5/2}}$$
$$-4\gamma(2,a_{1f})\ \frac{1}{a_{1f}^{3/2}}\ + 3\gamma(2.5,a_{1f})\ \frac{1}{a_{1f}^{2}}\ - \frac{1}{a_{1f}^{1/2}}\ - 3\gamma(1.5,a_{1f})\ \frac{1}{a_{1f}^{2}} - \frac{3}{5}\Gamma(-0.5,a_{1f})$$
$$+ \frac{12}{5}\gamma(2,a_{1f})\ \frac{1}{a_{1f}^{5/2}}\ + \frac{2}{a_{1f}^{3/2}}\ + 3\Gamma(-2.5,a_{1f})\}\ , \tag{A-15}$$

where

$$a_{1f} = \frac{(r-A)^2}{4\kappa(t_F - t_f)}\ . \tag{A-16}$$

For the sake of conciseness, the terms involving u_2 in Eq. (A-5) are omitted. The derivative of the polyharmonic function $T_f^*(P,q,t,\tau)$ and the normal derivative with respect to x_i are explicitly given by

$$\frac{\partial T_1^*}{\partial x_i} = \frac{-r}{16\pi^{3/2}[k(t-\tau)]^{5/2}} \exp(-a)\frac{\partial r}{\partial x_i} \tag{A-17}$$

$$\frac{\partial}{\partial n}\frac{\partial T_1^*}{\partial x_i}=\frac{1}{16\pi^{3/2}[k(t-\tau)]^{5/2}}\left(-n_i+2ur,_j n_j r,_i\right)\exp(-a) \tag{A-18}$$

$$\frac{\partial T_2^*}{\partial x_i}=\frac{1}{2\pi^{3/2}r^2}\gamma\left(\frac{3}{2},a\right)\frac{\partial r}{\partial x_i} \tag{A-19}$$

$$\frac{\partial}{\partial n}\frac{\partial T_2^*}{\partial x_i}=\frac{1}{2\pi^{3/2}r^3}\left[n_i\gamma\left(\frac{3}{2},a\right)-2\gamma\left(\frac{5}{2},a\right)r,_i n_j r,_j\right] \tag{A-20}$$

$$\frac{\partial T_3^*}{\partial x_i}=\frac{-1}{8\pi^{3/2}}\left\{\gamma\left(\frac{1}{2},a\right)-\frac{1}{u}\gamma\left(\frac{3}{2},a\right)\right\}\frac{\partial r}{\partial x_i} \tag{A-21}$$

$$\frac{\partial^2 T_3^*}{\partial n\partial x_i}=\frac{1}{8\pi^{3/2}r}\{n_i[-\gamma(0.5,a)+\frac{1}{u}\gamma(1.5,a)]+r,_i r,_j n_j[\gamma(0.5,a)-\frac{3}{u}\gamma(1.5,a)]\} \tag{A-22}$$

$$\frac{dT_{3B}^*}{dx_i}=-\frac{2a(kt)^{3/2}}{3\pi^{1/2}r}\frac{\partial r}{\partial x_i}[\frac{1}{r}\{2u_1^{3/2}\gamma(1.5,u_1)-2\gamma(3,u_1)-6u_1\gamma(2,u_1)+6u_1^{1/2}\gamma(2.5,u_1)$$
$$-\frac{1}{2}u_1^2-6u_1^{1/2}\gamma(1.5,u_1)+6\gamma(2,u_1)+3u_1+3\exp(-u_1)\}-\frac{2}{r_1}\{3u_1^{1/2}\gamma(1.5,u_1)$$
$$-6\gamma(2,u_1)+3u_1^{-3/2}\gamma(2.5,u_1)-1-3u_1^{-3/2}\gamma(1.5,u_1)+3u_1^{-1}-3u_1^{-1}\exp(-u_1)\}] \tag{A-23}$$

The time integrals of Eqs. (A-18), (A-20) and (A-22) can be obtained as follows:

$$\int_{t_f}^{t_F}\frac{\partial}{\partial n}\frac{\partial T_1^*}{\partial x_i}d\tau=\frac{1}{2k\pi^{3/2}r^3}\left\{2r,_i n_j r,_j\Gamma\left(\frac{5}{2},a_f\right)-n_i\Gamma\left(\frac{3}{2},a_f\right)\right\} \tag{A-24}$$

$$\int_{t_f}^{t_F}\frac{\partial}{\partial n}\frac{\partial T_2^*}{\partial x_i}d\tau=\frac{1}{8\pi^{3/2}kr}\left\{(n_i-3r,_i n_j r,_j)\frac{1}{a_f}\gamma\left(\frac{3}{2},a_f\right)+(n_i-r,_i n_j r,_j)[\Gamma\left(\frac{1}{2}\right)-\gamma\left(\frac{1}{2},a_f\right)]\right\} \tag{A-25}$$

$$\int_{t_f}^{t_F}\frac{\partial^2 T_3^*(p,q,t,\tau)}{\partial n\partial x_i}d\tau=\frac{r}{192\pi^{3/2}\kappa}\left\{n_i\left[3a_f^{-2}\gamma\left(\frac{3}{2},a_f\right)-6a_f^{-1}\gamma\left(\frac{1}{2},a_f\right)-3\Gamma\left(-\frac{1}{2},a_f\right)+16a_f^{-1/2}\right]\right.$$
$$\left.+r,_i r,_j n_j\left[-9a_f^{-2}\gamma\left(\frac{3}{2},a_f\right)+6a_f^{-1}\gamma\left(\frac{1}{2},a_f\right)-3\Gamma\left(-\frac{1}{2},a_f\right)\right]\right\} \tag{A-26}$$

$$\int_{t_f}^{t_F}\frac{\partial T_{3B}^*}{\partial x_i}d\tau=\frac{2a}{3\pi^{1/2}}\frac{\partial r}{\partial x_i}\frac{r^4}{32k}\{2\gamma\left(1.5,a_f\right)\frac{1}{a_f}+\frac{1}{5}\Gamma(0.5,a_f)-\frac{4}{5}\gamma\left(3,a_f\right)\frac{1}{a_f^{5/2}}$$
$$-4\gamma\left(2,a_f\right)\frac{1}{a_f^{3/2}}+3\gamma\left(2.5,a_f\right)\frac{1}{a_f^2}-\frac{1}{a_f^{1/2}}$$

$$-3\gamma\left(1.5,a_f\right)\frac{1}{a_f^{\;2}}-\frac{3}{5}\Gamma\left(-0.5,a_f\right)+\frac{12}{5}\gamma\left(2,a_f\right)\frac{1}{a_f^{\;5/2}}+\frac{2}{a_f^{\;3/2}}+3\ \Gamma(-2.5,a_f)\}\quad\text{(A-27)}$$

Appendix B (1D)

The functions for 1D unsteady heat conduction are

$$T_2^*(p,q,t,\tau)=\frac{r}{2\pi^{1/2}}\left[\gamma(a,0.5)+a^{-1/2}\exp(-a)\right]\quad\text{(B-1)}$$

$$\frac{\partial T_2^*(p,q,t,\tau)}{\partial n}=\frac{1}{2\pi^{1/2}}\gamma(0.5,a)\frac{\partial r}{\partial n}\quad\text{(B-2)}$$

$$T_3^*(p,q,t,\tau)=\frac{r^3}{12\pi^{1/2}}\{(a+1.5)a^{-1}\gamma(0.5,a)+a^{-1/2}\exp(-a)+2a^{-3/2}+a^{-3/2}\exp(-a)\}\quad\text{(B-3)}$$

$$\frac{\partial T_3^*(p,q,t,\tau)}{\partial n}=\frac{\partial r}{\partial n}\frac{r^2}{4\pi^{1/2}}[(1+a^{-1})\gamma(0.5,a)-a^{-1}\gamma(1.5,a)],\quad\text{(B-4)}$$

where $\gamma(,)$ is an incomplete gamma function of the first kind (Abramowitz, 1970). The time integral of Eqs. (49) and (B-1)-(B-4) can be obtained as follows:

$$\int_{t_f}^{t_F}T_1(p,q,t,\tau)d\tau=\frac{-r}{2\kappa\pi^{1/2}}[\Gamma(0.5,a_f)+\frac{1-\exp(-a_f)}{a_f^{\;0.5}}]\quad\text{(B-5)}$$

$$\int_{t_f}^{t_F}\frac{\partial T_1^*(p,q,t,\tau)}{\partial n}d\tau=\frac{\partial r}{\partial n}\frac{-1}{2\kappa\pi^{1/2}}\Gamma(0.5,a_f)\quad\text{(B-6)}$$

$$\int_{t_f}^{t_F}T_2^*(p,q,t,\tau)d\tau=\frac{r^3}{8\kappa\pi^{1/2}}\left\{\frac{1}{a_f}\gamma(0.5,a_f)-\frac{2}{3}\Gamma(0.5,a_f)+\frac{2}{3}\frac{(a_f^{1/2}+a_f^{3/2})}{a_f^2}\exp(-a_f)\right\}\quad\text{(B-7)}$$

$$\int_{t_f}^{t_F}\frac{\partial T_2^*(p,q,t,\tau)}{\partial n}d\tau=\frac{\partial r}{\partial n}\frac{r^2}{8\kappa\pi^{1/2}}\{-\frac{1}{a_f}\gamma(0.5,a_f)-2[\Gamma(0.5,a_f)-\frac{1}{a_f^{1/2}}\exp(-a_f)]\}\quad\text{(B-8)}$$

$$\begin{aligned}\int_{t_f}^{t_F}T_3^*(p,q,t,\tau)d\tau=\frac{r^5}{2880\kappa\pi^{1/2}}\{&15(\frac{3}{a_f^2}+\frac{4}{a_f})\gamma(0.5,a_f)-12\Gamma(0.5,a_f)\\&+12\gamma(0.5,a_f)+6(\frac{2}{a_f^{1/2}}+\frac{9}{a_f^{3/2}}+\frac{4}{a_f^{5/2}})\exp(-a_f)+\frac{48}{a_f^{\;5/2}}\}\end{aligned}\quad\text{(B-9)}$$

$$\int_{t_f}^{t_F}\frac{\partial T_3^*(p,q,t,\tau)}{\partial n}d\tau=\frac{\partial r}{\partial n}\frac{r^4}{16\kappa\pi^{1/2}}\{(\frac{1}{a_f}-\frac{1}{2a_f^{\;2}})\gamma(0.5,a_f)+\frac{1}{3}\Gamma(0.5,a_f)-\frac{\exp(-a_f)}{3a_f^{\;1/2}}+\frac{2\exp(-a_f)}{3a_f^{\;3/2}}\},\quad\text{(B-10)}$$

where

$$a_f = \frac{r^2}{4\kappa(t_F - t_f)}. \tag{B-11}$$

Appendix C (Linear time interpolation)

The time integrals of Eq. (62) using linear time interpolation in the two-dimensional case can be obtained as follows:

$$\int_{t_{f-1}}^{t_f} (t_f - \tau) T_1^* d\tau = \frac{1}{4\pi\kappa}\left[\frac{r^2}{4\kappa}\left\{\frac{\exp(a_{f-1})}{a_{f-1}} - \frac{\exp(a_f)}{a_f}\right\} - \left\{(t_F - t_f) + \frac{r^2}{4\kappa}\right\}\left\{E_1(a_{f-1}) - E_1(a_f)\right\}\right] \tag{C-1}$$

$$\int_{t_{f-1}}^{t_f} (\tau - t_{f-1}) T_1^* d\tau = \frac{1}{4\pi\kappa}\left[\frac{r^2}{4\kappa}\left\{\frac{\exp(a_{f-1})}{a_{f-1}} - \frac{\exp(a_f)}{a_f}\right\} - \left\{(t_F - t_{f-1}) + \frac{r^2}{4\kappa}\right\}\left\{E_1(a_{f-1}) - E_1(a_f)\right\}\right] \tag{C-2}$$

$$\int_{t_{f-1}}^{t_f} (t_f - \tau) \frac{\partial T_1^*}{\partial n} d\tau = \frac{1}{2\pi\kappa}\frac{\partial r}{\partial n}\left\{(t_F - t_f)\frac{1}{R^2}\left[\exp(-a_{f-1}) - \exp(-a_f)\right] - \frac{1}{4\kappa}\left[E_1(a_{f-1}) - E_1(a_f)\right]\right\} \tag{C-3}$$

$$\int_{t_{f-1}}^{t_f} (\tau - t_{f-1}) \frac{\partial T_1^*}{\partial n} d\tau$$
$$= \frac{1}{2\pi\kappa}\frac{\partial r}{\partial n}\left\{(t_F - t_{f-1})\frac{1}{R^2}\left[\exp(-a_{f-1}) - \exp(-a_f)\right] - \frac{1}{4\kappa}\left[E_1(a_{f-1}) - E_1(a_f)\right]\right\} \tag{C-4}$$

$$\int_{t_{f-1}}^{t_f} (t_f - \tau) T_2^*(p,q,t,\tau) d\tau = \frac{r^2}{256\kappa\pi}\left\langle 16(t_F - t_f)\left[E_1(a_f) + \frac{E_1(a_f) + \ln(a_f) + C + 1 - \exp(-a_f)}{a_f}\right]\right.$$
$$- \frac{r^2}{\kappa}\left\{E_1(a_f) - \frac{1 - \exp(-a_f)}{a_f} - \frac{2E_1(a_f) + 2\ln(a_f) + 2C + 1 - \exp(-a_f) + a_f}{a_f^2}\right\}$$
$$\left. + \frac{r^2}{\kappa}\left\{E_1(a_{f-1}) - \frac{1 - \exp(-a_{f-1})}{a_{f-1}} - \frac{2E_1(a_{f-1}) + 2\ln(a_{f-1}) + 2C + 1 - \exp(-a_{f-1}) + a_{f-1}}{a_{f-1}^2}\right\}\right\rangle \tag{C-5}$$

$$\int_{t_{f-1}}^{t_f} (\tau - t_{f-1}) T_2^*(p,q,t,\tau) d\tau = \frac{r^2}{256\kappa\pi}\left\langle 16(t_F - t_{f-1})\left[E_1(a_f) + \frac{E_1(a_f) + \ln(a_f) + C + 1 - \exp(-a_f)}{a_f}\right]\right.$$
$$- \frac{r^2}{\kappa}\left\{E_1(a_f) - \frac{1 - \exp(-a_f)}{a_f} - \frac{2E_1(a_f) + 2\ln(a_f) + 2C + 1 - \exp(-a_f) + a_f}{a_f^2}\right\}$$

$$+\frac{r^2}{\kappa}\left\{E_1(a_{f-1})-\frac{1-\exp(-a_{f-1})}{a_{f-1}}-\frac{2E_1(a_{f-1})+2\ln(a_{f-1})+2C+1-\exp(-a_{f-1})+a_{f-1}}{a_{f-1}^2}\right\}\Bigg\rangle \quad \text{(C-6)}$$

$$\int_{t_{f-1}}^{t_f}(t_f-\tau)\frac{\partial T_2^*}{\partial n}d\tau=\frac{r}{8\kappa\pi}\frac{\partial r}{\partial n}\Bigg\langle -(t-t_f)\left[E_1(a_{f-1})+\frac{1-\exp(-a_{f-1})}{a_{f-1}}\right]+(t-t_f)\left[E_1(a_f)+\frac{1-\exp(-a_f)}{a_f}\right]$$

$$+\frac{2\kappa(t-t_{f-1})^2}{r^2}\{1-\exp(-a_{f-1})+a_{f-1}\exp(-a_{f-1})-a_{f-1}^2\exp(-a_{f-1})\}$$
$$-\frac{2\kappa(t-t_f)^2}{r^2}\{1-\exp(-a_f)+a_f\exp(-a_f)-a_f^2\exp(-a_f)\}\ \Bigg\rangle \quad \text{(C-7)}$$

$$\int_{t_{f-1}}^{t_f}(t_f-\tau)T_3^*(p,q,t,\tau)d\tau=\frac{r^4}{9216\kappa\pi}\{36(t_F-t_f)\Bigg\langle\Bigg[E_1\left(a_f\right)+\frac{1}{a_f}\left\{4E_1(a_f)+4\ln(a_f)+4C-\exp(-a_f)+1\right\}$$

$$+\frac{2E_1(a_f)+2\ln(a_f)+2C-3\exp(-a_f)+3-5a_f}{{a_f}^2}\Bigg]$$

$$-\Bigg[E_1(a_{f-1})+\frac{1}{a_{f-1}}\left\{4E_1(a_{f-1})+4\ln(a_{f-1})+4C+1-\exp\left(-a_{f-1}\right)\right\}$$

$$+\frac{2E_1(a_f)+2\ln(a_f)+2C-3\exp(-a_f)+3-5a_f}{{a_{f-1}}^2}\Bigg]\Bigg\rangle$$

$$+\frac{r^2}{\kappa}\Bigg[\Bigg\{-E_1(a_{f-1})-\frac{1-\exp(-a_{f-1})}{a_f}+\frac{18E_1(a_{f-1})+18\ln(a_{f-1})+18C+a_{f-1}+1-\exp(-a_{f-1})}{a_{f-1}^2}$$

$$+\frac{12E_1(a_{f-1})+12\ln(a_{f-1})+12C+16-16\exp(-a_{f-1})-28a_{f-1}+11{a_{f-1}}^2}{a_{f-1}^3}\Bigg\}$$

$$-\Bigg\{-E_1(a_f)-\frac{9-\exp(-a_f)}{a_f}+\frac{18E_1(a_f)+18\ln(a_f)+18C+9a_f-\exp(-a_f)-27}{a_f^2}$$

$$+\frac{12E_1(a_f)+12\ln(a_f)+12C+16-16\exp(-a_f)}{a_f^3}\Bigg\}\Bigg]\} \quad \text{(C-8)}$$

$$\int_{t_{f-1}}^{t_f}(t_f-\tau)\frac{\partial T_3^*}{\partial n}d\tau=\frac{r^5}{1536\kappa^2\pi}\frac{\partial r}{\partial n}\left\{\left[-E_1(a_{f-1})+\frac{\exp(-a_{f-1})}{a_{f-1}}+6\frac{E_1(a_{f-1})+\ln(a_{f-1})+C}{a_{f-1}^2}\right.\right.$$

$$\left.+\frac{1-\exp(-a_{f-1})}{a_{f-1}^2}+4\frac{1-\exp(-a_{f-1})-a_{f-1}}{a_{f-1}^3}\right]$$

$$-\frac{24\kappa(t_F-t_f)}{r^2}\left[E_1(a_{f-1})+2\frac{E_1(a_{f-1})+\ln(a_{f-1})+C}{a_{f-1}}+\frac{1-\exp(-a_{f-1})}{a_{f-1}}+\frac{1-\exp(-a_{f-1})-a_{f-1}}{a_{f-1}^2}\right]$$

$$-\left[-E_1(a_f)+\frac{\exp(-a_f)}{a_f}+6\frac{E_1(a_f)+\ln(a_f)+C}{a_f^2}+\frac{1-\exp(-a_f)}{a_f^2}+4\frac{1-\exp(-a_f)-a_f}{a_f^3}\right]$$

$$\left.+\frac{24\kappa(t_F-t_f)}{r^2}\left[E_1(a_f)+2\frac{E_1(a_f)+\ln(a_f)+C}{a_f}+\frac{1-\exp(-a_f)}{a_f}+\frac{1-\exp(-a_f)-a_f}{a_f^2}\right]\right\}. \quad \text{(C-9)}$$

4. References

Abramowitz, M. and Stegun, A. Eds., (1970), *Handbook of Mathematical Functions*, pp. 255-263, Dover, New York.

Brebbia, C. A., Tells, J. C. F. and Wrobel, L. C., (1984), *Boundary Element Techniques-Theory and Applications in Engineering*, pp. 47-107, Springer-Verlag, Berlin.

Carslaw, H. S. and Jaeger, J. C., (1938), *Some Problems in the Mathematical Theory of the Conduction of Heat*, Phil. Mag., Vol. 26, pp. 473-495.

Dyn, N., (1987), Interpolation of Scattered Data by Radial Functions, in *Topics in Multivariate Approximation*, Eds. C. K. Chui, L. L. Schumaker and F. I. Utreras, pp. 47-61, Academic Press, London.

Micchelli, C. A., (1986), Interpolation of Scattered Data, *Constructive Approximation*, Vol. 2, pp. 12-22.

Nowak, A. J., (1989), The Multiple Reciprocity Method of Solving Transient Heat Conduction Problems, *Advances in Boundary Elements*, Vol. 2, pp. 81-93, Eds. C. A. Brebbia and J. J. Connors, Computational Mechanics Publication, Southampton, Springer-Verlag, Berlin.

Nowak, A. J. and Neves, A. C., (1994), *The Multiple Reciprocity Boundary Element Method*, Computational Mechanics Publication, Southampton, Boston.

Ochiai, Y. and Sekiya, T., (1995a), Steady Thermal Stress Analysis by Improved Multiple-Reciprocity Boundary Element Method, *Journal of Thermal Stresses*, Vol. 18, No. 6, pp. 603-620.

Ochiai, Y., (1995b), Axisymmetric Heat Conduction Analysis by Improved Multiple-Reciprocity Boundary Element Method, *Heat Transfer-Japanese Research*, Vol. 23, No. 6, pp. 498-512.

Ochiai, Y. and Sekiya, T., (1995c), Generation of Free-Form Surface in CAD for Dies, *Advances in Engineering Software*, Vol. 22, pp. 113-118.

Ochiai, Y., (1996a), Generation Method of Distributed Data for FEM Analysis, *JSME International Journal*, Vol. 39, No. 1, pp. 93-98.
Ochiai, Y. and Sekiya, T., (1996b), Steady Heat Conduction Analysis by Improved Multiple-Reciprocity Boundary Element Method, *Engineering Analysis with Boundary Elements*, Vol. 18, pp. 111-117.
Ochiai, Y. and Kobayashi, T., (1999), Initial Stress Formulation for Elastoplastic Analysis by Improved Multiple-Reciprocity Boundary Element Method, *Engineering Analysis with Boundary Elements*, Vol. 23, pp. 167-173.
Ochiai, Y. and Yasutomi, Z., (2000), Improved Method Generating a Free-Form Surface Using Integral Equations, Computer Aided Geometric Design, Vol. 17, No. 3, pp. 233-245.
Ochiai, Y., (2001), Two-Dimensional Unsteady Heat Conduction Analysis with Heat Generation by Triple-Reciprocity BEM, *International Journal of Numerical Methods in Engineering*, Vol. 51, No. 2, pp. 143-157.
Ochiai, Y., (2003a), Multidimensional Numerical Integration for Meshless BEM, *Engineering Analysis with Boundary Elements*, Vol. 27, No. 3, pp. 241-249.
Ochiai, Y., (2003b), The Multiple-Reciprocity Method for Elastic Problems with Arbitrary Body Force, *Transformation of Domain Effects to the Boundary*, Y. F. Rashed Ed., Chapter 5, ISBN 1-85312-896-1, WIT Press.
Ochiai, Y. and Sladek, V., (2005), Numerical Treatment of Domain Integrals without Internal Cells in Three-Dimensional BIEM Formulations, *CMES(Computer Modeling in Engineering & Sciences)* Vol. 6, No. 6, pp. 525-536.
Ochiai, Y., Sladek, V. and Sladek, S., (2006), Transient Heat Conduction Analysis by Triple-Reciprocity Boundary Element Method, *Engineering Analysis with Boundary Elements*, Vol. 30, No. 3, pp. 194-204.
Ochiai, Y and Takeda, S., (2009a), Meshless Convection-Diffusion Analysis by Triple-Reciprocity Boundary Element Method, *Engineering Analysis with Boundary Elements*, Vol.33, No.2, pp.168-175.
Ochiai, Y and Kitayama, Y., (2009b),Three-dimensional Unsteady Heat Conduction Analysis by Triple-Reciprocity Boundary Element Method, *Engineering Analysis with Boundary Elements*, Vol. 33, No. 6, pp. 789-795.
Partridge, P. W., Brebbia, C. A. and Wrobel, L. C., (1992), *The Dual Reciprocity Boundary Element Method*, Computational Mechanics Publications, pp. 223-253.
Sladek, J. and Sladek, V., (2003), Local Boundary Integral Equation Method for Heat Conduction Problem in an Anisotropic Medium, *Proceedings of ICCES2003*, Chap. 5.
Tanaka, M., Matsumoto, T. and Takakuwa, S., (2003), Dual Reciprocity BEM Based on Time-Stepping Scheme for the Solution of Transient Heat Conduction Problems, *Boundary Elements XXV*, WIT Press, pp. 299-308.
Wrobel, L. C., (2002), *The Boundary Element Method*, Vol. 1, John Wiley & Sons, West Sussex, pp. 97-117.

14

Particle Transport Monte Carlo Method for Heat Conduction Problems

Nam Zin Cho
Korea Advanced Institute of Science and Technology (KAIST), Daejeon, South Korea

1. Introduction

Heat conduction [1] is usually modeled as a diffusion process embodied in heat conduction equation. The traditional numerical methods [2, 3] for heat conduction problems such as the finite difference or finite element are well developed. However, these methods are based on discretized mesh systems, thus they are inherently limited in the geometry treatment. This chapter describes the Monte Carlo method that is based on particle transport simulation to solve heat conduction problems. The Monte Carlo method is "meshless" and thus can treat problems with very complicated geometries.

The method is applied to a pebble fuel to be used in very high temperature gas-cooled reactors (VHTGRs) [4], which is a next-generation nuclear reactor being developed. Typically, a single pebble contains ~10,000 particle fuels randomly dispersed in graphite-matrix. Each particle fuel is in turn comprised of a fuel kernel and four layers of coatings. Furthermore, a typical reactor would house several tens of thousands of pebbles in the core depending on the power rating of the reactor. See Fig. 1. Such a level of geometric complexity and material heterogeneity defies the conventional mesh-based computational methods for heat conduction analysis.

Among transport methods, the Monte Carlo method, that is based on stochastic particle simulation, is widely used in neutron and radiation particle transport problems such as nuclear reactor design. The Monte Carlo method described in this chapter is based on the observation that heat conduction is a diffusion process whose governing equation is analogous to the neutron diffusion equation [5] under no absorption, no fission and one speed condition, which is a special form of the particle transport equation. While neutron diffusion approximates the neutron transport phenomena, conversely it is applicable to solve diffusion problems by transport methods under certain conditions. Based on this idea, a new Monte Carlo method has been recently developed [6-8] to solve heat conduction problems. The method employs the MCNP code [9] as a major computational engine. MCNP is a widely used Monte Carlo particle transport code with versatile geometrical capabilities.

Monte Carlo techniques for heat conduction have been reported [10-13] in the past. But most of the earlier Monte Carlo methods for heat conduction are based on discretized mesh systems, thus they are limited in the capabilities of geometry treatment. Fraley et al[13] uses a "meshless" system like the method in this chapter but does not give proper treatment to the boundary conditions, nor considers the "diffusion-transport theory correspondence" to be described in Section 2.2 in this chapter. Thus, the method in this chapter is a transport theory treatment of the heat conduction equation with a methodical boundary correction. The transport theory treatment can easily incorporate anisotropic conduction, if necessary, in a future study.

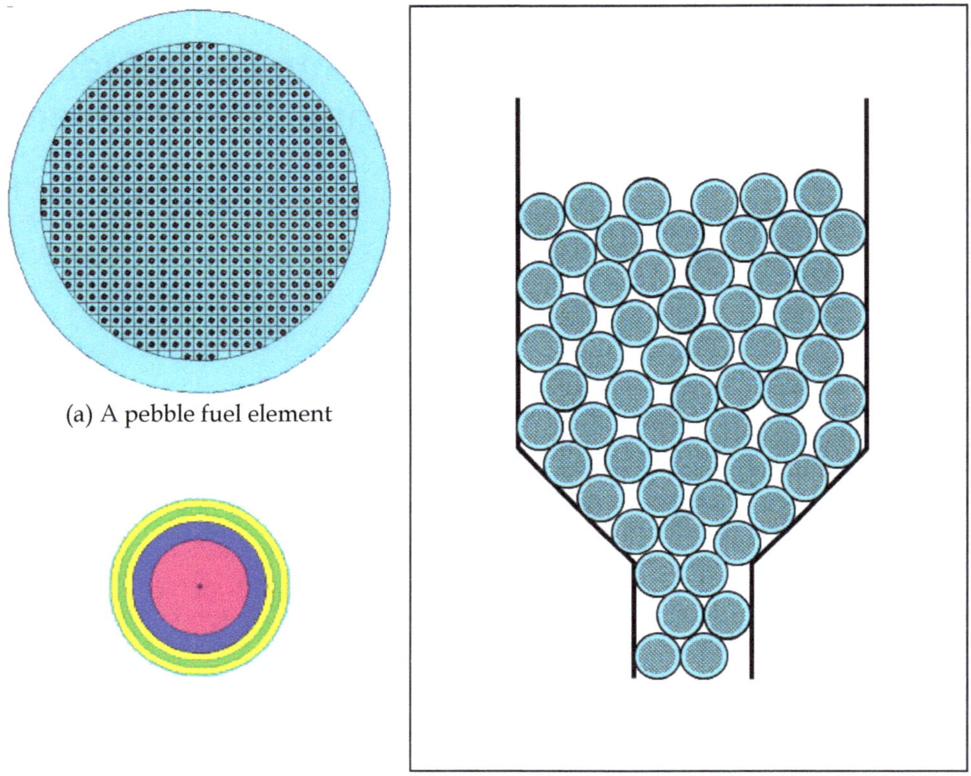

(a) A pebble fuel element

(b) A coated fuel particle

(c) A pebble-bed reactor core

Fig. 1. Cross-sectional view of a pebble fuel (a) consisting several thousands of coated fuel particles (b) in a reactor core (c)

2. Description of method

2.1 Neutron transport and diffusion equations

The transport equation that governs the neutron behavior in a medium with total cross section $\Sigma_t(\vec{r}, E)$ and differential scattering cross section $\Sigma_s(\vec{r}, E' \to E, \vec{\Omega}' \cdot \vec{\Omega})$ is given as [5]:

$$\vec{\Omega} \cdot \nabla \psi(\vec{r}, E, \vec{\Omega}) + \Sigma_t(\vec{r}, E)\psi(\vec{r}, E, \vec{\Omega}) = \int dE' \int d\vec{\Omega}' \Sigma_s(\vec{r}, E' \to E, \vec{\Omega}' \cdot \vec{\Omega})\psi(\vec{r}, E', \vec{\Omega}') + S(\vec{r}, E, \vec{\Omega}) \tag{1a}$$

with boundary condition, for $\vec{n} \cdot \vec{\Omega} < 0$,

$$\psi(\vec{r}_s, E, \vec{\Omega}) = \begin{cases} \psi_s(E, \vec{\Omega}), & \textit{given}, \\ 0, & \textit{if vacuum}, \end{cases} \tag{1b}$$

where

$$\vec{r} = neutron\ position,$$
$$E = neutron\ energy,$$
$$\vec{\Omega} = neutron\ direction,$$
$$S = neutron\ source,$$
$$\psi(\vec{r}, E, \vec{\Omega}) = neutron\ angular\ flux.$$

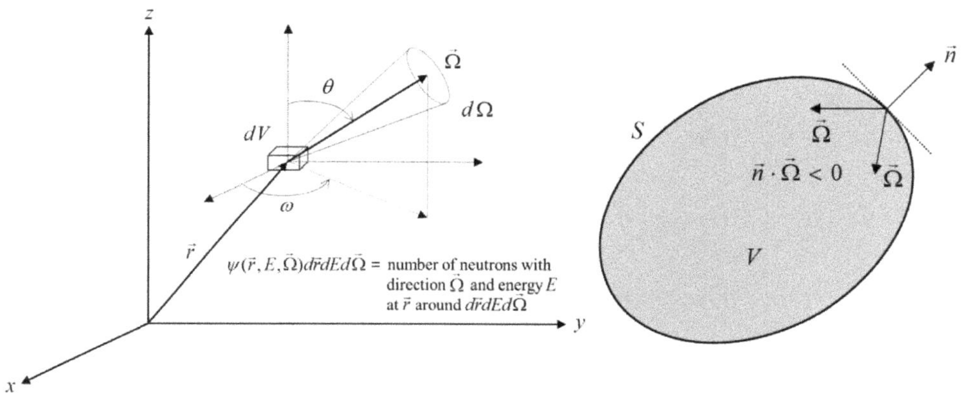

Fig. 2. Angular flux and boundary condition

Fig. 2 depicts the meaning of angular flux $\psi(\vec{r}, E, \vec{\Omega})$ and boundary condition. In the special case of no absorption, isotropic scattering, and mono-energy of neutrons, Eq. (1) becomes

$$\vec{\Omega} \cdot \nabla \psi(\vec{r}, \vec{\Omega}) + \Sigma_s(\vec{r})\psi(\vec{r}, \vec{\Omega}) = \frac{1}{4\pi}\Sigma_s(\vec{r})\phi(\vec{r}) + \frac{S(\vec{r})}{4\pi}, \tag{2a}$$

with vacuum boundary condition,

$$\psi(\vec{r}_s, \vec{\Omega}) = 0 \ for\ \vec{n} \cdot \vec{\Omega} < 0, \tag{2b}$$

where scalar flux is defined as

$$\phi(\vec{r}) = \int d\vec{\Omega}\psi(\vec{r}, \vec{\Omega}). \tag{2c}$$

Let us now consider a "scaled" equation of (2a),

$$\vec{\Omega} \cdot \nabla \psi(\vec{r}, \vec{\Omega}) + \beta\Sigma_s(\vec{r})\psi(\vec{r}, \vec{\Omega}) = \frac{1}{4\pi}\beta\Sigma_s(\vec{r})\phi(\vec{r}) + \frac{1}{\beta}\frac{S(\vec{r})}{4\pi}. \tag{3}$$

An important result of the asymptotic theory provides correspondence between the transport equation and the diffusion equation, i.e., the asymptotic $(\beta \to \infty)$ solution of Eq. (3) satisfies the following diffusion equation:

$$-\frac{1}{3\Sigma_s(\vec{r})}\nabla\phi(\vec{r}) = S(\vec{r}), \tag{4a}$$

with vacuum boundary condition

$$\phi(\vec{r}_s + d) = 0, \quad d = extrapolation\ distance. \tag{4b}$$

It is known that, between the two solutions from transport theory and from diffusion theory, a discrepancy appears near the boundary. Thus, the problem domain is extended using an extrapolated thickness (typically $d \simeq one\ mean\ free\ path = 1/\Sigma_t$) for boundary layer correction, as shown in Fig. 3.

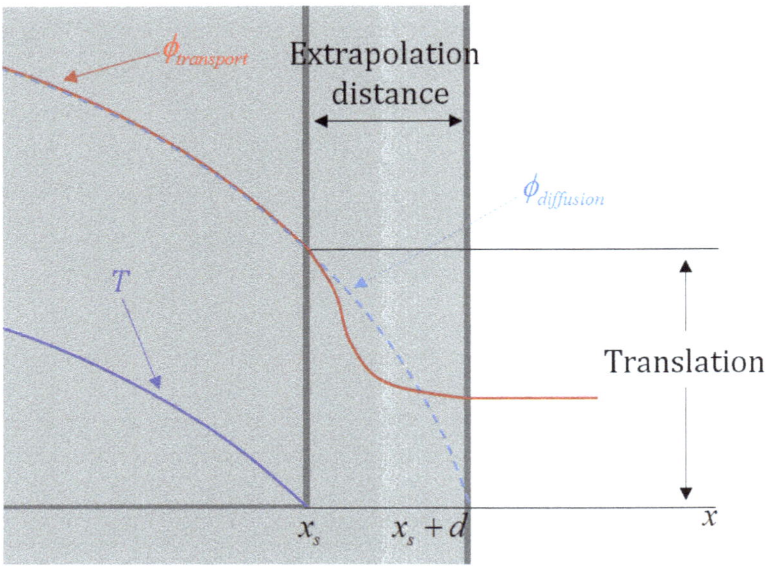

Fig. 3. Boundary correction with an extrapolated layer

2.2 Monte Carlo method for heat conduction equation

Correspondence

The steady state heat conduction equation for a stationary and isotropic solid is given by [1]:

$$\nabla \cdot k(\vec{r})\nabla T(\vec{r}) + q'''(\vec{r}) = 0, \tag{5a}$$

with boundary condition

$$T(\vec{r}_s) = 0, \tag{5b}$$

where $k(\vec{r})$ is the thermal conductivity and $q'''(\vec{r})$ is the internal heat source.

If we compare Eq. (5) with Eq. (4), it is easily ascertained that Eq. (4) becomes Eq. (5) by setting

$$\Sigma_s(\vec{r}) = \frac{1}{3k(\vec{r})}, \tag{6}$$

and

$$S = q'''(\vec{r}), \tag{7}$$

with a large β and the problem domain extended by d.

The Monte Carlo method is extremely versatile in solving Eqs. (1), (2) and (3) with very complicated geometry and strong heterogeneity of the medium. Thus, Eq. (3) is solved by the Monte Carlo method (with a large β) to obtain $\phi(\vec{r})$. The result of $\phi(\vec{r})$ is then translated to provide $T(\vec{r}) = \phi(\vec{r}) - \phi(\vec{r}_s)$ as the solution of Eq. (5) [See Fig. 3.]

Here, $\beta > 1$ is a scaling factor rendering the transport phenomena diffusion-like. A large scaling factor plays an additional role of reducing the extrapolation distance to the order of a mean free path. To choose a proper value for β, we introduce an adjoint problem to perform sensitivity studies, specific results for a pebble problem provided later in this section.

Proof of principles of the method

In order to confirm or provide proof of principles of the Monte Carlo method described in Section 2.2, first we consider a simple heat conduction problem which allows analytic solution. The problem consists of one-dimensional slab geometry, isotropic solid, and uniformly distributed internal heat source under steady state. The left side has reflective boundary condition and the right side has zero temperature boundary condition. Fig. 4(a) shows the original problem and Fig. 4(b) shows the extended problem to be solved by the Monte Carlo method, incorporating the boundary layer correction. Table 1 provides the calculational conditions.

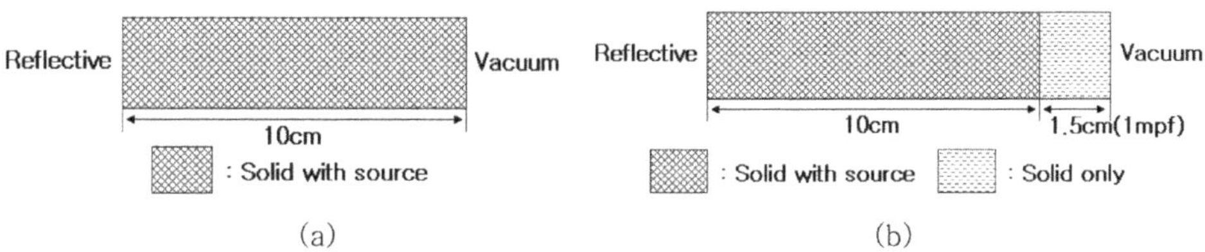

Fig. 4. A one-dimensional slab test problem

Thermal Conductivity ($W / cm \cdot °K$)	Internal Heat Source(W / cm^3)	Extrapolation Thickness (mfp)	Scaling Factor
0.5	0.01	1	1

Table 1. Calculation Conditions for Simple Problem

Figs. 5 and 6 show the Monte Carlo method results with and without the extension by extrapolation thickness in comparison with the analytic solution. Note that the result of the Monte Carlo method with boundary layer correction is in excellent agreement with the analytic solution.

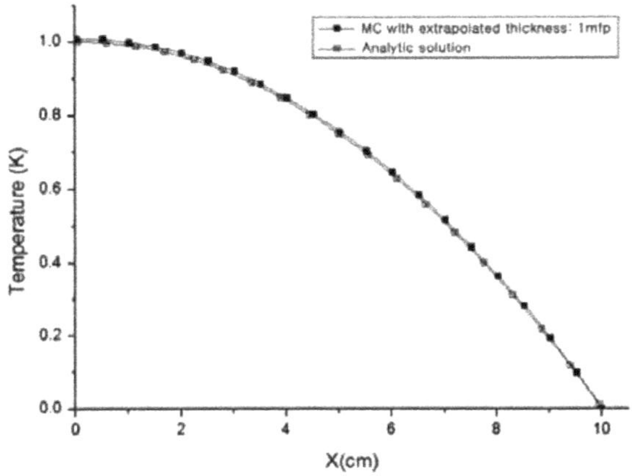

Fig. 5. Monte Carlo heat conduction solution with extrapolated layer

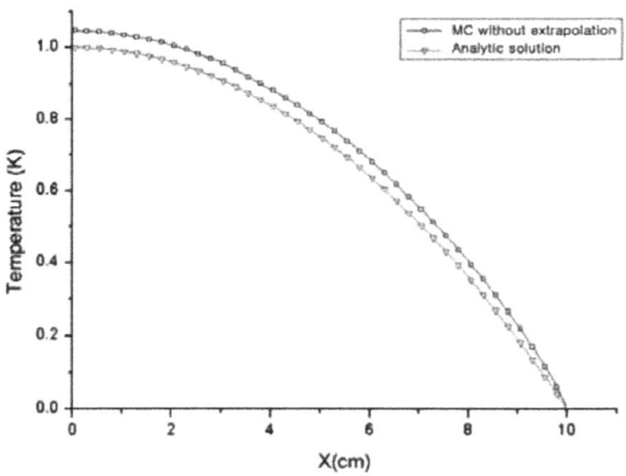

Fig. 6. Monte Carlo heat conduction solution without extrapolated layer

To test the method on a realistic problem, the FLS (Fine Lattice Stochastic) model and CLCS (Coarse Lattice with Centered Sphere) model [14] for the random distribution of fuel particles in a pebble are used to obtain the heat conduction solution by the Monte Carlo method. Details of this process are described in Table 2 and Fig. 7. The power distribution generated in a pebble is assumed uniform within a kernel and across the particle fuels. The pebble is surrounded by helium at 1173K with the convective heat transfer coefficient h=0.1006($W / cm^2 \cdot °K$). A Monte Carlo program HEATON [15] was written to solve heat conduction problems using the MCNP5 code as the major computational engine.

Material	Kernel	Buffer	Inner PyC	SiC
Thermal Conductivity ($W/cm\cdot{}^{\circ}K$)	0.0346	0.0100	0.0400	0.1830
Radius (cm)	0.02510	0.03425	0.03824	0.04177

Material	Outer PyC	Graphite-matrix	Graphite-shell
Thermal Conductivity ($W/cm\cdot{}^{\circ}K$)	0.0400	0.2500	0.2500
Radius (cm)	0.04576	2.5000	3.0000

Number of Triso Particles	9394
Power/pebble (W)	1893.95

Table 2. Problem Description for a Pebble

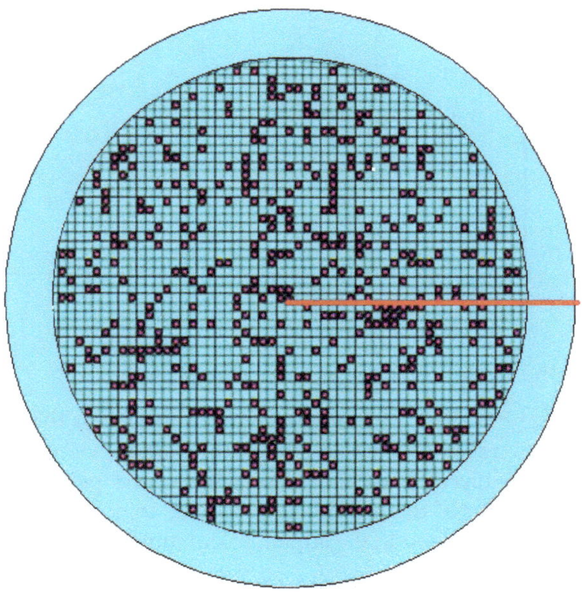

Fig. 7. A planar view of a particle random distribution for a pebble problem with the FLS model

Heat conduction solutions for the pebble problem with the data in Table 2 using the Monte Carlo method are shown in Table 3 and Fig. 8. The number of histories used was 10^7. Parallel computation with 60 CPUs (3.2GHz) was used. When the scaling factor β increases, the solution of the pebble problem approaches its asymptotic solution (diffusion solution). However, the computational time increases rapidly as the scaling factor increases. In Table 3 and Fig. 8, it is shown that a scaling factor of 10 or 20 is not large enough.

Scaling Factor	Maximum Temp. ($°K$)	Relative Error[a] (%)	Graphite Temp. Near Center($°K$)	Relative Error[a] (%)	Computing Time (sec)	Translation Temp. ($°K$)
1	1674.21	1.59	158.33	0.71	534	27.08
10	1556.96	1.14	1533.53	0.34	6,692	2.72
20	1558.54	1.12	1531.67	0.30	20,297	1.36
50	1553.22	1.11	1527.07	0.28	99,454	0.54

[a] One standard deviation in temperature / mean estimate of temperature by Monte Carlo $\times 100\%$

Table 3. Results of Fig. 7 Problem

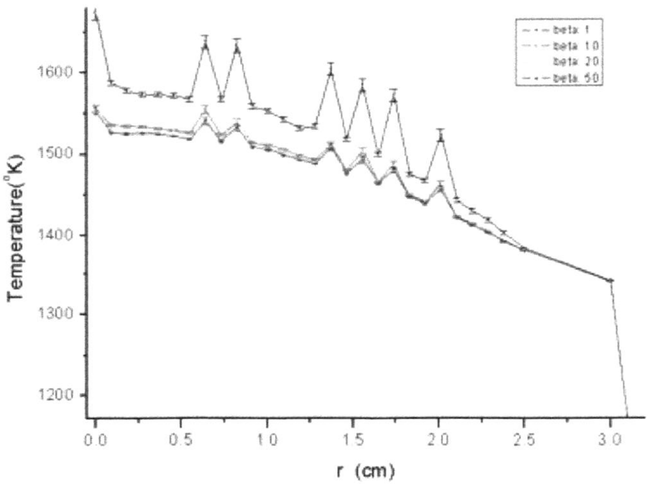

Fig. 8. Results along the red line of Fig. 7 vs the scaling factor

Therefore, it is necessary to determine an effective scaling factor that renders the problem more diffusive. This can be done using an adjoint calculation. Using an adjoint calculation, the computing time is reduced as the calculation transports particles backward from the detector region (at the center of the pebble) to the source region. Additionally, it is possible that the changed tally regions used in the adjoint calculation allow effective particle tallies.

Scaling Factor	Maximum Temp. ($°K$)	Standard Deviation($°K$)	Computing Time (sec)
1	1685.131	0.409	47
20	1558.817	0.308	1,427
50	1553.931	0.304	7,298
80	1553.586	0.304	17,976
100	1552.995	0.303	27,240
120	1552.713	0.303	39,435

Table 4. Maximum Temperature and Computing Time for Fig. 9

In order to confirm the appropriate scaling factor, the problem with the data of Table 2 and in Fig. 7 was again tested with a smaller number (10^6) of histories compared to the number used in the forward calculation. The results depending on the scaling factor are shown in Fig. 9 and Table 4.

Fig. 9 shows that the center temperature of a fuel pebble approaches its asymptotic solution (diffusion solution) as the scaling factor increases. Therefore, to obtain a diffusion solution, a scaling factor of > 30 (e.g., 50) is required.

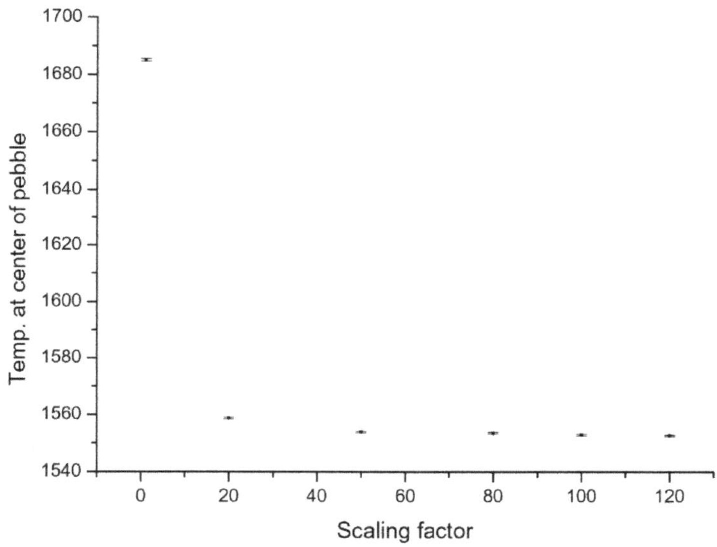

Fig. 9. Center temperature by the adjoint calculation

2.3 Heat conduction problems

Given varying-temperature boundary condition

The first kind of the boundary conditions is the prescribed surface temperature:

$$T(\vec{r}_s) = f(\vec{r}_s), \tag{8}$$

where $\vec{r}_s$ is on a boundary surface. Since the paradigm heat conduction problem that the Monte Carlo method can treat is a problem with zero temperature boundary condition (as described in Section 2.2), let T be decomposed into T^* and $\tilde{T}$ as follows:

$$T(\vec{r}) = T^*(\vec{r}) + \tilde{T}(\vec{r}), \tag{9}$$

where T^* satisfies the zero boundary condition, and $\tilde{T}$ is chosen such that it satisfies the given boundary condition (8). Eq. (5a) can then be rewritten as:

$$-\nabla \cdot k(\vec{r})\nabla T(\vec{r}) = -\nabla \cdot k(\vec{r})\nabla (T^* + \tilde{T}) = q'''(\vec{r}), \tag{10}$$

or

$$-\nabla \cdot k(\vec{r})\nabla T^{*}(\vec{r}) = q^{*'''}(\vec{r}), \tag{11a}$$

where the new source $q^{*'''}(\vec{r})$ is defined by

$$q^{*'''}(\vec{r}) = \nabla \cdot k(\vec{r})\nabla \tilde{T}(\vec{r}) + q'''(\vec{r}), \tag{11b}$$

Eq. (11a) is to be solved for T^{*} by the Monte Carlo method [6-8]. The Monte Carlo method cannot deal easily with the gradient term, $\nabla \cdot k(\vec{r})\nabla \tilde{T}$, in Eq. (11b) when the boundary condition temperature is not a constant and $k(\vec{r})$ is not smooth enough. In order to evaluate the new source term as simply as possible, let $\tilde{T}$ be zero in internally complicated thermal conductivity region as shown in Fig. 10. In addition, $\tilde{T}$ and $\nabla \tilde{T}$ must be continuous in the whole problem domain to render the $\nabla \cdot k(\vec{r})\nabla \tilde{T}$ term treatable.

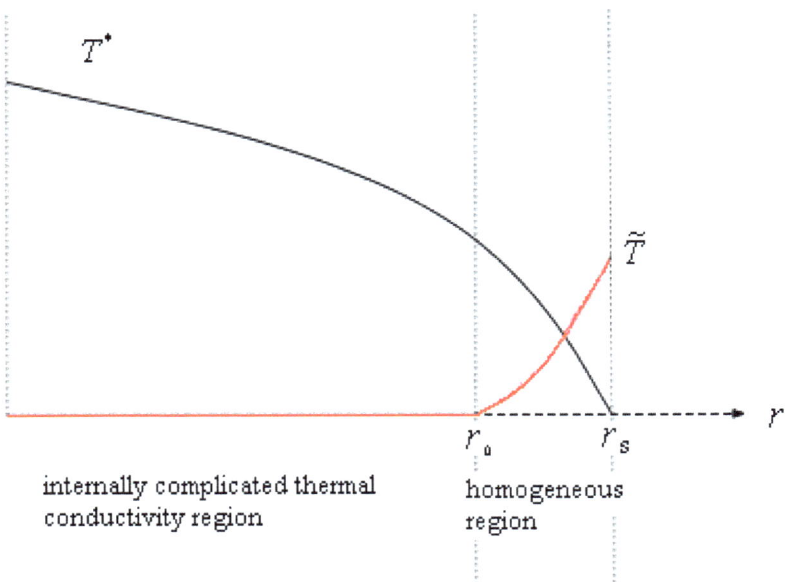

Fig. 10. Solution Decomposition $T = T^{*} + \tilde{T}$,

In Ref. [8], the following $\tilde{T}$ is chosen for a three-dimensional spherical model:

$$\tilde{T} = U(r)f(r_s,\theta,\omega)\frac{(r-r_0)^2}{(r_s-r_0)^2}, \tag{12}$$

where $f(r_s,\theta,\omega)$ is the given boundary condition (8), θ and ω indicate polar and azimuthal angle, respectively. r_s is radius to the boundary surface and there may be internally complicated thermal conductivity region inside r_0.

Convection boundary condition

A convection boundary condition is usually given by

$$k_1\frac{\partial T(\vec{r}_s)}{\partial n} = h(T_b - T(\vec{r}_s)), \tag{13}$$

where k_1 is the thermal conductivity of medium 1 (solid), h and T_b are the convective heat transfer coefficient and the bulk temperature of the convective medium, respectively. This condition can be equivalently transformed to a given temperature (T_b) boundary condition of a related problem, in which the convective medium is replaced by a hypothetical conduction medium with thermal conductivity

$$k_2 = h\Delta n\left(\frac{r_s}{r_b}\right), \tag{14}$$

where Δn is additional thickness beyond $r_s(\Delta n = r_b - r_s)$ in a spherical geometry. Here r_b is the radius where T_b occurs. k_2 involves a geometry factor and k_2's for several geometries are shown in Table 5 (see Appendix B for the derivation).

Geometry	k_2
Sphere	$h(r_b - r_s)\left(\frac{r_s}{r_b}\right)$
Cylinder	$hr_s\, ln\left(\frac{r_b}{r_s}\right)$
Slab	$h(x_b - x_s)$

Table 5. k_2 for Several Geometries

There is no approximation in the k_2 expressions for given h if there is no heat source in the fluid. The transformed problem can then be solved by the Monte Carlo method in Section 2.1 with replacement of r_0 by r_s and r_s by r_b, and T_b as the boundary condition. Eq. (13) with the right-hand side replaced by Eq. (14) is no more than a continuity expression of heat flux on the interface. Fig. 11 shows the concept in this transformation.

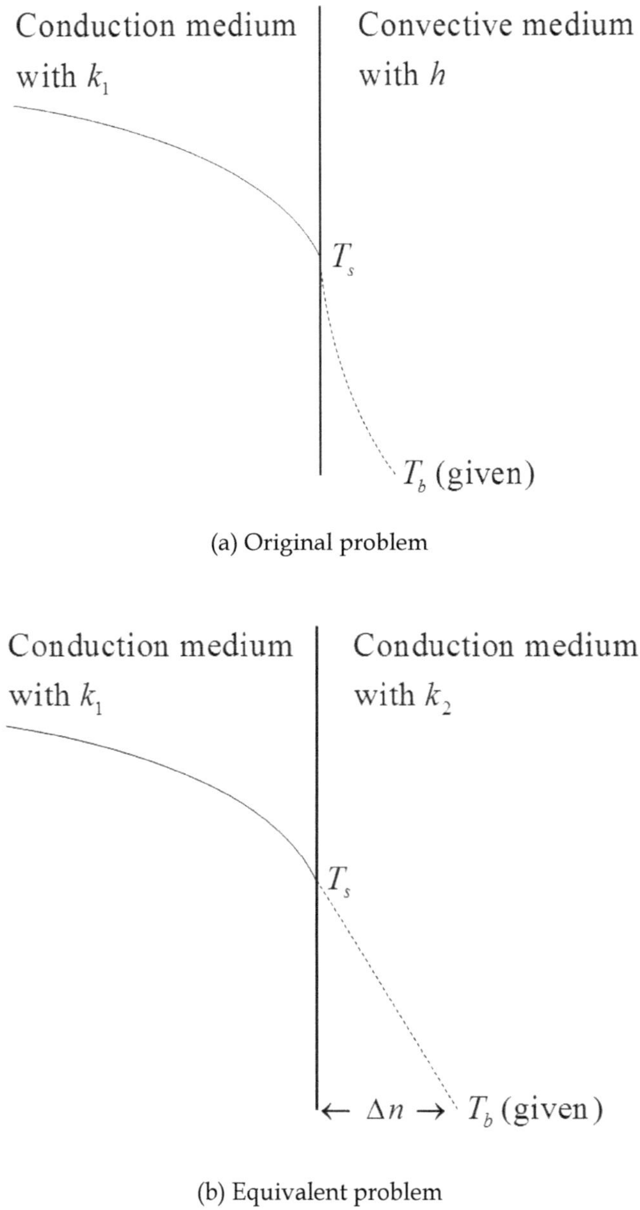

Fig. 11. Transformation of a convective medium to an equivalent conduction medium preserving heat flux

Examples

The method is applied to a pebble fuel with Coarse Lattice with Centered Sphere (CLCS) distribution of fuel particle [14]. The description of a pebble fuel is shown in Fig. 12 and Table 2. The pebble fuel is surrounded by helium at given bulk temperature with convective heat transfer coefficient $h = 0.1006(W / cm^2 \cdot {}^\circ K)$. The number of histories used in the Monte Carlo calculation was 10^7.

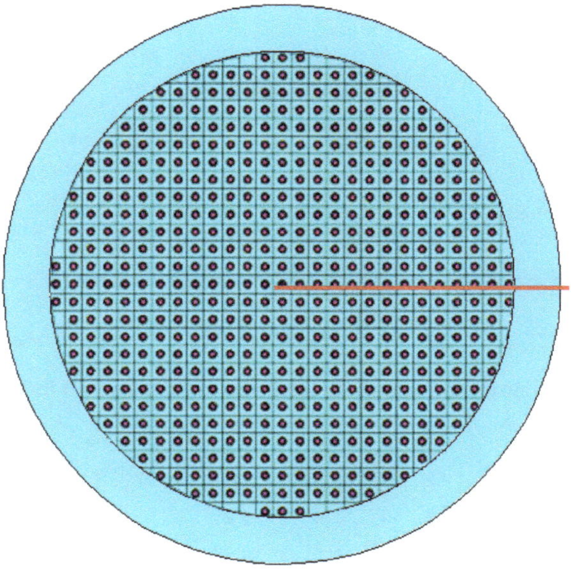

Fig. 12. CLCS distribution

Test Problem 1 is defined by the following non-constant bulk temperature of the helium coolant:

$$1173 + 10(1 + \cos\theta)^\circ K, \tag{15a}$$

where θ is the polar angle, or equivalently

$$1173 + 10\left(1 + \frac{z}{\sqrt{x^2 + y^2 + z^2}}\right), \tag{15b}$$

where

$$x^2 + y^2 + z^2 = r_b^2, \tag{15c}$$

with $r_b = 3.1$, x, y and z in centimeters.

The results are shown in Figs. 13, 14 and 15.

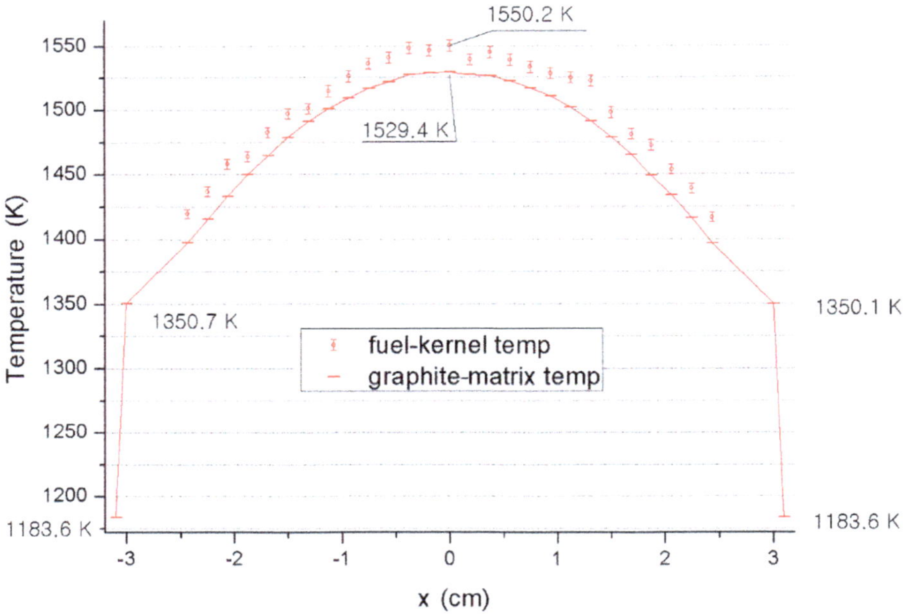

Fig. 13. Temperature distribution along x -direction with $y = z = 0$ in Test Problem 1

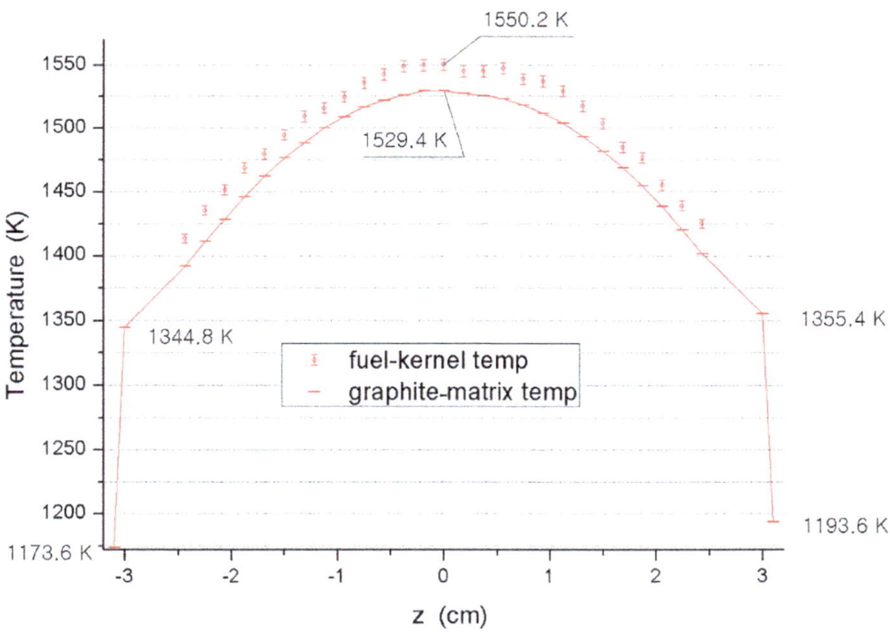

Fig. 14. Temperature distribution along z -direction with $x = y = 0$ in Test Problem 1

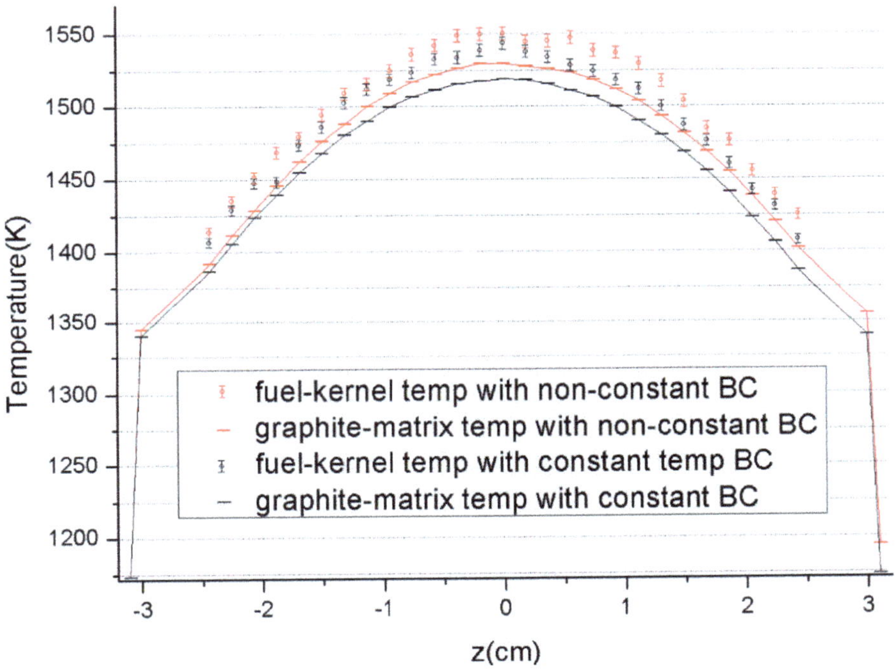

Fig. 15. Comparison of Test Problem 1 and a problem with constant helium bulk temperature (1173 °K)

Test Problem 2 is defined by the following non-constant bulk temperature of the helium coolant:

$$1173 + 10 + (x + y + z)°K \tag{16a}$$

where

$$x^2 + y^2 + z^2 = r_b^2, \tag{16b}$$

with $r_b = 3.1$, x, y and z in centimeters.

The results are shown in Figs. 16, 17, and 18.

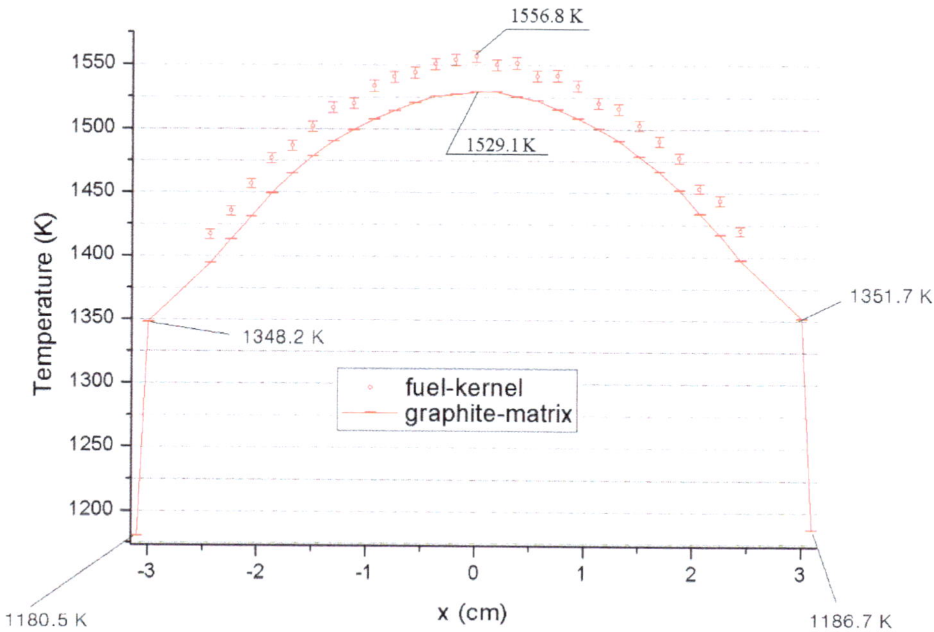

Fig. 16. Temperature distribution along x -direction with $y = z = 0$ in Test Problem 2

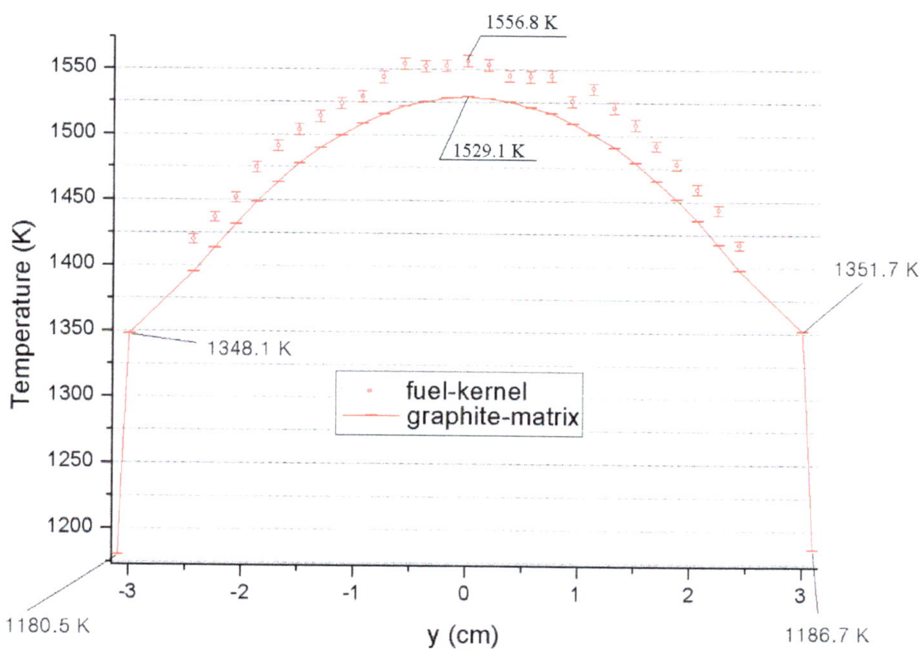

Fig. 17. Temperature distribution along z -direction with $x = y = 0$ in Test Problem 2

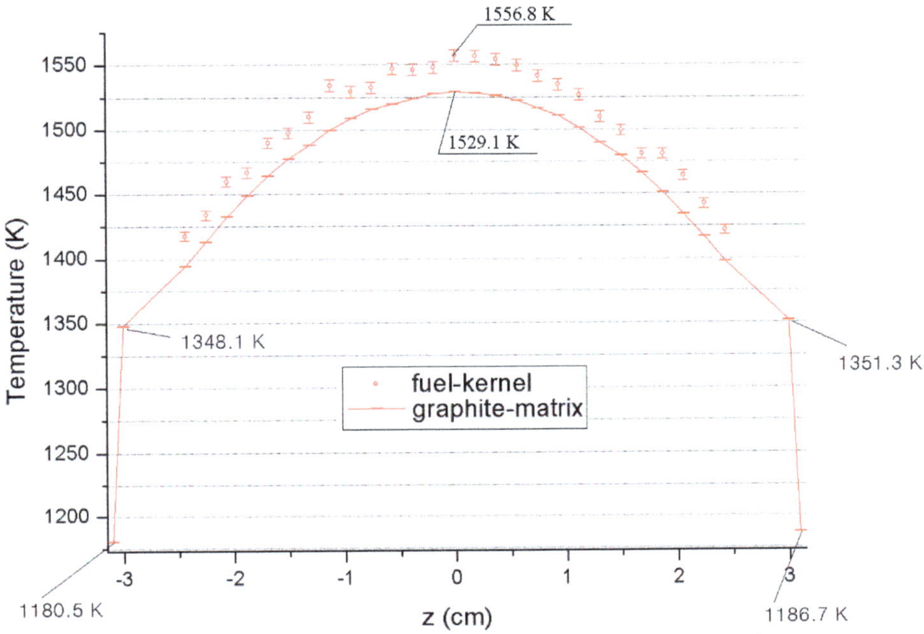

Fig. 18. Temperature distribution along y -direction with $x = z = 0$ in Test Problem 2

3. Applications

3.1 Comparison between the FLS (Fine Lattice Stochastic) model and analytic bound solutions

In this section, the data of the geometry information and thermal conductivity are identical to those in Table 2. Based on the results in the previous section, temperature distributions were calculated using a scaling factor of 50. Three triso particle configurations obtained by randomly distributed fuels in a pebble were considered (using the FLS model in Ref. 14). The tally regions as shown in Fig. 19 were chosen. If a (fine) lattice has a heat source, the tally is done over the kernel volume. If the lattice consists of only graphite, tally is done over the lattice cubical volume.

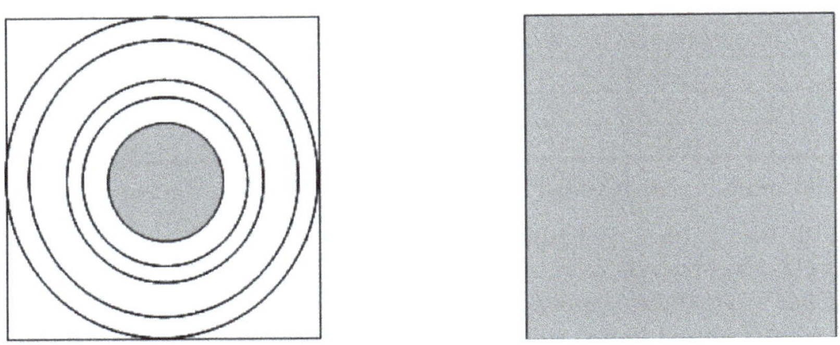

Fig. 19. Tally regions with and without a heat source

Fig. 20 shows the temperature distributions obtained from the Monte Carlo method compared to the two analytic bound solutions superimposed with a particle located at the center of the pebble based on commonly quoted homogenized models [16]. It is important to note that the volumetric analytic solution usually presented in the literature [17] predicts lower temperatures than those of
(thus underestimates) the Monte Carlo results. In the Monte Carlo results, the fuel-kernel temperature and graphite-matrix temperature are distinctly calculated. The results are summarized in Table 6.

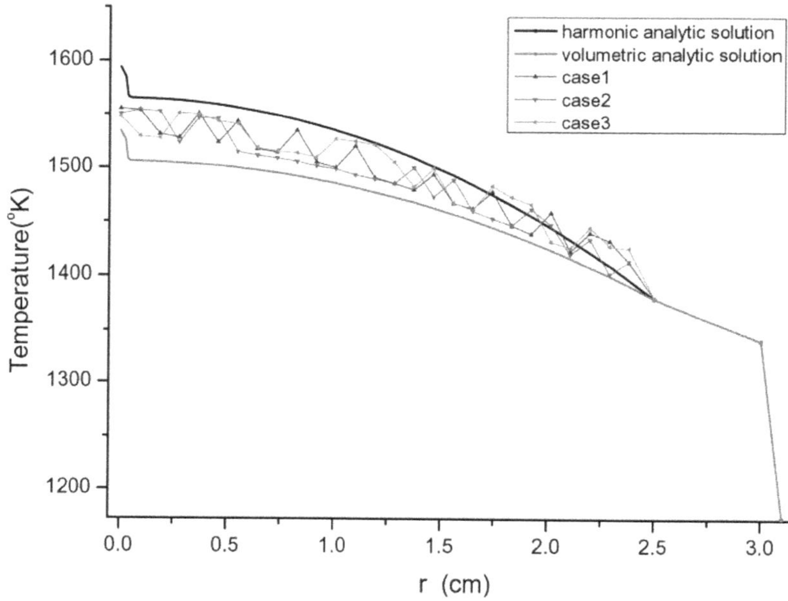

Fig. 20. Temperature profiles depending on the triso particle distribution configuration compared to two homogenized models

	Max. Temp. ($^\circ K$)	Average Kernel Temp. ($^\circ K$)	Average Graphite Temp. ($^\circ K$)
Case 1	1555.07	1497.84	1487.61
Case 2	1553.77	1499.63	1480.43
Case 3	1550.87	1501.89	1489.38
Average	1553.23	1499.79	1485.80

Table 6. Maximum, Average Kernel and Graphite Temperatures from Fig. 20

For a fourth triso particle configuration (Case 4), the tally region was further refined as shown in Fig. 21 to provide more accurate graphite-moderator temperature. Essentially, if the lattice has a kernel (heat source), the tally is done over the kernel volume and over the moderator (graphite and layers) volume separately. Otherwise, if the lattice consists of only graphite, the tally is done over the cubical volume.

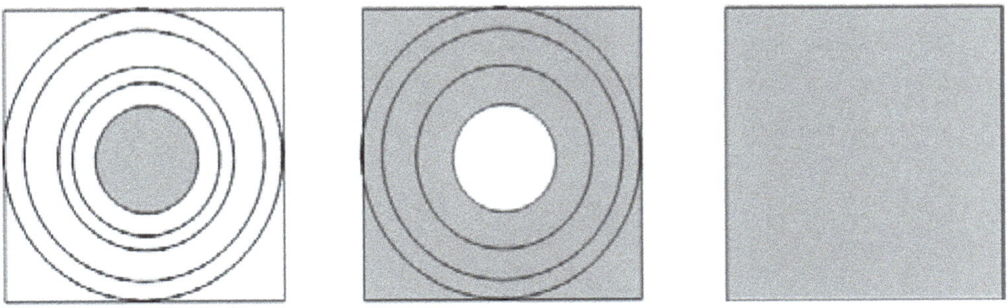

Fig. 21. Tally regions depending on the geometries

In this problem, geometry information is identical to those shown in Table 2. The distributed particle configuration is shown in Fig. 22. The kernel and graphite-moderator temperatures are shown in Fig. 23 and Table 7.

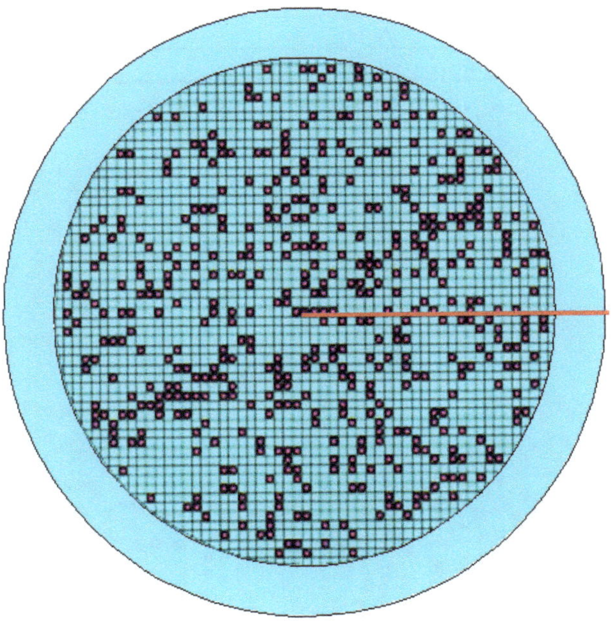

Fig. 22. A planar view of a fourth particle distribution configuration with the FLS model

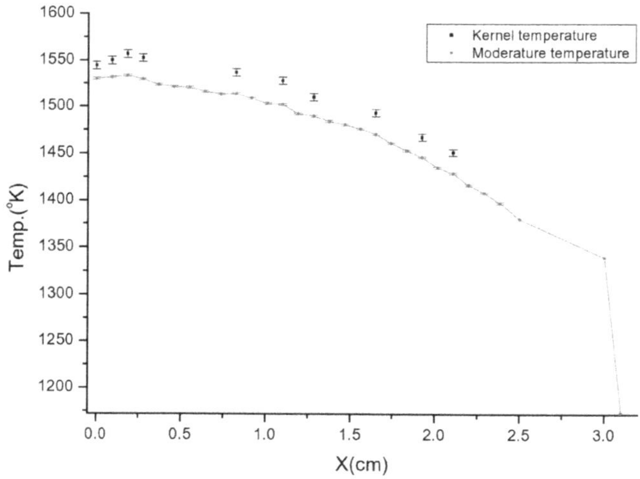

Fig. 23. Temperature distribution along red line for Fig. 22

Maximum temperature ($^\circ K$)	1556.70
Averaged kernel temperature ($^\circ K$)	1518.88
Averaged moderator temperature ($^\circ K$)	1484.61
Surface temperature at 2.5cm ($^\circ K$)	1379.82
Surface temperature at 3.0cm ($^\circ K$)	1339.65
Computing time	43h 35m 9s

Table 7. Results for the Fourth Configuration Shown in Fig. 23

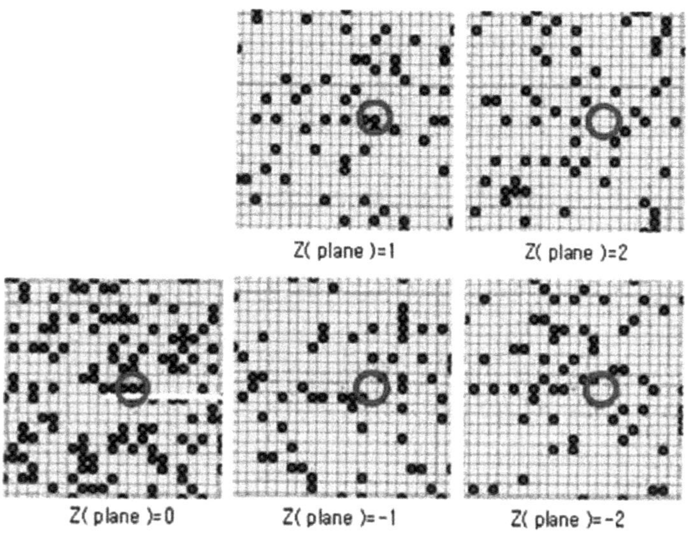

Fig. 24. Cross-sectional views for Fig. 22

The temperature profile on the $z = 0$ plane along red line is shown in Fig. 23 and Table 7. In this FLS model, the maximum fuel temperature appears not at the center point but near the central region, as the fuels are concentrated on the right side of the center point on the $z = 0$ plane, as shown in Fig. 24. Note that the red circle in Fig. 24 denotes particles with the dominant effect of the temperature increase on the $z = 0$ plane.

3.2 CLCS (Coarse Lattice with Centered Sphere) model

The temperature distribution was obtained again for the CLCS (Coarse Lattice with Centered Sphere) model [14]. In this model, the tally regions used are shown in Fig. 25. The general geometry
information is identical to that in Table 2, except that there are 9315 triso particles and each triso particle takes one lattice cube (and vice versa), as shown in Fig. 26. The resulting temperature distribution for the CLCS model is shown in Fig. 27.

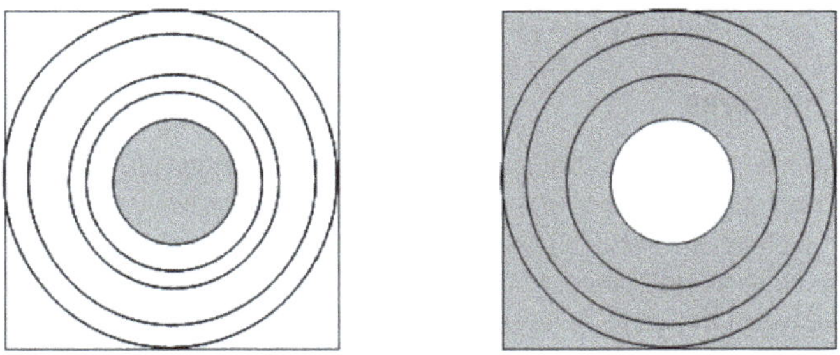

Fig. 25. Tally regions for the CLCS model

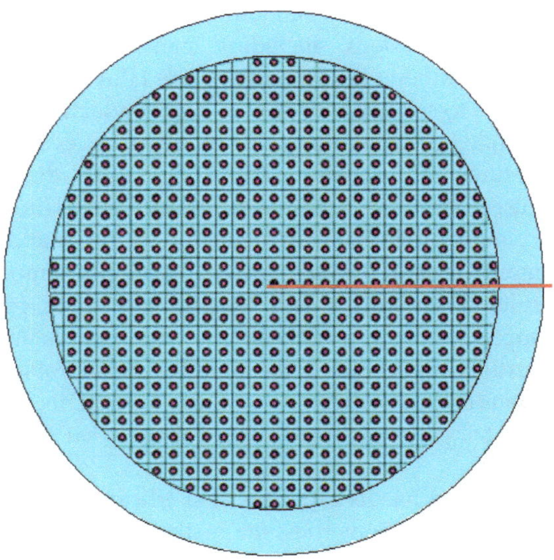

Fig. 26. Fuel particle configuration for the CLCS model

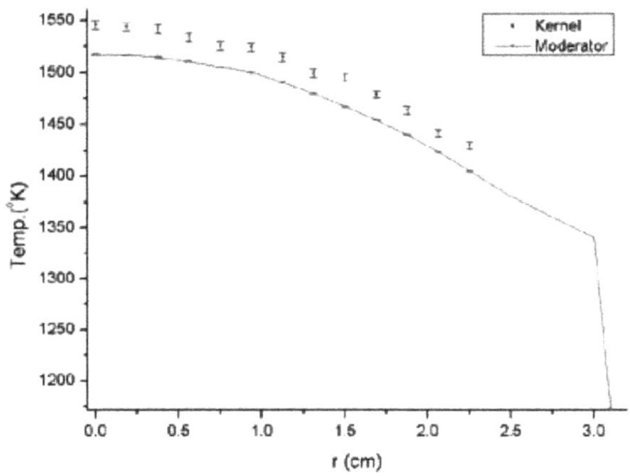

Fig. 27. Results of cubes along red line for Fig. 26

4. Concluding remarks

A Monte Carlo method for heat conduction problems was presented in this chapter. Based on the asymptotic theory correspondence between neutron transport and diffusion equations, it is shown that the particle transport Monte Carlo simulation can provide solutions to the heat conduction problems with two modeling devices introduced: i) boundary layer correction by the extended problem domain and ii) scaling factor to increase the diffusivity of the problem.

The Monte Carlo method can be used to solve heat conduction problems with complicated geometry (e.g. due to the extreme heterogeneity of a fuel pebble in a VHTGR, which houses many thousands of coated fuel particles randomly distributed in graphite matrix). It can handle typical boundary conditions, including non-constant temperature boundary condition and heat convection boundary condition. The HEATON code was written using MCNP as the major engine to solve these types of heat conduction problems. Monte Carlo results for randomly sampled configurations of triso fuel particles were presented, showing the fuel kernel temperatures and graphite matrix temperatures distinctly. The fuel kernel temperatures can be used for more accurate neutronics calculations in nuclear reactor design, such as incorporating the Doppler feedback. It was found that the volumetric analytic solution commonly used in the literature predicts lower temperatures than those of the Monte Carlo results. Therefore, it will lead to inaccurate prediction of the fuel temperature under Doppler feedback, which will have important safety implications.

An obvious area of further application is the time transient problem. The results of the steady-state heterogeneous calculations by Monte Carlo (as described in this chapter) can be used to construct a two-temperature homogenized model that is then used in transient analysis [18].

While the Monte Carlo method has its capability and efficacy of handling heat conduction problems with very complicated geometries, the method has its own shortcomings of the long computing time and variance due to the statistical results. It also has a limitation in that it provides temperatures at specific points rather than at the entire temperature field.

Appendix A: Elements of Monte Carlo method

A.1 Introduction

In a typical form of the particle transport Monte Carlo method [9,19], we simulate particle (e.g., neutron) behavior by following a finite number, say N, of particle histories and tallying the appropriate events needed to calculate the quantity of interest. The simulation is performed according to the physical events (expressed by each term in the transport equation) that a particle would encounter through the use of random numbers. These random numbers are usually generated by a pseudo random number generator, that provides uniform random number ξ between 0 and 1. In each particle history, the random numbers are generated and used to sample discrete events or continuous variables as the case may be according to the probability distribution functions. The results of tally are processed to provide estimates for the mean and variance of the quantity of interest, e.g., neutron flux, current, reaction rate, or some other quantities.

A.2 Basic operations of sampling

A.2.1 Sampling of random events

The discrete events such as the type of nuclides and collisions are simple to sample. For example, suppose that there are in the medium I nuclides with total macroscopic cross sections, $\Sigma_t^{(i)}, i=1,2,\cdots,I$. Let

$$\Sigma_t = \sum_{i=1}^{I} \Sigma_t^{(i)} , \tag{A1}$$

and

$$P_i = \frac{\Sigma_t^{(i)}}{\Sigma_t}, \quad i=1,2,\cdots,I. \tag{A2}$$

Now draw a random number ξ and if

$$P_1 + P_2 + \cdots + P_{i-1} \le \xi < P_1 + P_2 + \cdots + P_i , \tag{A3}$$

then the i -th nuclide is selected and the neutron collides with nuclide i . After determination of the nuclide, the type of collisions (absorption, fission, or scattering, etc.) is determined in a similar way. If the event is scattering, the energy and direction of the scattered neutron are sampled. In addition, the distance a neutron travels before suffering its next collision is sampled. These values are continuous variables and thus determined by sampling according to the appropriate probability density function $f(x)$. For example, the distance l to next collision (within the same medium) is distributed as

$$f(l)dl = e^{-\Sigma_t l}\Sigma_t dl , \tag{A4}$$

with its cumulative distribution function

$$F(l) = \int_0^l f(l')dl' = 1 - e^{-\Sigma_t l} . \tag{A5}$$

Since $F(l)$ is uniformly distributed between 0 and 1, we draw a random number ξ and let

$$F(l) = \xi \,, \tag{A6}$$

that in turn provides

$$l = -\frac{ln(1-\xi)}{\Sigma_t} = -\frac{ln(\xi)}{\Sigma_t} \,. \tag{A7}$$

A.2.2 Geometry tracking

In typical Monte Carlo codes, the geometries of the problem are created with intersection and union of surfaces. In turn, the surfaces are defined by a collection of elementary mathematical functions. For example, the geometry in Fig. A1 would be defined by functions that represent four straight lines and a circle.

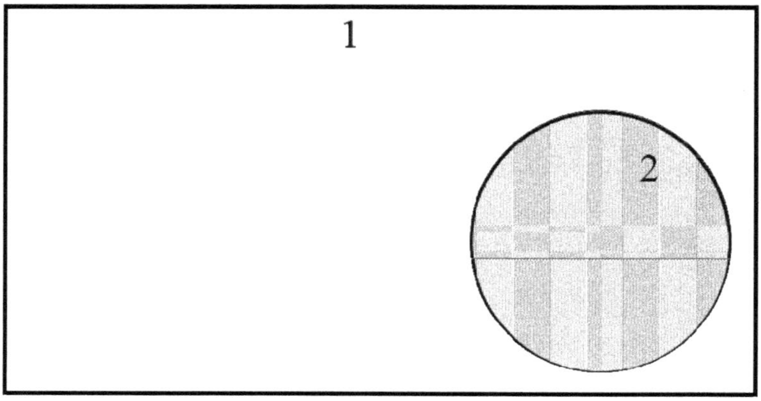

Fig. A1. An example of problem geometry with two material media

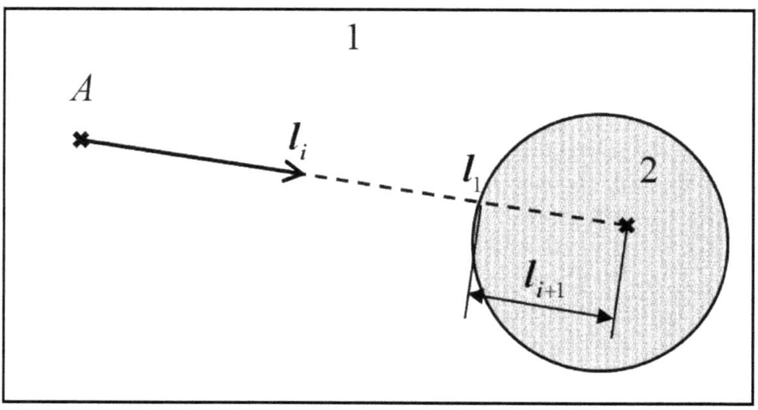

Fig. A2. Geometry tracking

Suppose that the neutron we follow is now at point A and heading to the direction as in Fig. A2. In order to determine next collision point, first we calculate the distance (l_1) to the nearest material interface and draw a random number ξ_i, then two cases occur; i) if $\xi_i \geq e^{-\Sigma_{t1}l_1}$, the collision is in region 1 at point $l_i = -\ln \xi_i / \Sigma_{t1}$, or ii) if $\xi_i < e^{-\Sigma_{t1}l_1}$, it says that the collision is beyond region 1, so draw another random number ξ_{i+1} to determine the collision point that may be in region 2 at $l_{i+1} = -\ln \xi_{i+1} / \Sigma_{t2}$ beyond l_1 along the same direction. This process continues until the neutron is absorbed or leaks out of the problem boundary.

A.2.3 Tally of events

To calculate neutron flux of a region, current through a surface, or reaction rate in a region, the events that are usually tallied are i) number of collisions, ii) total track length traveled, or iii) number of crossings through a surface. For example, suppose that we like to calculate average scalar flux $\overline{\phi}$ in a volume element V with total cross section Σ_t. From a well-known relation,

$$\overline{c} = V\Sigma_t\overline{\phi}\,, \tag{A8}$$

where c is the number of collisions made by neutrons in V, we can calculate $\overline{\phi}$ by tallying the number of collisions:

$$\overline{\phi} = \frac{1}{V\Sigma_t}\overline{c}\,. \tag{A9}$$

We provide sample estimate of $\overline{c}$ by

$$\hat{c} = \frac{1}{N}\sum_{n=1}^{N} c_n\,, \tag{A10}$$

where c_n is the number of collisions made in V during the n-th history and N is a large number. In addition, we also provide sample estimate of variance on c by

$$\begin{aligned} S^2 &= \frac{1}{N-1}\sum_{n=1}^{N}(c_n - \hat{c})^2 \\ &= \frac{N}{N-1}\sum_{n=1}^{N}(\widehat{c^2} - \hat{c}^2), \end{aligned} \tag{A11}$$

where

$$\widehat{c^2} = \frac{1}{N}\sum_{n=1}^{N} s_n^2\,. \tag{A12}$$

It can be easily shown that the sample standard deviation on $\hat{c}$ is

$$\sigma_{\hat{c}} = \frac{S}{\sqrt{N}}\,, \tag{A13}$$

which suggests to use a large N for accurate $\hat{c}$, since $\sigma_{\hat{c}}$ is a measure of uncertainty in the estimated $\hat{c}$.

Fig. A3 shows an example for c_n; in the shaded region,

$$c_1 = 0, c_2 = 1, c_3 = 1, and\, c_4 = 3,$$

thus

$$\hat{c} = \frac{1}{4} \times 5 = 1.25,$$
$$S^2 = \frac{3}{4} \times (\frac{11}{4} - 1.25^2) = 1.583,$$
$$\sigma_{\hat{c}} = \frac{\sqrt{1.583}}{\sqrt{4}} = 0.6291.$$

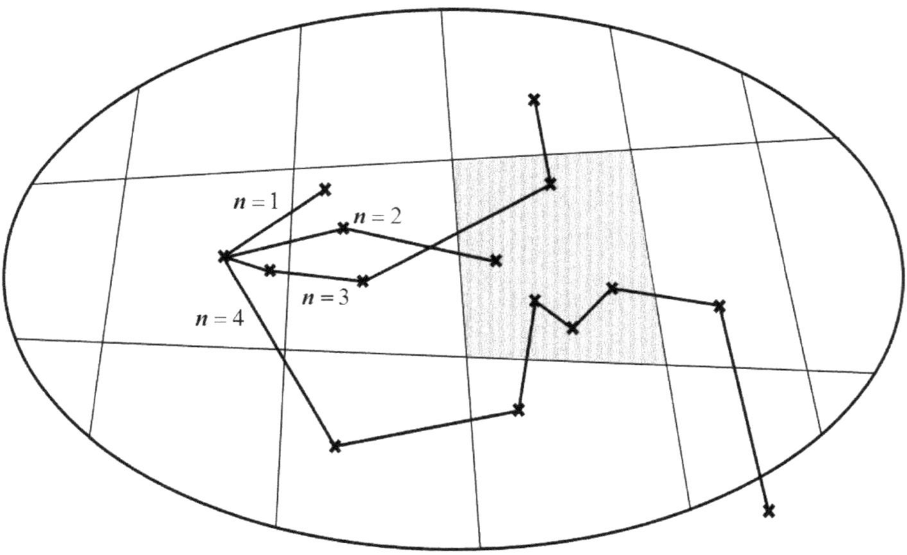

Fig. A3. Tally of number of collisions

Appendix B: Derivation of equivalent thermal conductivities

The expressions of k_2 (equivalent thermal conductivity) for the convective medium are derived in this Appendix for three (sphere, cylinder, slab) geometries.

B.1 Sphere geometry

The heat conduction equation in spherical coordinates is, in a region free of heat source,

$$\frac{k_2}{r^2} \frac{d}{dr} r^2 \frac{dT}{dr} = 0. \tag{B1}$$

Thus,

$$r^2 \frac{dT}{dr} = c_1, \tag{B2}$$

$$\frac{dT}{dr} = \frac{c_1}{r^2}, \tag{B3}$$

$$T = -\frac{c_1}{r} + c_2. \tag{B4}$$

From Eq. (B4),

$$T_s - T_b = c_1\left(\frac{1}{r_b} - \frac{1}{r_s}\right) = c_1 \frac{r_s - r_b}{r_s r_b}, \tag{B5}$$

and thus

$$c_1 = \frac{r_s r_b}{r_s - r_b}(T_s - T_b), \tag{B6}$$

The convective boundary condition equation for spherical geometry is,

$$-k_2 \left.\frac{dT}{dr}\right|_{r_s} = h(T_s - T_b). \tag{B7}$$

Substituting Eqs. (B3) and (B6) into (B7), we have

$$k_2 = h(r_b - r_s)\left(\frac{r_s}{r_b}\right). \tag{B8}$$

B.2 Cylinder geometry

The heat conduction equation in cylindrical coordinates is, in a region free of heat source,

$$\frac{k_2}{r}\frac{d}{dr} r \frac{dT}{dr} = 0. \tag{B9}$$

Thus,

$$r\frac{dT}{dr} = c_1, \tag{B10}$$

$$\frac{dT}{dr} = \frac{c_1}{r}, \tag{B11}$$

$$T = c_1 \ln r + c_2, \tag{B12}$$

From Eq. (B12),

$$T_s - T_b = c_1(ln\, r_s - ln\, r_b) = c_1\, ln\left(\frac{r_s}{r_b}\right), \tag{B13}$$

and thus

$$c_1 = \frac{T_s - T_b}{ln(r_s / r_b)}. \tag{B14}$$

The convective boundary condition equation for cylindrical geometry is,

$$-k_2 \left.\frac{dT}{dr}\right|_{r_s} = h(T_s - T_b). \tag{B15}$$

Substituting Eqs. (B11) and (B14) into (B15), we have

$$k_2 = h r_s\, ln\left(\frac{r_b}{r_s}\right). \tag{B16}$$

B.3 Slab geometry

The heat conduction equation in slab geometry is, in a region free of heat source,

$$k_2 \frac{d^2T}{dx^2} = 0. \tag{B17}$$

Thus,

$$\frac{dT}{dx} = c_1, \tag{B18}$$

$$T = c_1 x + c_2, \tag{B19}$$

From Eq. (B19),

$$T_s - T_b = c_1(x_s - x_b), \tag{B20}$$

and thus

$$c_1 = \frac{T_s - T_b}{x_s - x_b}, \tag{B21}$$

The convection boundary condition equation for slab geometry is,

$$-k_2 \left.\frac{dT}{dr}\right|_{r_s} = h(T_s - T_b). \tag{B22}$$

Substituting Eqs. (B18) and (B21) into (B22), we have

$$k_2 = h(x_b - x_s). \tag{B23}$$

5. References

[1] H.S. Carslaw and J.C. Jaeger, *Conduction of Heat in Solids*, 2nd ed., Oxford (1959).

[2] T.M. Shih, *Numerical Heat Transfer*, Hemisphere Pub. Corp., Washington, D.C. (1984).

[3] S.V. Patankar, *Numerical Heat Transfer and Fluid Flow*, McGraw-Hill, New York (1980).

[4] P.E. MacDonald, et al, "NGNP Point Design–Results of the Initial Neutronics and Thermal-Hydraulic Assessments During FY-03", Idaho Natural Engineering and Environmental Laboratory, INEEL/EXT-03-00870 Rev. 1, September (2003).

[5] James J. Duderstadt and Louis J. Hamilton, *Nuclear Reactor Analysis*, John Wiley & Sons, Inc. (1976).

[6] Jun Shentu, Sunghwan Yun, and Nam Zin Cho, "A Monte Carlo Method for Solving Heat Conduction Problems with Complicated Geometry," *Nuclear Engineering and Technology*, 39, 207 (2007).

[7] Jae Hoon Song and Nam Zin Cho, "An Improved Monte Carlo Method Applied to the Heat Conduction Analysis of a Pebble with Dispersed Fuel Particles," *Nuclear Engineering and Technology*, 41, 279 (2009).

[8] Bum Hee Cho and Nam Zin Cho, "Monte Carlo Method Extended to Heat Transfer Problems with Non-Constant Temperature and Convection Boundary Conditions," *Nuclear Engineering and Technology*, 42, 65 (2010).

[9] X-5 Monte Carlo Team, "MCNP - A General Monte Carlo N-Particle Transfer Code, Version 5(Revised)", Los Alamos National Laboratory, LA_UR-03-1987 (2008).

[10] T.J. Hoffman and N.E. Bands, "Monte Carlo Surface Density Solution to the Dirichlet Heat Transfer Problem", *Nuclear Science and Engineering*, 59, 205-214 (1976).

[11] A. Haji-Sheikh and E.M. Sparrow, "The Solution of Heat Conduction Problems by Probability Methods", *ASME Journal of Heat Transfer*, 89, 121 (1967).

[12] T.J. Hoffman, "Monte Carlo Solution to Heat Conduction Problems with Internal Source", *Transactions of the American Nuclear Society*, 24, 181 (1976).

[13] S.K. Fraley, T.J. Hoffman, and P.N. Stevens, "A Monte Carlo Method of Solving Heat Conduction Problems", *Journal of Heat Transfer*, 102, 121(1980).

[14] Hui Yu and Nam Zin Cho, "Comparison of Monte Carlo Simulation Models for Randomly Distributed Particle Fuels in VHTR Fuel Elements", *Transactions of the American Nuclear Society*, 95, 719 (2006).

[15] Jae Hoon Song and Nam Zin Cho, "An Improved Monte Carlo Method Applied to Heat Conduction Problem of a Fuel Pebble", *Transaction of the Korean Nuclear Society Autumn Meeting*, Pyeongchang, (CD-ROM), Oct. 25-26, 2007.

[16] J. K. Carson, et al, "Thermal conductivity bounds for isotropic, porous material", *International Journal of Heat and Mass Transfer*, 48, 2150 (2005).

[17] C. H. Oh, et al, "Development Safety Analysis Codes and Experimental Validation for a Very High Temperature Gas-Cooled Reactor", INL/EXT-06-01362, Idaho National Laboratory (2006).

[18] Nam Zin Cho, Hui Yu, and Jong Woon Kim, "Two-Temperature Homogenized Model for Steady-State and Transient Thermal Analyses of a Pebble with Distributed Fuel Particles," *Annals of Nuclear Energy*, 36, 448 (2009); see also "Corrigendum to: Two-Temperature Homogenized Model for Steady-State and Transient Thermal

Analyses of a Pebble with Distributed Fuel Particles," *Annals of Nuclear Energy*, 37, 293 (2010).
[19] E.E. Lewis and W.F. Miller, Jr., *Computational Methods of Neutron Transport*, Chapter 7, John Wiley & Sons, New York (1984).

Permissions

The contributors of this book come from diverse backgrounds, making this book a truly international effort. This book will bring forth new frontiers with its revolutionizing research information and detailed analysis of the nascent developments around the world.

We would like to thank Dr. Prof. Vyacheslav S. Vikhrenko, for lending his expertise to make the book truly unique. He has played a crucial role in the development of this book. Without his invaluable contribution this book wouldn't have been possible. He has made vital efforts to compile up to date information on the varied aspects of this subject to make this book a valuable addition to the collection of many professionals and students.

This book was conceptualized with the vision of imparting up-to-date information and advanced data in this field. To ensure the same, a matchless editorial board was set up. Every individual on the board went through rigorous rounds of assessment to prove their worth. After which they invested a large part of their time researching and compiling the most relevant data for our readers. Conferences and sessions were held from time to time between the editorial board and the contributing authors to present the data in the most comprehensible form. The editorial team has worked tirelessly to provide valuable and valid information to help people across the globe.

Every chapter published in this book has been scrutinized by our experts. Their significance has been extensively debated. The topics covered herein carry significant findings which will fuel the growth of the discipline. They may even be implemented as practical applications or may be referred to as a beginning point for another development. Chapters in this book were first published by InTech; hereby published with permission under the Creative Commons Attribution License or equivalent.

The editorial board has been involved in producing this book since its inception. They have spent rigorous hours researching and exploring the diverse topics which have resulted in the successful publishing of this book. They have passed on their knowledge of decades through this book. To expedite this challenging task, the publisher supported the team at every step. A small team of assistant editors was also appointed to further simplify the editing procedure and attain best results for the readers.

Our editorial team has been hand-picked from every corner of the world. Their multi-ethnicity adds dynamic inputs to the discussions which result in innovative outcomes. These outcomes are then further discussed with the researchers and contributors who give their valuable feedback and opinion regarding the same. The feedback is then collaborated with the researches and they are edited in a comprehensive manner to aid the understanding of the subject.

Apart from the editorial board, the designing team has also invested a significant amount of their time in understanding the subject and creating the most relevant covers. They scrutinized every image to scout for the most suitable representation of the subject and create an appropriate cover for the book.

The publishing team has been involved in this book since its early stages. They were actively engaged in every process, be it collecting the data, connecting with the contributors or procuring relevant information. The team has been an ardent support to the editorial, designing and production team. Their endless efforts to recruit the best for this project, has resulted in the accomplishment of this book. They are a veteran in the field of academics and their pool of knowledge is as vast as their experience in printing. Their expertise and guidance has proved useful at every step. Their uncompromising quality standards have made this book an exceptional effort. Their encouragement from time to time has been an inspiration for everyone.

The publisher and the editorial board hope that this book will prove to be a valuable piece of knowledge for researchers, students, practitioners and scholars across the globe.

List of Contributors

Krzysztof Grysa
Kielce University of Technology, Poland

M. S. Gadala and S. Vakili
Department of Mechanical Engineering, The University of British Columbia, Vancouver, BC, Canada

Hasna Louahlia Gualous
Caen Basse Normandie University/LUSAC, France

Semion Gutman
University of Oklahoma, USA

Junhong Ha
Korea University of Technology and Education, South Korea

Ernesto Marín Moares
Centro de Investigación en Ciencia Aplicada y Tecnología Avanzada (CICATA), Unidad Legaria, Instituto Politécnico Nacional (IPN), México

Mohammad Mehdi Kabir Najafi
Department of Engineering, Aliabad Katoul branch, Islamic Azad University, Golestan Province, Iran

Roman M. Kushnir and Vasyl S. Popovych
Pidstryhach Institute for Applied Problems of Mechanics and Mathematics, Ukrainian National Academy of Sciences, Ukraine

Ferenc Márkus
Department of Physics, Budapest University of Technology and Economics, Hungary

Xiaoguang Yuan
York University, Canada

Fengpeng Yang
Shanghai Jiao Tong University, China

A. Lazarian
Department of Astronomy, University of Wisconsin-Madison, USA

Yuriy Tokovyy
Pidstryhach Institute for Applied Problems of Mechanics and Mathematics, National Academy of Sciences of Ukraine, Lviv, Ukraine

Chien-Ching Ma
Department of Mechanical Engineering, National Taiwan University, Taipei, Taiwan ROC

Masakatsu Murakami
Institute of Laser Engineering, Osaka University, Japan

Yoshihiro Ochiai
Kinki University, Japan

Nam Zin Cho
Korea Advanced Institute of Science and Technology (KAIST), Daejeon, South Korea

www.ingramcontent.com/pod-product-compliance
Ingram Content Group UK Ltd.
Pitfield, Milton Keynes, MK11 3LW, UK
UKHW022206051026
15624UKWH00001B/22

9 781632 384416